AF784579

Vehicular and Automotive Engineering

Vehicular and Automotive Engineering

Edited by **Nicole Maden**

CLANRYE INTERNATIONAL

New Jersey

Published by Clanrye International,
55 Van Reypen Street,
Jersey City, NJ 07306, USA
www.clanryeinternational.com

Vehicular and Automotive Engineering
Edited by Nicole Maden

© 2015 Clanrye International

International Standard Book Number: 978-1-63240-511-1 (Hardback)

This book contains information obtained from authentic and highly regarded sources. Copyright for all individual chapters remain with the respective authors as indicated. A wide variety of references are listed. Permission and sources are indicated; for detailed attributions, please refer to the permissions page. Reasonable efforts have been made to publish reliable data and information, but the authors, editors and publisher cannot assume any responsibility for the validity of all materials or the consequences of their use.

The publisher's policy is to use permanent paper from mills that operate a sustainable forestry policy. Furthermore, the publisher ensures that the text paper and cover boards used have met acceptable environmental accreditation standards.

Trademark Notice: Registered trademark of products or corporate names are used only for explanation and identification without intent to infringe.

Printed in China.

Contents

Preface

Over the recent decade, advancements and applications have progressed exponentially. This has led to the increased interest in this field and projects are being conducted to enhance knowledge. The main objective of this book is to present some of the critical challenges and provide insights into possible solutions. This book will answer the varied questions that arise in the field and also provide an increased scope for furthering studies.

In this book, descriptive information regarding vehicular and automotive engineering has been presented. An automobile was seen as a luxurious accessory in the early years of the past century. However, in recent years the extent of technology and human effort applied by the vehicular industry for developing high quality vehicles has become evident. This book introduces new and advanced topics regarding vehicular technology and automotive engineering, and offers guidelines for future research. This book compiles contributions from various experts from around the globe working in automotive and related fields. The contributions were selected so as to offer a vast range of research. This book includes latest and cutting edge topics such as electronics, mechanics, materials, and manufacturing.

I hope that this book, with its visionary approach, will be a valuable addition and will promote interest among readers. Each of the authors has provided their extraordinary competence in their specific fields by providing different perspectives as they come from diverse nations and regions. I thank them for their contributions.

Editor

Section 1

Materials

Chapter 1

Nanocomposite Based Multifunctional Coatings

Horst Hintze-Bruening and Fabrice Leroux

Additional information is available at the end of the chapter

1. Introduction

The function of automotive coatings so far has been twofold – decoration and conservation. The former not only consists of providing a coloured and smooth surface but also to accentuate the shape of the car body via a viewing angle dependant brightness and / or colour. These optical effects mostly rely on tiny mirrors of metal flakes like aluminium or coated mica platelets that are more or less homogeneously dispersed within one coating layer; their surface normal predominantly being orthogonal to the substrates surface. The particles lateral dimensions typically are in the order of several hundred microns. The role of the coatings in the conservation of a metallic car body mainly consists in the protection of the substrate towards electrochemical degradation (= corrosion) either actively or via providing a barrier layer as well as a best possible limitation of the coatings defects produced in the wake of dramatic mechanical perturbations like impacting gravel stones. On the contrary mounted plastic parts or body panels in most cases have to be protected against chemical degradation triggered by exposure to UV light, physical erosion by swelling solvents like fuel or water depending on the nature of the polymers and the morphology of the bulk as well as towards catastrophic failure of the part upon mechanical impacts (= brittle crash behaviour). The means to achieve the latter are similar to those that are used in the design of stone chip resistant coatings for car bodies.

The present chapter describes the potential use of nano scaled particles based polymer composites as multifunctional constituents for novel protective as well as decorative automotive coatings. The focus lies on composites based on layered particles, namely layered double hydroxides (= LDH), their beneficial mechanical modes of action for energy dissipation, their diffusion barrier properties as well as their role in the active corrosion inhibition. For other coatings related applications of these materials we refer to a separate review article (Leroux et al., 2012).

2. Background

2.1. Automotive coatings

Since the industrialization of the car manufacturing process today's automotive coatings have been optimized in an evolutionary process over the past decades resulting in a typical stack of highly specialized individual coating layers (Fig. 1). This comprises an initially cathodic electrodeposited material that provides adhesion and active corrosion inhibition (ED coat) which then is covered by a spray coated layer (= primer) for the smoothening and protection of the ED coat towards UV light. On to the primer surface two layers are consecutively spray applied: first a "base coat" that provides the colour followed by a transparent top coat (= clear coat) which renders the whole system bright, smooth as well as resistant towards chemicals (bird droppings, tar, rosin, acid rain) and surface related mechanical impact like the scratching in the course of car washing. Except for the base coat itself all other layers are chemically cured after physical drying and film formation which implies that the car body has to pass three bake cycles up to 170°C (ED) or 140°C (combined base and clear layer).

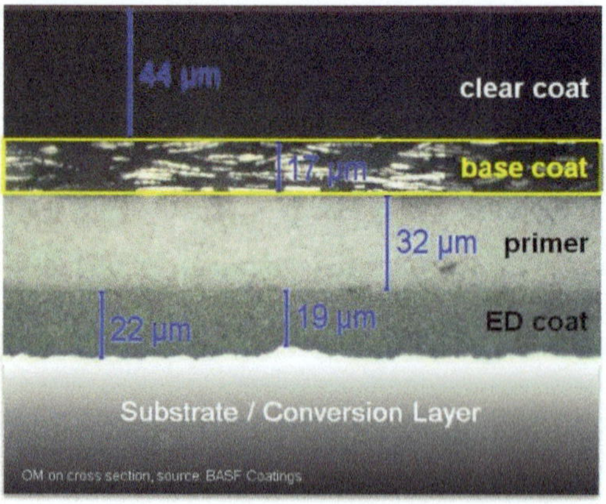

Figure 1. Automotive coating system applied on pretreated metal.

In order to render this coating process more eco-efficient some novel coating systems recently have been introduced which comprise stacks with a reduced number of layers, e.g. without the primer (Fig. 2), (Kreis, 2008). Under critical conditions such lean coatings reveal an insufficient capability to maintain the high level of performance of their technically more mature predecessors.

This is the setting for the development of new material concepts that may provide solutions capable to overcome the constraints inherent to these recent coatings systems which are still essentially based on the conventional technologies. With respect to the automotive coatings conservation functionality it has to be pointed out that – contrary to that what is sometimes suggested by textbooks and patent claims – some features are not attributed exclusively to one single layer, e.g. the relation of the primer layers properties and the stone chip resistance of the coating. Thus, although typically tested and introduced within one coating layer, the potential use of the described new materials concepts should not be considered to be restricted to these prototypes, substrates and applications respectively.

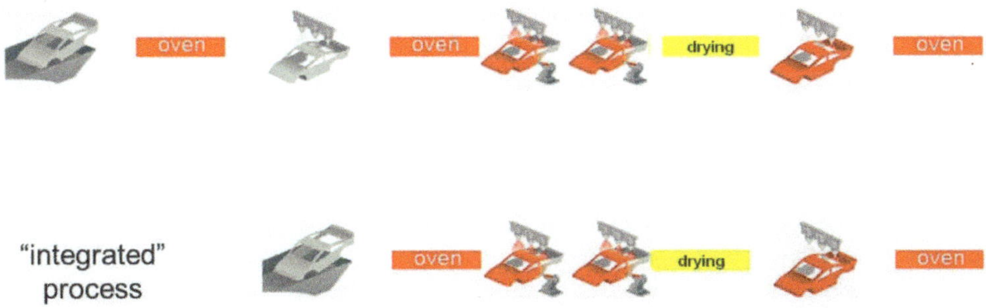

Figure 2. Conventional automotive coating process compared to novel integrated coating.

2.2. Stone impact

Gravel stone impingement is an occasional though dramatic event in the lifetime of an automotive body or mounted part and has intensively been studied (Zehnder et al., 1993; Ramamurthy et al., 1994; Zouari & Touratier, 2002; Lonyuk et al., 2007).

Regarding the physical processes occurring in a coating system there are three common basic features aside from all conceivable differences in terms of inclination angle, kinetic energy (up to 100 m/s, mass of projectiles up to several grams) as well as the shape (curvature radius) of the projectile:

1. Short time scales (micro seconds) and a very limited impact area smaller than one mm^2 are the features of an adiabatic process which yields a local heating (up to 10^2K) and a propagating shock wave.
2. A central compression area where the substrate plastically might deform and the coating material gets crunched is surrounded by a strained zone due to the lateral displacement of the coating material by the impinging projectile – thus shear stress in a transition zone causes coatings delamination.
3. The volume of laterally displaced coating material reaches its maximum towards the coatings surface – thus within the highly cross linked top coat which can least accommodate by plastic flow. Hence compressive and tensile stresses (depending on the inclination angle and shape of the object) yield crack initiation within this layer and crack propagation throughout the entire coating system.

This simplified scenario varies with different impact events as evidenced by comparing an orthogonal crush of a blunt object (infinite curvature radius) with the orthogonal "denting" of spiky or edged gravel or with the inclined "carving" by an edged impinging projectile (Dhar et al., 2005). Figure 3a shows the crater produced by a single spherical impact. Radial and concentric cracks as well as plastically flown coatings material encircle a central spot of crunched coating. The plastically deformed metal substrate and the coatings delamination at the interface with the conversion layer (bonder) is displayed after the removal of the damaged coating material using an adhesive tape (Figure 3b). In real life gravel stones are irregularly formed and they may impact under all conceivable inclination angles. Thus all coatings damage symptoms between "denting" and "carving" are summarized as "stone chipping".

Figure 3. Impact crater on a coated test panel after an orthogonal single spherical impact (steel, 0,03g, d= 2mm, v = 70 m/s, T = 273K) before (a) and after (b) removal of damaged coating. Scale bars are 200μm (a) and 1mm (b).

Standardized test methods have been established to assess the resistance of a coated work-piece against stone chipping. They consist of multiple impacts of accelerated gravels or steel grit particles on a defined area of a coated panel and a subsequent visual rating of the damage as well as a standardized image data processing of the exposed area. This latter procedure allows for the assessment of the area percentages of the uncovered substrate as well as the interfacial and cohesive fracture planes within the coating itself, provided appropriate thresholds can be defined to differentiate between these fracture sites. For the allocation of the uncovered metal surface the panel is immersed into a copper salt solution in order to get deposited elemental copper. Some tests specify that the impacted panels are exposed to aggressive environments in order to provoke the corrosion of the substrate which predominantly sets in at those damaged spots that bear bare metal surface.

The ratio of flow and cracking as well as the extent of the delamination zone depend on the film thickness and of the materials properties like the glass transition temperature (Lonyuk et al., 2008), the crosslink density (Bender, 1969), the interlayer adhesive strength (Ryntz et al, 1995) and finally the coatings morphology.

The importance of the latter has recently been highlighted with respect to the "conflicts between strength and toughness" in terms of the intrinsic fracture mechanics operating ahead the tip of a promoting crack (Ritchie, 2011). Making reference to hierarchical materials the author outlines the possibility to "attain both properties in a single material through the presence of multiple plasticity and toughening mechanisms acting on different length-scales". Within section 3.2. of this chapter we describe possibilities to translate these design principles into stone impact resistant coating materials that are based on multiple scale heterogeneity.

2.3. Automotive primer

Besides the smoothening and the protection of the ED coating layer against UV light the stone chip resistance of an automotive coating system is often attributed to the primer layer. Although the failure mode of any multilayer coating upon such a mechanical impact also depends on the characteristics of all individual layers (Dioh & Williams, 1994; Rutherford et al., 1997) the typical composition as well as the morphology of an automotive primer layer are somehow appropriate to mitigate the action of propagating cracks as well as the dislocation and crushing of the material.

In the first instance the organic fraction typically consists of weakly cross linked rather high molecular weight polymers that are bearing sites for hydrogen bonding. By a proper choice of the polyisocyanates as well as the polyols and / or the polyamines – the complementary reactants that yield polyurethanes or polyureas respectively via a step growth poly addition reaction – bulk materials can be obtained that segregate into rigid domains bearing hydrogen bonds between adjacent urethane / urea building blocks and soft domains linking the former. These soft segments typically consist of polyether diols or polyester diols that possess a high segment flexibility – which manifests in a low glass transition temperature (Tg). This flexibility in combination with the reversible breaking and formation of hydrogen bonds provides a pronounced toughness to the polymer phase. These polymers usually comprise rather low functionalities for chemical cross linking like hydroxyl groups. Typical values for "hydroxyl numbers"[1] are below 50 distributed over largely uneven chain lengths with mean molecular weights (Mn) in the order of 10^4 to 10^6 and typical polydispersity indices[2] (Mw/Mn) between 3 and 4. Therefore it is common to use them in admixture with more resinous hydroxyl functional polymers like polyesters and / or in combination with crosslinking agents like alkoxylated melamine formaldehyde resins. Under acid catalysis and with increasing temperature the latter either react with the hydroxyl groups of the former or they self-polymerize via transetherification. Depending on the overall formulation and the presence of internal interfaces due the presence of e.g. pigment particles the self-condensation of the MF resin may yield an additional heterogeneity of the polymer phase due to a pronounced anisotropic alignment of the aromatic melamine cores. This was also found to form hierarchical ordered structures over several length scales in the heterophasic self-condensation of hexamethoxy melamine formaldehyde resin in water (Weber et al., 2009). Besides the intrinsic toughness of the polymer phase stemming from the main chain segment flexibility and the intermolecular formation of reversible hydrogen bonds which favour plastic

---

[1] Hydroxyl- as well as acid numbers are defined as the amount of potassium hydroxide that would be consumed for the neutralization of these Brønstedt acids present within one gram of the solid (= nonvolatile) material. They are given as "mg KOH/g".

[2] Mean mass molecular weight (Mw) and mean number molecular weight (Mn) of polymers to be obtained from e.g. size exclusion chromatography.

deformation the heterogenization of the polymer phase due to segregation and polymer incompatibility might impede both crack initiation and propagation. This is well known from the impact behaviour of plastic parts where a more rigid mostly continuous phase imparts a soft second polymer phase as an impact modifier and the particular morphology of bicontinuous polymer phases was proposed as beneficial for stone impact resistant automotive primer layers (Makowski et al., 2005). The softer phase is attributed to release stress at the tip of cracks by stress reallocation and dissemination. Thus the softest conceivable material would be an included gaseous phase as nearly everyone's mundane experience with polymer foams or materials scientists work in the field of tough porous ceramics expressively attest (Studart et al., 2006). Indeed breakable hollow spheres have been proposed as impact modifying loci for stone chip resistant coatings (Roesler et al., 1997).

However these polymer matrices normally are filled with rather high amounts of inorganic pigments like titania (TiO_2), barite ($BaSO_4$) and fillers like precipitated calcium carbonate ($CaCO_3$) and talc (layered silicate minerals) – the particle sizes of the latter being in the micrometre range. In the course of a stone impact event such a composite mechanically dissipates energy via the damping of a traversing shock wave where the conservation of momentum commensurate to the mass, acts as a ballast bed via interlocking coarse particles thus disseminating stress or it fails at predetermined breaking points respectively. This latter manifests in the fracture surface of primer layers that comprise such clay minerals like talc or chlorite which can easily be cleaved upon shear stress or by the peak stress at a crack tip because just weak van der Waals forces are acting between adjacent silicate layers (fig. 4).

Figure 4. SEM picture shows a section of an impact crater. Smoothly cleaved talc particles led to cohesive failure of the conventionally pigmented automotive primer layer. Smaller white particles refer to titania, barite and calcite. Scale bar is 1 μm.

The physico-chemical principles described so far are all related to the properties of the solid coating layers. However with respect to the coatings design outlined in chapter 3 it has to be noted that the liquid coating materials – except for the ED coating which is exclusively a waterbone system – may either be solvent or water based. Whereas the vast majority of top clear coat materials still is based on organic polymer solutions (plus cross linker and additives) the base coat and the primer are increasingly formulated as aqueous coatings in order to minimize the amounts of volatile organic compounds and thus fulfilling ecological and environmental regulations. However the notion "aqueous" implies that such formulations still comprise certain levels of organic solvents which by the majority are partly or completely miscible with water at ambient conditions.

2.4. Metal corrosion

Metal corrosion results from a redox reaction system constituted by three elements: an anodic site with an excess of electrons produced by the release of an equivalent amount of metal cations into the environment (= metal oxidation), a cathode where electrons are reducing either ions or water (reactants being O_2, H_2O or H^+) thus releasing the corresponding reaction products like hydroxyl ions or molecular hydrogen respectively and an electrolyte as a conductive medium that enables an ionic current to flow thus ensuring electrical short-cut and a charge balance (fig. 5). The simultaneous presence of all three components and a perpetuate supply of the reactants is needed to maintain the currents and to degrade the anode material. Corrosion protection of non noble metal substrates requires that at least one of these elements becomes inhibited or at least significantly retarded. Active corrosion inhibition implies that a component from the environment (electrolyte, pretreatment, coating) interacts either with the substrate, the metal cations or the reduction products and that the reaction product forms an insoluble, stable layer or a deposit which impedes a further reaction at the respective electrode(s). On the contrary passive corrosion inhibition denotes the function of an applied coating to provide an effective diffusion barrier against water and ions from the environment thus isolating the corrosion sites from the electrolyte or retarding the formation of effective ion conductivity respectively.

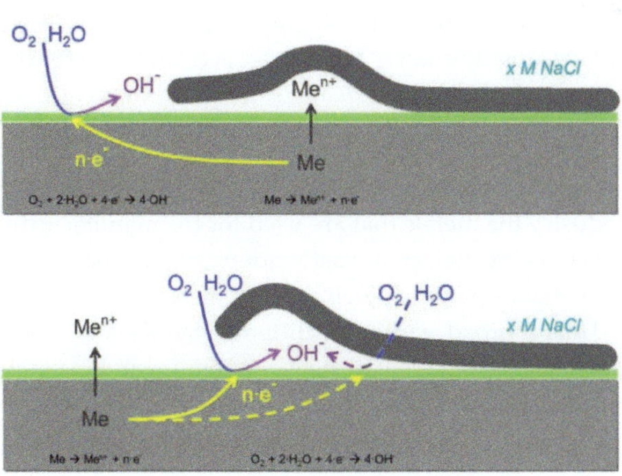

Figure 5. Anodic and cathodic corrosion process at defects on coated metal substrate (the green line symbolizes a conversion layer). Reaction of Me^{n+} with water may lead to precipitation of $Me(OH)_n$ and a local decrease of the pH value

There exist numerous techniques which have been used to investigate the electrochemical reactions occurring in the course of corrosion as well as to follow changes of the electrical resistance of a coated substrate. These analytical probes enable to determine electrochemical characteristics like the corrosion potential (E_{corr}), the corrosion current density (I_{corr}), the polarization resistance (R_p), the electrochemical noise resistance (R_n), the open circuit potential (E_{OCV}) by either polarizing the system (DC polarization, cyclic voltammetry), recording current and voltage fluctuations between two samples (electrochemical noise) or by following an open circuit potential of an immersed sample while the electrolyte is changed. Changing environmental conditions as well as progressing corrosion processes also alter the complex electrical resistivity of a coated sample which can be monitored with the electrochemical impedance spectroscopy which indirectly allows insights related to the evolution of the coatings barrier and self-healing properties or of the destabilization of the substrate coating interface. The locally resolved monitoring of corrosion allows the Scanning Vibrating Electrode Technique (SVET) by detecting gradients of the corrosion potential at the metal surface exposed to the corrosive medium. Similarly the Scanning Kelvin Probe (SKP) measures electrode potentials non-destructively at buried interfaces of a coated metal substrate and additionally displays the topography of the samples surface.

By coupling electrochemical techniques with chemical analysis of dissolved species within the electrolyte using e.g. ICP and / or the chemical nature of the exposed substrates surface during and after the exposure further valuable insights into the corrosion mechanism especially the role of the involved layers as well as the corrosion products can be obtained (Volovitch et al., 2011). Thus it could in principle be possible to quantitatively link stoichiometry (ratio of transferred electrons to converted chemical species) with corrosion rates.

Other tests evaluate and rate the corrosion resistance of coated samples by visual inspection of coated and in most cases marred (= scribed or stone impacted) specimen after an exposure to aggressive environments, either natural ones (e.g. maritime locations, industrialized districts) or artificial set-ups, the latter known as accelerated tests that are described by detailed protocols.

2.5. Corrosion protection

In the automotive industry the metals that are used for the manufacturing of the body or the body panels and framework reflect a vast variety of approaches towards lightweight construction, safety enhancement as well as an increased degree of freedom in the automotive design. Thus a broad variety of different metals and intermetallic alloys have been developed based on steel, zinc coated steel, aluminium and magnesium.

Most natural oxide or oxide-hydroxide passivation layers at the surface of these metals or alloys do not exhibit a sufficient resistance towards chemical degradation under those harsh environmental conditions often formed in the course of the corrosion processes. Another aspect is the fact that even the surface of a metal substrate consisting of only one single metal does not present a chemically homogeneous surface. First the solidification of the metal melt in the course of the substrates manufacturing proceeds via a seed growth mechanism yielding a final patchwork of metal grains separated by grain boundaries. Second the crystallized grains may expose different crystal planes with respect to their topological *hkl* systematic towards the surface as well as at defect structures and grain boundaries. These normally translate into different topological crystal planes of the oxide passivation layer. The different crystal planes expose a different metal coordination chemistry and hence a different chemical reactivity to the surface. Although not yet scaled to industrial processes a recently published approach proposes a pretreatment process that addresses the grain boundary selective deposition of polymer colloids from aqueous solutions on zinc coated and acid activated steel panels (Hintze-Bruening et al., 2011). The stabilizing anionic groups of the particles complex metal ions liberated from anodic reactions. Their concentration varies locally being highest in the vicinity of the grain boundaries thus yielding to a local selective destabilization and deposition of the polymer particles.

Another aspect relevant to corrosion protection stems from the fact that nearly every thick organic coating system may comprises some flaws like pores, pinholes and cracks on the micrometer scale as well as a locally varying cross-link density in the case of reactive systems. Additionally in highly filled pigmented coatings diffusion channels for aqueous solutions might be present, e.g. along polar interfaces or electric double layers at the surface of percolating pigment or filler particles, as well as capillary forces. Thus, before applying such coatings the metal surface or the passivation layer should be treated or replaced respectively in order to obtain a thin stable layer on the nano- to micro-meter scale, the conversion coating. These are formed on the substrates surface via a "pretreatment", a coating process normally comprising the cleansing (e.g. degreasing), an activation step (e.g. chemical dissolution to remove the native passivation layer) and finally the deposition of the conversion layer. As described above the common principle is to better protect the bare metal substrate against corrosion. However in many cases the conversion layer also provides a less polar and less smooth surface as the pristine substrate which in turn facilitates the adhesion of subsequently applied organic coating layers, either via mechanical anchoring or by specific molecular interactions or by both phenomena. Thus the bonderizing also known as phosphatization of steel and zinc coated steel comprises the precipitation of transition metal phosphates, e.g. the trication phosphatiszation forms zinc, nickel and manganese phosphates. These grow as coarse randomly oriented platelets from the metal surface leaving sufficient uncovered metal areas to ensure the electric conductivity which is needed for the cathodic reaction to take place in the subsequently applied cathodic electrodeposition paint process.[3] Indeed for this still most abundant metal substrate of automotive manufacturing the combined bonderizing and cathodic electrodeposition coating has become a mature effective corrosion preventing technology which is inter alia reflected in ever increasing time limits of the warranty issued by the car manufactures.

---

[3] The formed hydroxyl anions neutralize the organic acids that have been used for the formation of ammonium cationic groups by neutralizing tertiary amine bearing epoxy polymers. Hence the latter lose their colloidal stability and precipitate onto the metal surface of the automotive body. Aromatic moieties within the polymer backbone are supposed to interact effectively with the metal surface.

Such pretreatment may take place either as a continuous process in a coil coating line (e.g. zinc coating on steel, the passivation of aluminium with chromates) or subsequently on individual body panels or of the assembled automotive body respectively.

For nearly all other metal substrates but particularly for aluminium based parts the most effective corrosion inhibition system in the past was based on chromium associated in its hexavalent oxidation state (Cr^{6+}), the active species CrO_4^{2-} being slowly liberated from essentially insoluble strontium or barium salts up to their solubility limit under humid conditions, e.g. a coatings defect or water saturated coating. In the case of a starting corrosion process an insoluble and adherent layer of prevalently trivalent chromium comprising hydroxide forms on the substrate surface. Due to matching lattice parameters and the dielectric nature of the hydroxide layer both further metal dissolution as well as the cathodic reactions are effectively inhibited. However due its carcinogenic potential and environmental impact the use of this passivation has been banned by the Directive 2000/53/EC of the European Parliament and the Council of September 18, 2000. Therefore the development of chromate-free anti-corrosion coatings has become an ever growing research topic. Basically relying on the same principle, namely the slow release of oxoanions near a coatings damage and precipitation of a passivation or barrier layer as is the case of chromate and phosphate for aluminium and steel respectively, recent approaches to provide chromate substitutes have been introduced as "self-healing" concepts comprising the triggered release of corrosion inhibiting agents from depots imparted in the coating system.

2.6. Self-healing concepts

In the recent past the investigation of self-healing coatings has gained considerable attraction and even government-funded joint research initiatives like the European MUST program have been installed in order to provide the basis for future technical solutions. Common to all is the notion to furnish a coating system with depots of chemical active compounds which are released upon either a physical impact like mechanical stress, temperature or UV light or a chemical stimulus like a change of the pH value of the surrounding or the exposure to other species like water or ions. The chemicals imparted within the depots might either be active corrosion inhibiting compounds (e.g. ligands for substrate metals, precursor for passivation layers) or chemical reactants like monomers optionally along with cross-linkers or initiators which are thought to refill a void in the coating layer formed upon e.g. mechanical impact. However concerning the design of the host material a plethora of different approaches have been proposed. They span from sol-gel chemistry over micellar particles, Pickering particles, hard polymer capsules, sand-wiched shells of oppositely charged polyelectrolytes, porous minerals or oxides, tubular and layered clay minerals to hierarchically assembled frameworks. Regarding the different length scales covered as well as the different chemical stabilities of these materials it is obvious that some of these host architectures may be prone to different or even complementary trigger for release than others (e.g. hard polymer versus polyelectrolyte based capsules) or may provide additional benefits like an efficient diffusion barrier as it is the case of layered clay particles (table 1).

System / Trigger	Cl^-	Δ pH	H_2O	Δ T	mech.	light	chem.
no host		■	■				
sol-gel	■	■	■				■
layered particles	■	■	■				■
rigid polymers				■	■	■	
polyelectrolytes	■	■		■	■		■
micelles		■	■	■			■
Pickering particles		■			■		■
hierarchical	■	■	■				■

Table 1. Encapsulation systems and responsiveness towards stimuli compared to the direct addition of the inhibitor species into the matrix (= no host). This survey is loosely based on: Hughes et al., 2010.

In terms of container geometry the prevalent conception comprised discrete capsules dispersed in a polymer matrix. However with respect to rather thin layers in automotive coatings the overall dimensions of such depots and thus the amount of releasable components is very limited. Less relevant for corrosion inhibiting species which have to address a limited and in most cases moving interface or surface this aspect in particular affects the potential use for the self-repair of voids especially if larger defects from stone chipping are considered. In this context other coatings morphologies might be worth to be pursued where the depot extends over – theoretically – the whole coatings volume, e.g. by micro vascular networks which could either supply larger amounts of the repair system to one location or smaller amounts to the very same location if it comes to repeated damage. Filled fibres (Pang & Bond, 2005) and micro lithographic techniques (Toohey et al., 2009) have been proposed for the realization of such extended reservoirs in coatings.

Other self-repair systems target the curability (mendability) of smaller defects like fissures and cracks while maintaining an in principle cross-linked material via "photo-plasticity". This concept implies the presence of photoactive polymer groups which – attached to the polymer backbone – might cleave upon UV exposure, rearrange and form new covalent bonds with other "dangling" reactive polymer ends or side groups. This can also be coupled with other functional groups attached to the polymer chains which are prone to mechanical stress and yield reactive groups complementary to those from the photo induced scission (Gosh & Urban, 2009). Such photoactive systems could react to tiny fissures that are formed within the organic coating due to residual stress present from the coatings processing in combination with a stress build-up over time via e.g. cyclic exposure to different environments with respect to temperature, humidity, chemicals and UV radiation[4]. Regular exposure of an automotive to sunlight thus could help to maintain the coatings barrier function over time by impeding the formation of larger crack or fracture.

[4] Shorter wave length UV radiation is known to induce the formation of radicals as well as aggressive oxygen species (O_3, $^1O_2$) that trigger cascades of reactions leading to cross linking as well as chain scission. Hence embrittlement as well as polymer degradation may occur contemporaneously and/or successively and optionally spatially separated depending on the formulation, physical properties and morphology as well as exposure parameters.

Within the present chapter we will in the following focus on the aspect self-healing of active electrochemical sites on the substrate by using lateral extended inorganic particles capable to form nano composites as well as to release active corrosion inhibiting species via ion exchange.

2.7. Polymer nano composites

Following pioneering research on clay-organic intercalates (Lagaly et al., 1973, 1975; Lagaly, 1981) it was in the late 1980s that Toyota's R&D work on Montmorillonite polyamide composites (Fukushima et al., 1988) triggered extensive research and application related work in the field of polymer intercalated and exfoliated layered particles. The morphologies compared to a composite comprising the aggregated counterpart are shown in figure 6. These were supposed to impart strength and thermal stability as well as a diffusion barrier to the polymer at comparatively low loadings with the inorganic particles of 5 – 10 weight per cent. Consequently they have also been proposed as new materials for automotive applications, however not in the context of coatings (Garcés et al., 2000). The vast majority of composites have been based on Smectite clays and Montmorillonite in particular. Each platelet of this natural clay that is essentially constituting the mineral Bentonite consists of a central layer of aluminium and magnesium cations octahedral coordinated with hydroxyl anions covered on both sides by a layer of silica tetrahedrons: $\{Al_{1,67}Mg_{0,33}(OH)_2[Si_4O_{10}]\}$. The net negative charge of the overall structure which is equivalent to the proportion of Mg(II) in the central layer is compensated by cations like sodium or calcium that are intercalated. Typical charge densities are in the order of 100 meq/100g and the lateral dimensions of the 1 nm thick platelets vary between several hundred nanometers and one micrometer. Such a polar, charged stacking readily incorporates water molecules. Therefore the interlayer distance which is in the order of one nanometer depends on the level of environmental humidity. Additionally natural intercalated cations may easily be exchanged by organic cations like quarternized fatty acid derived amines. Indeed "organo-clays" have been the workhorse for the preparation of polmer nano-composites in most cases due to a given tunable compatibility with organic hydrocarbon based structures and an increased interlayer distance. Methods comprise in-situ polymerization, solution blending (polymer is dissolved in solvent) as well as polymer melt processing. Besides rendering polar clay

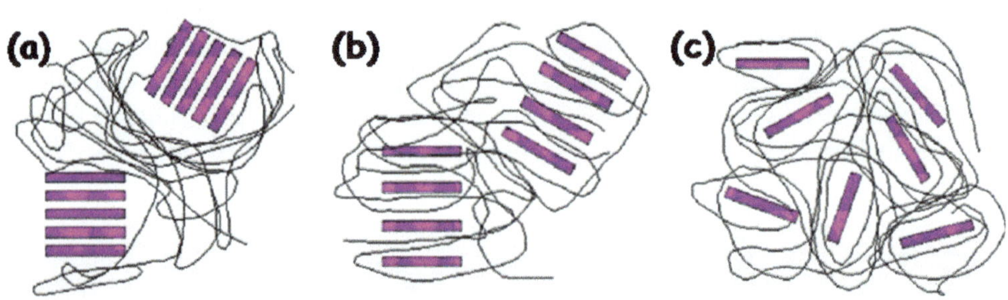

Figure 6. Aggregated (a), intercalated (b) and exfoliated (c) structures for nano composite materials based on layered fillers dispersed in polymer matrices.

"organophilic" reactive clays have been proposed for numerous in-situ polymerization systems, e.g. of epoxy resins via acidic ammonium (Lan et al., 1996), lactone via transition metal compound (Lepoittevin et al., 2002) or in combination with radically polymerized olefinic unsaturated monomers (Di & Sogah, 2006). Other approaches make use of pristine clays in heterophase polymerizations, either emulsion polymerization (Huang & Brittain, 2001) or mini emulsion polymerization (Cauvin et al., 2005), where the mineral platelets are found at the interface between the aqueous phase and the polymer particles to yield Pickering like armored particles.

Regarding the mechanical properties the stiffening effect of intercalated and to some extend exfoliated composites has been attributed to an immobilization of the polymer matrix. Thus polymer chains confined within the platelets interstitial space or tethered to the platelets surface may extend into the surrounding matrix and entangle with each other (Rao & Pochan, 2007). Hence the effective volume of the platelets exceeds that of the neat particles. Consequently a huge effect in the mechanical response to external stress can be observed as reinforcement accompanied with a decrease of polymer chain flexibility expressed as a reduced loss modulus in dynamic mechanical analysis. Obviously the stress transfer over the volume of a work piece scales with the platelets aspect ratio and from theoretical considerations as well as from biological systems it is known that the probability to dramatic failure due to flaws inversely scales with the thickness of the particles (Gao et al., 2003). However numerous studies have shown that polymer clay nano composites tend to show brittle failure in impact tests or to provide a reduced toughness in tensile tests in comparison with the unfilled polymers. Reasons might be local stress build up and thus crack initiation at aggregated inclusions. However in first instance it is due to the particles restricted mobility in a glassy matrix system. By shifting the time-temperature parameter frame of the test method relative to the glass temperature of the polymer a shift from brittle to tough failure can be observed (Shah et al., 2005). This was corroborated by simulation results which show that the particles mobility in the matrix is a decisive prerequisite for tough failure (Gersappe, 2002). In semi-crystalline polymer systems the nano-particles impact on the crystallization of the polymer phase was also shown to dramatically increase the failure toughness by changing the overall matrix morphology (Shah et al., 2004).

With respect to the diffusion barrier of clay based composites this is mainly attributed to the tortuosity for the diffusing species (Triantafyllidis et al., 2006). It was particularly shown that penetration rates inversely scale with the aspect ratio of the clay platelets (Gatos & Karger-Kocsis, 2007; Kugge et al., 2011).

3. Layered Double Hydroxide – Polymer composite based coatings

3.1. Layered Double Hydroxides (LDH)

LDH structure is described with the ideal formula, $[M^{II}_{x}M^{III}_{1-x}(OH)_2]_{intra}[A^{m-}_{x/m}.nH_2O]_{inter}$, where M^{II} and M^{III} are metal cations, A the anions and *intra* and *inter* denote the intralayer domain and the interlayer space, respectively. The structure deviates from the layered morphology of magnesium hydroxide (Brucite, $Mg(OH)_2$) – consisting in edge sharing octahedrons of hydroxyl anion coordinated metal cations. Compared to the neutral brucite framework, partial M^{II} to M^{III} substitution induces a positive charge for the layers, counter-balanced with the presence of the interlayered anions. Edge sharing $M(OH)_6$ octahedra yield inorganic sheets in the lateral dimension while the anions induce a stacking of the sheets as illustrated in Fig. 7. Structurally, polytypes based on two-, three- and six-layer polytypes presenting either rhombohedral (2R, 3R, 6R) or hexagonal symmetry (2H, 3H, 6H) are determined from the powder X-ray diffraction diagrams (Bookin et al., 1993). However the LDH and the hybrid derivative materials are usually poorly crystallized and the more common three-layer polytype $3R_1$ with rhombohedral symmetry (space group R-3m) is often adopted. From a long range order, the stacking of LDH sheets is viewed as layers translation of ($2/3a + 1/3b$) between consecutive layers.

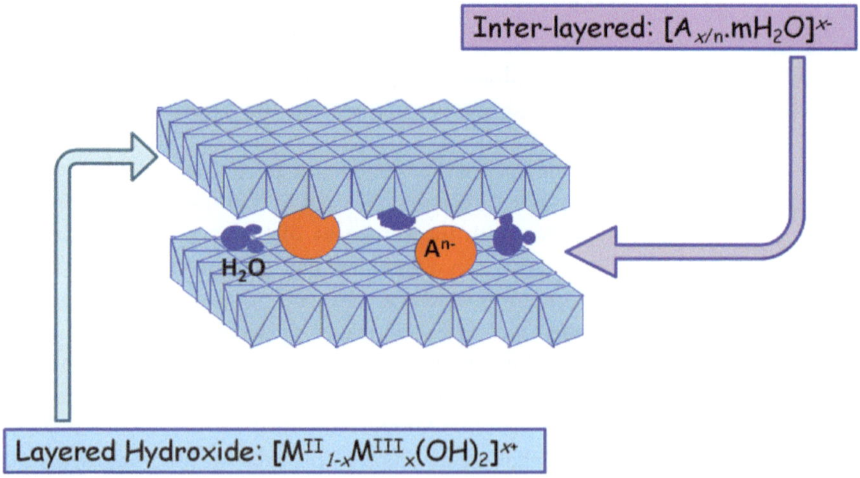

Figure 7. Structure of Layered Double Hydroxide material

In term of crystal identification and natural abundance, hydrotalcite of composition $Mg_6Al_2(OH)_{16}(CO_3)\bullet4H_2O$ was first identified in Sweden in the year 1842. Many other minerals presenting different chemical natures are belonging to this large family, such as manasseite, $Mg_3Al(OH)_8(CO_3)_{0.5}\bullet2H_2O$, meixnerite, $Mg_3Al(OH)_8(CO_3)_{0.5}\bullet2H_2O$, sjögrenite $Mg_3Fe(OH)_8(CO_3)_{0.5}\bullet2.25H_2O$, stichtite, $Mg_3Cr(OH)_8(CO_3)_{0.5}\bullet2H_2O$, takovite $Ni_3Al(OH)_8$- $(CO_3)_{0.5}\bullet2H_2O$, pyroaurite, $Mg_3Fe(OH)_8(CO_3)_{0.5}\bullet2.25H_2O$, wermlandite, $Mg(Al,Fe)_{0.5}SO_4\bullet2H_2O$, hydrocalumite, $Ca_2Al(OH)_6(CO_3)_{0.11}(OH)_{0.78}\bullet2.38H_2O$, $Fe^{II}_4Fe^{III}_2(OH)_{12}SO_4\bullet nH_2O$, an iron hydroxysulphate commonly called greenrust, and others. Naturally not abundant, synthetic LDH materials nowadays are, however, produced in relative large amounts, this promoted by a high global demand of their use in ethanol steam reforming, biodiesel production and in plastics processing industry to replace heavy-metal additives for PVC as well as bromide-based fire retardants in other polymers.

Simple, reproducible and rather inexpensive syntheses have been developed to produce LDH materials. The co-precipitation process at a fixed pH value is largely employed to yield materials with chemical homogeneity. Standard LDH products bearing anions such as nitrate or chloride may subsequently be used to yield LDH phases comprising other anion compositions through ion exchange topotactic reaction. Other procedures such as metal salt and metal oxide addition and reconstruction from amorphous calcined LDH precusors yield materials of lower crystallinity and potentially accompanied with non LDH phases. During a co-precipitation both nucleation and growth process steps occur simultaneously, resulting in particles presenting a relative broad size distribution. Size distribution may be limited at the expense of a lower crystallinity by using short addition times and by avoiding any ageing step after the co-precipitation. For a better control of the particle size distribution, a rapid coprecipitation reaction is associated either to a separate aging step (Zhao et al., 2002) or to a subsequent controlled hydrothermal treatment (Xu et al. 2006), both methods yielding stable homogenous LDH suspensions with controllable particle sizes of interest for the elaboration of thin films or nanostructured materials. Another option is the so called "homogeneous precipitation" through urea hydrolysis and decomposition at slightly elevated temperatures (Ogawa & Kaiho, 2002). Urea decomposes into ammonium cyanate and successively into ammonium carbonate and ammonia upon hydrolysis, the rising pH value provokes the precipitation of the metal salts yielding micrometre sized, monodisperse and well-shaped carbonate bearing LDH particles.

Aqueous medium is largely preferred for the coprecipitation process, but other solvents maybe employed. For instance the polyol method performed by the hydrolysis of the metal acetate salts in ethylene glycol or di ethylene glycol under heating produces pure acetate intercalated LDH, this in absence of pH and atmosphere control and without an excess of a base (Prevot et al., 2005). Alternatively the sol-gel route in an organic medium yields LDH materials after hydrolysis and condensation polymerization reactions of metal alkoxide precursors (Wang et al., 1999). Of lower relevance for large scale production and from application related aspects, reversed micelles or micro emulsions may be employed to drastically change the LDH platelets morphology into fiber-like particles (Hu & O'Hare, 2005), and a sophisticated method using inverse opals is reported to yield three dimensional ordered macroporous LDH materials, the polystyrene beads based opal array acting as a sacrificial template (Géraud et al., 2006).

As evidenced from the natural occurrence, the chemical variety of hydrotalcite class of materials is rooted in a broad range of compositional tolerance with respect to the nature and ratio of divalent and trivalent cations according to their associated oxidation state and ionic radius as well as from the nature of the interleaved anion. The ratio of divalent and trivalent cations induces a net positive charge layer, resulting in a variable of interleaved anions. The anionic exchange capacity usually ranges between 450 to 200 meq/100 mg that corresponds to an average area per charge between 25 and 40 $Å^2/e^-$. It is importance to tune the platelets interface in order to enable the accommodation of cumbersome guest molecules as such high charge densities picture layers that are tightly stacked by electrostatic attraction translated via the interlayer anions and / or by attractive forces acting laterally between the densely stacked anions. Indeed the usual $M^{II}_2M^{III}$ cation repartition results in LDH stacking being reluctant to exfoliate. However stable colloidal suspensions of exfoliated platelets can be obtained by the use of specific solvents such as formamide in addition to ultrasound for nitrate-based LDH (Liu et al., 2006), *N,N*-dimethylformamide-ethanol for carbonate-based LDH (Okamoto et al., 2006) or even water for alkoxide (Gurski et al., 2006) or lactate (Hibino & Kobayashi, 2005) intercalated LDH. It was observed that LDH nano sheets gradually dissolve in organic solvents.

With regard to their size – whether atomic or polymeric – and their valence from one to polyvalent a vast variety of species can be intercalated and exchanged by others depending on their characteristics. However driven by the intense research devoted to polymer-clay nanocomposite (PCN) (see section 2.7), surfactant molecules are commonly used to render the LDH sheets organophilic. It is comparable to the lipophilization of the smectite-type clays by alkyl ammonium cation to reduce the surface polarity of the silicate layers, and thus to supply a satisfactory affinity between them and a polymer. Another aspect is to consider LDH open-structure as an available container to embark and deliver functional species. This aspect is developed in more detail in section 3.3.

In addition to the tunable intra- and inter- chemical composition that permits to adapt LDH platelets for multiple applications and specific requirements (ecology, toxicity, functionality) bi-dimensional assembly of LDH platelets presents mesophasic ordering in suspension and in bulk polymer. Evidenced by birefringence observation, colloidal platelets of LDH show isotropic (I) to nematic (N) phase transition as a function of the concentration (Zhang et al., 2007). However high platelets concentration as high as 20% (wt.wt.) is needeed to realize the I-N phase transition with the co-existence of a sol-gel transition. At lower concentrations in toluene I-N transition was observed with LDH platelets grafted with amino-modified polyisobutylene (Mourad et al., 2008) as well as with LDH particles dispersed in high-molecular weight polyvinylpyrrolidone (Luan et al., 2009). More recently the possibility to use such organized hierarchy was applied to polymer coatings (Troutier-Thuilliez et al., 2011), more developed in section 3.2.

Another interesting aspect for the application of LDH in coatings is their ability to arrange themselves in the form of thin films on a (metal) substrate or to be generated onto a metal substrate. Academically the formation of M^{II}-Al-NO_3 LDH films is reported on the surface of Al_2O_3 (Paulhiac & Clause, 1993) when contacting with M^{II} (M^{II} = Zn, Mg, Co and Ni). Optimized adsorption onto Al_2O_3 surface arises near its isoelectric point (i.e.p.) at pH 7.3 demonstrating that the formation of the LDH film proceeds by surface adsorption rather than electrostatic impregnation. Onto Al metal substrate highly porous Zn-Al LDH films with thicknesses on several micrometers are grown by reacting zinc salts and ammonia (Gao et al., 2006), while Zincite (ZnO) precipitates onto Al-free surfaces (Si, glass, plastic). The cross-section of a LDH film obtained after 12h at 50°C using a NH_3 to Zn ratio of three is shown in figure 8.

Figure 8. SEM images of LDH film grown on Al-bearing glass substrate after (Gao et al., 2006).

Alternatively this technology appears appealing to protect Mg and Zn rich alloy substrates. The combined deposition of LDH with surfactants yields superhydrophobic barrier films that are of interest in coatings science (Chen et al., 2006). Finally thin and ultrathin LDH films have been developed for energy management, thus via electrodeposition highly exposed electrode materials can be prepared (Gupta et al., 2008). For instance Ni or Co LDH films are cathodically electrodeposited through the reduction of nitrate ions; the increasing OH^- concentration promoting the LDH precipitation at the surface (Indira & Kamath, 1994). However published work on electrodeposition on the laboratory scale suggests that this method shall be deemed to be challenging to scale-up due to the observed vulnerability of the coatings to form defects like cracks and uncovered domains despite carefully controlled parameters (current density, concentration, time, temperature).

3.2. Aqueous LDH based coatings

As outlined in chapter 2.2 automotive primers for instance are based on tough polymers capable to form hydrogen bonds (e.g. polyurethanes) that are highly filled with inorganic micron-sized particles like titania, barite, and easily cleavable talc, which primarily cause cohesive failure of the coating upon impact. With respect to integrated coating systems (see chapter 2.1.) without such a primer layer the mechanical functionality should be imparted into the remaining coating layers which have to provide additional functionalities like colour effect (aesthetics) or corrosion inhibition (diffusion barrier, inhibitor release). Thus it would be highly desirable to increase the scope for formulation by reducing the amount of loading of the polymer phase with filler particles. One attractive strategy is the use of exfoliated as well as polymer intercalated layered inorganic particles which are known to reinforce polymer matrices at low loadings through coordinate stress transfer. As described in chapter 2.4. large aspect ratios of these particles are preferred in view of the mechanical properties. However dimensions of up to several hundred nanometer surpass typical sizes of polymer colloidal particles that are in the order of 20 to 100 nm. Accordingly they usually have to be dispersed within the continuous aqueous phase of the liquid coating. In order to maintain a low initial as well as storage stable viscosity of this material allowing for spray application particle-particle interactions have to be controlled. This concerns in particular the interactions between the layered inorganic particles – thus the gel formation of quite diluted aqueous dispersions of phyllosilicates like Laponite and Montmorillonite resulting from the aggregation of clusters formed by exfoliated platelets or small staples of platelets has been a widely studied phenomenon (Mourchid et al., 1995; Dijkstra et al., 1997; Pignon et al., 1997; Shalkevich et al., 2007; Joensson et al., 2008; Cousin et al., 2009). Also specific interactions between polymers and the inorganic particles have to be considered (Balazs et al. 1998; Milczewska et al., 2003; Alemdar & Buetuen, 2004; Wang et al., 2006). In this respect combining oppositely charged components like positively charged platelets and anionic stabilized polymers or colloidal polymer particles appears to be adventurous regarding possible heterocoagulation as it was observed with oppositely charged inorganic particles (Ji et al., 2004). However within the following chapters it is shown that, within a given toolbox of waterborne polymer colloids, water soluble polymers and water dispersible LDH particles, by tuning the compatibility between the polymers as well as the interfacial properties between the polymers and the LDH particles, storage stable aqueous coating materials can be formulated. Furthermore upon spray application and film formation of several of these formulations a broad range of different coatings morphologies – all providing excellent impact resistance – are obtained.

3.2.1. Intrinsically incompatible matrix

For stone impact resistant coatings, dispersed soft phases, in particular bicontinuous phases, have been claimed, since heterogeneous (polymer) film morphologies are known to be beneficial for stress relaxation and for the limitation of unstable crack growth as outlined in section 2.2. In this context, a ternary composition was designed, comprising polyester (PES), blocked polyiso-cyanate (PURx) and hydrophobic polyurethane (PURh) (Hintze-Bruening et al., 2009). It has to be emphasized that contrary to the crystalline LDH phases, the polymers both of the reference system and of the mixtures described in this and the following sections are technical, structurally non uniform products with rather broad molecular weight distributions obtained from step growth polymerizations. However as visualized by CLSM using fluorescent labeled versions of PES and PURh, the mixture shows a spinodal like polymer phase separation upon water flash off. This produces a heterogeneous film structure in the baked coating with large dispersed areas of a less dense phase which primarily is constituted by PURh.

In the presence of the carbonate (CO_3^{2-}) and 4-amino benzene sulfonate (4-ABSA) comprising $Zn_2Al(OH)_6$ LDH particles, the phase separation becomes either retarded or accelerated, respectively, as was deduced from the resulting film morphologies. With the organic modified particles, a dispersed softer PURh phase of significant smaller droplet size was found within a continuous more rigid and elastic composite phase. Morphologies where both phases are continuous were also obtained using LDH particles with smaller lateral dimensions, as with the 3-ABSA intercalated $Mg_2Al(OH)_6$ phase (Fig. 9).

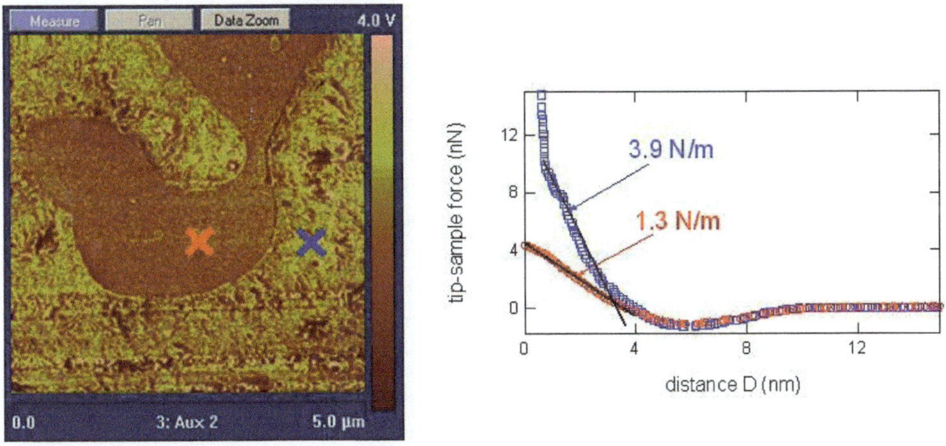

Figure 9. Multiple heterogeneous morphology of a baked film comprising polymer intercalated $Mg_2Al(OH)_6$ – 3,ABSA as rigid phase (blue) besides PURh based soft phase (red). AFM phase contrast image (5•5 µm) and force distance curves. The colour codes are also valid for Fig. 10.

It is conceivable that besides a positive impact of the particles on the viscosity development of the continuous polymer phase (as observed in the rheological behavior of a reference system, see below), the LDH particles might simply form physical obstacles for the moving interface between the demixing polymer phases. This latter effect was simulated for binary polymer systems comprising spherical hard particles (Ginzburg et al., 1999) and corroborating morphologies were observed with several silica beads filled (Karim et al., 1999) and smectite filled non crystalline polymer blends (Yurekli et al., 2003, 2004). Indeed CLSM and XRD studies revealed that the PES (and possibly both the polar components PURx and MF resin) readily intercalates the LDH galleries in the initial waterborne coating material while the particular shape of the micron sized aggregates of polymer intercalated LDH particles slowly disappears during the air drying process. An inverse impact on the kinetics of the polymer phase separation was found with the carbonate bearing LDH particles which enables a stratification of the film where a continuous PURh layer covers a composite phase that comprises the aggregated, not polymer intercalated LDH particles. Due to their basic nature, it is conceivable that the carbonate bearing LDH particles retard the acid catalyzed transetherification of the melamine formaldehyde resin.[5]

It is noteworthy that the particle impact on the kinetics of the polymer phase separation follows the same trend which was observed for the viscosity development of a reference system that was based on an aqueous PUR dispersion (Troutier-Thuilliez et al., 2009). This commercial grade polymer was shown by XRD studies and TEM imaging to readily intercalate organic anion bearing LDH phases whereas no specific interactions and intimate blending occurs in the presence of the carbonate comprising LDH particles. However, the 4-ABSA anions render the LDH particles organophilic but still incompatible with the hydrophobic

[5] Importantly, curing reactions are not generally be hampered by the presence of LDH phases as demonstrated with cross-linked polysiloxanes on a metal substrate (Kanai & Nomura, 2009) and in a sol-gel based coating (Alvarez et al., 2010).

polyurethane PURh of the ternary mixture which was demonstrated with another reference coating material that was exclusively based on PURh as the polymer part. This incompatibility might not be fully attributed to the hydrophobic nature of the polymer chains stemming from the incorporated dimer fatty acid moieties. This was shown with another LDH phase that beared intercalated fatty acid carboxylates (C12) instead of 4-ABSA. Although favourable interactions with the PURh chains are expected no intercalation could be observed. A possible explanation might be the lack in entropic gain. Assumed that the polymer intercalation proceeds via anion exchange the system would gain entropy by the released small anions 4-ABSA or dodecyl carboxylate. However due to distinct van der Waals attractions and as the conjugated base of a weak acid the latter are only slightly water soluble under ambient conditions.

As with the baked films of the reference system, typically stacks of tennish polymer intercalated and approximately 10-15 nm separated platelets of the initially 4-ABSA bearing LDH phase are found to be homogeneously and isotropically distributed throughout the continuous phase besides minor amounts of singular platelets and some stacks of platelets which retained their original interlayer distance. The stacks of polymer intercalated platelets or polymer separated stacks of non-intercalated platelets appear as alternating rigid soft layers which micromechanically could be considered as sliding planes in the case of shear stress and as springs or attenuators under compressive load. Additionally as described in section 2.7. and known in the context of smectite filled polymer system, the confinement of polymers strengthens the continuous polymer phase near and above the glass transition temperature. In combination with the presence of the dispersed soft PURh phase as well as the tortuous path for growing cracks these features were attributed to be the main causes for the excellent stone chip performance obtained with model automotive coatings comprising these materials as a primer layer which is visualized by the fracture surface that is shown in figure 10.

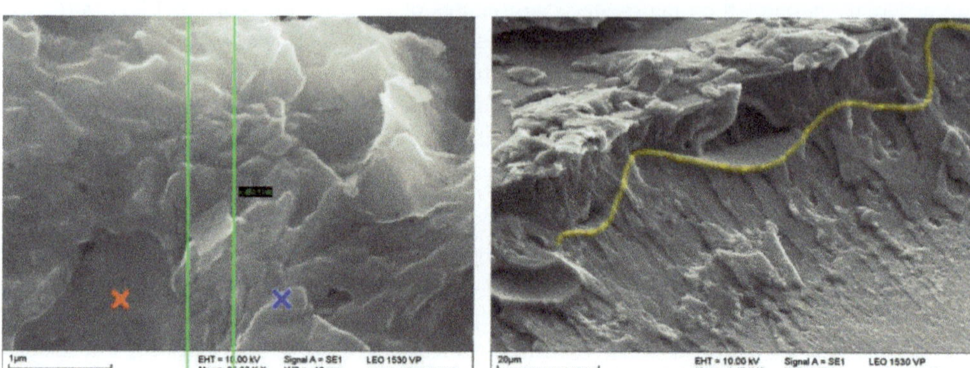

Figure 10. Sections of impact crater produced from single spherical impacts at 273K. Fracture surfaces of the primer layer are shown. Left image shows the composite of Fig. 9., scale bar is 1 µm. Tortuous cracking took place in the rigid composite leaving individual LDH stacks behind. Right image is a stratified primer (yellow line = interface) comprising aggregated carbonated LDH particles in the layer beneath the upper PURh phase, scale bar is 20 µm.

3.2.2. *Compatible matrix & ordered films*

Another approach to achieve high impact resistance in coatings could be the formation of textured composites including platelets that are aligned parallel to the substrate surface. Thus a high level of ordering at the scale of building blocks is reminiscent of bio-related frameworks, inspired by aragonite protein brick mortar assemblies found in e.g. nacre and reported in the literature as high strength materials. However these artificial materials are typically obtained by tedious application techniques, e.g. via layer by layer deposition that are economically unaffordable for industrial processing (Bonderer et al., 2008). In contrast LDH particle-based composites with distinctly oriented thin stacks of LDH platelets were obtained from a single application step of a $Zn_2Al(OH)_6$ 4-ABSA LDH particle comprising aqueous binary mixture of compatible polymers (Hintze-Bruening et al., 2011). Both polymers exhibit some amphiphilic features by combining hydrophobic backbones with anionic stabilizing groups: a polyurethane (PURh)[6] and a polyester (PESh).

It could be shown that the formation of this texture is due to some intermediate colloidal stage of a lyotropic liquid crystal like (LC) ordered liquid phase which arises from ionic interactions between protonated amino groups of 4-ABSA molecules on the LDH platelets surface and the carboxylate moieties of the PESh. Besides the well-known attempt of charged layered particles to gain entropy via reducing their excluded volume if dispersed in water, an additional enthalpy related driving force might be an effective lamellar shielding of the hydrophobic PESh backbone against the aqueous phase if adsorbed between adjacent LDH platelets. Thus a lamellar texture indicated by the fluorescent labelled PESh within an intermediate preparation stage of the coating was visualized with CLSM. Besides a nematic phase of poly(isobutylene) stabilized LDH particles in toluene (Mourad et al., 2008) a lyotropic mesophase of polymer stabilized LDH in water had not been described before.

Upon stirring the LDH particle slurry into the aqueous PESh dispersion, a stable viscous paste showing "schlieren" is obtained. These could be assigned to a nematic ordering as was shown by textural domains visible under cross polarized light as well as in cryo SEM pictures. Ultra small angle X-ray scattering (USAXS) confirms the mean spacing between the lamellae as observed in the cryo SEM picture by exhibiting a peak corresponding to a distance of 75 nm. Such a distance is not revealed by cryo TEM images of individual lamellae that are predominantly constituted by parallel aligned LDH platelets (Fig. 11). Despite their rather non-uniformly stacking in terms of the number as well as with regard to the interlayer distance of the individual LDH layers, this varying morphology still gives rise to distinct peaks in the SAXS curve over four harmonics corresponding to a mean spacing of 21 nm. The scattering intensity follows an $I(q) = q^{-2.01}$ power law decay which is that expected for individual platelets.

[6] Note that the hydrophobic nature of the polymer backbones is stemming from dimerized fatty acid that is used as a monomer in the synthesis of the polyester intermediate polyols. With this respect PURh is similar to that PURh described in the preceding section, both only differing by the type of diisocyanate that was used for the polyaddition step in the synthesis of the polyurethanes.

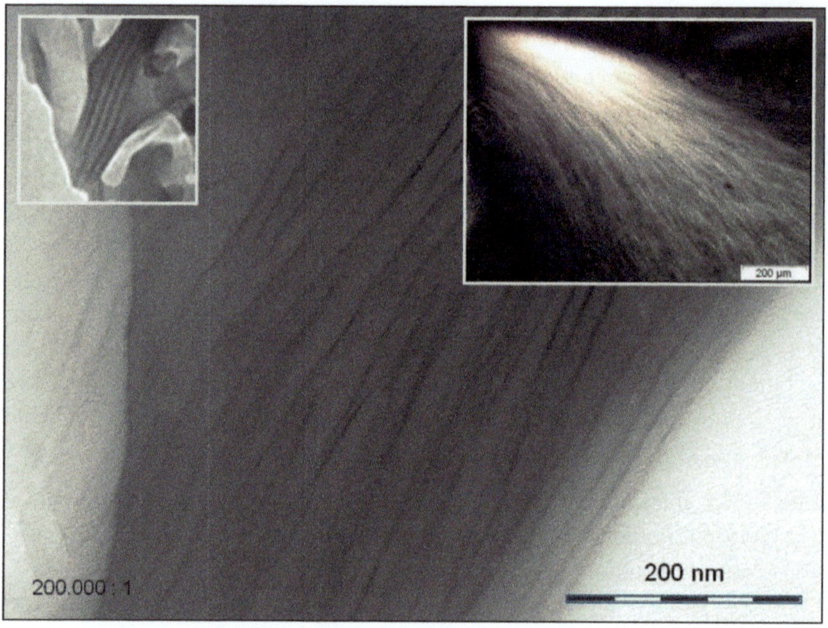

Figure 11. Cross polarized OM image (right insert, scale bar 200 µm) shows an ordered liquid composed of aligned lamellae (TEM image, left insert) which comprise irregularly parallel aligned individual LDH platelets with a mean spacing of 20 nm (cryo TEM images).

Combined with the aqueous PURh dispersion and the melamine formaldehyde resin (MF), smooth highly transparent films can be obtained after drying and baking.[7] It could be shown that within the liquid coating material the ordered nematic like phase of the intermediate disrupts into micron sized fragments which are still constituted of LDH platelets separated by 20 nm according to cryo SEM images and peaks in the SAXS curve. These lamellar fragments still comprise the major portion of the PESh which was visualized by the fluorescent labelled polymer in confocal laser scanning microscopy (CLSM). The morphology of free films was investigated by TEM and XRD (Fig. 12).

The LDH platelets' interstitial space from the liquid coatings LDH bearing fragments has been compacting during film formation and baking yielding aligned tiny stacks with an interlayer spacing of 10 nm. These stacks comprise LDH platelets that are separated by 1.76 nm. The alignment parallel to the substrates surface is far from being perfect as revealed by stitched TEM pictures but essential enough to give rise to either the 003 or the 010 and 100 reflections in the SWAXS curves depending on the inclination angle of the specimen.

The small interstitial spacing of 1.76 nm corresponds to that of the pristine LDH phase bearing intercalated 4,ABSA anions. However these were found to be liberated from the intermediate PESh LDH preparation into the aqueous phase. Hence such a restacking would

[7] The ordered nematic like liquid of the intermediate preparation does not yield a film. After drying and baking extended, quasi "bicontinuous" phases are found: a LDH rich solid phase besides a transparent LDH free polyester based (viscous resinous) phase.

be a remarkable finding and deserves further investigations as well as the role of the film forming polyurethane colloids. The latter was found to be incompatible with the (pristine) LDH phase like the one of the incompatible polymer system described in section 3.2.1.

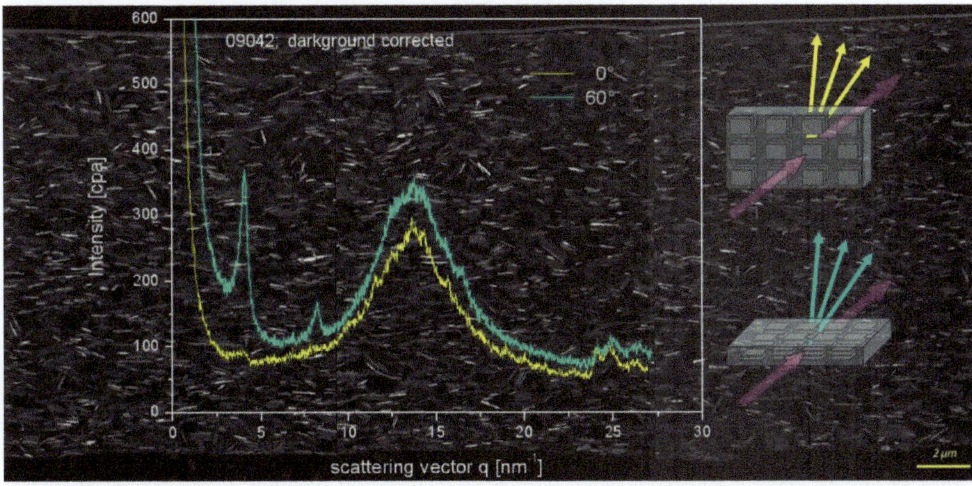

Figure 12. Stitched HAADF-TEM images and corresponding SWAXS diffraction curves from different inclination angles of a baked film of the aqueous, PURh diluted nematic like meso-phase of PESh stabilized LDH platelets. Black spheres with diameters of ~ 1 µm in the lower part of the film are discussed in the text and emphasized in Fig. 13.

If such a liquid formulation of PURh fragmented ordered LDH – PESh intermediate liquids was sprayed as a primer layer in the same model automotive coating system already introduced in the preceding section an outstanding stone chip resistance was observed. SEM fracture analysis of an impact crater produced by a single spherical impact test that was applied on the very same test panel revealed that remnant substrate near pieces of the fractured layer expose – besides some LDH platelets – the replica of numerous spherical "inclusions" with a typical diameter of one micro-meter or less (Fig. 13).

These witness a foamy morphology of the lower part of the primer layer. The gaseous inclusions are also visible in the TEM image of the free film as dark spots and they are most probably constituted by either accumulated air from the application or (and) by liberated methanol from the trans-etherification cross linking reaction of the melamine formaldehyde resin. Thus the excellent impact resistance may not (exclusively) be attributed to the ordered LDH composite since a porous morphology is also known to be beneficial for stress dissipation and limitation of crack growth (see also sections 2.2 and 2.3). The use of breakable hollow spheres as filler particles for stone chip resistant coatings accordingly was an early proposal albeit studies were limited on systems comprising glass spheres on a much bigger length scale. In contrast the present model coating demonstrates the possibility of an in-situ formation of such porous structures on an appropriate scale by providing an efficient diffusion barrier within the coating layer. Such a combination of mechanical performance and barrier properties might be the basis of multifunctional coatings in terms of corrosion protection of the metal substrate.

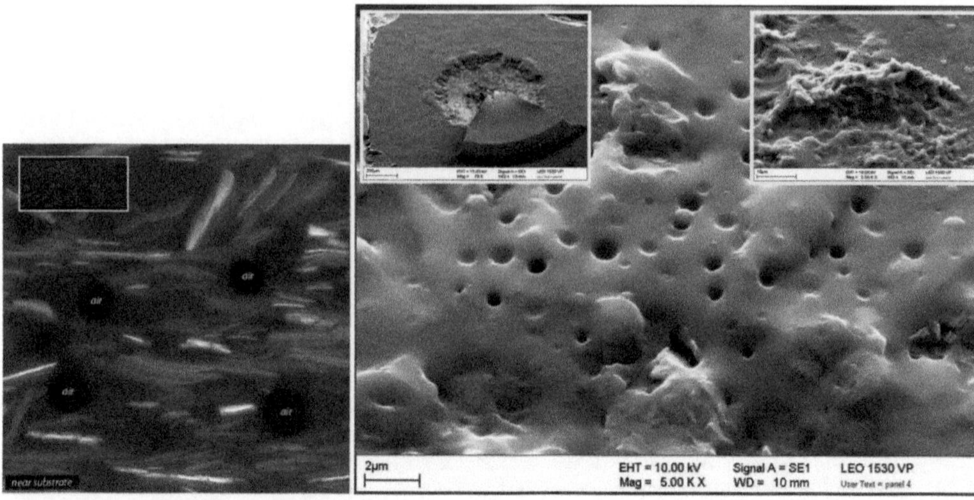

Figure 13. Impact crater from single spherical impact with increasing magnifications (inserts) showing spherical gaseous inclusions in the remnants of the primer layer obtained from the formulation also used to prepare the film shown in Fig. 12. (scale bar 2μm). A zoomed section of image Fig. 12 is shown for comparison. The term "air" is discussed in the text.

3.3. LDH based anti-corrosion coatings

As mentioned in section 2.5., the most effective corrosion inhibitor chromium ion Cr^{6+} used for years was banned from automotive coatings (Directive 2000/53/EC of the European Parliament and the council of 18 September 2000). In more details, it was declared in the European director that new vehicles after 1 July 2003 should not contain any lead, mercury, cadmium or hexavalent chromium, the latter being poisonous to the environment as well as being a potent carcinogen. Many chemical additives are also blacklisted by international regulations through REACH "Registration, Evaluation, Authorisation and Restriction of Chemicals" applied by the European Commission for enterprise and industry. With respect to metal substrates, such as aluminum and aluminum alloys like AA 2024-T3, commonly used in aeronautic, aerospace and automotive transportation, for the protective coatings these regulations apply. This prompted developments of chromate-free coatings anti-corrosion (Twite & Bierwagen, 1998; Sinko, 2001; Buchheit et al., 2003). To protect industrial ferrous and non-ferrous alloys as well as modern upcoming substrates from corrosion, applied coating technologies (spin coating, plasma, electrodeposition, anodization) and/or specific inhibitors (cerium, phosphate, molybdate, phosphate compounds, conducting polymers) may be selected (section 2.5.).

However the observation that a corrosion inhibitor dispersed into a coating may result in a weakening in its structure, leading to greater permeation trough diffusion mechanism and finally to its deterioration led to the idea to prevent such deleterious phenomena by loading the inhibitor in some sort of "passive" container. Among possible containers LDH-type materials are of great interest due to their high lateral aspect ratio allowing efficient barrier properties as well as due to their ion exchange ability (see section 3.1.). The former arising from the strong structural platelets anisotropy is also to a great extend dependent on the degree of the particles dispersion within the coating. Once provided, reduction in gas (*e.g.* O_2) and electrolyte (*e.g.* H_2O and Cl^-) permeability through the polymer film should be expected through the creation of tortuous paths, thus decelerating the corrosion process (see section 2.4). With respect to the latter, the LDH host structure is acting as a scavenger towards aggressive ions (*e.g.* Cl^-), *vice et versa* Cl^- ions present a rather good affinity with LDH materials. Such ion entrapment is associated by displacing the interleaved anion, i.e. the corrosion inhibitor, if the ion exchange reaction proceeds. Reminding that LDH host structure is able to load very diverse anions, inhibition efficiency of intercalated inorganic or organic anions such as carbonate (Lin et al., 2007), molybdate (Yu et al., 2008), vanadate (Mahajanam & Buchheit, 2008; Zheludkevich et al., 2010) and laurate (Zhang et al., 2008a), quinaldate and thiazole derivative (Poznyak et al., 2009) has been reported recently.

Coating systems usually present pores and / or bear heterogeneities in the crosslink density from the sub-micrometer to the nano scale that facilitates the corrosion process to migrate along the metal coatings interface. Both cathodic and anodic corrosion mechanisms drastically increase or lower local pH values (see section 2.4.) respectively. Practically, an anodic corrosion process will induce a low pH value that would promote the dissolution of the LDH layer, thus buffering the medium as well as releasing the interleaved anion. On the contrary a cathodic corrosion process (i.e. oxygen evolution) imposes a high pH value that stabilizes the LDH phase that consequently would exchange either Cl^- or OH^- with the anti-corrosion agent. Hence such desorption triggered by the corrosion process may be viewed as a "smart" release of active molecules able to decrease the damaging on demand. It is also commonly designated as a "self-healing" behavior although the polymer phase is not repaired by the anticorrosion agent. In the anodic case, one should also mention that the dissolution of an appropriate LDH host framework could release Zn^{2+} ions into the media that are reported to act as an inhibitor to chloride-induced corrosion (Veleva et al., 1999; Williams & McMurray, 2003). In the following, one should distinguish between the barrier effect (passive inhibition) and the release of chemical agents (active inhibition). For the former, it is interesting to note that LDH-type barrier layers may be generated either chemically or electrochemically (section 3.1.). After prolonged immersion times in seawater, LDH-type material was found on the surface of composites and of monolithic 6092- and 6061-T6 aluminum panels (Ding et al., 2009). LDH / alumina bilayer film was fabricated on an Al foil via a single hydrothermal crystallization step (Guo et al., 2009) and topological growth of LDH platelets interleaved with surfactant molecules onto an aluminum substrate was reported to yield a super-hydrophobic protective film (Zhang et al., 2008). Due to the high cathodic corrosion activity of magnesium, a corrosion-resistant barrier forms on its surface in alkaline environments. In the presence of a trivalent metal in a basic medium, a LDH precursor is deposited on the Mg alloy sheet (Buchheit & Martinez, 1998). Similarly, chemical conversion yielding a hydrotalcite (Mg,Al-LDH) was reported for a Mg-Al-Zn (AZ91D) alloy surface (Uan et al., 2008). No corrosion spot was observed after a 72 h salt spray test which was attributed to the formed protective barrier layer (Lin & Uan, 2009). Such a chemically-generation process may be further optimized using the separate nucleation and aging steps (SNAS) method which permits to form uniform and compact LDH films (Wang et al., 2010). Spin coating process was reported for AZ31 Mg alloy substrate (Zhang et al., 2008b), and spray-coated LDH formulation $Mg_3Al(OH)_8NO_2{\cdot}2H_2O$ bearing nitrite ions onto a steel surface was claimed to result in a good corrosion resistance when immersed for 30-days into a 3% brine solution (Abe & Yasuda, 2005). Additionally to protect steel from corrosion during outdoor exposure over a long period of time, LDH barrier acts as a rust-inhibiting pigment when dispersed into an adhesion-promoting primer (Nagai et al., 2003; Sato et al., 2002) and such a "rustproofing" pretreatment of steel surfaces with a "paint-like" coating comprising a LDH formulation prior to painting was also reported (Yung et al., 2002; Sato et al., 2003) and related to overall coatings efficiency in the way that the procedure is enabling to reduce the number of coating stages (Nagai et al., 2004). For some coatings comprising zinc powder, formulations with hydrocalumite (Ca_2Al-LDH) as a fixing agent for corrosive ions provide a good adhesion to blast-treated steel plates as well as a high corrosion resistance in salt spray test (Kihira et al., 2007). Electrogeneration of LDH particles reported for magnesium-based substrate was also found to improve the adhesion of subsequently applied paintings in addition to an enhanced corrosion resistance. Intercalated with corrosion inhibitor anions electrodeposited hydrotalcite films were claimed to adsorb corrosive ions and release corrosion inhibitor, and in the same time the LDH layer being firmly bound to the magnesium alloy (Yu et al., 2009). Similarly cathodic deposited coatings comprising LDH-like materials with an organic acid salt were claimed to provide excellent edge corrosion prevention, a high throwing power and chemical resistance along with other beneficial properties for film formation (Nakau et al., 2000).

For the release aspect, the beneficial effect to load the anti-corrosion agent into a LDH host structure is illustrated on Al 2024 as a substrate using nitrate Zn_2Al-LDH ($Zn_2Al(OH)_6(NO_3)$ $\bullet 2H_2O$) that is a well-known model system for its anticorrosion efficiency (Miyata, 1988) (Fig. 14. & Fig. 15.). Figure 14 displays photographed specimens of epoxy based primer coated panels after an immersion test. The deleterious effect of the free nitrate is obvious if the extended corroded zones all over the exposed substrate area are compared to the much smaller domains that were formed when the equivalent amount of 2000 ppm nitrate had been embarked within the interstitial space of LDH platelets. This demonstrates the difference in (mobile) ionic strength throughout the test duration that facilitates diffusion of corrosive electrolyte in combination with a barrier effect imposed by the layered LDH particles. Additionally the beneficial effect of providing the inhibitor via an efficient triggered release from LDH hosts was observed on scratched and immersed panels which had been coated with the very same primer formulations and subsequently with a top coat (Fig. 15).

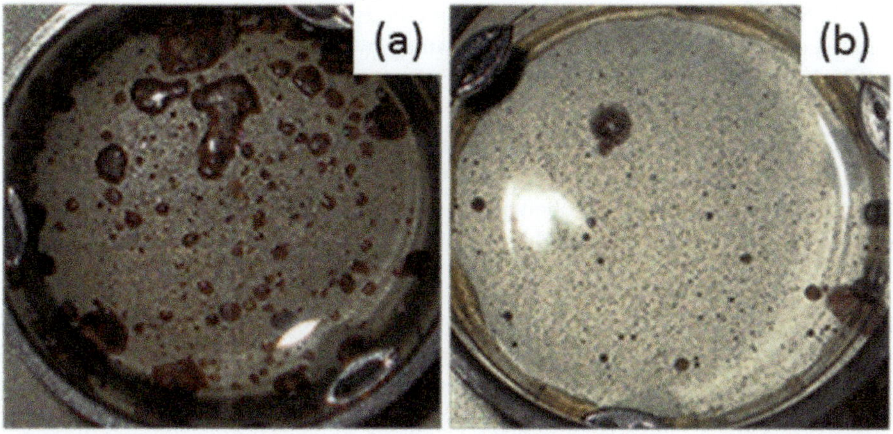

Figure 14. Epoxy primer coated panel, the primer comprising $NaNO_3$ (a) or LDH/NO_3 (b) respectively. Panels were exposed to 0.005 M NaCl for 720 hours, the diameter of the exposed area being 4,6 cm. (Courtesy of Dr. Thomas Stimpfling)

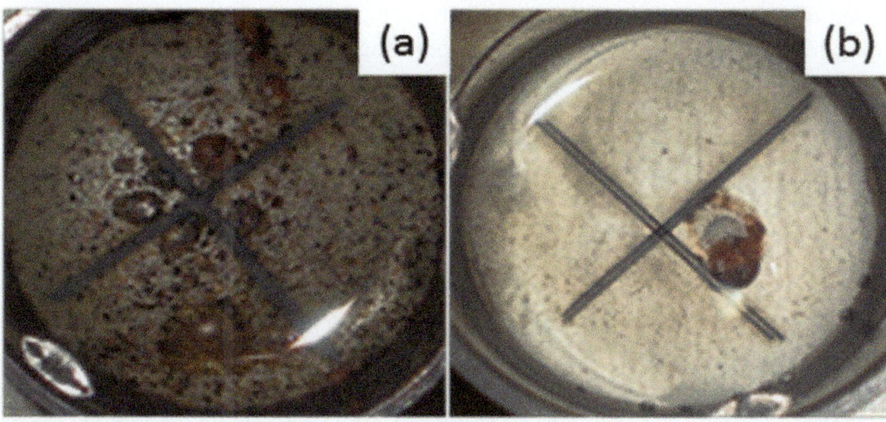

Figure 15. The same experimental set-up as described for figures 3a and 3b, except for an additional top coat being applied on the primer layer and that the coatings were scratched before exposure. (Courtesy of Dr. Thomas Stimpfling)

From pioneering work underlining the ability of LDH to act as a container for corrosion inhibiting species and in particular against filiform corrosion (Miyata, 1988), a lot of literature is now available. Most of them report the interest of LDH-based anti-corrosion coatings against blistering, pitting corrosion, visible and non-visible (sub layer electrolyte accumulation) damage zones involved in filiform corrosion and other corrosive reactions or coating failures. Similarly numerous patents report the relevance of LDH additives to prevent metal substrate corrosion; this is particularly well documented for Al-based alloys and steel substrates. Thus for coating AA2024-T3 alloy, a suitable composition was proposed that uses 2,5-dimercapto-1,3,4-thiadiazole (DMTD) as guest anion intercalated between LDHs layers via the reconstruction method (Sinko & Kendig, 2005). DMTD does not form insoluble compounds with Al^{3+} but rather interacts locally at copper inclusions due to the high chemical affinity between the thiol and copper that results in a stable chemisorbed layer of the corresponding complexes. Deposited on the electrochemically active copper rich domains, it inhibits the oxygen reduction reaction. In succession LDH phases loaded with different thiazole derivatives like mercaptobenzothiazole, mercaptobenzimidazole, mercaptobenzotriazole, mercaptobenzoxazole, benzotriazole, or toloyltriazole were claimed to act in this vein (Price et al., 2010).

For steel-based substrates like low carbon steel plates vanadate bearing LDH hybrids as ion-exchangeable pigments in alkyd paints provide anticorrosive performance (Chico et al., 2008, however they were found to be less effective than zinc chromate. A special treatment providing a good corrosion resistance was reported for SPCC-SB (cold rolled carbon steel sheets or coils with dull or bright finish associated with standard temperature grade) sheets that is used for automotive modules, body panels and interior parts (Okumiya T, Ikematsu, 2007): LDH phases modified with saturated aliphatic monocarboxylic acids in admixture with magnesium acetate yield gels that can be applied on the rolled sheets in the form of the pulverized material. Finally synergistic corrosion inhibition for metal substrates in general and based on LDH was claimed (Gichuhi & Novelli, 2005): organic or inorganic corrosion inhibiting interleaved agents were selected according to the criteria of non-toxicity and to their ability to form stable, insoluble metal-ligand complexes, thus resulting in the protection of metal substrates and utilizing synergistic corrosion inhibitive mechanisms.

4. Conclusions

Within this chapter it was shown that modern and future automotive coatings might be realized which could either enhance resistance towards impact and corrosion to traditional systems or maintain their accustomed performance in more eco-efficient systems comprising less individual layers or in significantly thinner coatings by making use of layered particle based polymer composites. Following design principles for materials that combine tough failure with reinforcement and translating them into heterogeneous morphologies on varied length scales coatings damage due to stone impact can effectively be limited. By tuning the compatibility between the components of the polymer matrix as well as the affinity between these and layered inorganic particles of a toolbox of some polyesters, polyurethanes and layered double hydroxides such different coatings morphologies like disperse, bi-continuous, stratified as well as textured structures can be obtained, all of them performing outstandingly as a primer layer of an automotive coating system in stone impact tests.

Contrary to some other layered inorganic particles like Montmorillonite as well as compared to other encapsulation materials like polyelectrolytes or hard polymer capsules it is the chemical and structural variability in combination with both their chemical stability under moderate conditions and their structural lability in more acidic as well as basic environments that renders layered double hydroxides versatile host materials for corrosion inhibiting species. Such hybrid particles thus act on different time scales and in different modes in a polymer matrix which yields at low loadings multifunctional organic coatings:

1. Mechanical reinforcement and stress dissipation.
2. Diffusion barrier towards oxygen and humidity.
3. Triggered release of corrosion inhibitors.
4. Buffering local changes in pH value via structural decomposition.
5. Supply of chemical components for the formation of passivation layers and in the case of deposited LDH-like phases an adhesion promoting layer might be of relevance for repair coatings thus acting similar to a conversion layer.

The latter aspect deserves further investigations with respect to zinc, aluminium and magnesium based alloys that are progressively used as materials in especially light weight automotive body construction. Recent investigations of corrosion processes have shown that layered double hydroxides are deposited as passivating surface layers in the course of the anodic reaction.

Additionally related to high aspect ratios and the ability to assemble into ordered lyotropic mesophases layered double hydroxide particles based formulations could yield textured films that are attractive regarding automotive effect coatings which demand highly oriented layered effect pigments like aluminium flakes or mica platelets in order to provide highest possible viewing angle dependent optical properties of the coating. Thus it was recently proposed that the LDH based aqueous formulation that has been described in the context of impact resistance in this chapter might be used to impose its tendency to form ordered films on the pigment orientation parallel to the substrates surface (Hintze-Bruening et al., 2010).

Author details

Horst Hintze-Bruening
BASF Coatings GmbH, Germany

Fabrice Leroux
Clermont University, CNRS, France

5. References

Abe, Y & Yasuda, M. (2005). Hydrotalcite-containing inorganic coatings showing long-lasting anticorrosive property, JP2005112998.

Alemdar, A. & Buetuen, V. (2005). Interaction between a Tertiary Amine Methacrylate Based Polyelectrolyte and a Sodium Montmorillonite Dispersion and Its Rheolo-gical and Colloidal Properties. *Journal of Applied Polymer Science,* Vol.95,pp. 300–306.

Alvarez, D. ; Collazo, A. ; Hernandez, M. ; Novoa, XR. & Perez, C. (2010). Corrosion protective properties of hydrotalcites doped hybrid sol-gel coatings on aluminium substrates, *Materials Science Forum,* Vol. 636-637, pp. 996-1003.

Balazs, A.C.; Singh, C. & Zhulina, E. (1998). Modeling the Interactions between Polymers and Clay Surfaces through Self-Consistent Field Theory. *Macromolecules,* Vol. 31, pp. 8370-8381.

Bender, H. (1969). The mechanical properties of films and their relation to paint chipping. *Journal of Applied Polymer Science*, Vol. 13, pp. 1253-1264.

Bonderer, L.J.; Studart, A.R. & Gauckler, L.J. (2008). Bioinspired Design and Assembly of Platelet Reinforced Polymer Films. *Science*, Vol. 319, pp. 1069-1073.

Bookin, S.; Cherkashin, V.I. & Drits, V.A. (1993). Polytype diversity of the hydrotalcite-like minerals. II.

Determination of the polytypes of experimentally studied varieties. *Clays and Clay Minerals*, Vol.41, pp. 558-564.

Buchheit, RG.; Guan, H.; Mahajanam, S. & Wong, F. (2003). Active corrosion protection and corrosion sensing in chromate-free organic coatings. *Progress in Organic Coatings*, vol.47, pp. 174-182.

Buchheit, RG. &Martinez, MA. (1998). Formation of corrosion-resistant oxide coating on metal or alloy surface in alkaline bath, US985756218.

Cauvin, S.; Colver, P.J. & Bon, S.A.F. (2005). Pickering Stabilized Miniemulsion Polymerization: Preparation of Clay Armored Latexes. Macromolecules, Vol. 38, pp. 7887-7889

Chen, H.; Zhang, F.; Fu, S. & Duan, X. (2006). In situ microstructure control of oriented layered double hydroxide monolayer films with curved hexagonal crystals as superhydrophobic materials. *Advanced Materials*, Vol.18, pp. 3089-3093.

Chico, B.; Simancas, J. ; Vega, JM.; Granio, N.; Diaz, I.; de la Fuente, D. & Morcillo, M. (2008). Anticorrosive behavior of alkyd paints formulated with ion-exchange pigments. *Progress in Organic Coatings*, Vol.61, pp. 283-290.

Cousin, F.; Cabuil, V. & Levitz, P. (2002). Magnetic Colloidal Particles as Probes for the

Determination of the Structure of Laponite Suspensions. *Langmuir*, Vol. 18, pp. 1466-1473.

Dhar, S.; Krajac, T.; Ciampini, D. & Papini, M. (2005). Erosion mechanisms due to impact of single angular particles. *Wear*, Vol. 258, pp. 567-579.

Di, J. & Sogah, D.Y. (2006). Exfoliated Block Copolymer/Silicate Nanocomposites by One-Pot, One-Step in-Situ Living Polymerization from Silicate-Anchored Multifunc-tional Initiator. *Macromolecules*, Vol. 39, pp. 5052-5057.

Dijkstra,M.; Hansen, J.-P. & Madden, P.A. (1997). Statistical model for the structure and gelation of smectite clay suspensions. *Physical Review E*, Vol. 55, pp. 3044-3053.

Ding, H.; Hawthorn, GA. & Hihara, LH. (2009). Inhibitive effect of seawater on the corrosion of particulate-reinforced aluminum-matrix composites and monolithic aluminum alloy. *Journal of Electrochemical Society*, Vol.156, pp. C352-C359.

Dioh, N.N. & Williams J.G. (1994). The impact behavior of paints. *Journal of Materials Science*, Vol. 29, pp. 6091-6096.

Fukushima, Y.; Okada, A.; Kawasumi, M; Kurauchi, T. & Kamigaito, O. (1988). Swelling behavior of Montmorillonite by Poly-6-amide. *Clay Minerals*, Vol. 23, pp. 27-34.

Gao, H.; Ji, B.; Jaeger, I.L.; Arzt, E. & Fratzl, P. (2003). Materials become insensitive to flaws at nanoscale: Lessons from nature. *Proceedings of the National Academic Society*, Vol. 100, pp. 5597-5600.

Gao, YF.; Nagai, M.; Masuda, Y.; Sato, F.; Seo WS. & Koumoto, K. (2006). Surface precipitation of highly porous hydrotalcite-like film on Al from a zinc aqueous solution. *Langmuir*, Vol.22, pp. 3521-3527.

Garcés, J.M.; Moll, D.J.; Bicerano, J.; Fibiger, R. & McLeod, D.G. (2000). Polymeric Nanocomposites for Automotive Applications. *Advanced Materials,* Vol. 12, pp. 1835-1839.
Gatos, K.G. & Karger-Kocsis, J. (2007). Effect of the aspect ratio of silicate platelets on the mechanical and barrier properties of hydrogenated acrylonitrile butadiene rubber (HNBR)/layered silicate nanocomposites. *European Polymer Journal,* Vol. 43, pp. 1097-1104.
Géraud, E.; Prevot, V.; Ghanbaja J. & Leroux, F. (2006). Macroscopically ordered hydrotalcite-type materials using self-assembled colloidal crystal template. *Chemistry of Materials,* Vol.18, pp. 238-240.
Gersappe, D. (2002). Molecular Mechanisms of Failure in Polymer Nanocomposites. *Physical Review Letters,* Vol. 89, pp. 058301-1 – 058301-4.
Gichuhi, T. & Novelli W. (2005). Synergistic corrosion inhibitor for metal substrates based on hydrotalcite or layered double hydroxide complexing agents, US20050235873.
Ginzburg, V.V.; Qiu, F.; Paniconi, M.; Peng, G.; Jasnow, D. & Balazs, A.C. (1999). Kinetic model of phase separation in binary mixtures with hard mobile impurities. *Physical Review E,* Vol. 60, pp. 4352-4359.
Ginzburg, V.V.; Qiu, F.; Paniconi, M.; Peng, G.; Jasnow, D. & Balazs, A.C. (1999). Simualtion of hard particles in a phase-separating binary mixture. Physical Review Letters, Vol. 82, pp. 4026-4029.
Gosh, B. & Urban, M. (2009). Self-repairing oxetane substituted chitosan polyurethane networks. *Sciene,* Vol.323, pp. 1458-1460
Guo, X. ; Xu, S. ; Zhao, L. ; Lu, W. ; Zhang, F. ; Evans, DG. & Duan, X. (2009). One-Step Hydrothermal Crystallization of a Layered Double Hydroxide/Alumina Bilayer Film on Aluminum and Its Corrosion Resistance Properties. *Langmuir,* Vol. 25, pp. 9894-9897.
Gupta, V.; Gupta, S. & Miura, N. (2008). Potentiostatically deposited nanostructured Co_xNi_{1-x} layered double hydroxides as electrode materials for redox-supercapacitors. *Journal of Power Sources,* Vol. 175, pp. 680- 685.
Gurski, JA.; Blough, SD.; Luna, C.; Gomez, C.; Luevano A.N. & Gardner, EA. (2006). Particle-particle interactions between layered doublé hydroxide nanoparticles. *Journal of American Chemical Society,* Vol. 128, pp. 8376-8377.
Hibino T. & Kobayashi, M. (2005). Delamination of layered double hydroxides in water. *Journal of Materials Chemistry,* Vol.15, pp. 653-656.
Hintze-Bruening, H. ; Troutier-Thuilliez, A.-L. & Leroux, F. (2009). Layered particle-based polymer composites for coatings: Part II—Stone chip resistant automotive coatings. *Progress in Organic Coatings,* Vol. 64, pp. 193–204.
Hintze-Bruening, H. ; Steiner, H.-P.; Leroux, F.; Troutier-Thuilliez, A.-L. & Stimpfling, T. (2010). WO2010130308.
Hintze-Bruening, H. ; Troutier-Thuilliez, A.-L. & Leroux, F. (2011). Layered particle-based polymer composites for coatings: Part III— Textured coatings obtained via lyotropic liquid Crystals. *Progress in Organic Coatings,* Vol. 70, pp. 240-244.
Hintze-Bruening, H.; Dornbusch, M.; Toews, S. & Bremser, W. (2011). Method for autophoretic coating, coating agent and multilayer paint finish. WO2011138290.
Hu G. & O'Hare, D. (2005). Unique layered double hydroxide morphologies using reverse microemulsion synthesis. *Journal of American Chemical Society,* Vol.127, pp. 17808-17813.

Huang, X. & Brittain, W.J. (2001). Synthesis and Characterization of PMMA Nanocomposites by Suspension and Emulsion Polymerization. *Macromolecules*, Vol. 34, pp. 3255-3260.

Hughes, A.E.; Cole, I.S.; Muster, T.H. & Varley, R.J. (2010). Designing green, self-healing coatings for metal protection. *NPG Asia Materials*, Vol.2, pp. 143-151.

Indira, L. & Kamath, PV. (1994). Electrogeneration of base by cathodic reduction of anions: novel one-step route to unary and layered double hydroxides (LDHs). *Journal of Materials Chemistry*, Vol.4, pp. 1487-1490.

Ji, Y.-Q.; Black,L.; Weidler, P.G. & Janek, M. (2004). Preparation of Nanostructured Materials by Heterocoagulations Interaction of Montmorillonite with Synthetic Hematite Particles. *Langmuir*, Vol. 20, pp. 9796-9806.

Joensson, B.; Labbez, C. & Cabane, B. (2008). Interaction of Nanometric Clay Platelets. *Langmuir*, Vol. 24, pp. 11406-11413.

Jung, C. ; Schimakura, T. ; Maurus, N. & Domes, H. (2002). Method for pretreating and/or coating metallic surfaces with a paint-like coating prior to forming and use of substrates coated in this way, WO02031064.

Kanai, T. & Nomura H. (2009). Metals surface-treated with crosslinked polysiloxanes, their manufacture, and surface treating liquids containing alkoxysilanes or their hydrolyzates, JP2009262402.

Karim, A.; Douglas, J.F.; Nisato, G.; Liu, D.-W. & Amis, E.J. (1999). Transient Target Patterns in Phase Separating Filled Polymer Blends. *Macromolecules*, Vol. 32, pp. 5917-5924.

Kihira, H.; Aiga, T.; Imai, A.; Hiramatsu, K.; Mitsutsuka, Y.; Nagai, M.; Sato, T. & Matsumoto, T. (2007). Anticorrosive zinc dust-containing coating compositions, JP2007284600.

Kreis, W.; 26th Eurpean Car Body Conference, Bad Nauheim, 27 Febr. 2008

Kugge, C.; Vanderhoek, N. & Bousfield, D.W. (2011). Oscillatory shear response of moisture barrier coatings containing clay of different shape factor. *Journal of Colloid and Interface Science*, Vol. 358, pp. 25-31.

Lagaly, G.; Stange, H. & Weiss, A. (1973). Adsorption of long chain molecules onto aromatic swelling liquids in mica type layer silicate. *Proceedings of the International Clay Conference, Madrid*, pp. 693-704.

Lagaly, G.; Beneke, K. & Weiss, A. (1975). Magadiite and H-Magadiite: II. H-Magadiite and its intercalation compounds. *American Mineralogist*, Vol.60, pp. 650-658

Lagaly, G. (1981). Characterization of clays by organic compounds. *Clay Minerals*, Vol. 16, pp. 1-21.

Lan, T.; Kaviratna, P.D. & Pinnavaia, T.J. (1996). Epoxy self-polymerization in smectite clays. *Journal of Physics and Chemistry of solids*, Vol.57, pp. 1005-1010.

Lepoittevin, B.; Pantoustier, N.; Devalckenaere, M.; Alexandre, M.; Kubies, D.; Calberg, C. ; Jérôme, R. & Dubois, P. (2002). Poly(e-caprolactone)/Clay Nanocomposites by in-Situ Intercalative Polymerization Catalyzed by Dibutyltin Dimethoxide. *Macromolecules*, Vol. 35, pp. 8385-8390.

Leroux, F.; Stimpfling, T. & Hintze-Bruening, H. (2012). Relevance and Performance of LDH Platelets in Coatings. *Recent Patents on Nanotechnology*, Vol. 6, accepted

Lin, J. K.; Hsia, C. L. & Uan, J. Y. (2007). Characterization of Mg,Al-hydrotalcite conversion film on Mg alloy and Cl- and anion-exchangeability of the film in a corrosive environment. *Scripta Materialia,* Vol.56, pp. 927-930.

Lin, JK. & Uan, JY. (2009). Formation of Mg,Al-hydrotalcite conversion coating on Mg alloy in aqueous HCO_3^-/CO_3^{2-} and corresponding protection against corrosion by the coating. *Corrosion Science,* vol.51, pp. 1181- 1188.

Liu, Z.; Ma, R.; Osada, M.; Iyi, N.; Ebina, Y.; Takada, K. & Sasaki, T. (2006). Synthesis, anion exchange, and delamination of Co-Al layered double hydroxide: assembly of the exfoliated nanosheet/polyanion composite films and magneto-optical studies. *Journal of American Chemical Society,* Vol.128, pp. 4872-4880.

Lonyuk, M.; Bosmab, M.; Riemslag, A.C.; Zuidema, J.; Bakker, A. & Janssen, M. (2007). Stone-impact damage of automotive coatings: A laboratory single-impact tester. *Progress in Organic Coatings,* Vol.58, pp. 241-247.

Lonyuk, M.; Bosmab, M.; Vijverberg, C.A.M.; Bakker, A. & Janssen, M. (2008). Relation between chip resistance and mechanical properties of automotive coatings. *Progress in Organic Coatings,* Vol.61, pp. 308-315.

Luan, L.; Li, W.; Liu, S. & Sun, D. (2009). Phase behavior of mixtures of positively charged colloidal platelets and nonadsorbing polymer. *Langmuir,* Vol. 25, pp. 6349-6356.

Mahajanam, S. P. V. & Buchheit, R. G. (2008). Characterization of Inhibitor Release from Zn-Al-$[V_{10}O_{28}]^{6-}$ Hydrotalcite Pigments and Corrosion Protection from Hydrotalcite-Pigmented Epoxy Coatings. *Corrosion,* Vol.64, pp. 230-240.

Makowski, M.P.; Martz, J.T.; Novak, C.A. & Verardi, C.A. (2005). Coatings with improved chip resistance and methods of making the same. WO2005052077.

Milczewska, K.; Voekel, A. & Jeczalik, J. (2003). The use of Flory-Huggins parameter to characterization of polymer/filler interaction. *Macromolecular Symposia,* Vol. 194, pp. 305-311.

Miyata S. (1988). Filiform corrosion-resistant primer coating composition and preventing filiform corrosion with hydrotalcite inhibitors. US884761188.

Mourad, M.C.D. Devid, E.J.; van Schooneveld, M.M.; Vonk, C. & Lekkerkerker, H.N.W. (2008). Formation of nematic liquid crystals of sterically stabilized layered double hydroxide platelets. *Journal of Physical Chemistry B,* Vol.112, pp. 10142-10152.

Mourchid, A.; Delville, A.; Lambard, J.; LeColier, E. & Levitz, P. (1995). Phase Diagram of Colloidal Dispersions of Anisotropic Charged Particles: Equilibrium Properties, Structure, and Rheology of Laponite Suspensions. *Langmuir,* Vol. 11, pp. 1942-1950.

Nagai, M.; Sato, T.; Taki, T.; Yamazaki, A. & Tanabe H. (2003). Corrosion protection of weather-resistant steels with moisture-curable primers. JP2003082477.

Nagai, M.; Taki, T.; Sato, T.; Matsuno, H. & Yamazaki, A. (2004). Corrosion prevention of weather- and rust-resistant steels with less number of coating stages. JP2004097945.

Nakao, F.; Sugisaki, K. & Tominaga, A. (2000). Cationically electrodepositable coating material. EP001046684.

Ogawa M. & Kaiho, H. (2002). Homogeneous precipitation of uniform hydrotalcite particles. *Langmuir,* Vol.18, pp. 4240-4242.

Okamoto, K.; Sasaki, T.; Fujita, T. & Iyi, N. (2006). Preparation of highly oriented organic-LDH hybrid films by combining the decarbonation, anion-exchange, and delamination processes. *Journal of Materials Chemistry,* Vol.16, pp. 1608-1616.

Okumiya, T. & Ikematsu, D. (2007). Aqueous anticorrosive coatings containing layered double hydroxides exfoliated in water and metalis substrates coated therewith. JP2007039549.

Pang, J.W.C. & Bond, I.P. (2005). 'Bleeding composites' – damage detection and self-repair using a biomimetic approach. *Composites: Part A,* Vol. 36, pp. 183-188

Paulhiac, JL. & Clause, O. (1993). Surface coprecipitation of cobalt(II), nickel(II), or zinc(II) with aluminium(III) ions during impregnation of γ-alumina at neutral pH. *Journal of American Chemical Society,* Vol.115, pp. 11602-11603.

Pignon, F.; Magnin, A.; Piau, J.-M.; Cabane, B.; Lindner, P. & Diat, O. (1997). Yield stress thixotropic clay suspension: investigations of structure by light, neutron and X-ray scattering. *Physical Review E,* Vol. 56, pp. 3281-3289.

Poznyak, S. K.; Tedim, J.; Rodrigues, L. M.; Salak, A. N.; Zheludkevich, M. L.; Dick, L. F. P. & Ferreira, M. G. S. (2009). Novel Inorganic Host Layered Double Hydroxides Intercalated with Guest Organic Inhibitors for Anticorrosion Applications. *ACS Applied Materials & Interfaces* Vol.1, pp. 2353-2362.

Prevot, V.; Forano, F. & Besse, JP. (2005). Hydrolysis in polyol: new route for hybrid-layered double hydroxides preparation. *Chemistry of Materials,* Vol.17, pp. 6695-6701.

Price, C.; Matzdorf, C.; Nickerson, W. & Lipnickas, E. (2010). Magnesium-rich anticorrosive coating compositions. US20100197836.

Ramamurthy, A.C.; Lorenzen, W.I. & Bless, S.J. (1994). Stone impact damages to automotive paint finishes: an introduction to impact physics and impact induced corrosion. *Progress in Organic Coatings,* Vol.25, pp. 43-71.

Rao, Y.Q. & Pochan, J.M. (2007). Mechanics of Polymer-Clay Nanocomposites. *Macromolecules,* Vol. 40, pp. 290-296.

Ritchie, R.O. (2011). The conflicts between strength and toughness. *Nature Materials,* Vol. 10, pp. 817-822.

Roesler, M.; Klinke, E. & Kunz, G. (1997). Glashohlkugeln als stossabsorbierende Fuellstoffe. Farbe und Lack, Vol.103, pp. 49-54.

Rutherford, K.L.; Trezona, R.I.; Ramamurthy, A.C. & Hutchings, I.M. (1997). The abrasive and erosive wear of polymeric paint films. *Wear,* Vol. 203-204, pp. 325-334.

Ryntz, R.A.; Ramamurthy, A.C. & Holubka, J.W. (1995). Stone impact damage to painted plastic substrates. *Journal of Coatings Technology,* Vol. 67, pp. 23-31.

Ryntz, R.A.; Ramamurthy, A.C. & Mihora, D.J. (1995). Thermal and impact induced stress failure in painted TPO: the role of surface morphology. Journal of Coatings Technology, Vol. 67, pp. 35-46.

Sato, T.; Nomura, R.; Maekawa, S. & Yamazaki, A. (2002). Agent for steel surface preparation prior to painting. JP2002285362.

Sato, T.; Yamazaki, A. & Matsuno, H. (2003). Surface conditioning agents of steel for paint coating. JP2003328162.

Shah, D.; Maiti, P.; Gunn, E.; Schmidt, D.F.; Jiang, D.D.; Batt, C.A. & Giannelis, P. (2004). Dramatic enhancements in toughness of polyvinylidenefluoride nanocomposites via nano-clay directed crystal structure and morphology. *Advanced Materials,* Vol. 16, pp. 1173-1177.

Shah, D.; Maiti, P.; Jiang, D.D.; Batt, C.A. & Giannelis, P. (2005). Effect of nanoparticle mobility on toughness of polymer nanocomposites. *Advanced Materials,* Vol. 17, pp. 525-528.

Shalkevich, A.; Stradner, A.; Bhat, S.K.; Muller, F. & Schurtenberger, P. (2007). Cluster, Glass, and Gel Formation and Viscoelastic Phase Separation in Aqueous Clay Suspensions. *Langmuir,* Vol. 23, pp. 3570-3580.

Sinko, J. (2001). Challenges of chromate inhibitor pigments replacement in organic coatings. *Progress in Organic Coatings,* Vol.42, pp. 267-282.

Sinko, J. & Kendig, MW. (2005). Pigment-grade corrosion inhibitors for protective coating on metal substrates. WO2005003408.

Studart, A.R.; Gonzenbach, U.T.; Tervoort, E. & Gauckler, L.J. (2006). Processing Routes to Macroporous Ceramics: A Review. Journal of the American Ceramic Society, Vol.89, pp. 1771–1789.

Toohey, K.S.; Sottos, N.R. & White, S.R. (2009). Characterization of Microvascular-Based Self-healing Coatings. *Experimental Mechanics,* Vol.49, pp. 707-717

Triantafyllidis, K.S.; LeBaron, P.C.; Park, I. & Pinnavaia, T.J. (2006). Epoxy-Clay Fabric Film Composites with Unprecedented Oxygen-Barrier Properties. *Chemistry of Materials,* Vol. 18, pp. 4393-4398.

Troutier-Thuilliez, A.-L. ; Taviot-Guéhoa, C. ; Cellier, J. ; Hintze-Bruening, H. & Leroux, F. (2009). Layered particle-based polymer composites for coatings: Part I. Evaluation of layered double hydroxides. *Progress in Organic Coatings,* Vol. 64, pp. 182–192.

Troutier-Thuilliez, A.-L.; Hintze-Bruening, H.; Taviot-Guého, C.; Verney, V. & Leroux, F. (2011). Exfoliation and liquid crystal phase formation of layered double hydroxide into waterborne polyurethane coatings. *Soft Matter,* Vol.7, pp. 4242-4251.

Twite, RL. & Bierwagen, GP. (1998). Review of alternatives to chromate for corrosion protection of aluminium aerospace alloys. *Progress in Organic Coatings,* Vol.33, pp. 91-100.

Uan, J-Y.; Yu, B-L. & Pan X-L. (2008). Morphological and microstructural characterization of the aragonitic $CaCO_3$/Mg,Al-hydrotalcite coating on Mg - 9 Wt Pct Al - 1 Wt Pct Zn alloy to protect against corrosion. *Metal Materials Transaction A,* Vol.39, pp. 3233-3245.

Veleva, L.; Chin, J. & del Amo, B. (1999). Corrosion electrochemical behavior of epoxy anticorrosive paints based on zinc molybdenum phosphate and zinc oxide. *Progress in Organic Coatings,* Vol.36, pp. 211-216.

Volovitch, P.; Vu, T.N.; Allély, C. ; Aal, A.A. & Ogle K. (2011). Understanding corrosion via corrosion product characterization: II. Role of alloying elements in improving the corrosion resistance of Zn–Al–Mg coatings on steel. Corrosion Science, Vol.53, pp. 2437-2445

Wang, JA.; Morales, A.; Bokhimi, X.; Novaro, O.; Lopez, T. & Gomez, R. (1999). Cationic an anionic vacancies in the crystalline phases of sol-gel magnesia-alumina catalysts. *Chemistry of Materials,* Vol.11, pp. 308-313.

Wang, J. ; Li, D. ; Yu, X.; Jing, X.; Zhang, M. & Jiang, Z. (2010). Hydrotalcite conversion coating on Mg alloy and its corrosion resistance. *Journal of Alloys Compounds,* Vol. 494, pp. 271-274.

Wang, X.; Gao, Y.; Mao, K.; Xue, G.; Chen, T.; Zhu, J.; Li, B.; Sun, P.; Jin, Q. & Ding, D. (2006). Unusual Rheological Behavior of Liquid Polybutadiene Rubber/Clay Nanocomposite Gels: The Role of Polymer-Clay Interaction, Clay Exfoliation, and Clay Orientation and Disorientation. *Macromolecules*, Vol. 39, pp. 6653-6660.

Weber, B.; Bremser, W. & Hiltrop, K. (2009). Creating new materials with melamine resins. *Progress in Organic Coatings*, Vol.64, pp. 150-155.

Williams, G. & McMurray, H. N. (2003). The mechanism of group (I) chloride initiated filiform corrosion on iron. *Electrochemistry Communications*, Vol.5, pp. 871-877.

Xu, ZP.; Stevenson, G.; Lu, CQ.; Lu, GQ.; Bartlett, P. & Gray, P. (2006). Stable suspension of layered double hydroxide nanoparticles in aqueous solution. *Journal of American Chemical Society*, Vol.128, pp. 36-37.

Yu, X.; Wang, J.; Zhang, M.; Yang, L.; Li, J.; Yang, P. & Cao, D. (2008). Synthesis, characterization and anticorrosion performance of molybdate pillared hydrotalcite/in situ created ZnO composite as pigment for Mg-Li alloy protection. *Surface and Coatings Technology*, Vol.203, pp. 250-255.

Yu, X.; Li, J.; Wang, J.; Jing, X. & Zhang, M. (2009). Electrodeposition method for forming corrosion inhibitor anion-intercalated hydrotalcite film on magnesium alloy surface. CN200910072431.

Yurekli, K.; Karim, A.; Amis, E.J. & Krishnamoorti, R. (2003). Influence of Layered Silicates on the Phase-Separated Morphology of PS-PVME Blends. *Macromolecules*, Vol. 36, pp. 7256-7267.

Yurekli, K.; Karim, A.; Amis, E.J. & Krishnamoorti, R. (2004). Phase Behavior of PS-PVME Nanocomposites. *Macromolecules*, Vol. 37, pp. 507-515.

Zang, J.; Luan, L.; Zhu, W.; Liu, S. & Sun, D. (2007). Phase behavior of aqueous suspensions of Mg_2Al layered double hydroxide : the competition among nematic ordering, sedimentation, and gelation. *Langmuir*, Vol.23, pp. 5331-5337.

Zehnder, A.T.; Ramamurthy, A.C.; Bless, S.J. & Brar, N.S. (1993). Stone impact damages to automotive paint finishes: measurement of temperature rise due to impact. *International Journal of Impact Engineering*, Vol.13, pp. 133-143.

Zhang, F.; Zhao, L.; Chen, H.; Xu, S.; Evans, D. G. & Duan, X. (2008a). Corrosion Resistance of Superhydrophobic Layered Double Hydroxide Films on Aluminum. *Angewandte Chemie International Edition*, Vol.47, pp. 2466-2469.

Zhang, F.; Sun, M.; Xu, S.; Zhao, L. & Zhang, B. (2008b). Fabrication of oriented layered double hydroxide films by spin coating and their use in corrosion protection. *Chemical Engineering Journal*, Vol.141, pp. 362-367.

Zhao, Y.; Li, F.; Zhang, R.; Evans, DF. & Duan, X. (2002). Preparation of layered double-hydroxide nanomaterials with a uniform crystallite size using a new method involving separate nucleation and aging steps. *Chemistry of Materials*, Vol.14, pp. 4286-4291.

Zheludkevich, M. L.; Poznyak, S. K.; Rodrigues, L. M.; Raps, D.; Hack, T.; Dick, L. F.; Nunes, T. & Ferreira, M. G. S. (2010). Active protection coatings with layered double hydroxide nanocontainers of corrosion inhibitor. *Corrosion Science*, Vol.52, pp. 602-611.

Zouari, B. & Touratier, M. (2002). Simulation of organic coating removal by particle impact. *Wear*, Vol. 253, pp. 488-497.

Chapter 2

Lubricating Aspects of Automotive Fuels

Evripidis Lois and Panagiotis Arkoudeas

Additional information is available at the end of the chapter

1. Introduction

Back in the 1960s, the term "lubricity" was defined by Appeldorn and Dukek as: "If two liquids have the same viscosity, and one gives lower friction, wear or scuffing, it is said to have better lubricity". It should be noted, however, that this definition was not strictly applied and many researchers carried out lubricity experiments on fuels based on their own understanding of the concept. The lubricating ability of fuels, because of their very low viscosity, depends mostly on their boundary film-forming properties. Some moving parts of diesel fuel pumps and injectors are protected from wear by the fuel. To avoid excessive wear, the fuel must have some minimum level of lubricity. Lubricity is the ability to reduce friction between solid surfaces in relative motion. The lubrication mechanism is a combination of hydrodynamic lubrication and boundary lubrication. In hydrodynamic lubrication, a layer of liquid prevents contact between the opposing metal surfaces. For diesel fuel pumps and injectors, the liquid is the fuel itself and viscosity is not the key fuel property as one could profoundly expect. The history of fuel lubricity is associated with problems in engine performance as liquid-hydrocarbon based fuels must possess a minimum of lubricating ability to be able to protect high-pressure injection pumps and related fuel supply equipment from wear. The topic of gasoline lubricity has recently become more urgent with the introduction of direct-injection gasoline engines, which will necessitate high-pressure gasoline injection pumps, a development that is most likely to place considerably more emphasis on the lubricating ability of gasoline, accelerating wear especially in rotary distributor fuel pumps. According to pump manufacturers this loss of lubricity may be the difference between fuels from a controlled laboratory environment and a cost-conscious production environment. [1-7].

In the last decades, the demand for both gasoline and automotive diesel fuel has increased rapidly and strongly. In the early 1990s, world gasoline production rose to about 800 billion litres, about half of which was consumed in the United States. The world demand for gasoline is estimated to be an average 20 million barrels a day. The United States is the largest consumer with an average consumption of around 8.9 million barrels a day in 2008, accounting for over 40% of global consumption. This was over 9 million barrels a day in 2007. However, the production of gasoline in any country depends on the type of economy it follows. For example, while US have adopted a gasoline based economy, India is largely a diesel based economy, leading to more production and consumption of gasoline in US and High-Speed Diesel (HSD) in India. Moreover, the light sweet crude oil used by US, yields more gasoline. So, there are considerable differences in the relative demand for gasoline and diesel fuels from region to region. The refinery industry has met these challenges by improving efficiency of crude oil utilisation, increasing the depth of crude oil processing and reducing product losses as well as adjusting refining processes to maximise production of either gasoline or diesel fuel [1-4, 8].

Fuel composition is a key factor in determining the lubricity of fuels, since it depends not only on the crude oil the fuel is prepared from, but also on the refinery process, finishing process, and blending method. The gradual increase in severity of refinement in recent years to meet tightening environmental regulations has simultaneously reduced the concentration of many potential lubricity agents and thus made fuel lubricity poorer. Gasoline is the lightest liquid fraction of petroleum, boiling between about 30°C and 200°C, i.e. containing mainly C5 to C12 hydrocarbons. It is reasonable to infer that the inherent lubricity of gasoline will be poorer than that of aviation fuel and diesel fuel due to the lighter distillation cut, in which natural antiwear impurity concentration will be lower. Fortunately and till now, the lubricity requirements of gasoline are generally much lower than for diesel since gasoline fuel injection systems inject fuel upstream of the inlet valves and thus operate at much lower pressures than diesel fuel pumps.

In 1990s, the amount of sulphur, nitrogen and aromatics in diesel fuels was reduced by severe hydrotreating to minimize SO_x emissions from diesel powered vehicles. The use of low sulphur diesel fuels led to numerous pump failures. To combat the loss of this lubrication, packages of additives that increase lubricity could be blended with the fuel prior to distribution [9]. The lubricity characteristics of diesel fuel are similar to aviation turbine fuels, up to the middle of the 1980s, but the lubricity of diesel fuels was not considered a significant factor that could lead to serious problems and little work concerning diesel lubricity had been carried out. There was not widely accepted test method existed to determine the lubricity of diesel fuels. In the 1990s, Sweden and United States introduced low sulphur, low aromatic diesel fuels and this was followed by other countries, including Canada, Switzerland, Austria, and Germany. Soon after the introduction of these environmental diesel fuels in the Scandinavian and Californian markets in the early 1990s, a number of injector equipment failures were reported from all manufacturers. These failures took place in passenger cars working with Bosch rotary pumps after only 3000 to 10,000 km. In Europe and the USA, such fuels have been shown to reduce the life of distributor type pumps by up to 95%. Field trials and pump rig durability testing of both Swedish Class 1 and 2 showed that their inherent lubricity was unacceptable [9-15]. Diesel fuel work has revealed that humidity, which reflects environmental water vapour pressure, can have an important influence on the friction and wear, although this was not taken into account in test work until recently [1-7, 9-15]. It is possible to eliminate, at least to a large extent, the influence of humidity on test repeatability of friction, wear, and film formation by carefully controlling humidity in a relative narrow range

[1-7]. A survey of low sulphur diesel fuels (sulphur content ranged from 1 to 498 ppm) has showed that in spite of high refinement most low sulphur diesels still contain considerable residual polyaromatics (0.3 to 2.2 %wt) and diaromatics (2 to 11% wt.). A detailed analysis of data has suggested that even in low sulphur diesel fuels, polyaromatics may still play a more important role than diaromatics in determining diesel lubricity.

Gasoline lubricity is a complex phenomenon, involving many complicated and interconnecting factors, such as the presence of water, concentration of sulphur, nitrogen, oxygenates, diolefins, diaromatics, the effect of viscosity and the synergistic effect of different wear mechanisms. The lubricity mechanism of gasoline is quite different from that of diesel fuels that leads to severe adhesive wear. With low-sulphur fuels, adhesive wear is observed instead of corrosive and mild oxidative wear, and deposits build up on top land [1-4]. Metallurgy and mechanical properties of test specimens used to study wear have important effects on the lubricating mechanisms of fuels. When the hardness of the lower specimen in an HFRR test is not enough to support the generated oxide films formed by the reaction between surfaces and dissolved oxygen and the adsorption films formed on top of the oxide films by gasoline polar impurities, severe adhesion and metal transfer occur [1-7, 9-15]. The wear behaviour of some gasolines was found to be sensitive to the time of exposure to air, in that the wear values obtained fell slightly after the fuel container had been opened several times. This may be due to the oxidation of gasoline components. Gasolines containing olefins, and dienes, in particular, which have very poor oxidation stability [1-7]. However, polyaromatics in gasolines are absent due to the lower boiling range and only a few thousandths by volume of diaromatics, i.e. naphthalenes are present. More than 99% of aromatics in gasoline are monoaromatic, i.e. benzene, toluene and xylenes [1-7, 16-19].

In this chapter, the effect of various compositional and physicochemical characteristics of automotive fuels will be examined, with respect to their lubrication mechanisms. Additionally, tribological aspects (e.g. wear scar analysis, Scan Electron microscopy, etc) of automotive fuels and their mixtures with biofuels or/and bio-additives like essential oils will be presented and discussed.

2. Diesel fuel lubricity and addition of essential oils

For diesel fuel, the 1980s was an important transition period from high lubricity to moderate lubricity due to the increase of severity of refinement. Hydrotreating processes were widely used and these doubtless caused a great reduction of natural lubricity agents in diesel fuels. However, unlike aviation kerosene, no major lubricity problems were encountered in diesel fuels until the late 1980s. This may be because:

a. Diesel fuel has a higher boiling temperature range than kerosene and thus contains a larger proportion of naturally-occurring lubricity agents.
b. The severity of the refinement used in the production of early- and middle-1980s diesel fuels was moderate and this allowed enough naturally-occurring lubricity agent to survive during refining and maintain adequate lubricity.
c. In general, diesel fuels have higher viscosity, which is beneficial to film formation.

Recent concerns over the environmental impact of diesel powered equipment have driven various countries to legislate reductions in vehicle exhaust emission levels and changes to diesel fuel quality. These reductions in exhaust emissions have caused changes in engine design such as increased fuel injection pressure and control of the fuel injection. Hardware changes tend to require improved diesel fuel lubricity to avoid excessive wear of the fuel injection system [20-22].

Past studies showed that diesel lubricity is largely provided by trace levels of naturally occurring polar compounds which form a protective layer on the metal surface. Typical sulfur compounds do not confer this wear protection themselves; rather it is the heterocyclic aromatics and nitrogen and oxygen compounds that are the most important [23-25]. A complex mixture of polar compounds is found in diesel, and some are more active than others. The process of hydrotreating to reduce sulfur levels also destroys some of this natural lubricant. Other refinery processes also influence the concentration of the lubricity agents in the final fuel blend [26]. Lubricity additives have been developed to compensate for the deterioration in natural lubricity observed in low sulfur diesels. A moderate dosage of chemically suitable additive is beneficial in most cases, but if the dosage is too high, some common diesel-fuel additives can cause fuel injector deposits, water separation problems, or premature filter plugging. These problems are not always identified in the standard fuel specification tests, and result in field problems [27-29].

In this chapter, results are presented on the lubricating properties of low sulfur diesel fuels additized with ten different essential oils. Data were generated to identify the minimum concentration of the above oxygen containing compounds, which provide lubricity improvement down to the 460 μm wear scar diameter (WSD) level. The value of 460 μm was proposed by the European Committee for standardization (CEN) in February 1997, and generally adopted by the industry, as the minimum requirement for an acceptable field performance [30].

Oxygen containing compounds such as fatty acids are superior friction reducing agents. These compounds adsorb or react on rubbing surfaces to reduce adhesion between contacting asperities and limit friction, wear and seizure [31-34]. Wei and Spikes considered that the significant wear reduction was produced by oxygen compounds with phenolic-type or carboxylic acid groups and occurred at a concentration of just a few parts per million [35]. Essential oils contain omega-3 and omega-6 fatty acids like α-Linolenic acid (18:3), Linoleic acid (18:2), eicosapentaenoic acid (20:5), docosahexaenoic acid (22:6), gamma-linolenic acid (18:3), dihomo-gamma-linolenic acid (20:3) and arachidonic acid (20:4). They also contain ethers, esters and terpenes [36].

Although the lubricating efficiency of fatty acids and their derivatives has been closely examined, the impact of adding other oxygenates such as essential oils has not been examined in detail. On the other hand, the addition of these oxygenates to diesel fuel has been proposed as a method to help complete the oxidation of carbonaceous particulate matter and associated hydrocarbons, thereby reducing particulate matter PM emissions [37-39].

2.1. Experimental procedure focused on enhancing diesel lubricity by addition essential oils

To assess the impact of the selected oxygen components on the lubrication properties of low sulfur automotive diesel, two fuels (A1 and A2) that comprised distillates of the hydrodesulfurization process were obtained by a Greek refinery and were used for all the tribological experiments as base fuels. The fuel properties are presented in Table 1, along with the standard methods that were used for their determination.

Fuel code	A1	A2	Test method
Density (kg/m^3, 15°C)	0.832	0.838	ISO 12185
Viscosity (cSt, 40°C)	2.86	2.78	ISO 3104
CFPP (°C)	-8	-11	ISO 116
Flash point (°C)	61	88	ISO 2719
Cetane number	57.9	52.2	ISO 5165
Cetane Index	55.8	49.0	ASTM D 4737
Sulphur (ppm)	31	51	ASTM D 5453
Nitrogen (ppm)	13	26	ASTM D 5762
Water (ppm)	117	154	ISO 12937
Total Acid Number (mg KOH/ g)	0.12	0.15	ISO 7537
Refractive Index	1.4595	1.4745	ISO 5661
Conductivity (pS/m, 25 °C)	48	299	ISO 6297
Calorific value (Kcal/kg)	10120	9953	ASTM D-240
Residue (% m/m)	1.32	0.92	ISO 3405
Mono-aromatics (% v/v)	22.3	23.5	ASTM D-6591
Di-aromatics (% v/v)	3.8	3.7	ASTM D-6591
Poly-aromatics (% v/v)	0.10	0.16	ASTM D-6591
Distillation (°C)			
IBP	168	168	ISO 3405
10%	213	198	ISO 3405
50%	278	268	ISO 3405
90%	334	325	ISO 3405
FBP	358	349	ISO 3405
Lubricity (μm , average)	425	555	ISO 12156

Table 1. Properties of the base fuels

All tribological measurements were carried out using the HFRR apparatus, according to the CEC F-06-A-96 method. The test temperature was 60 °C and the volume of the fuel sample used was 2 ml. Relative humidity was kept between 55 and 59%, while the mean ambient temperature was 24 °C. The lubricating efficiency of the fuels was estimated by measuring the average wear scar diameter WSD of the spherical specimen by using a photomicroscope. The wear scars quoted were corrected to give WS 1.4 values. The HFRR WS 1.4 parameter is the mean WSD normalized to a standard vapour pressure of 1.4 kPa. The repeatability was calculated using the following equation (1) [17]:

$$R = 139 - (0.1648 \times WSD1.4) \quad (1)$$

The ten essential oils used in this series of experiments, included kernel peach oil, grape seed oil, pine oil, carrot seed oil, castor oil, camomile oil, laurel oil, eucalyptus oil, lavender oil and rosemary oil. The carrot seed oil, castor oil, camomile oil, laurel oil, eucalyptus oil, lavender oil, rosemary oil were obtained from Aldrich Chemical Company and they were used as received, without further purification. However, kernel peach oil, grape seed oil and pine oil that were not commercially available and were prepared by extracting their seeds with alcohols and hexane. Afterward they were vacuum distilled to receive the final products, whose properties were similar to those reported in the relevant literature. In an effort to establish the purity of the prepared compounds, the density d_4^{20} and the refractive index n_D^{20} (Table 2) were compared to the corresponding data found in literature, the purity level of the compounds eventually used in this study was estimated to be at least 95% [40-42]. The chemical constitution of the essential oils used in this study is complex and are mixtures of many compounds, as shown in Table 3 and 4.

Nomenclature	Measured density, d_4^{20}	Density of pure compounds, d_4^{20}	Measured Refractive Index, n_D^{20}	Refractive Index of pure compounds, n_D^{20}
Kernel peach oil	0.921	0.918	1.470	1.471
Grape seed oil	0.924	0.921	1.473	1.475
Pine oil	0.812	0.825	1.478	1.480

Table 2. Chemical characteristics of pure essential oils

Chemical structure	Grape seed oil	Kernel peach oil	Camomile oil	Laurel oil	Carrot seed oil
Monoterpenes			α-terpene 0-10%, α-pinene 0-10%, β-pinene 0-10%	α-pinene 4-10%, β-pinene 3-8%, sabinene 4-12%	α-pinene <13%, β-pinene <18%
Sesquiterpenes			Sabinene 0-10%, caryophyllene 0-10%		
Alcohols			Trans-pinocarveol 5%, farnesol and nerolidod (5-6%)	Linalool 4-13.5%, methyl eugenol 12%	Carotol 18%
Esters			2-methylbutyl 2-methyl propionate 0.5-25%, 2-	α-terpinyl acetate 10-	Geranyl acetate

Chemical structure	Grape seed oil	Kernel peach oil	Camomile oil	Laurel oil	Carrot seed oil
			methylpropyl butyrate 0.5-10%, 2-methylbutyl 2-methyl butyrate 0.5-25%, 2-methylpropyl 3-methyl butyrate 0-10%, propyl angelate 0.5-10%, 2-methylpropyl angelate 0.5-25%, butyl angelate 0.5-10%, 3-methylpentyl angelate 0-10%, isobutyl angelate 36-40%, isobutyl isobutyrate 4%, 2-methylbutyl methyl-2-butyrate 3%, isoamyl methyl-2-butyrate 3%, propyl angelate 1%, hexyl acetate 0.5-10%	18%	10%
Phenols					
Ketones			Pinocarvone 13%		
Aldehydes			Myrtenal 0-10%		
Oxides			1,8-cineole 0-25%	1,8-cineole 30-50%	
Acids	Oleic acid 15-20%, linoleic acid 69-78%, palmitic acid 5-11%, Stearic acid 3-6%, α-linolenic acid 0.3-1%, palmitoleic acid 0.5-0.7%	Oleic acid 55-70%, linoleic acid 17-30%, palmitic acid 4-7%, Stearic acid 1.5-3%, α-linolenic acid <1.5%, palmitoleic acid <1%, arachidic acid <0.5%, eicosenoic acid <0.5%, behenic acid 0.3%, myristic acid <0.1%, margaric acid <0.1%, margaroleic acid <0.1%			

Table 3. Chemical structure characteristics of pure essential oils used [37]

Essential oil	Castor oil	Eucalyptus oil	Pine oil	Lavender oil	Rosemary oil
Monoterpenes		α-pinene 3.7%, β-pinene 1.0%, myrcene 2.0%	α-pinene 22-43%, β-pinene 3-33%, limonene 0.7-4.1%, δ-3-carene 0.4-31%, β-caryophyllene 0.7-5.5%, camphene 1.6-3.3%, sabinene 0.2-0.6%, γ-terpinene 0.1-0.5%, trans-ocimene 0.7-1.4%, β-phellandrene 1-2.7%, p-cumene 0-0.2%, terpinolene 0.3-3%	α-pinene 0.02-1.1%, β-pinene 0.1-0.2%, cis-ocimene 1.3-10.9%, trans-ocimene 0.8-5.8%, limonene 0.2-7%	α-pinene 15-34%
Sesquiterpenes			γ-cadinene 0.5-5.4%, α-copaene 0-0.2%, longifolene 0-0.2%, β-guaiene 0.2-0.7%, γ-muurolene trace-0.4%, α-humelene trace-0.5%, γ-patchoulene trace-0.4%, γ-cadinene trace-0.3%, α-muurolene trace-1%	β-caryophyllene 2.6-7.6%, β-farnesene 1%	
Alcohols		Linalool 0.4%, geraniol 2.6%, α-terpineol 14.0%, isoterpineol-4 2.0%	Borneol 2%, terpinene-4-ol 1%, epi-α-cadinol <1%, epi-α-muurolol <1%, α-cadinol 0-0.2%	Linalool 26-49%, terpinen-4-ol 0.03-6.4%, α-terpineol 0.1-1.4%, borneol 0.8-1.4%, geraniol 1%, lavandulol 0.5-1.5%	Borneol trace-7%
Esters			Bornyl acetate 0-3%	Linalyl acetate 36-53%, lavandulyl acetate 0.2-5.9%, terpenyl acetate 0.5%, geranyl acetate 0.5%	
Phenols					
Ketones				Octanone-3 0.5-3%	Verbenone 15-37%, camphor 1-15%
Aldehydes			Citronellal 0-0.2%	Myrtenal 0.1%, cuminal 0.4%, benzaldehyde	

Essential oil	Castor oil	Eucalyptus oil	Pine oil	Lavender oil	Rosemary oil
				0.2%, neral and genarial 0.4%, trans-22-hexanal 0.4%	
Oxides		1,8 cineole 62-72%		1,8 cineole 0.5-2.5%	1,8 cineole trace-20%
Acids	Ricinoleic acid 85-95%, oleic acid 2-6%, linoleic acid 1-5%, linolenic acid 0.5-1%, stearic acid o.5-1%, palmitic 0.5-1%, dihydroxystearic acid 0.3-0.5%				

Table 4. Chemical structure characteristics of pure essential oils used [37]

The ten essential oils were examined for their lubricating performance using the base fuels A1 and A2. Base fuel A1 had an average wear scar diameter (WS 1.4) of less than 460 μm, whereas base fuel A2 showed an increased wear scar diameter (WS 1.4) of more than 460 μm. All oils were dissolved in the base fuels at the same concentration levels, i.e. 200, 500, 1000 and 5000 ppmw. The fuel properties of the pure essential oils, including their WS 1.4 values, are presented in Tables 5a and 5b.

Essential oil	Grape seed oil	Kernel peach oil	Camomile oil	Laurel oil	Carrot seed oil
Density (kg/m³, 15°C)	0.924	0.921	0.929	0.940	0.923
Viscosity (cSt, 40°C)	29.90	35.56	28.15	52.05	32.51
Potassium (ppm)	8	4	4	3	2
Sodium (ppm)	1	2	2	2	1
Flash point (°C)	165	326	55	56	51
Calorific value (Kcal/kg)	10084	10268	10378	10227	10029
Residue (% m/m)	0.067	0.025	0.037	0.086	0.035
Sulphur (ppm)	5	8	3	5	8
Nitrogen (ppm)	7	17	9	18	11
Water (ppm)	741	421	693	1698	334
Ash (% m/m)	0	0	0	0	0
Lubricity (μm , average)	152	117	169	217	237

(a)

Essential oil	Castor oil	Eucalyptus oil	Pine oil	Lavender oil	Rosemary oil
Density (kg/m^3, 15°C)	0.966	0.798	0.812	0.789	0.774
Viscosity (cSt, 40°C)	207.33	1.32	1.90	1.12	1.23
Potassium (ppm)	10	5	6	7	3
Sodium (ppm)	0	2	0	1	1
Flash point (°C)	228	48	38	75	49
Calorific value (Kcal/kg)	9854	9841	9826	9564	9339
Residue (% m/m)	0.003	0.001	0.002	0.001	0.001
Sulphur (ppm)	9	0	0	0	0
Nitrogen (ppm)	3	0	3	0	2
Water (ppm)	3035	126	57	201	167
Ash (% m/m)	0.010	0	0	0	0
Lubricity (μm , average)	151	406	487	350	462

(b)

Table 5. Properties of pure essential oils.

2.2. Results and discussion

The two base fuels were initially analyzed to determine their lubrication effectiveness. The average wear scar diameter WSD values for the two base fuels are given in Table 1. It is evident that base fuel A1 has WSD value under the acceptable limit of 460 μm, and is characterized as fuel with good lubricating properties. Base fuel A2 has WSD value over the acceptable limit of 460 μm, and is characterized as fuel with poor lubricating properties. Consequently, these fuels were well suited to determine the response of essential oils on their lubrication properties.

Figure 1 shows the impact of adding essential oils on the lubricity of base fuel A1. On the basis of the HFRR test results, grape seed oil increases the lubricity of the base fuel. The WSD value decreased from 425 μm to approximately 365 μm at the concentration range 200-5000 ppm, the optimum value being at 1000 ppm. In the case of kernel peach oil, the WSD value decreased from 425 μm to approximately 335 μm, the lowest value obtained again at the concentration of 1000 ppm. In the case of camomile oil, the WSD value decreased from 425 μm to approximately 335 μm, the minimum value being exhibited at 1000 ppm. Similar behaviour if followed by carrot oil, the minimum value for lubricity being 353 μm. Laurel oil, decreases the WSD value from 425 μm to approximately 267 μm at the concentration of 5000 ppm. The addition of eucalyptus oil, causes WSD value to increase from 425 μm to approximately 581 μm at the concentration range 200-5000 ppm. Similarly pine oil, increases the WSD value from 425 μm to approximately 543 μm, the minimum value being observed at 200 ppm. Lavender oil, increases the WSD value for all the concentrations examined,

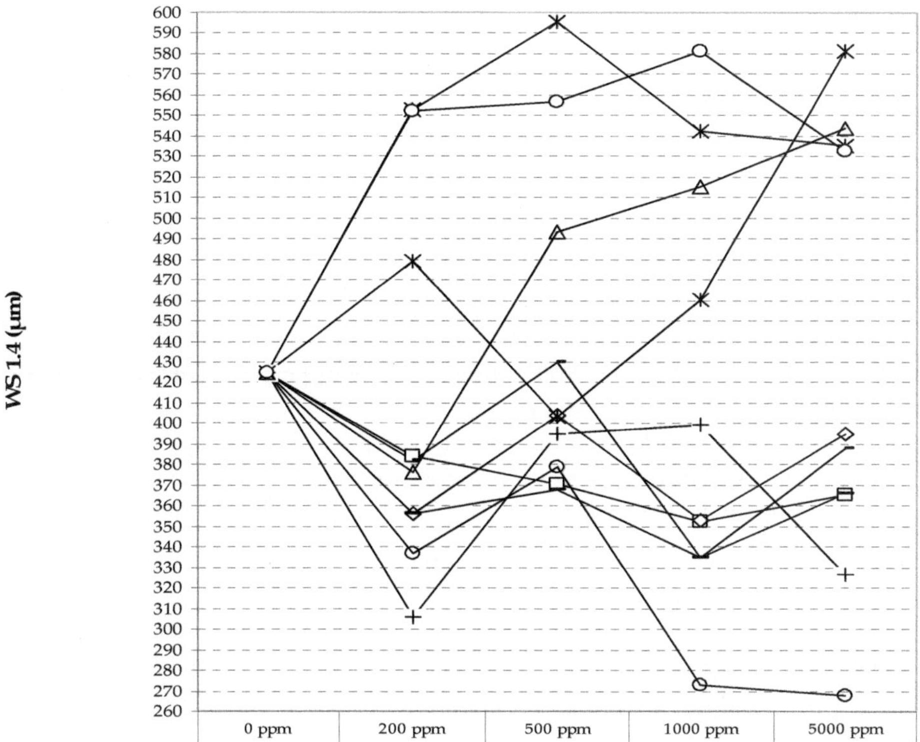

	0 ppm	200 ppm	500 ppm	1000 ppm	5000 ppm
Grape seed oil	425	384	370	352	365
Kernel peach oil	425	382	430	335	388
Camomile oil	425	357	368	335	366
Laurel oil	425	337	379	273	267
Carrot seed oil	425	357	404	353	395
Castor oil	425	306	395	399	327
Eucalyptus oil	425	479	404	460	581
Pine oil	425	376	493	515	543
Lavender oil	425	553	595	542	535
Rosemary oil	425	552	557	581	532

Concentration (ppm)

Figure 1. Impact of essential oils addition on the lubrication properties of fuel A1 reaching a maximum at 595 μm. Similarly, rosemary oil increases the WSD value for all the concentrations examined. In contrast to eucalyptus oil, pine oil, lavender oil and rosemary oil, the addition of grape seed oil, kernel peach oil, camomile oil, laurel oil and carrot seed oil improved the base fuel lubricity. On the other hand, small concentrations of the pine oil and the eucalyptus oil, 200 and 500 ppm respectively, were sufficient to set the WS 1.4 value well within the required limit of 460 μm. However, a further increase in the concentration to 5000 ppm for the grape seed oil, kernel peach oil, camomile oil, carrot seed oil, eucalyptus and pine oil, led to an increase in the WSD value. The analysis of the trend curves of the eucalyptus oil, pine oil, lavender oil and rosemary oil show that the required treat rate to obtain a satisfactory WSD (WS 1.4) of 460 μm was 200 ppm for the pine oil and 500 ppm for the eucalyptus oil, while the other two oils could not obtain a satisfactory WSD of 460 μm for any of the concentrations tested. If one examines the experimental data closely, it appears that the essential oils as they increase their density and viscosity, they improve their lubrication performance in the range of 200-5000 ppm.

Figure 2 gives the effect of essential oils on the tribological properties of fuel A2. It can be seen that grape seed oil, kernel peach oil, camomile oil, laurel oil, carrot seed oil and castor oil provide satisfactory HFRR mean WSD (WS 1.4) of less than 460 μm at the concentration level of 200 ppm. Any further increase in the concentration of these oils led to a slight increase or decrease in the WSD values.

It should be noted that when 200 ppm of pine oil was added to the base fuel, the tribological results showed a significant improvement in WSD value of 450 μm. Further addition of pine oil in concentrations between 500 and 5000 ppm increased the WSD significantly.

It is evident that in order to improve the lubrication properties of low sulfur diesel fuels, small concentration levels of grape seed oil, kernel peach oil, camomile oil, laurel oil, carrot seed oil and castor oil ranging from 200 to 500 ppm are sufficient to bring the WSD value within the required limit. In the Tables 5a and 5b, it can be seen that lavender oil and eucalyptus oil provide satisfactory HFRR mean WSD (WS 1.4) of less than 460 μm as pure essential oils, but as additives have worse lubricating properties. Similar conclusions may be drawn if one considers the chemical constitutions of eucalyptus oil, pine oil, lavender oil and rosemary oil where the terpenes (monoterpenes and sesquiterpenes) and oxides (1,8-cineole) do not help in the direction of improving the lubricating ability of essential oils, Table 3. In contrast, essential fatty acids play an important role providing better lubricating performance (castor oil, kernel peach oil and grape seed oil). In general, esters seem to appear a better lubricating performance as their density increases.

Overall, it appears that the essential oils having higher density and viscosity improved their lubrication performance in the range of 200-5000 ppm.

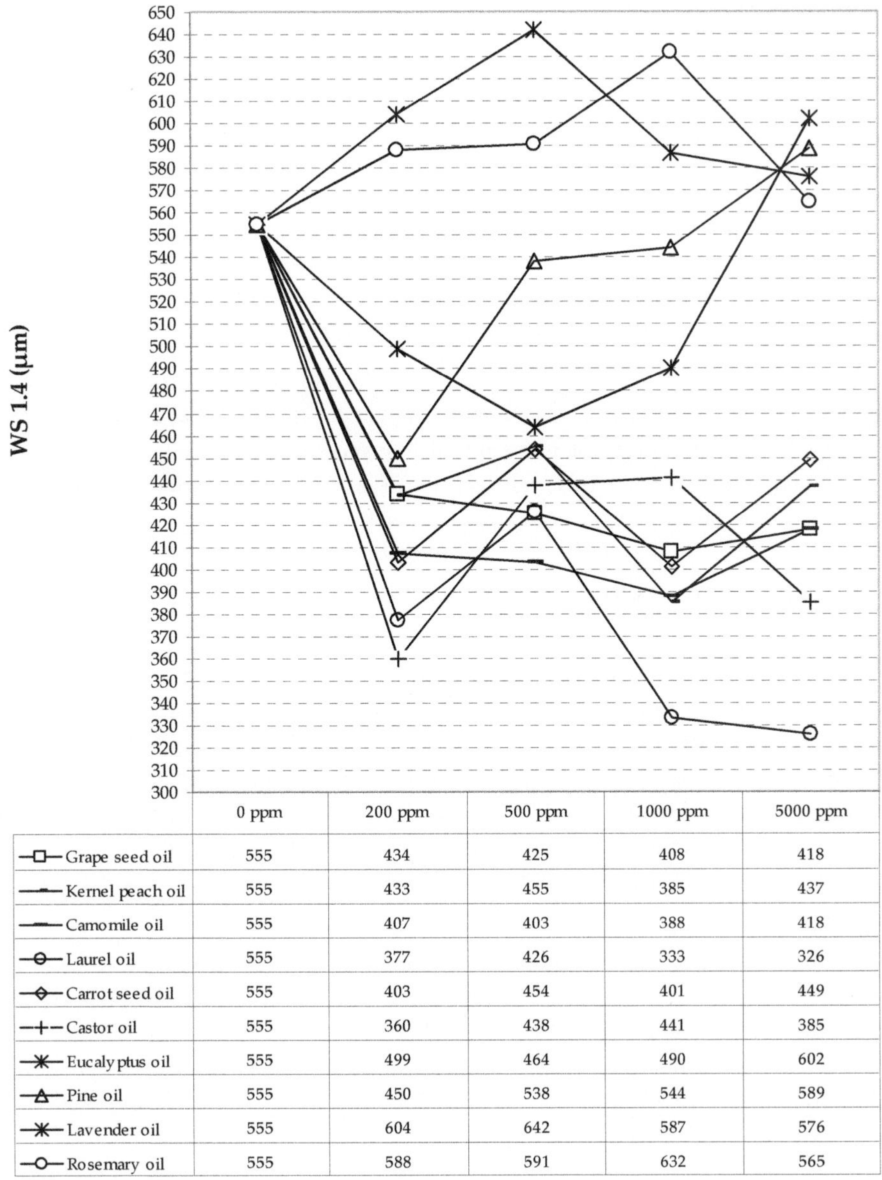

	0 ppm	200 ppm	500 ppm	1000 ppm	5000 ppm
Grape seed oil	555	434	425	408	418
Kernel peach oil	555	433	455	385	437
Camomile oil	555	407	403	388	418
Laurel oil	555	377	426	333	326
Carrot seed oil	555	403	454	401	449
Castor oil	555	360	438	441	385
Eucalyptus oil	555	499	464	490	602
Pine oil	555	450	538	544	589
Lavender oil	555	604	642	587	576
Rosemary oil	555	588	591	632	565

Concentration (ppm)

Figure 2. Impact of essential oils addition on the lubrication properties of fuel A2

3. Gasoline fuel lubricity

Gasoline lubricity is a complex phenomenon, involving many complicated and interconnecting factors, such as the presence of water, oxygenates diolefins, diaromatics, the effect of viscosity and the synergistic effect of different wear mechanisms. The lubricity mechanism of gasoline is quite different from that of diesel fuels that leads to severe adhesive wear. With low-sulphur fuels, adhesive wear is observed instead of corrosive and mild oxidative wear, and deposits build up on top land.

Metallurgy and mechanical properties of test specimens have important effects on the lubricating mechanisms of fuels. When the hardness of the lower specimen in an HFRR test is not enough to support the generated oxide films formed by the reaction between surfaces and dissolved oxygen and the adsorption films formed on top of the oxide films by gasoline polar impurities, severe adhesion and metal transfer occur.

Fuel quality in recent years became increasingly important, not only for its role in the actual performance of the vehicles, but also for its impact on the emissions. However, the fuel pump at the service stations is the point at which the actual specifications of the fuels should be ascertained [43]. In this chapter, results are presented on the gasoline properties impact on lubricity, based on the study of numerous petrol samples.

3.1. Experimental procedure

The two principal problems in testing gasoline regarding the lubricity, are evaporation of gasoline fuel due to its very high volatility and the extreme sensitivity of gasoline lubricity to tiny amounts of contaminant. Researchers have recently reached to the solution to modify the conventional HFRR test method for studying diesel fuels, principally by deepening the fuel holder so that a larger sample of fuel could be accommodated and by covering the lubricant test chamber with a close-fitting lid. The test rig is also completely enclosed in a plastic box from Polytetrafluoroethylene (PTFE). This enabled the humidity of the test to be controlled, a factor that has been shown to influence wear of fuels, and also helped retain gasoline vapors.

The test conditions used for the gasoline tests were chosen to be identical to those specified for diesel fuel tests according to ISO 12515-1 except for the fuel temperature. A fuel temperature of 25^{0}C was employed in all gasoline tests. It is important to mention that fuel tests were carried out in compliance with the standard ASTM G 133 as there is not as yet a standard for gasoline lubricity.

The following gasoline properties were determined since they are directly related to the exhaust emissions: Research Octane Number (RON), Benzene, Toluene, Xylene, Olefins, Saturates, MTBE and Total Aromatics were determined using the mid-IR method, while Sulfur and Nitrogen content was measured using the ANTEK 9000NS elemental analyser according to ASTM D 5453 and ASTM D 5762 respectively. Gasoline vapour pressure measurements were conducted with a Setavap Vapour Pressure tester 22420-3. The Setavap Vapour Pressure results were converted to DVPE in strict conformance with the requirements of ISO 3007 method, using the appropriate conversion equation. Potassium content (an additive used in lead replaced gasoline (LRP) to protect valve seat recession) was measured using the Atomic Absorption Spectrometry (AAS) according to IP 456. Viscosity was determined at 15^{0}C and 25^{0}C using the Anton Paar viscometer by ASTM D 7042. Distillation data including the value of residue were obtained according to the procedure of ISO 3405. Also, the ISO 10370 method was used for the calculation of residue and the examination of adulteration with heavier distillates or solvents. It should be mentioned that sulphur content is a physical marker concerning the matter of gasoline adulteration with heavier distillates of petroleum. Water content and conductivity (at 20^{0}C) were also measured according to the standards ISO 12937 and ISO 6297, respectively. Also, Total Acid Number (TAN) was measured according to the standard ISO 7537. The TAN value might indicate the potential of corrosion problems.

One hundred twenty six (126) samples of commercial gasolines were collected from service stations located all over Athens and its suburbs. They consisted of 40 samples of LRP gasoline, 46 samples of unleaded gasoline and 40 samples of super unleaded. The analysis of the samples showed that 19 samples were found to be adulterated (3 unleaded, 10 LRP and 6 super unleaded). Also, thirty six (36) samples of non-additized gasolines were tested, which were produced by mixing in different proportions the following refinery streams: FCC, Isomerate, Alkylate, Dimate, Reformate, MTBE and ETBE which were obtained by the Hellenic Petroleum refinery installations in 2005. The above samples were not randomly classified during their preparation but they stemmed from a multivariate analysis of variance and an appropriate empirical and statistical process to evaluate and predict the values of Research Octane Number, density, vapour pressure and benzene content (< 1% v/v) of the final mixtures.

Gas chromatography was applied to the quantitative determination of paraffins, isoparaffins, olefins, naphthenes and aromatics with 3 to 12 carbon atoms according to the standard methodology described by CAN/CGSB-3.0 No. 14.3M93.

Emphasis was given to the experimental procedure because of the amount of samples and properties measured, in order to ensure that no contamination or lighter substances loss would influence the final result of the measured values. The values of the properties were statistically analyzed and compared as a completely randomized factorial experiment to assess whether and how the different type of gasoline fuel and the measured properties affect the lubricity using analysis of variance and complex neural networks [35, 36]. The lubricating properties of gasolines were expressed from the value of mean wear scar diameter (MWSD1.4) of the spherical specimen, detected using a photomicroscope to an accuracy ± 1μm and were corrected at the absolute water pressure 1.4 kPa at the temperature of 25^{0}C. It was found that the sensitivity to humidity of friction, wear and film formation of gasoline and diesel is quite different in different humidity ranges. In the range of water vapour pressure less than about 0.8 kPa, wear increases rapidly with humidity and then remains insensitive to humidity with increasing humidity. This phenomenon may suggest the presence of a transition concentration of dissolved water in gasolines.

The appearance of the transition may be explained by different effects of dissolved water on the lubricity performance of fuels themselves, oxide film formation in rubbing surfaces, as well as the interfacial chemistry of protective film formation by naturally-occurred antiwear agents in fuels. The results of previous studies suggest that the influence of humidity on the measurement of friction, wear and film formation of fuels can be, at least to a large extent, eliminated by carefully controlling humidity in a certain range, say from 0.9 to 1.2 kPa.

The optical assessment of the wear and the possible wear mechanisms that took place during the experimental process, were evaluated with back scattered electron imaging, quantitative x-ray analysis, and x-ray mapping of the surface of the metal specimens using SEM technique (Scanning Electronic Microscope). Additionally, the samples were photographed using scanning probe microscopy for the metal surfaces.

3.2. Results and discussion

3.2.1. Gasoline lubricity evaluation

Examination of the gasoline lubricity has shown that the majority of the samples for the 36 non-additized gasoline fuels were above the acceptance limit of diesel lubricity, the 460-μm limit. The values ranged from 711 μm to 1064 μm as shown in Figure 3. The preliminary results on a non-additized gasoline showed that the repeatability of the modified HFRR test is quite good. The tested samples were evaluated three repeated times, in order to obtain the mean lubricity value. The commercial gasolines were evaluated twice, to obtain their mean lubricity value. Their values ranged from 279 μm to 846 μm as shown in Figure 4. On the contrary, most of the samples of LRP gasoline were near the limit of 460-μm indicating that the presence of the potassium additive had a main effect on the lubricating properties of fuels. The limit of 460-limit is even lower if we consider the reduction of temperature to 25°C (about 400 μm). Adulterated new super gasolines with unleaded gasoline have poorer lubricating properties, as shown in Figure 4. It is obvious that unleaded and super unleaded gasolines have much higher lubricity values than LRP gasolines. Especially, samples with sulphur content below 50 ppm and nitrogen content below 10 ppm, exhibit extremely high lubricity values above the limit of 700 μm after appropriate statistical analysis (Factor Analysis, Two-step cluster analysis and Neural network approach).

There was no linear or other type of correlation between the concentration of potassium and the lubricity, but after statistical approach, emerges that with a concentration above 5 ppm K there may be a significant reduction of MWSD1.4 value near the limit of 460 μm, as shown in Figure 5. The factors most likely to cause the observed differences in lubricity are the bulk fuel composition, the use of additives and the use of oxygenates.

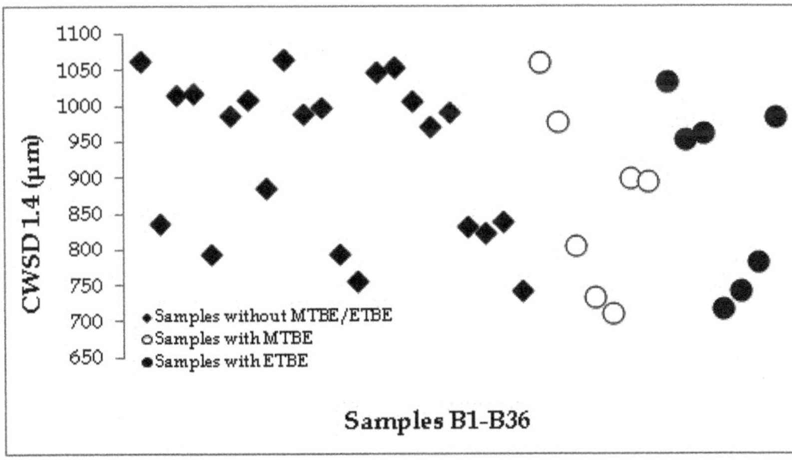

Figure 3. Lubricity mean values (CWSD1.4 – Corrected Wear Scar Diameter at 1.4 kPa water pressure) for the non-additized gasoline samples B1-B36.

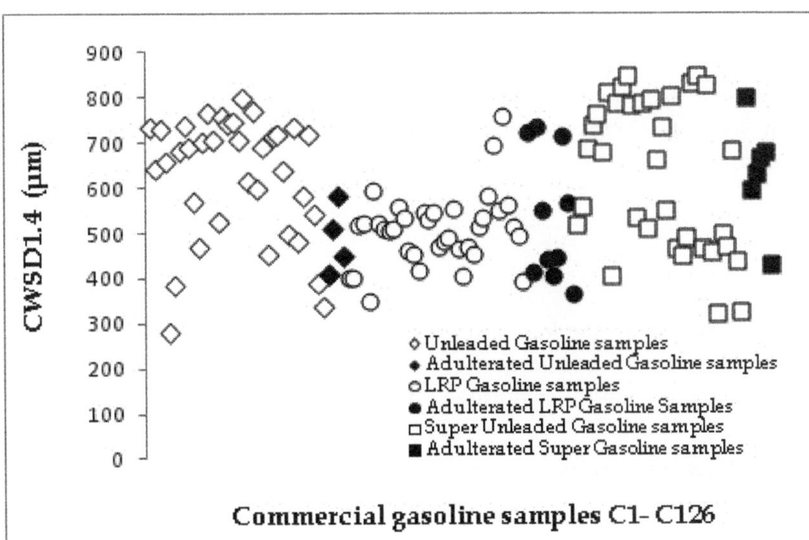

Figure 4. Gasoline lubricity mean values (CWSD1.4 – Corrected Wear Scar Diameter at 1.4 kPa water pressure) for the three commercial types (Unleaded, Super Unleaded and LRP).

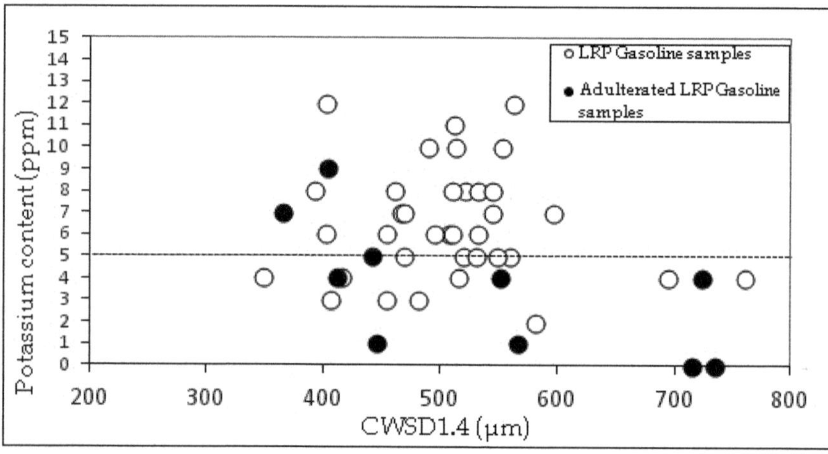

Figure 5. Potassium content of LRP gasoline fuels versus lubricity (CWSD1.4)

The diagrams for the variation of the lubricity values with the coefficient of friction and film are shown in Figures 6 and 7 respectively. In Figure 5, there appears a linear correlation between the gasoline lubricity and the coefficient of friction for the whole series of the samples, non-additized and commercial samples (R^2=0.85 and 0.88 respectively). This strong linear relationship between the coefficient of friction and lubricity is due to the wear mechanism of the gasoline fuel and especially adhesive wear which is the most basic sub-category observed. In Figure 7, two regions were observed according to the variation of the film in relation to the lubricity values. The repeatability limits (R) for these two regions can be calculated according to the following equations (2, 3) which resulted from statistical and mathematical analysis, while the limit in which this difference was observed at the variation of the film with the lubricity values concure to the mean value of lubricity for the commercial samples:

$$\text{If MWSD1.4} > 589\ \mu\text{m, then R} = 137 - 0.0854 \times \text{MWSD1.4} \tag{2}$$

$$\text{If MWSD1.4} \leq 589\ \mu\text{m, then R} = 137 - 0.1094 \times \text{MWSD1.4} \tag{3}$$

A possible explanation for the formation of two film categories, maybe the severe adhesive wear and metal transfer occurring in the unleaded, super unleaded and adulterated gasolines, exhibited in Figure 7 as the lower plateau.

3.2.2. Gasoline fuel comparison

The adulterated fuel samples were isolated and two statistical computations were carried out each time, one with these adulterated samples and the other without them. The spread of the values can be depicted using box-plots. Figure 8 shows the median, quartiles, and extreme values of lubricity for each type of gasoline fuel. Each box plot displays the 50% percentage of samples' population in the square area, the 75% percentage of them within the

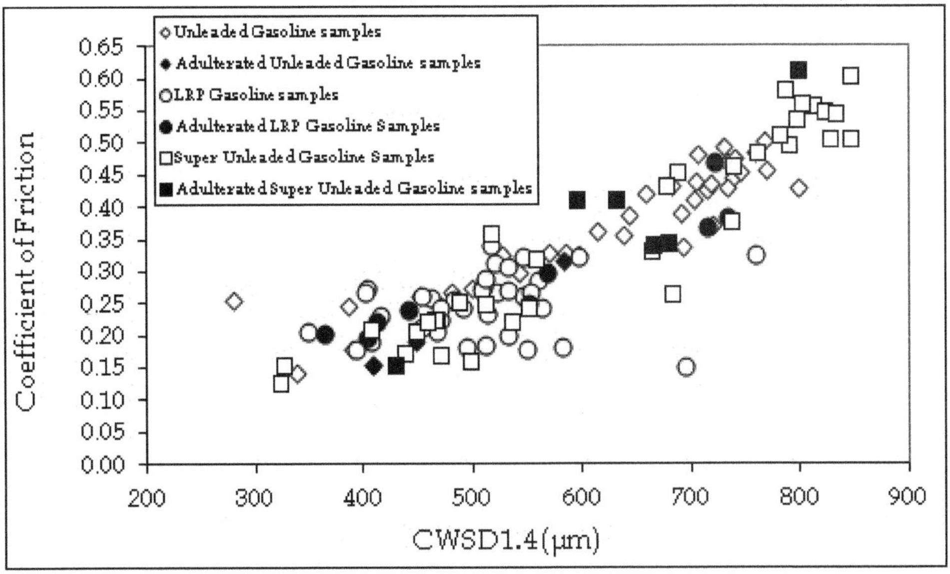

Figure 6. Gasoline lubricity mean values (CWSD1.4) versus coefficient of friction for 126 commercial gasoline fuels.

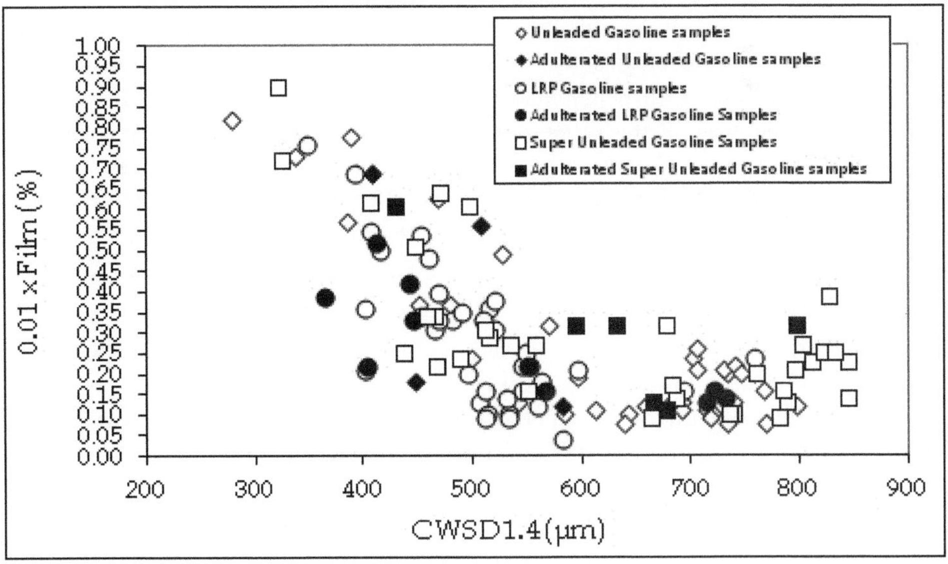

Figure 7. Gasoline lubricity mean values (CWSD1.4) versus film for 126 commercial gasoline fuels.

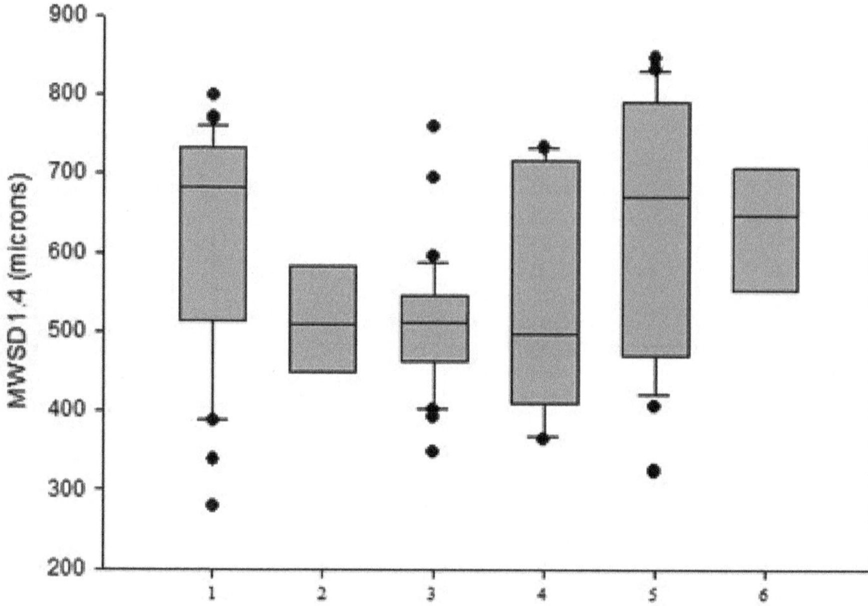

Figure 8. Box plot analysis – First statistical graphic approach to the data (1=Unleaded, 2=Adulterated Unleaded, 3=LRP, 4=Adulterated LRP, 5=Super Unleaded, 6= Adulterated Super Unleaded).

upper and lower limit and the extreme values which are cases with values more than 3 box lengths from the upper or lower edge of the box. It is shown that LRP gasolines have a much better representative sample population indicating good lubricating properties compared with the other two types of gasoline. One unleaded gasoline has shown extreme good lubricity value, 279μm, but it is assumed to be caused by the use of special anti-wear or other additives. Because all the properties were not normally distributed for correlation analysis with Pearson correlation coefficient, the correlation coefficients of Spearman and Kendall's tau-b were chosen to be computed. The effect of the properties on the gasoline lubricity is different for each type of gasoline. The chemical structure and the related individual physical properties seemed to inter-correlate in their effect on lubricity in different degree for each type of gasoline.

More specifically, the statistically significant coefficients showed that unleaded gasoline samples appear to have lower values of wear diameter, as sulphur and nitrogen content, saturates and viscosity increased. On the contrary, unleaded gasolines appear to have greater values as toluene, oxygen, MTBE and vapor pressure increased.

LRP gasolines appear to have lower values of wear diameter as sulphur, potassium and nitrogen content, conductivity (non-adulterated samples), saturates and viscosity increased. On the contrary, LRP gasolines appear to have greater values as the total acid number, benzene, aromatics and xylene is increased.

Finally, super unleaded gasolines appear to have lower values of wear as sulphur content, nitrogen content and olefins increased. On the contrary, super unleaded gasolines seem to have greater values as toluene, xylene, water, benzene, aromatics and oxygen increased.

Linear regression and categorial regression was applied for each type of commercial gasoline fuel and was concluded that the gasoline lubricity values could be predicted accurately by models if we know at least 15 to 20 physicochemical and constituent characteristics of the fuels measured for the commercial and non-additized gasoline samples, excluding the tribological characteristics such as the coefficient of friction, film and absolute water pressure.

Neural network approach showed that the concentration of cyclic olefins with 6,7,8,9 and 11 carbon atoms and normal olefins with 5 and 8 carbon atoms increases with the lubricity, while the concentration of normal paraffins with 4,7,9,10 and 11 carbon atoms decreases with the lubricity. Also, it was confirmed that the percentage of dimate stream, the concentration of total normal olefins (% w/w) and the concentration of olefins (% v/v) increases with the lubricity, as shown in Figure 9.

Numerical analysis for the refinery streams that were blended together in order to evaluate the gasoline lubricity of non-additized gasoline fuels showed the optimal proportions for each stream which lead to a minimum lubricity as shown in Table 6. We conclude that a percentage of 8 to 9% v/v of oxygenate such as MTBE and ETBE with the an optimum composition for the rest streams can give us a minimum lubricity and an increase in the percentage of isomerate stream in contrast with FCC and reformate stream proportions which would lead to a minimum lubricity, taking into account the modern refinery practice used.

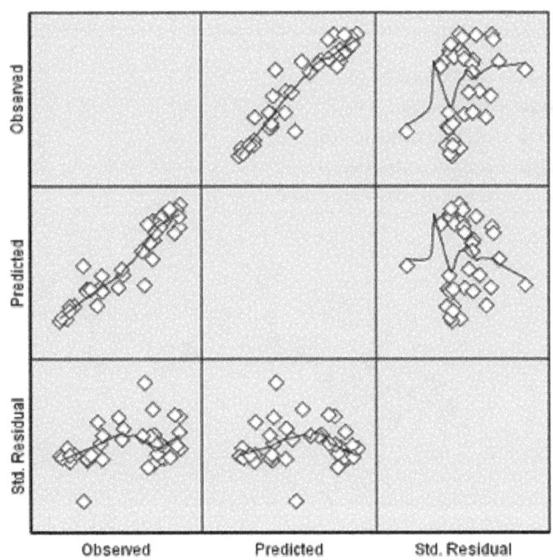

Figure 9. Predicted and observed values showing a linear effect of the percentage of dimate stream, the concentration of total normal olefins (% w/w) and the concentration of olefins (% v/v) on lubricity mean values (CWSD1.4).

	Optimum percentage of addition	Addition limits	
		max	**min**
Reformate	0.26	0.50	0.11
FCC	0.35	0.50	0.20
Alkylate	0.10	0.15	0.00
Dimate	0.00	0.05	0.00
Isomerate	0.21	0.30	0.10
MTBE or ETBE	0.09 or 0.08	0.13	0.00

Table 6. Data of numerical analysis for optimum percentage of addition.

3.2.3. Viscosity and density effect

There is no specification limit for the viscosity of gasoline fuels. It was decided to test all the samples at the temperatures of 15ºC and 25ºC. During the statistical process, a linear correlation between the viscosity and density appeared (R^2= 0.92, 0.98 and 0.90 for unleaded, LRP and super unleaded samples respectively). Figure 10 shows the correlation between density and viscosity for all the gasoline samples for each type separately. Both these properties are greatly influenced by the composition of the fuel.

This enhances the fact that the compositional characteristics of the fuel do influence the gasoline lubricity to a considerable degree, but there is not any linear or other correlation between density and viscosity with lubricity.

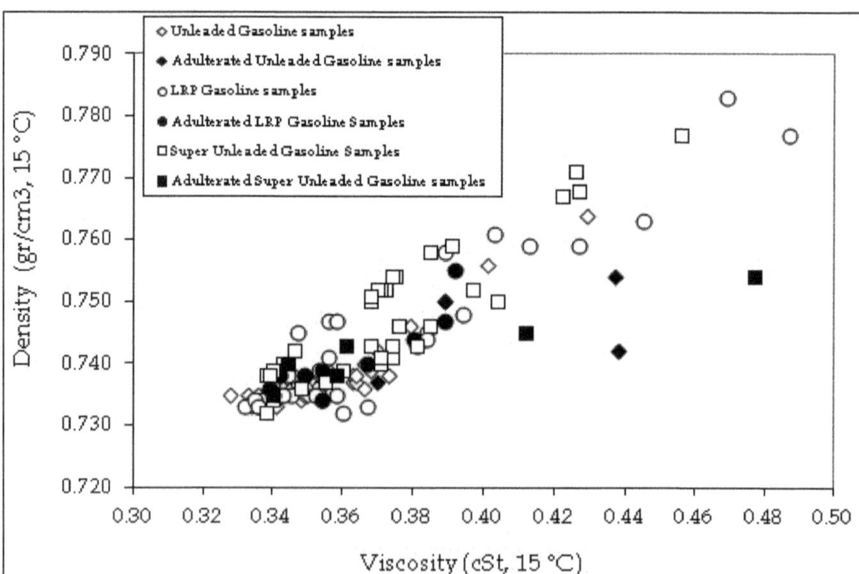

Figure 10. Graphs indicating linear correlation between viscosity and density at 15°C.

3.2.4. Scanning electron microscopy results

The quantitative x-ray analysis of SEM showed the existence of the elemental chlorine in six of the nine cases. The level of the chlorine content was greater than the potassium content which was detected using the same technique. There is strong evidence that the level of concentration for the chlorine in the gasoline fuels varies from 0.1 to few ppm as potentiometric titrations revealed (ASTM D 4929). Only very small quantities of impurities containing chlorine are present in gasolines but they can chemically react with the metal surface under high pressure conditions. The activity of halogenated hydrocarbons increases with decreasing stability of the carbon-halogen bond. At local contact temperatures ranging between 305-330 °C, the additive thermally decomposes and the reactive halogen atoms form a surface layer of iron halogenides on the part surface. Eventual failure of the contact point comes when the contact temperature exceeds the melting point of the iron halide layer. Under such conditions, small particles of carbon are generated as well.

More elements were detected, such as K, Fe, S, Si, Cl, Cu, Cr and Mn in accordance with the preliminary data of the elemental analysis for the specific batch of specimens that were used apart from the sulfur concentration which originated from the fuel constituents.

It was observed that the material first transferred to the ring was a disc grey layer which then oxidized and detached as fine brown powder, either haematite (α-Fe_2O_3) or hydrohaematite (α-$Fe_2O_3.nH_2O$). Three wear processes were suggested: (a) transfer of metal from ball to disc, (b) oxidation of transferred layer and (c) removal of the oxide as detached debris.

In the HFRR tests in the current study, a ball-on-flat contact geometry was used, similar to those described above. However the upper specimen is hard, 750~850 VPN (kg/mm^2) and lower specimen, soft 190~210 VPN. In such a system, strong adhesion transfers material from soft to hard specimen. Oxidation, or probably severe cold working, will then transform the layers into hard abrasive lump or debris.

In HFRR tests, strong adhesion resulted in transfer of the material of soft flat specimen to the hard ball specimen and the transferred layers then formed wear particles. Therefore, abrasive wear is a wear process secondary to adhesive wear. The key property a material needs to resist abrasive wear is hardness. Moisture also has a strong influence on abrasive wear rates. Usually, abrasive wear rate increases with increasing moisture content.

In the case of gasolines, the corrosive medium can consist of gasoline components, additives, dissolved water or dissolved oxygen for corrosive and oxidative wear. Except for trace amount of polar impurities the most chemically-active gasoline components are olefins, including monoolefin and diolefins.

The severe adhesive wear in gasolines can be depressed by adding moderately reactive additives, such as corrosion inhibitors. The minimum gasoline wear can probably be obtained by carefully selecting the corrosivity of antiwear additives to balance adhesive wear/corrosive wear. In the case of gasoline, three different types of wear, i.e. adhesive wear, abrasive wear and corrosive wear persist together and probably interact synergistically.

The amount of oxygen and water dissolved in gasoline fuels is quite important. The mean value for water content was 208 ppm (maximum value 618 ppm and minimum 64 ppm) and the mean value for oxygen content was 1.07 % w/w (maximum value 2.87 % w/w and minimum 0.14 % w/w). In most cases, the water absorbed by gasolines from the atmosphere reached the water solubility in gasoline. Therefore if humidity increases further, the water content in gasolines does not increase, so neither does the wear.

Figures 11 and 12 depict the wear mechanisms mentioned above.

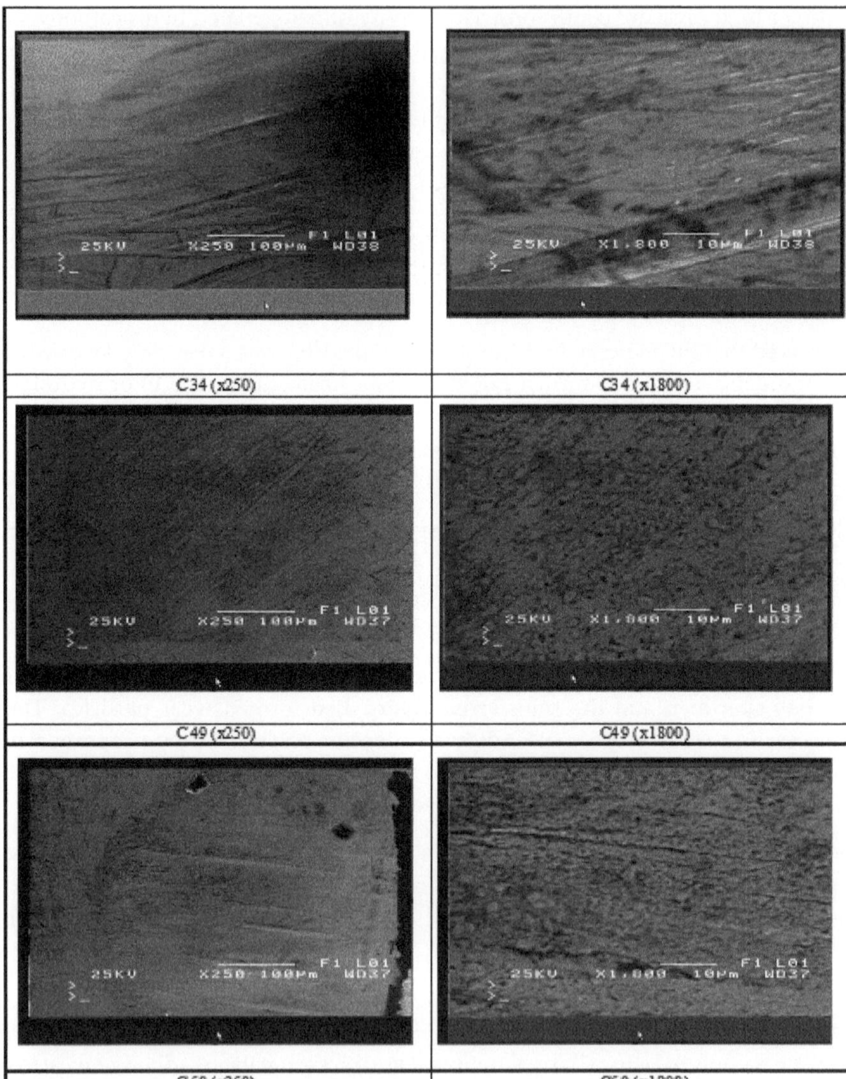

Figure 11. SEM images for commercial gasoline samples.

Figure 12. SEM images for commercial gasoline samples.

4. Conclusions

In an effort to investigate the impact of essential oils on the tribological properties of low sulfur diesel fuels, ten essential oils were added to low sulfur fuels. The following conclusions can be drawn from this study:

1. Six of the ten essential oils used, i.e. grape seed oil, kernel peach oil, camomile oil, laurel oil, carrot seed oil and castor oil provided satisfactory HFRR mean WSD (WS 1.4) of less than 460 μm, for concentration levels between 200 and 5000 ppm.

2. The other four essential oils, i.e. eucalyptus oil, pine oil, lavender oil and rosemary oil , could not obtain a satisfactory WSD of 460 μm, even for the base fuel A1 which has WSD value under the acceptable limit of 460 μm, except from the pine oil at the concentration level of 200 ppm and the eucalyptus oil at the concentration level of 500 ppm.
3. Lubricating mechanisms of essential oils are probably controlled by some of the constituents present in essential oils and their polar constituent contribution but this needs further investigation.
4. Overall, it appears that the essential oils having increased density and viscosity exhibit improved lubricating performance in the range of 200-5000 ppm.

From the study on gasoline lubricity it was concluded that to a large extent, gasoline lubrication has to rely on its bulk components to provide good film forming lubricating ability, except the inherent ability of tiny polar amounts or other impurities to provide film-forming characteristics during an applicant load. Conductivity values of LRP gasolines indicate the influence of such polar compounds as potassium additives and their ability to be activated to form chemical bonds in the metal surface above 5 ppm limit concentration.

Nineteen gasoline samples were found to be adulterated based on the quinizarin tracing and the sulphur content. Also, some of these samples were found to be mixed up with aromatic solvents. But, most of the key properties of the gasoline fuels were found to comply with the current EU legislation. The degree of adulteration does influence the lubricity especially for additized LRP gasoline samples altering the final values.

The findings of this research, verified the poorer lubricating properties of gasoline fuels compared with that of diesel fuels. Different type of gasoline fuel is affected in different degree from the compositional characteristics of the fuel and its physico-chemical properties.

Potassium concentration seems to play a significant role even in very low concentrations protecting satisfactorily from wear under boundary conditions.

The amount of water that could be absorbed during handling must be taking into account.

It is known that certain alkali compounds may accelerate the oxidation of certain organic compounds which are found in the gasoline fuels. However, further research on the oxidation stability of gasoline fuels and its effect on gasoline lubricity must be initiated in this direction.

Author details

Evripidis Lois and Panagiotis Arkoudeas
National Technical University of Athens, School of Chemical Engineering, Greece

5. References

[1] P. Arkoudeas, F. Zannikos, D. Karonis, E. Lois, Lubricity assessment of gasoline fuels Fuel Processing Technology (In process)

[2] P. Arkoudeas, E. Lois, F. Zannikos (2008) The tribological behaviour of essential oils in low sulphur automotive diesel Fuel Volume 87 Issues 17-18 Pages 3648-3654
[3] G. Aanastopoulos, P. Arkoudeas, E. Lois, F. Zannikos, S. Kalligeros, P. Shoinas (2003) Study of the tribological properties of automotive diesel – effect of sulphur and nitrogen compounds 4th Greek National Congress of Chemical Engineering in Patras
[4] Arkoudeas P., Karonis D., Lois E., Korres D., Karavalakis G., (2005) Vegetable oils, essential oils and their derivatives as substitutes and anti-wear additives of diesel engines 1st National Greek Congress of Alternative Fuels in Athens
[5] Spikes HA, Wei DP. (2000) Fuel lubricity—fundamentals and review Fuels International 1:45-65.
[6] Hadley JW, Owen GC, Mills B. (1993) Evaluation of high frequency reciprocating wear test for measuring diesel fuel lubricity SAE paper 932692.
[7] Wei D., (2000) Thirty years of research on fuel lubricity Shigou Xuebao Shigou Jiagong.
[8] Miura M, Ikeda T, Takizawa H, Yoshida H, Ikebe H. (1997) Study on lubricity of low sulfur diesel fuels SAE Paper 972895.
[9] Lacey PI, Lestz SJ. (1992) Effect of low lubricity fuels on diesel injection pumps—Part I: field performance SAE Paper 920823.
[10] Lacey PI, Lestz SJ. (1992) Effect of low lubricity fuels on diesel injection pumps—Part II: laboratory evaluation SAE Paper 920824.
[11] Barbour RH, Rickeard DJ, Elliott NG. (2000) Understanding diesel lubricity SAE Paper 2000-01-1918.
[12] Wei DP. (1986) The lubricity of fuels I. Wear studies on diesel fuel components Acta Petrolei Sinica 2:79-87.
[13] Wei DP. (1990) The lubricity of fuels II. Wear studies using model compounds Acta Petrolei Sinica 4:90-95.
[14] Wei DP. (1990) The lubricity of fuels. III. Wear studies on diesel fuels Acta Petrolei Sinica 6:15-19.
[15] Tucker RF, Stradling RJ, Wolveridge PE, Rivers KJ, Unbbens A. (1994) The lubricity of deeply hydrogenated diesel fuels—the Swedish experience SAE Paper 942016.
[16] Anastopoulos G, Lois E, Karonis D, Zanikos F, Stournas S, Kalligeros S. (2001) A preliminary evaluation of esters of mono-carboxylic acids on the lubrication properties of low sulfur diesel fuels Ind Engng Chem Res 40:452-6.
[17] CEC F-06-A-96. Measurement of diesel fuel lubricity—approved test method. HFRR Fuel Lubricity Test.
[18] Wei D-P, Spikes H.A. (2000) Fuel Lubricity – Fundamentals and Review, Fuels International p-45-65.
[19] P.I. Lacey, R.L. Mason. (2000) Fuel Lubricity: Statistical Analysis of Literature Data SAE paper 2000-01-1917.
[20] Miura M, Ikeda T, Takizawa H, Yoshida H, Ikebe H. (1997) Study on lubricity of low sulfur diesel fuels SAE Paper 972895.
[21] Cameron F. (1998) Lubricity of California diesel fuel SAE Paper 981362.
[22] Nikanjam M. (1993) Development of the first CARB certified California alternative diesel fuel SAE Paper 930728.

[23] Wei DP. (1986) The lubricity of fuels I. Wear studies on diesel fuel components Acta Petrolei Sinica 2:79-87.
[24] Wei DP. (1990) The lubricity of fuels II. Wear studies using model compounds Acta Petrolei Sinica 4:90-95.
[25] Wei DP. (1990) The lubricity of fuels III. Wear studies on diesel fuels Acta Petrolei Sinica 6:15-19.
[26] Tucker RF, Stradling RJ, Wolveridge PE, Rivers KJ, Unbbens A. (1994) The lubricity of deeply hydrogenated diesel fuels—the Swedish experience SAE Paper 942016.
[27] Caprotti R, Bovington C, Fowler WJ, Taylor MG. (1992) Additive technology as a way to improve diesel fuel quality SAE Paper 922183.
[28] Batt RJ, McMillan JA, Bradbury IP. (1996) Lubricity additives performance and no harm effects in low sulfur fuels SAE Paper 961943.
[29] Nikanjam M., Burk E. (1994) Diesel fuel lubricity additive study SAE Paper 942014.
[30] EN 590 (2003) Automotive fuels-Diesel-Requirements and test methods.
[31] Galbraith RM, Hertz PB (1997) The Bocle test for diesel and biodiesel fuel lubricity SAE Paper 972862.
[32] Hertz PB. (1995) Winter engine wear comparisons with a canola biodiesel fuel blend Saskatchewan Canola Commission Report.
[33] Anastopoulos G, Lois E, Serdari A, Zanikos F, Stournas S, Kalligeros S.(2001) Lubrication properties of low-sulfur diesel fuels in the presence of specific types of fatty acid derivatives Energy Fuels 15:106-12.
[34] Anastopoulos G, Lois E, Karonis D, Zanikos F, Stournas S, Kalligeros S. (2001) A preliminary evaluation of esters of mono-carboxylic acids on the lubrication properties of low sulfur diesel fuels Ind Engng Chem Res 40:452-6.
[35] Wei D, Spikes H. (1986) The lubricity of diesel fuels. Wear 111:217- 35.
[36] Wanda Sellar. (2001) The Directory of Essential Oils Reprint, Essex: The C.W. Daniel Company, Ltd.
[37] Tsurutani K, Takei Y, Fujimoto Y, Matsudaira J, Kumamoto M. (1995) The effects of fuel properties and oxygenates on diesel exhaust emissions SAE Paper 952349.
[38] Akasaka Y, Sukurai Y. (1994) Effects of oxygenated fuel and Cetane improver on exhaust emissions from heavy-duty DI diesel engines SAE Paper 942023.
[39] Bruce A. Buchholz, Charles J. Mueller, Glen C. Martin, A. S. Cheng, Robert W. Dibble and Brian R. Frantz. (2004) Tracing fuel component carbon in the emissions from diesel engines Nuclear Instruments and Methods in Physics Research Section B: Beam Interactions with Materials and Atoms 223-224:837-841
[40] Catalog handbook of fine chemicals (2005-2006) Aldrich Chemical Company, Inc.
[41] Vogel's textbook of practical organic chemistry (1989) 5th ed. Bath Press.
[42] Wanda Sellar, (2001) The Directory of Essential Oils Reprint, Essex: The C.W. Daniel Company Ltd.
[43] A.J. Von Wielligh, N.D.L. Burger, T.L. Wilcocks (2003) Diesel Engine Failures due to Combustion Disturbances, caused by Fuel with Insufficient Lubricity Industrial Lubrication and Tribology Vol. 55, p-65-75.

Chapter 3

The Role of Nanotechnology in Automotive Industries

Mohsen Mohseni, Bahram Ramezanzadeh,
Hossein Yari and Mohsen Moazzami Gudarzi

Additional information is available at the end of the chapter

1. Introduction

Nanotechnology involves the production and application of physical, chemical, and biological systems at atomic or molecular scale to submicron dimensions and also the integration of the resulting nanostructures into larger systems. Therefore, nanotechnology deals with the large set of materials and products which rely on a change in their physical properties as their sizes are so small. Nanotechnology promises breakthroughs in areas such as materials and manufacturing. Nanoparticles, for example, take advantage of their huge surface area to volume ratio, so their optical properties become a function of the particle diameter. When incorporated into a bulk material, these can strongly influence the mechanical properties such stiffness or elasticity. For example, traditional polymers can be reinforced by nanoparticles leading to novel materials to be used as lightweight replacements for metals. Such enhanced materials will enable a weight reduction together with an increase in durability and enhanced functionality.

There are different reasons why this length scale is so important. The wavelike behavior of materials predominates when the size lies in the atomic scale. This changes the fundamental properties of materials such as melting temperature, magnetization and charge capacity without changing the chemical composition. The increased surface area of nano materials make them ideal for use in composites, reacting systems and energy storage. By increasing the surface area the number of surface atoms increases dramatically, making surface phenomena play a vital role in materials performance. This is because a greater amount of a substance comes in contact with surrounding material. This results in better catalysts, since a greater proportion of the material is exposed for potential reaction. At nanoscale the gravitational forces become negligible and electromagnetic forces dominate. At nano scale surface and interface forces become dominant. From optical point of view, when the size of materials is comparatively smaller than the wavelength of visible light they do not scatter light and can be used in applications where transparency is of great importance. The automotive sector is a major consumer of material technologies. It is expected that nanotechnologies improve the performance of existing technologies for car industries. significantly. Applications range from already existing paint quality, fuel cells, batteries, wear-resistant tires, lighter but stronger materials, ultra-thin anti-glare layers for windows and mirrors to the futuristic energy-harvesting bodywork, fully self-repairing paint and switchable colors. The basic trends that nanotechnology enables for the automobile are : lighter but stronger materials (for better fuel consumption and increased safety); improved engine efficiency and fuel consumption for gasoline-powered cars (catalysts; fuel additives; lubricants); reduced environmental impact from hydrogen and fuel cell-powered cars; improved and miniaturized electronic systems; better economies (longer service life; lower component failure rate; smart materials for self-repair).

This chapter attempts to discuss the applications of nanotechnology in automotive sector and bring some examples of each set of products being used in car industries.

2. Exterior applications

2.1. Nano-clearcoats with high scratch and wear resistance

2.1.1. An introduction on scratch/mar

In a multilayer automotive coating system (basecoat/clearcoat), the main responsibility of the clearcoat layer is to protect the pleasing appearance of the metallic underneath layer from environmental factors. However, the clearcoat's appearance may be vulnerable to degradation in exposure to harsh environmental conditions, especially weathering and mechanical damages (Bautista et al., 2011, Barletta et al., 2010, Courter et al., 1997, Ramezanzadeh et al, 2011d, Ramezanzadeh et al, 2011e). Scratch and mar are the most important types of mechanical damages which impose serious challenges for the coatings formulators. Depending on the size and morphology of the scratch/mar, the appearance changes of the clearcoat may vary. Based on the viscoelastic properties of the clearcoat and scratching condition (the cause of scratch, scratches force, scratch velocity and environmental temperature) scratch can be produced by two primary mechanisms, i.e. plastic and fracture flow. The fracture type scratch has sharp edges and irregular shapes having high capability of light scattering (Ramezanzadeh et al, 2011f, Ramezanzadeh et al, 2011g, Yari et al., 2009a, Shen et al., 2004). On the other hand, plastic type scratch has smoother surface and less ability to light scattering (Fig. 1).

The plastic type scratches are deeper than fracture types and have greater tendency to self-healing at temperatures around clearcoat 's Tg.

2.1.2. Approaches to improve scratch resistance

Two main strategies can be sought in order to produce highly scratch resistant clearcoats: the first is optimizing cross-linking behavior of the clearcoat utilizing appropriate components and the second is introducing reinforcing inorganic fillers into the clearcoat formulation. The first approach deals with low enough Tg-clearcoats showing the reflow behavior or extraordinary high cross-linking density (Bautista et al., 2011, Barletta et al., 2010, Courter et al., 1997, Ramezanzadeh et al, 2011, Ramezanzadeh et al, 2011d, Ramezanzadeh et al, 2011e, Yari et al., 2009a, Shen et al., 2004). The clearcoats scratch resistance can be highly improved by these two ways. However, there exist disadvantages for each of these strategies alone. Producing low-Tg clearcoats needs changing clearcoat chemical composition. This may negatively influence other properties of the clearcoat such as reduced chemical resistance. A highly cross-linked clearcoat can be obtained by the reaction of melamine based resins and polyols to form etheric bonds. Although this system may appropriately resist against scratch, the coating will be susceptible to acid etching and performs weakly in weathering. One alternative way to improve scratch resistance of the coating while the lowest weathering performance is maintained is the use of so called hybrid materials including both organic and inorganic domains simultaneously. In this system, the inorganic domains improve clearcoat scratch resistance and organic domain guarantees the stability in weathering. The hybrid materials can be obtained by direct embedding inorganic fillers into them or by in-situ production of inorganic domain in a method called sol-gel processing. The micro-sized inorganic fillers cannot be used due to their effects on clearcoat transparency. By using inorganic fillers in nano sized form, the mechanical properties of the clearcoat will be improved even at low loadings mainly due to their small particle size and huge surface area. Unlike conventional micron-sized fillers, they do not affect the transparency of the clearcoat. The advantages and disadvantages of incorporating nano-fillers into the clearcoat matrix or in-situ creation of inorganic domains in the clearcoat matrix will be discussed below (Shen et al., 2004, Schulz et al., 2001, Hara et al., 2001, Jardret et al., 2000, Weidian et al., 2001, Thorstenson et al., 1994, Ramezanzadeh et al., 2010d).

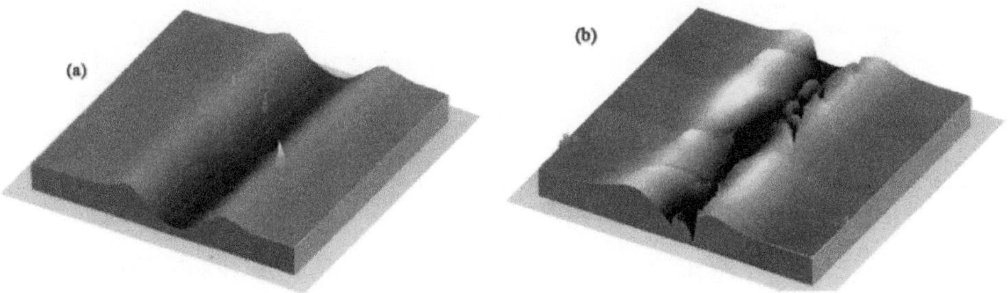

Figure 1. Visual illustrations of (a) plastic type and (b) fracture type scratches.

2.1.3. *Highly scratch resistant clearcoat containing inorganic nano fillers*

It has been found that incorporation of nanoparticles such as Al_2O_3, SiO_2, ZrO_2 and TiO_2 into a clearcoat matrix could significantly enhance the scratch resistance (Bautista et al., 2011, Amerio et al., 2008, Tahmassebi et al., 2010, Groenewolt et al., 2008, Sangermano et al., 2010). Ceramic nanoparticles have been found as appropriate hardening materials to significantly improve clearcoat hardness and therefore scratch resistance. However, the improvement cannot be easily obtainable when the particles are poorly dispersed. The inorganic fillers do not have intrinsic affinity to organic phase. These lead to phase separation and aggregate formation. The aggregated particles (>100 nm) depreciate clearcoat properties especially the optical clarity. Attempts have been carried out to solve this problem by surface modification of fillers with organosilanes to render them hydrophobic and thereby improve their dispersibility into the polymeric matrix. The surface modification not only can influence dispersibility but also can result in stronger physical/chemical interfacial adhesion between particles and the matrix (Tahmassebi et al., 2010). Different factors may be influential for the effects of nano fillers on the scratch resistance of a clearcoat: the particles chemistry, size, shape and surface modification. It has been demonstrated that nanoparticles could improve clearcoat properties in different ways. The most important of which will be discussed here (Tahmassebi et al., 2010).

Inorganic nanoparticles have hardness and elastic modulus greater than organic polymers. Incorporation of these particles to the clearcoat matrix increases hardness and elasticity (Fig. 2). This depends on the content, the intrinsic hardness and the dispersion degree of the inorganic filler. Increased hardness and elasticity may result in better clearcoat resistance against sharp scratching objects penetrating into the surface.

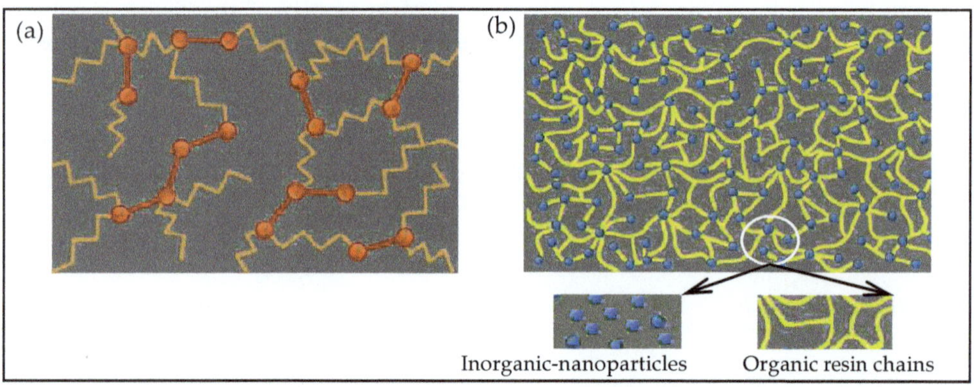

Figure 2. Schematic illustrations of the chemical structures of the conventional coating consist of resin/cross-linker (a) and inorganic-nanoparticles loaded paint (b).

However, it has been shown that greater hardness does not necessarily guarantee clearcoat scratch resistance. There are problems with highly increased clearcoat hardness. For examples, when the applied forces are greater than the critical force, it leads to fracture type scratches. Increasing coating hardness can also result in an increase in clearcoat brittleness and therefore reduction of other properties like flexibility. To overcome this problem, attempts have been carried out to obtain tough clearcoat in presence of nanoparticles. Results obtained in recent researches show that nanoparticles could influence cross-linking density of the clearcoat by affecting curing reaction. Nanoparticles with organosilane modifications include functional groups with high capability of reacting with functional groups of resins. As a result, some chemical bonds between resin and hardener (curing agent) will be replaced by the bonds created between particle/hardener and/or particle/resin.

This results in a decrease in the cross-linking density of the clearcoat. On the other hand, nano fillers enhance the hardness and elasticity. These two phenomena result in clearcoat toughness improvement in presence of nanoparticles. A tough clearcoat can resist abrasive condition and show less fracture behavior (Amerio et al., 2008, Tahmassebi et al., 2010, Groenewolt et al., 2008, Sangermano et al., 2010).

2.1.4. Highly scratch resistant clearcoat using sol-gel method

Nanofiller embedded clearcoats show enhanced scratch and wear resistance. However, the clearcoat transparency will be influenced as a result of nanoparticles aggregation. Obtaining appropriate dispersion needs surface modification as well as using different dispersing techniques. In-situ process of inorganic phase formation inside organic matrix using sol-gel technique has been considered (Ramezanzadeh et al., 2011d, Presting et al., 2003, Hernandez-Padron et al., Hernandez-Padron et al., 2003). Organic/inorganic precursors can be used to produce in- situ silica network in the matrix. These precursors, either as network former, such as tetraethyl orthosilicate (TEOS) or network modifier such as methacryloxy propyl trimethoxysilane (MEMO) and glycidoxy propyl trimethoxysilane (GPTS), can be introduced to the main polymeric film former to obtain a so-called hybrid nanocomposite films. This process includes precursor hydrolysis and self-condensation reactions. The hydrolyzed precursors could be cross-linked with the organic coating matrix by reacting with polyol and other curing cross-linkes such as amino or isocyanate compounds in the automotive coating formulation. In this way, a hybrid nanocomposite containing organic/inorganic phases can be obtained (Fig. 3). The organic phase presented in the hybrid nanocomposit can be responsible for the adhesion and flexibility and the inorganic phase can help coating resists mechanical damages (Ramezanzadeh et al. 2010).

2.2. Scratch resistant polymer glasses

Nowadays, fuel consumption of a car is an important factor for both car manufacturers and consumers. Request for producing cars with lower fuel consumption has been enormously developed in recent years. Reducing the weight of the cars is one way reaching this target. The car weight can be significantly reduced by replacing heavy glass parts (i.e. head lights and windows) by light polymeric glass sheets (Fig. 4) (Yahyaei et al., 2011).

One of the most used kinds of glass polymers is polycarbonate which has excellent impact strength, high toughness and light weight. Polycarbonates have been already used in light covers and lenses. However, polycarbonate has limited scratch/abrasion and chemical resistance together with the tendency to yellowing when it is exposed to UV light in long term. Glass is a hard material having excellent scratch resistance. However, it has higher weight and lower impact strength compared with polymeric glasses. Washing (both automatic carwash and hand washing) and sand/dust particles presented in air are main causes of scratching polycarbonates glass parts. This may result in a significant reduction in head lights transparency and therefore light scattering. Attempts have been carried out to solve the problem. Two methods have been sought for this purpose. Producing polycarbonate polymeric glass parts by embedding nanoparticles into it and/or using acrylate or polysiloxane paints over the head light. Aluminum oxide nanoparticles are also used in the coatings composition in order to make it hard enough to resist scratch and abrasion. This coating is highly transparent due to the small size of the filler particles and their fine distribution (Yahyaei et al., 2011, Pang et al., 2006, Brinker et al., 1990). Embedding nano-sized silica particles into an organic modified siloxane based coating results in nano-coating for automotive glazing application. This coating can produce various properties for the plastic glazing like hydrophobic/anti-smudge, infra-red (IR) and ultra-violet (UV) shielding and anti-fogging behavior. The schematic illustration of a nano-enhanced automotive plastic glazing is shown in Fig. 5.

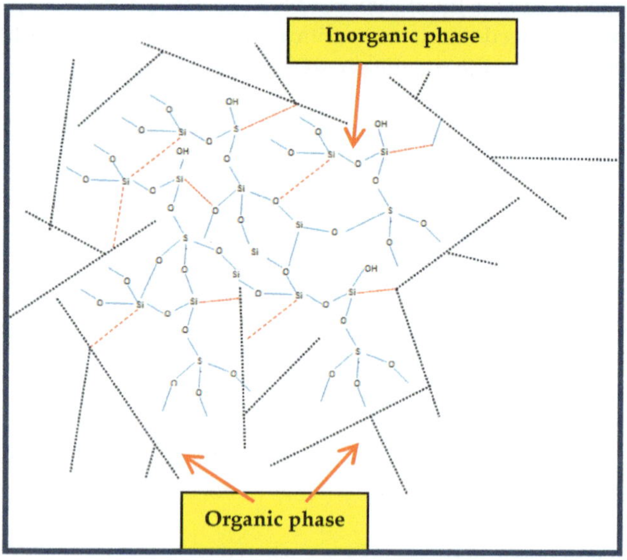

Figure 3. Schematic illustration of a sol-gel based automotive clearcoat containing organic/inorganic precursors (Ramezanzadeh et al. 2010).

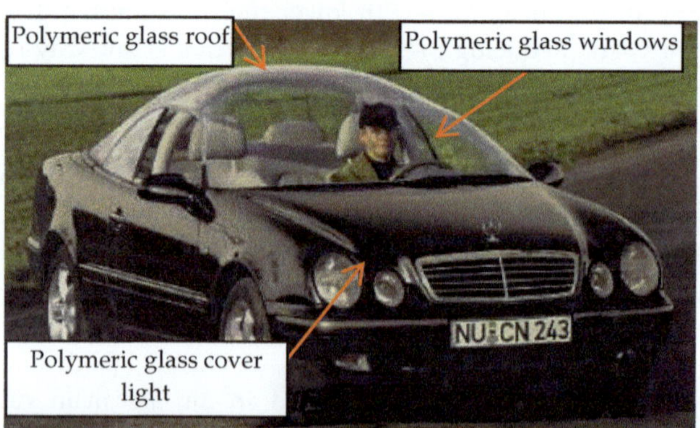

Figure 4. Modern automobiles equipped with nanostructure polymeric glasses for roof, windows and cover light.

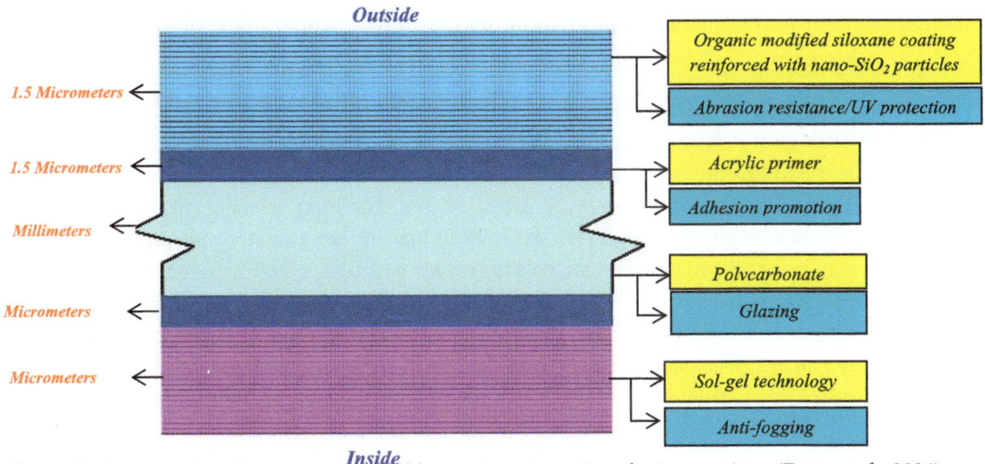

Figure 5. Cross-section of a nano-enhanced layers in automotive glazing coatings (Pang et al., 2006).

The average thickness of the nano-embedded coating used for polycarbonate is approximately 1 mm. Different nanocoating layers (as shown in Fig. 5) are responsible for anti-scratch/easy-to-clean/anti-fogging and UV stabilization of polycarbonate plastic glazing. To this end, nanoparticles such as TiO_2, SiO_2 and Al_2O_3 for abrasion resistance improvement, TiO2 and ZnO for UV protection, sol-gel based TiO for anti-fogging behavior and TiO_2 for easy-clean properties are used (Yahyaei et al., 2011, Pang et al., 2006, Brinker et al., 1990).

In premium optical glazing like glass panes, using coatings with extremely high scratch resistance is necessary. To this end, attempts have been carried out to apply hard materials over the polymer glass from gaseous phase. Using physical vapor deposition (PVD) and chemical vapor deposition (CVD) procedures as well as plasma polymerization, a highly cross-linked nanometric polymeric layer containing inorganic components can be obtained. Producing highly scratch-resistant polymer glass using these techniques opens new possibilities for designing transparent roof tops (Fig. 5) and car body shell parts.

2.3. Nano-coatings with anti-corrosion performance for car bodies

Anti-corrosive coatings both in form of conversion and organic coatings are used to protect metal body against corrosive materials. The most important of these coatings are Cr(VI) and phosphate conversion coatings together with electrodeposition coating (ED). Cr(VI) due to its excellent anticorrosion performance has been widely used to protect car bodies from corrosion in the last decades. The high anticorrosion performance of this coating is related to its high self-healing behavior in corrosive environment. However, the toxic and hazardous nature of chromium compounds are well documented and their uses have been banned in recent years. Phosphate coating is another kind of conversion coating which has appropriate anticorrosion properties and is less toxic compared with Cr(VI). However, the bath containing these materials leave huge amounts of sludge (Nobel et al., 2007, Dhoke et al., 2009, Shchukin et al., 2007, Brooman, 2002, Kasten et al., 2001).

It has been shown that Cr(III) is less toxic compared with Cr(VI). However, compared to Cr(VI), Cr(III) does not have long-term protection. Nanotechnology has been employed to eliminate this disadvantage. A three layer system including zinc layer, Cr(III) enriched layer and nano-SiO_2 particles containing layer are used for this purpose (Fig. 6). Each layer has specific role for corrosion protection of steel. Zinc has higher negative potential than iron. And when it exposes to corrosive electrolyte, it can produce electron needed for cathodic reaction and prevents iron from oxidation. As a result, Zn^{2+} cations produce positive charge at surface. On the other hand, SiO_2 nanoparticles have negative charges. Therefore, nanoparticles can migrate to corroded area and cover it. In fact, nanoparticles produce a layer with approximate thickness of 400 nm. This phenomenon is called self-healing by nano passivation.

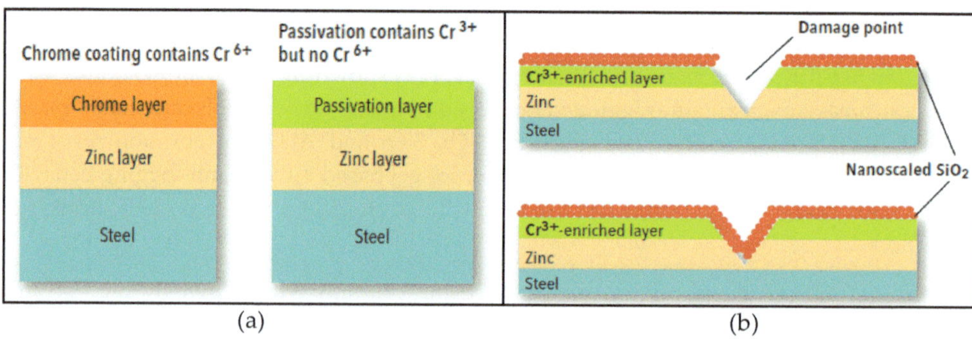

Figure 6. Conventional anti-corrosion coatings (a) and nanostructured anticorrosion system (b).

Incorporating nanoparticles into electrodeposition coating formulation is another approach to improve the anti-corrosion performance of car body. Nanoparticles such as Nano-SiO2, Nano-TiO_2, Nano-Clay, Nano Carbon Tube etc. are used to improve electrocoating properties. What is important here is that electrocoatings are waterborne coatings. Therefore, the nanoparticles used for this system must be compatible with coating formulation. Hydrophilic surface modifications are used to produce nanoparticles compatible with this kind of coating. Nanoparticles due to their very small size and high surface area could improve barrier properties of the organic electrocoating against corrosive electrolyte penetration. These particles increase electrolyte pathways through the coating (Nobel et al., 2007, Dhoke et al., 2009).

2.4. Smart nano-scale container anticorrosive coating

New generation of self-repairing coatings are developed to further enhance anticorrosive properties of metal substrates. In conventional systems, the barrier property of the coating is the main mechanism for metal protection against corrosion. However, the barrier performance of a coating will be damaged soon and corrosive electrolyte comes into contact with the metal substrate. Use of corrosion inhibitors is another approach to produce active coatings in exposure to corrosive electrolytes. These active agents are soluble in corrosive electrolytes and protect metal surface by passivation mechanism. There are different kinds of corrosion inhibitors which can be divided to three main types based on the mechanism controlling corrosion process. Anodic inhibitor (only reduces anodic reaction rate), cathodic inhibitor (only reduces cathodic reaction rate) and mixed inhibitors (both cathodic and anodic reactions can be influenced) (Nobel et al., 2007, Dhoke et al., 2009, Shchukin et al., 2007, Brooman, 2002, Kasten et al., 2001, Sheffer et al., 2004, Garcia-Heras et al., 2004, Osborne et al., 2001, Vreugdenhil et al., 2005). The solubility of the inhibitors is found an important factor affecting its corrosion inhibiting efficiency. Very low solubility leads to low passivating behavior at metal substrate. There are disadvantages for very high solubility: first, the inhibitor will be rapidly leached out from the coating and second, the active surface blistering and delamination may occur due to osmotic pressure effect. As a result of osmotic pressure, water transportation into the coating matrix and passive layer destruction may occur. Because of this, adding active inhibitors at high concentration is not possible in conventional processes. This problem has been solved in modern coatings using nanoscale container (carrying active agents like inhibitors). In this approach, active inhibitor is loaded into nanocontainer. The nanocontainers have a permeable shell which could release active agents in coating matrix. In fact, the shell is designed in a way which release active agent in a controlled process. There is another approach which instead of nanocontainer in which the passive layer is doped with active agents. However, interaction of active materials with coating matrix leads to short-coming in the stability and self-repairing activity of the coating. The disadvantages cannot be seen for the system loaded with nanocontainers as the coating matrix does not directly contact with active agents. The system is schematically shown in Fig. 7 (Shchukin et al., 2007).

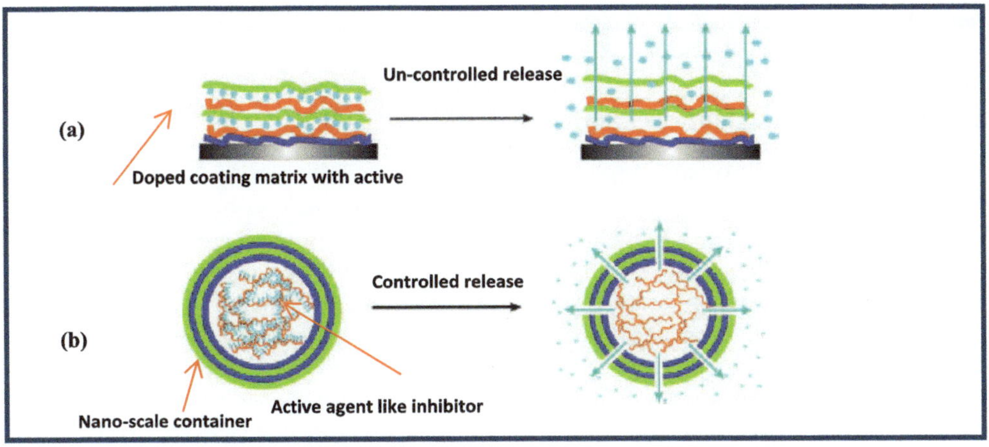

Figure 7. Active material is embedded in the "passive" matrix of the coating (a) and active material is encapsulated into nanocontainers (b) (Shchukin et al., 2007).

The nanocontainers will be uniformly distributed in coating matrix keeping active materials in a trapped state. The nanocontainers respond to any signal or when the environment undergoes changes they release encapsulated active materials. Controlling nanocapsuls permeability and nanocontainers compatibility with coating matrix are the most important parameters affecting their anticorrosion performance. The optimum nanocontainers size range is 300-400 nm. Using nanocontainers with larger sizes may lead to large hollow cavities formation inside coating matrix resulting in significant reduction of the passive protective properties of the coating. Depending on the sensitive components presented in nanocontainers (i.e polyelectrolytes or metal nanoparticles) different parameters like ionic strength, pH changes, temperature, ultrasonic treatment, magnetic field alteration may influence shell permeability. The mechanism in which nanocontainers release active agents and protect attacked areas of metal surface by corrosive electrolyte forming a passive layer is schematically shown in Fig. 8 (Shchukin et al., 2007).

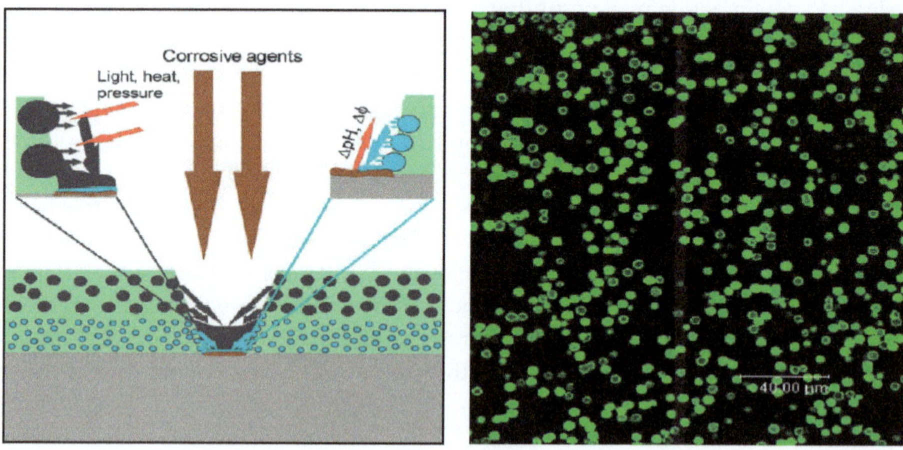

Figure 8. Schematic illustration of self-repairing mechanism of nanocontainers when metal is exposed to corrosive electrolyte (Shchukin et al., 2007).

2.5. Weathering resistant automotive coatings

Two main purposes of coating application in automotive industries are protecting the car body against environmental conditions and imparting desirable esthetic appearance. To fulfill these functions, the coatings themselves have to remain intact for a long time in a harsh environment. Photo and hydrolytic degradations respectively caused by sunlight and humidity are two common processes occurred, resulting in changes in all aspects and properties of automotive coatings (Yari, et al., 2009a; 2009b; Ramezanzadeh, et al., 2012a). These chemical alterations may greatly influence all aspects of the coating. Therefore, automotive coatings are required to be highly resistant against weathering condition. To enhance coating resistance against sunlight, HALS (hindered amine light stabilizer) and/or organic UV-absorbers has been added to clearcoat formulation. Although these ingredients considerably enhanced weathering performance, in addition to having high prices, they can migrate to other layers and are also prone to undergo decomposition during service life. To fight weathering, nanotechnology has offered new solutions that have no drawbacks as mentioned above for organic UV stabilizers.

In recent researches, various nanoparticles such as zinc oxide, iron oxide, cerium oxide, titanium oxide and silica have been incorporated into conventional polymeric coatings to enhance their resistance against sunlight. Nanoparticles, possessing a high surface area for absorbing the harmful part of sunlight (ultraviolet part), prevent the coatings from weathering degradation. Since they are inorganic and particulate, they are more stable and non-migratory within an applied coating. So, they present better effectiveness and longer protection.

As mentioned before, TiO_2 nanoparticles are effective to fight against UV rays and can protect the coating against weathering. However, these nanoparticles especially can exert strong oxidizing power and produce highly reactive free radicals and degrade the coating in which has been incorporated. Thus, photocatalytic activity of TiO_2 nanoparticles has to be controlled. For this purpose, treatment of nanoparticles by different techniques such as silane agents not only suppresses photocatalytic activity of TiO_2 nanoparticles, also offers clear advantages like simplicity and low cost and processing temperatures. It has also been demonstrated that surface modification of TiO_2 nanoparticles with aminopropyl trimethoxy silane (APS), considerably has reduced photocatalytic activity of nanoparticles and enhanced the weathering resistance of a polyurethane coating(Mirabedini, et al., 2011).

In various researches, it has been shown that zinc oxide nanoparticle can be an effective option to nearly completely screen the UV rays and protect the coating(Lowry, et al., 2008; Ramezanzadeh & 2011a). In an attempt to improve the UV resistance of an aromatic polyurethane-based automotive electro-coating nano-ZnO particles were used. The results obviously illustrated that the presence of nano-zinc oxide particles could decrease the photodegradation tendency of the film and protect it against deterioration (Rashvand, et al., 2011).

2.5.1. Weathering due to biologicals

Although, humidity and sunlight are the two main factors which degrade automotive coatings, other environmental factors, i.e. those originated from the biological sources can have a spoiling impact on the appearance of a car body during its service life (Ramezanzadeh et al., 2009) . In a systematic manner, the degradation mechanism and influence of various biological substances such as bird droppings, tree gums and insect gums on automotive coatings have been thoroughly studied(Yari et al., 2009c; Ramezanzadeh et al., 2010a; 2010b; 2011b) . It was revealed that both natural gum and bird dropping seem to affect the coating physically (by imposing stress to coating while they are being dried) and chemically(by catalyzing the hydrolytic degradation) (Yari et al., 2011; Ramezanzadeh et al., 2010c; 2010b; 2011b). While natural gum has extensively created large cracks with scattered etched areas (Ramezanzadeh et al., 2010c; 2010b), the influence of bird droppings was formation of numerous etched regions with some small cracks (Yari et al., 2011). It was also found that the most important factors governing the degradation are the coating chemical structure at surface and adhesion between coating surface and biological materials. Therefore, it was thought that any modification which could be able to alter both surface chemistry of the clearcoat and adhesion would be an ideal option to fight bio attacks. This idea was proved by a series of experiments. It was demonstrated that modification of clearcoat with a functional silicone additive significantly improved the coating performance against bird droppings and tree saps(Yari et al., 2012a, Ramezanzadeh et al., 2012b)(Fig. 9).

According to these new findings, creating a clearcoat with non-stick and superior water-repellency properties would significantly reduce the failure of coatings caused by biological materials. Ultra-hydrophobic self-cleaning coatings which are produced by nanotechnology is a powerful approach for this purpose. Contaminants on such surfaces are swept by water droplets or adhere to the water droplet and are removed from the surface when the water droplets roll off. Although these types of coatings for automotive glasses have been already commercialized, their development for automotive paints is in progress.

2.6. Smart windows based on electrochromism

As stated before, providing a secure and comfortable condition for driver and passengers has become an important task in automotive industry. To this end, automotive experts strongly believe that all types of energy like sound, light and heat which enter the car body have to be controlled. Recent progresses in polymer and different types of dichromic technology have allowed the development of smart glasses which intelligently control solar radiation transmission and modulate glare, increase passenger comfort and safety. Among different kinds of smart glasses, electrochromic (EC) ones are very important.

EC materials alter their optical characteristics (darkness/lightness) when a small electric potential difference is applied. They are suitable for a wide range of applications. They can be employed in different parts of an automobile like for energy-efficient windows, antiglare rear-view mirrors, sunroofs and displays.

A typical ECD composition has a complicated structure. As shown in Fig.10, it usually consists of a five-layer sandwiched-structure which are applied between two glass substrates. This structure includes transparent conductor, an electrochromic coating, ion conductor and ions storage coating and another transparent Conductor(Pawlicka , 2009).

Since the layers in this structure are very thin, the technology used for this assembly can be considered in nanotechnology domain. The thicknesses of transparent conductor, EC, electrolyte and counter electrode (ion storage) layers in a typical EC structure are 1500A°, 4000A°, 100μm and 1250A°, respectively. These layers can be deposited by different techniques such as sputtering, CVD, spin- or dip-coating from sol-gel precursors, etc.

The EC devices operate based on the reversible electrochemical double injection of positive ions i.e. H^+, Li^+, Na^+ and electrons into the host lattice of EC materials. Diffusion of mentioned ion and electrons into EC layer initiate some redox reactions, leading to a change in electrochemical state of EC material and therefore its resultant color. This variation in color of EC layer alters the color of the whole structure (for more details, see (Monk et al., 2007).

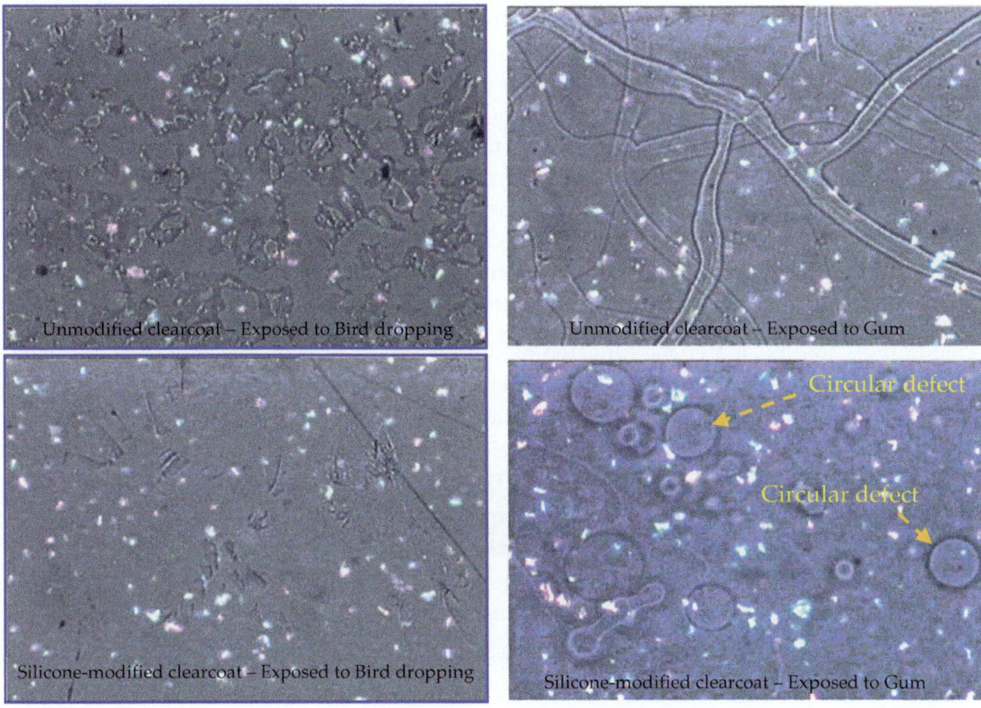

Figure 9. Microscopic images pure and silicone-modified clearcoats exposed to bird droppings and tree gums.

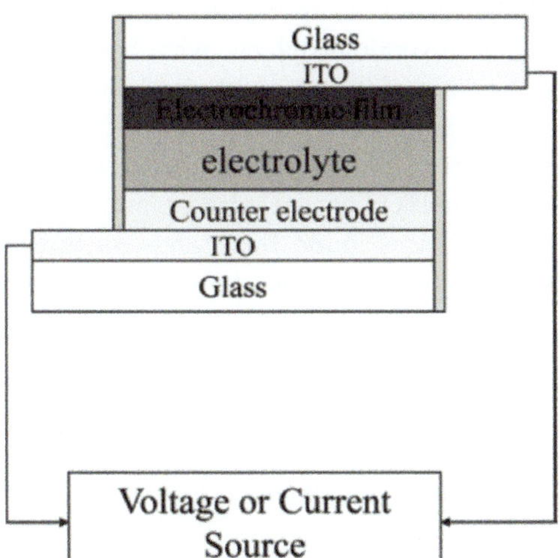

Figure 10. A typical EC system consists of different layers (Pawlicka , 2009).

EC technology becomes more and more important because it possesses low power consumption. However, due to slow diffusion of ions, response time (the time that a perceivable change in color occurs) in conventional EC systems is relatively slow and this drawback limits application of EC systems where a fast response is needed like in automotive rear-view mirrors. A significant portion of studies related to EC systems is devoted to new methods or materials to reduce the response time. In recent years, although scientists have achieved successes using new materials like hydrogen ions instead of lithium ions, nanotechnology has opened new rooms in this field and has triggered plenty of academic and industrial enthusiasms. In an EC process, if the surface area of the EC materials increases by producing nano-structured oxide films, migration of ions will be improved and consequently redox reactions will occur faster. Here, a few of nanotechnology-involved studies are briefly presented.

Among inorganic EC materials, tungsten oxide films have the highest coloration efficiency in the visible region and, therefore, have been most extensively studied so far. In a recent study, EC films from crystalline WO_3 nanoparticles have been fabricated (Lee et al., 2006). Porous films of crystalline WO_3 nanoparticles were grown by hot-wire chemical-vapor deposition and electrophoresis techniques. The nano-scale porosity of the films not only increases the surface area and ion-insertion kinetics, but also diminishes the overall material cost. It was also revealed that compared to conventional amorphous WO3 films, nanoparticle films demonstrated superior electrochemical-cycling stability in acidic electrolytes, a higher charge density, and comparable CE. It seems that these findings will eventually revolutionize current EC technology.

The first commercial EC product was based on a patented document fom Gentex Corp. in 1992. It was a solution-phase EC rear-view mirror for automotive vehicles which had a reflectance greater than 70% to less than 10%. This technology has been installed in many cars. In addition, in 2007 Donnelly Corporation designed an EC system for use as automotive mirrors. In this invention, which was based on polymerization of an electrochromic monomer, the color of the mirror varies uniformly from a silver appearance to bluish purple, and its reflectance changes from 60% to 20%. In a similar study, Thin mesoporous films of nickel oxides and nanotube manganese oxides were electrochemically produced on indium tin oxide(ITO) coated glasses and compared with conventional structure ones(Yoshino, 2012). It was found that nano-structured films exhibited marked changes in optical transmittance and electric charge with respect to the electrochromic reactions.

Coating EC material on different types of nano-particles are much more novel approaches to take advantage of large surface area granted by nano-materials to overcome the drawback of long switching time. Coating Viologn on TiO2 nanoparicles (Cummins, et al., 2000) or preparation of Poly(3,4- ethylenedioxythiophene) (PEDOT) nanotubes (Kim, & Lee, 2005) or arrangement of PEDOT films (Kimura, & Yamada, 2009) on Au nano-brush electrodes are of the most important published activities.

In a same research, in order to enhance the contrast and switching time of regular Prussian Blue (PB), which is widely used as a sole electrochrome in EC devices, the nano-technology concept has been applied(Cheng et al., 2007). Fig. 11 clearly describes this research.

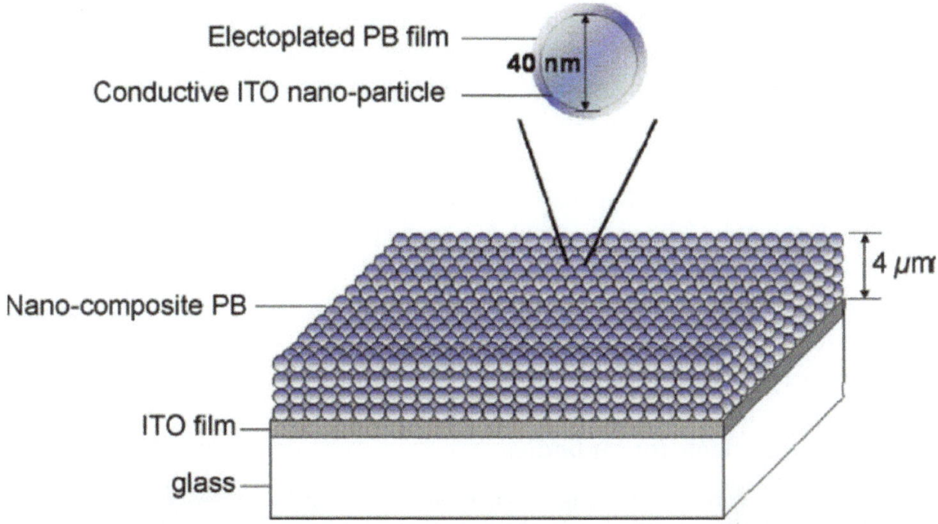

Figure 11. Conceptual structure of nano-composite PB film.

In the naocomposite shown in Fig. 11, indium tin oxide (ITO) nano-particles was applied as a medium layer for PB to gain larger operative reaction surface area. It was prepared by spraying a well-dispersed ITO nano-particle solution onto an ITO-coated glass and followed by electroplating PB on pre-sprayed ITO nano-particles. Due to proper covering of ITO nano-particles with PB, the final film produce a nano-porous electrochromic layer. Fig. 12 shows the SEM photograph of this nanocomposite from top and cross-sectional view. It was also revealed that switching speed and contrast of nano-structured film exhibit much better performances than traditional PB thin films. It was explained by this fact that Nanocomposite PB offers much larger operative reaction surface area than traditional PB film does.

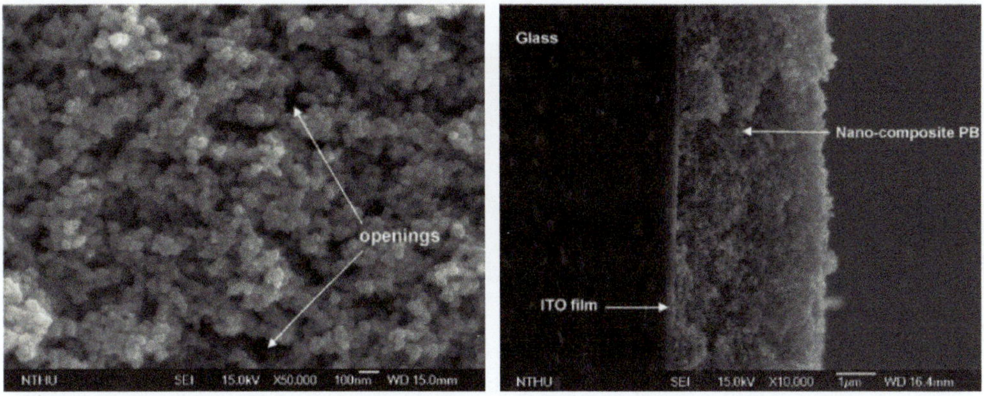

Figure 12. SEM images of final PB nanocomposite film (a) top and (b) cross-sectional view.

DuPont has developed an EC device based on an organic polymer technology to control light transmission in automotive applications. In comparison with current EC technologies, this not only is less complicated, but also it can be used in rigid and flexible forms, large sizes, and curved shapes. Target markets of this technology in automotive include sunroofs, mirrors, instrument clusters, windshield shade bands, sidelights, and backlights.

It is predicted that the market for smart windows will become a billion-dollar one by 2015 and will be doubling by 2018. The automotive market provides the next largest source of smart window opportunities for glass suppliers, after the architectural markets.

The Ferrari 575 M Superamerica had an electrochromic roof as standard, and the Maybach has a PDLC roof as option. Some Polyvision Privacy Glass has been applied in the Maybach 62 car for privacy protection purposes.

2.7. High-strength steels for car bodies

In order to enhance cars and passengers safety at crashes, the automotive producers have attempted to use high-strength steels in car bodies. However, it is difficult recasting of high-strength steel parts in cold state as the size accuracy changes and undesirable spring-back effects may occurs. Recasting in a hot state (at 1000 °C) helps us to avoid such disadvantages during recasting of high-strength steel parts. However, the scaling of this kind of steel is difficult at high temperatures. Nanotechnology is utilized to solve this problem. To this end, a multifunctional protective coating produced based on nanotechnological approach with the principles of conventional paint technologies. This multifunctional coating is produced using bonded and connected nano sized vitreous and plastic like materials together with aluminum particles. This system is schematically shown in Fig. 13 (21-22).

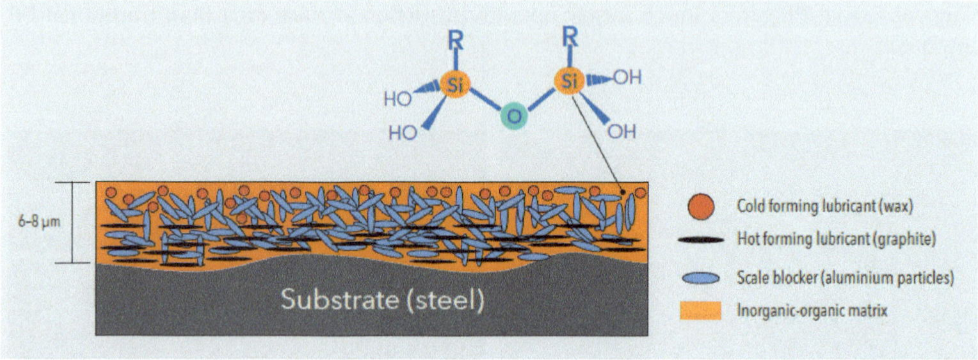

Figure 13. Nanostructured high-strength steel for car body.

2.8. Nanostructured tyres

Tyres performance extremely depends on their cover composition which is continuously in contact with road. So the rubber composition of the cover plays an important role on its properties. Different properties like abrasion resistance, grip and resistance against tear propagation are important. Incorporation 30% filler can improve such properties. Type and loading of filler as well as chemical and physical interactions between filler and rubber are influential parameters affecting its properties (Das et al., 2008, Zhou et al., 2010).

While the tyre resistance against grip should be high, its rolling resistance has to be low. Tyres need to resist abrasion but they should have slip-proof properties to reduce the car slide. In fact, there are three main factors which necessarily should be considered for a desired car tyre. These are reducing fuel efficiency by improving rolling resistance, increasing tyre life time by improving its abrasion resistance and reducing car fuel consumption by reducing friction. However, reducing friction can negatively influence car safety. The modern tyres consist of a mixture of synthetic and natural rubber, carbon black and silica, additives and steel/textile or nylon rod as reinforcement.

Soot (carbon black), silica and organosilane are the important examples of materials used to enhance rubber properties. Adding such materials to rubber composition at nanometric dimensions can significantly improve tyre properties. The size and surface modification of the particles can affect their chemical and/or physical interactions with rubber matrix. This varies the particles cross-linking with natural rubber molecules, affecting its properties. Nano sized soot particles can significantly enhance tyres durability as well as higher fuel efficiency. These particles have courser surface compared with traditional ones and due to their higher surface energy, they could produce stronger interactions with rubber matrix (Fig. 14). As a result, inner friction can be reduced which results in better rolling properties (Das et al., 2008, Zhou et al., 2010).

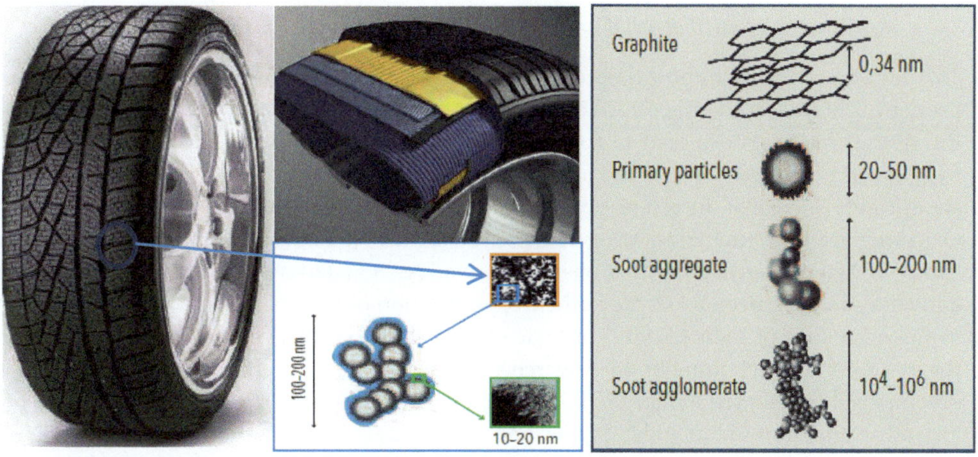

Figure 14. Schematic illustration of a modern nanostructured based tyre for cars (Das et al., 2008, Zhou et al., 2010).

It is well known that strain vibration will occur within tyre material at high car speed. Nanoparticles can reduce this strain vibration and results in superior traction, especially on wet roads. The surface modification of the particles is important which will affect their interaction with rubber matrix and its final properties. It has been found that carbon nanotube (CNT) can improve mechanical properties such as tensile strength (600%), tear strength (250%) and hardness (70%) of styrene-butadien rubbers. Tyres with higher stiffness and better thermoplastic stability can be produced using lamellar nano-sized organoclays like montmorillonite. The other nanoparticles used to enhance car tyre properties are nano-alumina, carbon nano fibers (CNF) and graphene. The rolling resistance of tyres can be significantly improved using silane-treated silica compared with traditional carbon black based tyres. Using nanoparticles, tyres with better traction on wet and icy roads can be produced. As a result, the stopping distance of car can be reduced by 15-20 % and 5% in fuel consumption (Das et al., 2008, Zhou et al., 2010).

3. Interior applications

3.1. Automotive fabrics

Car industry's commercial strategy today is to improve the safety and convenience aspects of automobiles. Textiles, especially fabrics, as the main substances in designing of interior parts of a vehicle, are very important. They are utilized in various parts such as interior panels for doors, pillars, seats coverings and paddings, parts of the dashboard, cabin roof and boot carpets, headliner, safety belts, airbags. Nanotechnology as a powerful tool has aided the auto-manufacturers to reach their goals in a short period of time. The most important properties of automotive fabrics which have been modified by the aid of nanotechnology include: a) anti-microbial b) self-cleaning c) fire-retardancy.

3.1.1. Antimicrobial/antibacterial and Anti-odour properties

Textiles can grant an appropriate environment for micro-organisms growth especially at proper humidity and temperature in contact to human body. Rapid and uncontrolled fast thriving of microbes can lead to some serious problems. Because of public concern about hygiene, the number of studies about anti-microbial modification of textiles has been significantly increased in recent years. To this end, various anti-microbial agents such as Oxidising agents (aldehydes, halogens), Radical formers (halogens, isothiazones and peroxo compounds), diphenyl ether (bis-phenyl) derivatives, Quaternary ammonium compounds and chitosan have been used. Nevertheless, application of many of these materials has been avoided because of their harmful or toxic effects. More recently, nanotechnology has been the basis of a great number of researches to produce novel anti-microbial textiles. As schematically presented in Fig. 15, the most important nano-structured anti-microbial agents are silver, titanium oxide, gold, copper and zinc oxide and chitosan nano-particles, silver-based nano-structured materials, titania nanotubes (TNTs), carbon nanotubes (CNTs), nano-clay, gallium, liposomes loaded nano-particles (Dastjerdi & Montazer, 2010). These nanoparticles can be coated directly on textiles or via a vehicle (incorporated nanoparticles in a matrix such as silica network). Various techniques can be utilized for coating of these antimicrobial agents on textiles like sol-gel processes and chemical vapor deposition.

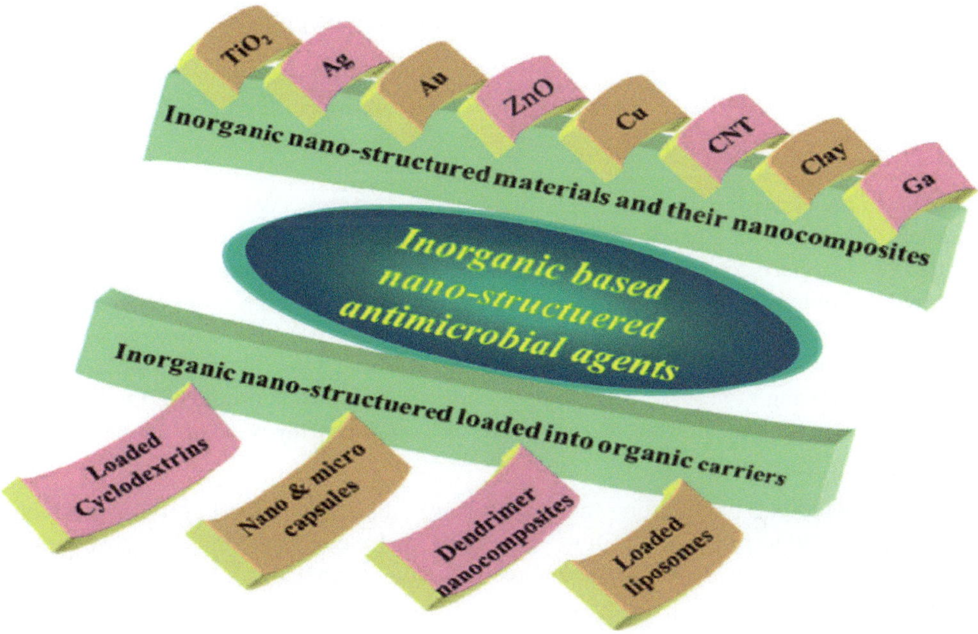

Figure 15. Classification of inorganic based nano-structured anti-microbial agents(Dastjerdi & Montazer, 2010).

The anti-bacterial action in these agents is caused via either a photo-catalytic reactions or biocidical processes. An example of former type of anti-bacterials is titania-based agents that act through the absorption of light, photo-catalytic reactions. As a result of these reactions, excited charge carriers (an electron and a positively charged electron-hole) are produced. While the positively charged holes induce the oxidation of organic molecules, the electrons can react with oxygen, leading to formation of hyperoxide radicals. These radicals attack and oxidize the cell membranes of microorganisms. The described photo-catalytic process is the cleaning mechanism of superhydrophilic self-cleaning surfaces which leads to the degradation of stains (Banerjee, 2011; Fujishima et al., 2008). Silver and gold are examples of the latter type of anti-bacterial materials. In biocidical action, the antibacterial effect happens via interaction between the positively charged biocide and the negatively charged cell membranes of microorganisms which damages the microorganism. In the majority of researches a combination of both mechanisms (photo-catalytic and biocidal processes) are used to achieve an efficient anti-bacterial effect (Yuranova, et al., 2006; Yeo et al., 2003).

Among different anti-bacterial agents, silver has received the most attention because of potential advantages(Montazer et al., 2012a; Montazer, et al., 2012 b). Besides possessing a high degree of biocompatibility, silver is highly resistant to sterilization conditions and has a long-term antibacterial efficiency against different bacteria.

In commercial viewpoint, anti-bacterial automotive textiles based on nanotechnology are beginning to enter the market. For example, Tencel™ material based on nanofibrils of cellulose was produced by Lenzing. It has a combination of properties and in particular antibacterial properties which reduces growth of bacteria. This product has been introduced to the market as a good candidate for seat car covers.

3.1.2. Hydrophobic surfaces and anti-stain textiles

Lotus leaf is a natural model for super-hydrophobic surfaces. Very low surface energy materials (like fluoro- or silicone- containing polymers) and nano-scale roughness structures (created by nanoparticles or nanotechnology-based procedures) are required for creating a superhydrophobic self-cleaning surface. A schematic picture of such surfaces is shown in Fig. 16. On these surfaces the distance between summits of such nano-roughnesses is around few hundreds nanometer and they are so close together that a speckle of dirt would not fit between them(Wang et al., 2011). Therefore, a non-stick surface is produced. On the other hand, low surface energy substances make water roll off and easily wash off unattached dirt from surface.

Different methods based on nanotechnology like Layer-by-Layer Deposition, Electrodeposition/Electropolymerisation, Plasma and Laser Treatment, Electrospinning, Casting and Molding can be employed for creating nano-roughness.

Among researches to make super hydrophobic surfaces, carbon nanotube, silica and fluoro containing polymer nanoparticles were applied to the nylon, cotton and polyester fabrics in form of a coating (46-48). In these works, they could achieve artificial lotus leaf structures.

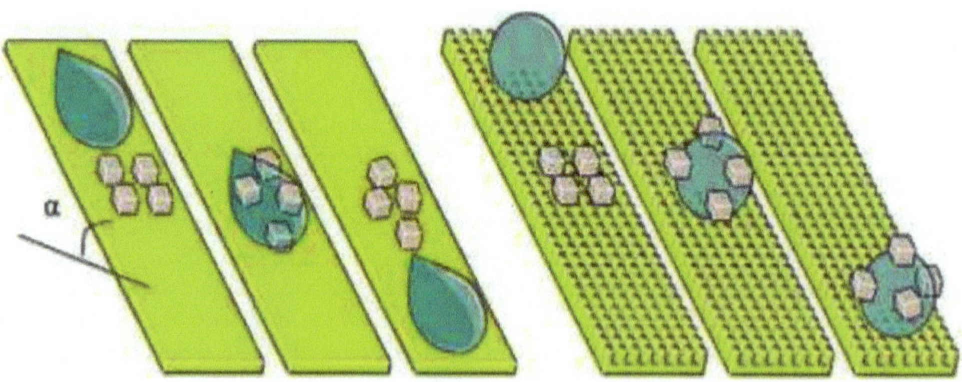

Figure 16. Self-clean action on a conventional and on a nano-structured textiles by removing dirt with water (lotus effect).

Opel Co. was the first manufacturer in the world to equip seating upholstery of Insignia (Car of the Year 2009) with the Nanogate coatings that repel dirt and liquid staining.

3.1.3. Flame retardant fabrics

For the last half a century, various compounds have been employed to improve the fire resistance performance of textiles. Inorganic chemicals such as antimony, aluminum and tin as well as Bromine, Chlorine- and Phosphorus- based compounds are the main chemical families of flame retardants (Horrocks, 2011). These conventional chemicals are not usually harmless. It has been proved that halogen–antimony and phosphorus–bromine combinations, besides having limited performance have environmental concerns. Environmental regulations have restricted the use of these flame-retardant additives, initiating a search for replacing toxic flame retardants in polymer formulations with safer and more environmentally-friendly alternatives. This has sparked the interest of nanoscientists.

Recently, polymer nanocomposites offering significant advantages over conventional formulations have received many attentions in the field of flame retardancy. Nanoparticle fillers are highly attractive for this purpose, because they can simultaneously modify both the physical and flammability properties of the polymeric matrixes. Layered silicates (clay) and carbon nanotubes (CNTs) are two main nanostructured materials that have attracted the attention of scientists to promote fire performance of polymeric substrates like textiles (Bellayer et al., 2004; Kiliaris, & Papaspyrides, 2010). The nano-materials make fabrics less ignitable and self-extinguishable when the flame is removed.

Since flame retarding mechanisms of clay and CNTs are different, significant synergism happens when they are introduced to textile together, leading to a much more efficient approach to improve the flame retardancy.

In recent studies, polyhedral oligomeric silsesquioxane(POSS) compounds have been utilized as fire-retardant agents. In a series of experiments, Bourbigot and coworkers introduced POSS nanoparticles in polypropylene yarns, cotton and knitted polyester and showed that the time-to-ignition increased significantly as a result of presence of nanoparticles (Bourbigot, et al., 2005).

Since clays, CNTs and POSS nanoparticles are more expensive than traditional fire retardants, their uses are currently hampered even if they are more environmentally friendly. Therefore, cost reduction would likely change this situation.

3.2. Nano-coatings for engine application

Coatings plays an important role in improving efficiency and life of the car engine. These are listed below (Dahotre et al., 2005, MacLean et al., 2003, Lin., et al 1993):

Lubrication (reduced frictional loss)

Thermal insulation (higher operating temperature)

Reduced friction (surface finish and affinity or oil)

Reduce dimension weight (replaces cast iron block/liner)

It is well known that engine of a car can operate at higher temperatures by reducing external heat removal. Using lightweight materials in engine to reduce load, heat losses and frictional losses is another approach to improve fuel efficiency. One of the most important factors to improve fuel efficiency is reducing weight of engine. Replacing cast iron (with density of 7.8 g/cm^3) used in engine blocks with low-cast aluminum-silicone (with density of 2.79 g/cm^3) is one possibility for engine weight reduction. However, aluminum alloys do not have adequate wear resistance and high seizure loads to be used in the cylinder bores. Because of these, cylinder bores are made of cast iron liners which have good wear resistance. Therefore, attempts have been carried out to improve aluminum bars properties using new composites and/or monolithic coatings (Dahotre et al., 2005, MacLean et al., 2003, Lin et al., 1993, Venkataraman et al., 1996, Rao et al., 1997).

Nanomaterials can be employed to achieve extraordinary properties for aluminum bars. Schematic illustration showing the variations of hardness versus grain size is depicted in Fig. 17.

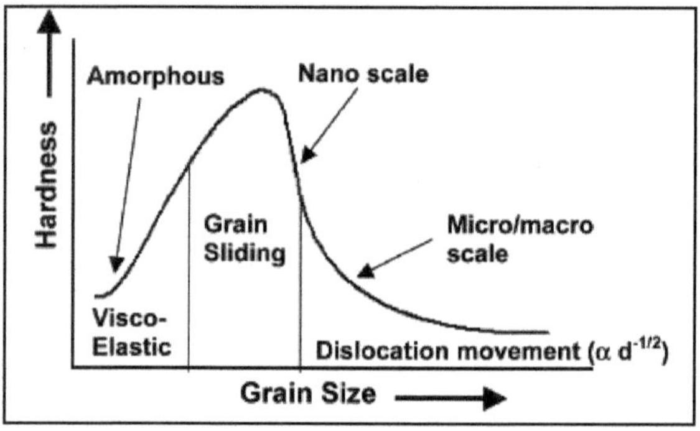

Figure 17. The effects of grain size of a metal on its hardness and other properties (Dahotre et al., 2005).

Fig. 17 clearly shows increase of hardness and flow stress as the grain size decreases (<100 nm). At grain sizes smaller than 100 nm, the deformation mechanism will be changed from dislocation-controlled slip to grain boundary sliding whilst the plasticity is increased simultaneously. Different parameters including toughness, flow stress, ductility and thermal insulation of the aluminum will be intensified when the grain size is in nano scale. Nanocoatings have been utilized in order to improve engine efficiency as described below:

3.3. Wear resistant nano-coatings for engines

Scratch and wear are criteria parameters which will be considered for the metal parts used in automobile engines. Electrodeposited hard chrome and microstructure ceramic coatings are the most used kinds of protective coatings for engine parts. The ceramic coatings are frequently applied on metal parts using thermal spray. In plasma spraying, the coating powder reinforced with ceramic particles is injected into a plasma stream following by heating and accelerating toward the metal substrate. The ceramic rapidly cools and produce a coating layer over the substrate. However, there are limitations for the use of microstructure ceramic and eletrodeposited chrome coatings. Chrome coatings include hazardous materials influencing the environment and are also expensive. The conventional microstructure ceramic coatings are less expensive than chrome coating but are brittle and show low adhesion to the substrate. Attempts have been carried out to find other replacements. Nanostructured containing ceramic coatings have been utilized to improve metal parts of engine against abrasion and wear. Reducing the scale of materials microstructure like grain size, particle size or layer thickness can significantly alter its properties (Fig. 18) (Rao et al., 1997, Wuest et al., 1997, Rastegar et al., 1997, Cole et al., 1997, Ebisawa et al., 1991, Kabacoff et al., 2002, Sanchez et al., 2007).

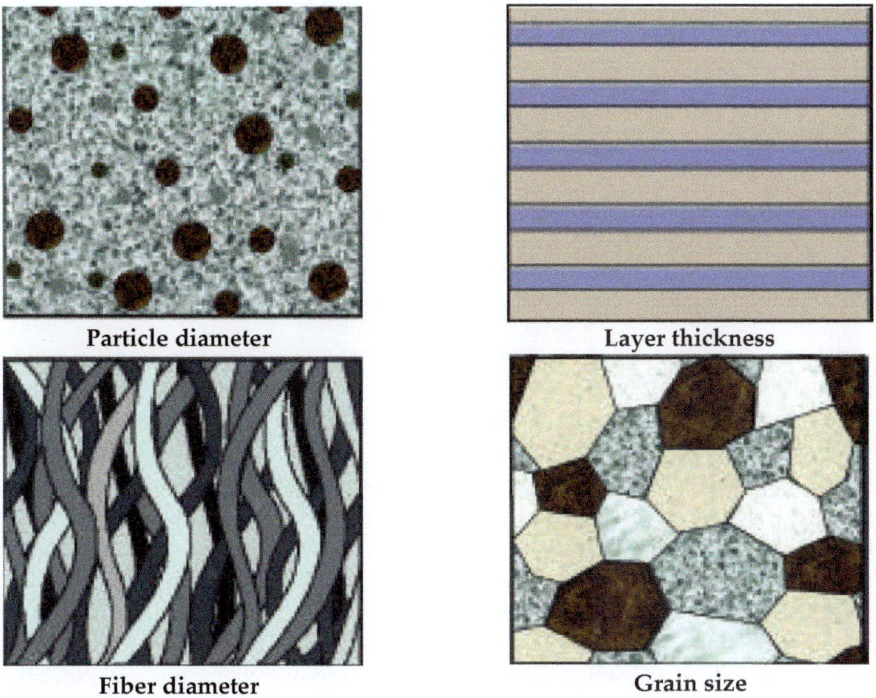

Figure 18. Different states of nanostructured materials used in order to improve car body properties (Kabacoff et al., 2002).

Recently, new nanoceramic composites (alumina-titania ceramic coatings) have exhibited excellent wear resistance. However, there are problems with spraying nanoparticles because of their low mass and poor fluidity. One way to solve this problem is agglomerating nanoparticles into micrometer-sized aggregates. However, in order to obtain initial nanostructure after spraying, the process must be carefully controlled. Alumina, alumina-titania, cemented tungsten carbides, or zirconia powders are examples for the nanostructured coatings obtained from agglomerated particles plasma spraying. Both wear and abrasion resistance of the metal substrate will be considerably improved after the nanostructured materials spraying. Recent findings have revealed that atmospheric plasma spraying of the nanostructured materials results in better nanostructure formation in final coating layer and better tribological properties of the coating (Dahotre et al., 2005, Kabacoff et al., 2002, Sanchez et al., 2007).

3.4. Nano-coatings with good lubrication for engine application

It is well known that mechanical friction could significantly influence the internal combustion (IC) engine fuel economy. Valve train, piston system crank and bearing system are the most important sources of frictions (Fig. 19) (Dahotre et al., 2005, Kabacoff et al., 2002, Sanchez et al., 2007).

Figure 19. New coatings used to improve (right) engine body structure and (left) cylinders.

These friction sources could reduce engine life and increase oil consumption. Coatings could reduce frictions and result in lower oil usage. Examples of these coatings are Ni-Mo-MoS2, Ni-BN, graphite-Ni, etc. Recently, nano-structured materials have been utilized to improve friction properties of piston rings. Zirconium ceramic coatings can modify surface properties. Nano-size zirconium powder can be dispersed in a mineral oil. The nanoparticles can reach working surface of the engine when the piston moves. The nano-size zirconium help ceramic particles better bond to the metal surface. Heat generated during engine operation would be enough to cure ceramic powder attached to the engine surface. After curing, ceramic coating produces hard and smooth surface at different parts of the engine including cylinder walls, piston rings, piston top, valve tops and bearing surfaces. The nano-size zirconium particles can also improve fuel economy, power output, oil burning and reduce noise, vibration of engine and pollution discharge (Rao et al., 1997, Wuest et al., 1997, Rastegar et al., 1997, Cole et al., 1997, Ebisawa et al., 1991, Kabacoff et al., 2002, Sanchez et al., 2007).

3.5. Nanofluids and nanolubricants

3.5.1. Nanofluids: Properties and application in automotive industry

Adding nano sized materials like nanofibers, nanotubes, nanowires, nanorods and nanosheets to fluids results in producing new generation of fluids having superior properties in comparison with conventional fluids. In fact, nanoscale colloidal suspensions loaded with condensed nanomaterials are named nanofluids. This system consists of two phases: liquid phase (base fluid) and solid phase (nanoparticles). Using nanoparticles, the thermoplastic properties such as thermal diffusivity, thermal conductivity, viscosity and convective heat transfer coefficient of the fluid will be significantly enhanced. Achieving such properties need making stable nanofluids which has shown serious challenge in recent years (Yu et al., 2011, Trisaksri et al., 2007, Wang et al., 2007, Wang et al., 2008). Using nanofluids, cooling systems with higher efficiency have been designed for cars. Decreasing cooling system weight and reducing its complexity are the most important advantages of using nanofluids. In this way, compact cooling system with smaller size and weight can be designed for cars' radiator. Improving thermal conductivity of ethylene glycol-based fluids using nanomaterials has attracted much attention as engine coolant. In conventional cooling systems, a ratio of 50:50 of water and ethylene glycol is used as coolant. However, there are advantages of using ethylene glycol based nanofluids such as low pressure operation compared with mixture of water and ethylene glycol. Nanofluids based coolants have boiling point higher than conventional ones helping it reject more heat through coolant system. It has been shown that using nanofluids in cars' radiator could reduce frontal area of radiator up to 10%. In this way, nanofluids could reduce aerodynamic drag and fuel saving up to 5%. Nanofluids could also reduce friction and wear in pumps and compressors, leading to fuel saving up to 6%. These all reveal that nanofluids are suitable materials which not only could improve cars cooling system performance but also can greatly influence the structure design of cars (Wang et al., 2008, Li et al., 2009, Kakac et al., 2009, Xie et al., 2009, Yu et al., 2009, Yu et al., 2007, Kole et al., 2007, Tzeng et al., 2005).

3.5.2. Heat transfer improvement using nanofluids

Maxwell's model reveals that increase in volume fraction of spherical nanoparticles results in thermal conductivity improvement of a liquid. Moreover, increase in surface area-to-volume ratio of the particles leads to an increase of the conductivity of the liquid. In addition to particles size and particles loading, the particles sphericity (defined as the ratio of the surface area of a perfect spherical particle to that of non-spherical particle at the same volume) is another parameter influencing thermal conductivity of a suspension. Hamilton and Crosser's (Yu et al., 2009, Yu et al., 2007, Kole et al., 2007, Tzeng et al., 2005) revealed that decreasing particle sphericity from 1.0 to 3.0 results in significant increase in thermal conductivity more than two times. Particle with 10 nm diameters has surface-area to volume ratio of 1000 times greater than a particle with 10 μm size. Consequently, it has been expected to enhance thermal conductivity using nanometer sized particles much greater than micrometer sized particles.

Attempts have been carried out to improve heat transfer ability of water/ethylene glycol liquids (used in a car radiator) using nanoparticles. Nano CuO and Al_2O_3 particles are added to these liquids. Results showed significantly enhanced thermal conductivity of the liquids using these nanometric materials. It is shown that using 4 vol% nano-CuO (30 nm diameter) can increase thermal conductivity of the ethylene glychol by 20%. The same observation was seen in case of using nano-Al2O3 particles in water. It has been found that reducing nano-CuO particles' size results in further increase in thermal conductivity of the liquid. In another research, the effects of addition of nano sized ZnO, Al_2O_3 and TiO_2 particles (at 5 vol%) to an ethylene glycol on its thermal conductivity and viscosity was studied. The highest thermal conductivity and the lowest viscosity were observed for the liquid loaded with MgO nanoparticles. Carbon nanotube is found effective nanoparticle to enhance thermal conductivity of water and ethylene glycol. Using 1 vol% carbon nanotube can improve water/ethylene glycol mixture conductivity up to 175% (Kakac et al., 2009, Xie et al., 2009, Yu et al., 2009, Yu et al., 2007, Kole et al., 2007, Tzeng et al., 2005).

3.6. Lubricating oils for cars using nanoparticle additives

Lubricants like mineral oil are used to reduce friction and wear in automobile engine. The pistons movement in cylinder of an engine produces frictions as a result of metals wear. This may lead to reduced engine efficiency as well as lowered engine life. Oils are used as lubricant to reduce friction. The conventional oils need to be exchanged after a special engine working time. In fact, the oil lubricant properties will be gradually reduced. Researches to produce better oils with longer life are developed in recent years. Nanotechnology is one of the most effective ways of fulfilling this target (Wu et al., 2007, Chinas-Castillo et al., 2003). It has been shown that nanoparticles could improve lubricant behavior of conventional oils. Particles shape, size and concentration are influential parameters affecting wear and friction reduction. It has been shown that gold particles having particle size of 20 nm has the best lubricating effects. Dialkyldithiophosphate modified copper nanoparticles is shown as an effective nanoparticle with high ability of improving anti-wear ability of metal surface by producing an anti-friction film. Diamond and inorganic fullerene-like (IF) particles are other examples of anti-wear nanoparticles being used as additives for lubrication. The most important mechanisms which result in friction reduction are colloidal effects, rolling effects, protective film and third body. Diamond nanoparticles were added to oil to improve its anti-wear ability. This nanoparticle has found to improve oil lubricant behavior via various mechanisms including: (a) ball bearing effects of the spherical particles existed between rubbing surfaces, (b) the surface polishing and (c) increasing surface hardness. Adding CuO nanoparticles to oil could significantly reduce friction coefficient. Ball bearing at high temperature and viscous effect at low temperature are the reasons CuO nanoparticles can improve anti-wear behavior of oil. The nanoparticles depositions at worn surfaces would be responsible for shear stress reduction leading to tribological properties improvement of the surface (Wu et al., 2007, Chinas-Castillo et al., 2003, Zhou et al., 1999, Rapoport et al., 1999, Chen et al., 1998).

3.7. Energy criterion in cars

To replace combustion engines, different strategies and methods have been developed. Among them, electrochemical energy production/storage is the most important option owing to sustainability and being environmentally friendly (Schlapbach & Zuttel, 2001). The so-called electrochemical energy storage and conversion systems include fuel cells, batteries and supercapacitors. Although batteries have found their way in marketplace in different applications and fuel cells and supercapacitors are competing to establish promising applications, there are still many challenges to be solved to have energy conversion/storage systems which could surpass combustion engines in terms of power/energy performance and cost (Winter & Brodd, 2004). Nanomaterials are finding great contribution to overcome these challenges (Arico et al., 2005; Serrano et al., 2009).

3.7.1. Fuel cells

Automobiles powered by fuel cells are believed to have considerable market in near future and it is envisioned about 80 million fuel cell vehicles will be on the road by 2020 (Serrano et al., 2003). High efficient energy conversion, safety, high energy density, nonpolluting are the advantages of employing fuel cells as energy source for driving a car (Jacobson et al., 2005). However, high cost, low volumetric power density, low durability and cell life plus high sensitivity to purity gas stream and complex operation are disadvantage of using fuel cells. Thus, hybrid configuration of fuel cells with batteries or supercapacitors is being developed in order to supply power for peak-power demands such as acceleration and start-up and also recovering braking energy (Kötz et al., 2001). In fact, despite of comparable energy density of fuel cells (100-1000 $Wh.kg^{-1}$) relative to combustion engines, they have few order of magnitude lower power densities, rendering them as steady energy source. On the other hand, supercapacitors possess high power density comparable to combustion engines (Fig. 20). Therefore, hybrid systems of fuel cells and supercapacitors or batteries are an efficient configuration for replacing the combustion engines. In general there are six types of fuel cells systems including i) alkaline fuel cells ii) polymer electrolyte membrane fuel cells (PEFCs) iii) direct methanol fuel cells (DMFC) iv) phosphoric acid fuel cells v) molten carbonate fuel cells and vi) solid oxide fuel cells (Winter & Brodd, 2004).

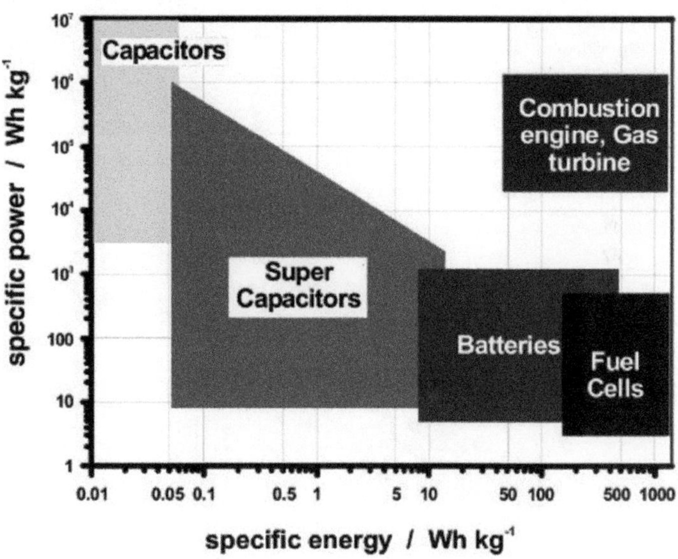

Figure 20. Ragone plot of the energy storage domains for the various electrochemical energy conversion systems compared to an internal combustion engine and turbines and conventional capacitors (Winter & Brodd, 2004).

Although one can drive a car powered by some classes (PEFCs) of fuel cells, there are still some challenges associated with employing them which are mostly high cost, fuel supply/storage and life time. Generally the fuel choice is hydrogen and oxygen which finally exhaust water. Hydrogen supplication, refuelling infrastructure and storage of hydrogen are still ongoing challenges (Chan, 2007). Although hydrogen has very high gravimetric energy density, its' application hindered due to low volumetric energy density. Storing hydrogen in liquid state require employing highly expensive cryogenic tanks (Fig. 21). Compressing hydrogen gas also urges using costly storage facilities. There has been considerable research to develop new materials enabling storage of hydrogen at high enough concentration at not too high pressure and too low temperature. Initial interest was focused on metal hydrides through chemisorptions of hydrogen (Fig. 22). However, efforts were quite unsuccessful to synthesis metal alloys reaching to theoretical limit (~8wt%) unless some promising results were reported for nanosized metal (oxide) composites (Arico et al., 2005). Metal alloys such as LaNi, TiFe, MgNi are generally expensive and in all cases are heavy which makes commercialization of products dealing with mobile applications problematical.

Figure 21. GM ElectroVAN, the first hydrogen fuel cell powered car introduced at 1966. The hydrogen and oxygen stored in super-cooled liquid in cryogenic tank.

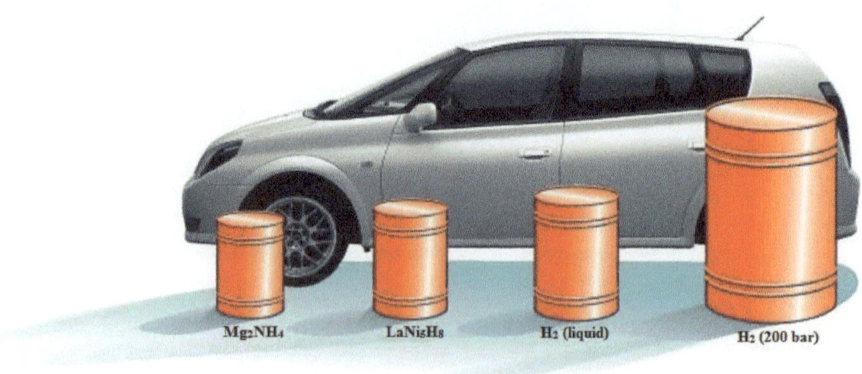

Figure 22. Comparison of volumetric density of different systems to store hydrogen (Schlapbach & Zuttel, 2001)

In another set of research, the efforts are focused on physical entrapping of hydrogen in porous materials. Physisorption of H_2 allows fast loading and unloading. Nanostructured materials are the sole candidates for this purpose. Carbon nanomaterials (Dillon et al., 1997) and metal organic framework (MOFs) (Ce′te et al., 2005; Zhou et al., 2007) are two dominate studied classes of materials with the goal of solid-phase storage of H_2. By the advent of carbon nanotubes at early 90s (Iijima, 1991; Iijima & Ichihashi, 1993), great deal of attention was attracted to this novel nanomaterial. Dillon *et. al.* (Dillon et al., 1997) for the first time reported high capability of single-walled carbon nanotubes (SWNTs) as porous media for H_2 uptake, reporting 5 to 10wt% of H_2 adsorbtion at ambient conditions. Chen *et. al.* (Chen et al., 1999) later reported significant enhancement of H_2 uptake (20wt%) of multi-walled carbon nanotubes (MWNTs) upon doping with alkali metals at 300K and 0.1MPa. These results were beyond the requirements that have been established by Department of Energy (DOE) of US where gravimetric density of H_2 must be at least 6wt%. However, subsequent studies revealed that CNTs are not promising candidates for H_2 uptake. The results indicate metal doping of CNTs can enhance the H_2 uptake of nanotubes, but hardly exceeds 4wt% (Liu et al., 1999; Béguin & Frackowiak, 2006).

Novel nano-porous materials, MOFs, have been centre of attention for gas adsorbtion. These materials are product of reaction of metals ions with rigid organic molecules (Ce′te et al., 2005). Due to exceptionally high surface area and tunable chemical structure of MOFs, high potential for high enough H_2 uptake is envisioned. In September 2011, Daimler introduced a concept vehicle, Mercedes-Benz F125!, which was pictured to be derived by 2025 (http://media.daimler.com). The most interesting technology of this conceptual car is its source of energy which is hybrid of Li-S battery (See below for more details) and hydrogen fuel cell. Despite current technology used in available fuel cell vehicles, liquid or compress hydrogen, storage facility of hydrogen is based on MOFs materials in this concept vehicle. The manufacturer claims one will be able to derive up to 1000km with maximum speed of 220km.h^{-1} before it is needed to be refilled.

Although interest to fabricate solid-phase hydrogen reservoir using CNTs are now quenched, emergence of the carbon-based thinnest materials, *i.e.* graphene, again revived hopes to have carbon-based hydrogen storage tanks (Park et al., 2007; Dimitrakakis et al., 2008; Burress et al., 2010). In fact, as the H_2 storage mechanism in carbon nanostructures relies on physisorption on graphenic surface, the hydrogen uptake is proportional to specific surface area of nanostructure which reaches ultimate value for carbon nanostructures in graphene (2630 m^2.g^{-1}) (Dimitrakakis et al., 2008). Researches on developing high surface area graphene-based materials are ongoing and more time requires confirming whether graphene-based nanoporous materials are able to solve the mystery of the hydrogen storage or not (Subrahmanyam et al., 2011).

Another major issue which hinder widespread application of fuel cell vehicles, is high cost associated with manufacturing them (Fig. 23). The expensive constituents of fuel cells including the catalyst and electrolyte membrane are the origin of high cost of fuel cells (Steele & Heinzel, 2001). In addition, performance and life time of these classes of fuel cells (PEFC, DMFC) can be remarkably improved by nano-engineering of the catalyst and electrolyte membrane. In fact, the commercialization of fuel cells will very much rely on the ability to reduce the cost and improve the performance of catalyst, membrane and other expensive parts to launch a fuel cell powered vehicle at a competitive cost and driving capabilities (Arico et al., 2005). To reach this goal, nanotechnology plays a dominant role.

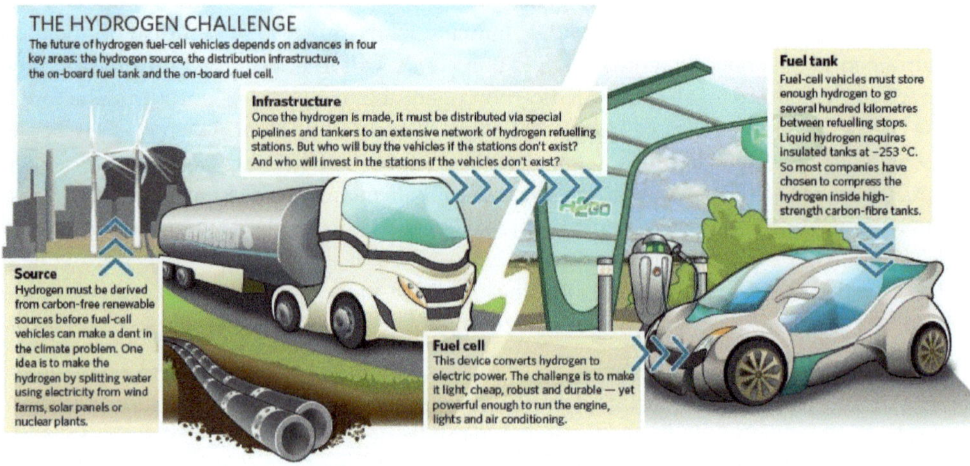

Figure 23. The challenges associated with implementing hydrogen as the next generation green fuel including hydrogen source and production, storage infrastructure, fuel tank and high efficient fuel cells (Tollefson, 2010).

Principally, fuel cells (PEFCs and DMFCs) operate with a polymer electrolyte membrane which is sandwiched between cathode and anode that separate the fuel (hydrogen) from the oxidant (air or oxygen) (Fig. 24). The performance of these low-temperature fuel cells is mostly limited by oxygen reduction reaction (ORR) (Arico et al., 2005). The current generations of fuel cells are utilized platinum (Pt)-based catalyst for both the oxidization of fuel and reduction of the oxygen (Greeley et al., 2009). Several factors are motivating researchers to replaces Pt which the high cost of this precious metal (40-70 $\$.g^{-1}$) is one them. In addition, this metal is so scarce (less than 0.005 p.p.m in the Earth's crust) and about 90% of the world's Pt supply comes from just two countries, South Africa and Russia (http://en.wikipedia.org/wiki/Platinum). A few approach are being actively pursued with the goal of improving the electro-catalyst activity plus lowering the overall cost. However, these efforts have dominantly followed three major strategies including: i) improving the Pt-based catalyst ii) developing new class of non-precious catalyst using other transition metals and finally iii) metal-free catalyst materials.

Today Pt-carbon catalyst which are widely used in PEFCs and DMFCs, are nanoparticles of Pt decorated on carbon support (*e.g.* carbon black) (Greeley et al., 2009). Pt nanoparticles' activity increases as the particle size decreases reaching a minimum of ~3nm (Arico et al., 2005). Rational nano-engineering of Pt alone or with other metal atoms into specific arrangement of nanostructured alloys such as core-shell nanoparticles or nanowire (Koenigsmann et al., 2011; Koenigsmann et al., 2012; Hong et al., 2012; Lim et al., 2009) is a highly effective tool to synthesis new generation of catalyst if fundamentals of governing the performance of catalyst are understood. Core-shell nanoparticles of Pt-Cu, for example, are more active than Pt which consists of cores made of a Cu-Pt alloy and Pt-rich outer shell (Mayrhofer et al., 2009). Such structure is obtained through a controlled dealloying the hybrid nanoparticles. However, many challenges associated with using precious Pt-based catalyst still remain untouched through these strategies. For instance, General Motor's fuel-cell set-up used around 80 grams of platinum as electrocatalyst to split hydrogen into electrons and protons which costs roughly 5000$ (Tollefson, 2010). Although General Motors's officials hope to reduce the Pt loading to 30 grams in near future and less than 10 grams in the next decade, the price of catalyst would still be high considering the fact that the Pt is so scarce.

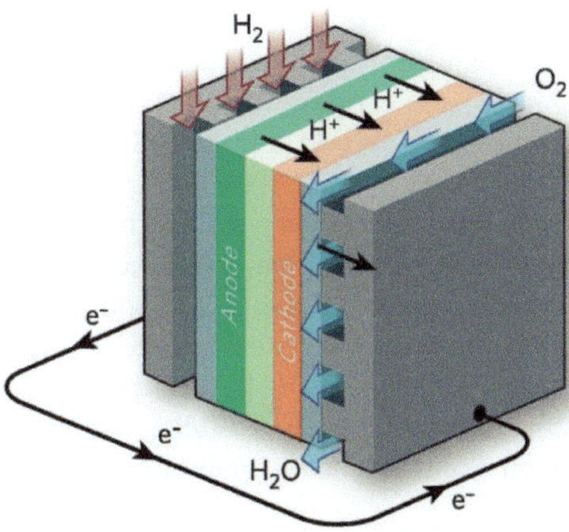

Figure 24. Basic structure of hydrogen fuel cell in which hydrogen split into electron and proton at anode and at cathode protons reduce oxygen exhausting water finally (Tollefson, 2010).

With the goal of achieving inexpensive catalyst for fuel cells' application, tremendous attempts have been done since 60s decade (Jasinski, 1964). Despite quite ineffective early catalysts, recent developments through nanoscale engineering of nanostructured catalysts revive the hopes to have non-precious replacement of Pt catalysts. Catalysts based on thermally annealed precursors comprising nitrogen, carbon and transition metals, especially Fe and Co (Fe (or Co)/N/C), have attracted more attention due to high activity and performance (Lefevre et al., 2009; Bashyam & Zelenay, 2006; Proietti et al., 2011). This class of catalysts consists of metal nanoparticles embedding in nanostructured nitrogen-doped graphenic carbon (Wu et al., 2011). Metal free catalysts have been also synthesized and evaluated as catalyst. Among them, carbon-based nanostructures are the most studied systems. At 2009, Dai's group (Gong et al., 2009) at Case Western University showed gas phase N-doping of CNTs would result in a metal free electrocatalyst. After that, many studies revealed promising performance of N-doped carbon nanomaterials such as SWNTs (Zhang & Dai, 2012), graphene (Qu et al., 2010), mesoporous graphitic array (Liu et al., 2010) and carbon quantum dots (Li et al., 2011), for replacing Pt-based catalysts (Gong et al., 2009; Chen et al., 2012; Wang et al., 2011). Findings in Fe (or Co)/N/C systems and N-doped carbon nanomaterials may be combined possibly through using CNTs or graphene as support in Fe (or Co)/N/C catalysts instead of carbon black. However, it seems that carbon nanomaterials would have much more contribution in the next generation of catalyst in fuel cells than what they have in current Pt/C commercial catalysts. Rational design of nanostructure of the so-called upcoming catalyst would be the key issue.

3.7.2. Batteries

Although fuel cell powered cars have been driven since more than 45 years ago (Fig. 21), the current efficiency as replacing the internal combustion engines hinders widespread use of FCV (additionally lack of hydrogen storage infrastructures and cost are real barriers) (Turner, 2004). In addition, from practical point of view, because of low power density of FCs, FC powered automobiles require an energy storage device to deliver required energy in power-peak demands. As a result, currently automakers have come to this decision to launch new versions of hybrid cars before FC cars, which has resulted to introduction of new generation of automobiles with considerably low gasoline combustion reaching to record less than one litre per 100 km (Demirdeven & Deutch, 2004). Therefore, at the heart of the upcoming automobiles, energy storage devices play a key role. Batteries are blooming in different markets and automobile industry is not an exemption. Among different classes of batteries, lithium ion batteries have higher potential for employing in the next generation of cars due to higher energy density, unless in the first generation of electric cars even lead-acid batteries (for example the EV1, GM introduced 1996) and Ni-MeH (Prius, Toyota) were used (http://www.economist.com/sciencetechnology/tq/PrinterFriendly.cfm?story_id=10789409).

Similar to other batteries, Li-ion batteries also consist of cathode, electrolyte and anode (Fig. 25). Due to principles governing the electrochemistry of Li-ion batteries wide ranges of materials can be used in this class of devices which significantly affect the cell potential, energy density and safety of batteries (http://www.economist.com/sciencetechnology/tq/PrinterFriendly.cfm?story_id=10789409). Therefore, huge amount of attention has been attracted towards developing high performance Li-ion batteries from both academic communities and industrial firms. The efforts are focused on improving the capacity, safety and the charging rate. Nanoscopic materials are presumed to have great contribution in the world's $56 billion battery market in near future (Serrano et al., 2009).

The cathode and anode materials must be able to be intercalated with Li ions having high Li hosting capacity and also high electron conduction (Tarascon & Armand, 2001). Among the various materials employed as cathode for Li-ion batteries, iron (or other metal) phosphate is a promising and safe replacing candidate for conventional cathode material, cobalt oxide (Padhi et al., 1997). This case is interesting as nanoengineering helped a lot to see marketable version of this material. In fact, due to the low electrical conductivity of $FePO_4$, the energy charging rate is very low. Unless it is found that doping with other transition metals can enhance the material's conductivity, $FePO_4$-based materials found their way in marketplace when Chiang's group at MIT uncovered that nanosized $FePO_4$ particles can store and deliver energy much faster than usual size (~10 μm) due to higher surface area which facilitate intercalation of Li ions (Chung et al., 2002; Kang & Ceder, 2009). In fact, conventional Li-ion batteries equipped with cobalt oxide which are prone to catch fire due to thermal runaway, phosphates can be used to fabricate larger Li-ion batteries much more suitable for automobiles. For example, Volvo 3CC concept car, because of limitations associated with using cobalt oxide based batteries, is equipped with 3000 Li-ion cells with AA size. The safety issue can be overcome by employing phosphate-based cathode, the technology which is being commercialized by A123 Systems (co-founded by Dr. Chiang) which is collaborating with GM on a plug-in hybrid car, Chevy Volt (Fig. 26) (http://www.economist.com/sciencetechnology/tq/PrinterFriendly.cfm?story_id=10789409).

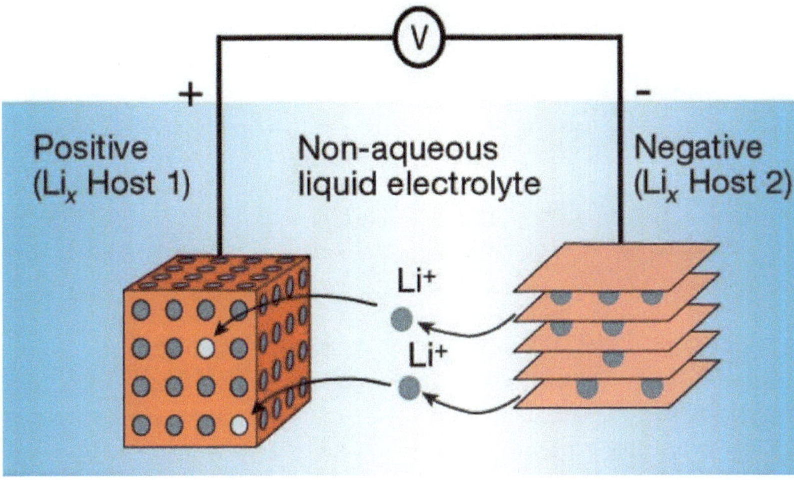

Figure 25. Basic structure of a Li-ion battery in which lithium ion intercalation into anode and cathode during charge and discharge process, respectively, is employed to store electrochemical energy (Tarascon & Armand, 2001).

Nanosizing the cathode and anode materials are now tremendously followed in different battery materials (Fig. 27). Silicon, one of the most promising anode materials, may find somewhere in market if researchers could overcome instability of this materials during charge-discharge process through the nanostructuring of this element (Armand & Tarascon, 2008). Different nanostructures of Si such as nanoparticles (Lee et al., 2010), nanowire (Chan et al., 2008), nanotube (Wu et al., 2012), hierarchical nanoporous structures (Magasinski et al., 2010) and their composites with nanocarbons (Lee et al., 2010; Cui et al., 2009), have shown to have exceptionally high capacity and stability raising hopes to have commercial batteries with Si-based anode.

In the realm of cathode materials for Li-based batteries, sulfur boosts the capacity of Li-ion batteries with one order of magnitude higher theoretical capacitance (Peramunage & Licht, 1993). Therefore, the Li-S batteries may succeed Li-ion batteries as their energy density is extremely high plus low cost and density of sulfur (Kang et al., 2006). However, its capacity fades away during charge-discharge of the cell due to polysulfide ions (the reaction intermediates) dissolution in electrolyte causes irreversible loss of active materials diminishing the capacitance few times lower than theoretical value. At 2009, by employing an innovative technique through the nanostructuring the sulfur inside the mesoporous carbon, capacitances near the theoretical limits were attained (Ji et al., 2009). After that, different carbon nanostructures such as hollow carbon nanofibers (Zheng et al., 2011), graphene oxide (Ji et al., 2011) and pyrolzed PAN/graphene (Yin et al., 2012) were used to immobilize sulfur. This class of batteries (Li-S) would find market in automobile industries as also claimed by Daimler in its concept vehicles, Mercedes-Benz F125! having exceptional high range of 1000km (Fig. 28) (http://media.daimler.com).

Figure 26. Chery Volt, a plug-in hybrid car, was introduced by GM. The automaker is trying to use safer nanophosphate-based batteries (http://www.economist.com/sciencetechnology/tq/PrinterFriendly.cfm?story_id=10789409).

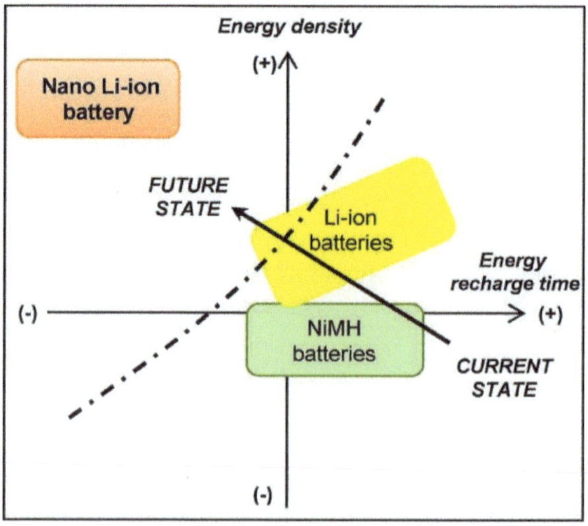

Figure 27. Nanotechnology will make batteries with higher energy density which charge faster than current Li-ion batteries (Serrano et al., 2009).

Figure 28. Mercedes-Benz F125! the concept vehicle introduce by Daimler at 2011 equipped with advanced technology in energy storage including high performance Li-S batteries having power density of 350 $Wh.kg^{-1}$ (energy capacity 10 kWh) and revolutionary ultra-porous MOFs technology (104 $m^2.g^{-1}$) making it possible to store 7.5 kg of H_2 in a flexible framework. All these technologies owe a lot to nanoscience and technology (http://media.daimler.com).

3.7.3. *Supercapacitors*

Despite of advantages of using batteries as energy storage device in hybrid cars including high energy density, challenges associated with employing batteries especially timely recharging, safety and lifetime bring the another electrochemical storage device as candidate for the same purpose, i.e. supercapacitors. Supercapacitors store energy by forming a double layer of electrolyte ions on the surface of conductive electrodes, called EDLCs (Miller & Simon, 2008). The widespread applications of supercapacitors are limited by their low energy density (1-5 Wh/kg) comparing to batteries (10-500 $Wh.kg^{-1}$) and as a result high cost of energy storage. But the fact that supercapacitors can be charged and discharged in less than a minute over a million cycles motivates scientific communities to enhance energy density of supercapacitors (Simon & Gogotsi, 2008; Chmiola et al., 2006). It is envisioned at energy density of 40 Wh/kg, the supercapacitors would be an improvement over the batteries used in some hybrid vehicles. In addition, the concept of supercapacitor powered urban bus which recharge at each bus stop in a minute is another intriguing idea (Fig. 29, left side).

At the heart of the current EDLCs, nanoporous carbon acts as electrode. To improve the energy density of supercapacitors, different nanomaterials such as MWNTs (Frackowiak et al., 2000), SWNTs (Kaempgen et al., 2009), metal oxide nanoparticles (Hu et al., 2006) and conducting polymers (Zhang et al., 2010) (two later cases are classified as pseudo-capacitors), have been used but high cost of nanotubes and some metal oxide (*e.g.* RuO_2) and poor stability of pseudo-capacitors render them not enough efficient storage device.

Emergence of graphene has revolutionized this field, as this material is the thinnest imaginable carbon allotrope (Fig. 29, right side) (Stoller et al., 2008). The graphene-based pseudo-capacitors are still in infancy stage, but initial results confirm high capacity of graphene-based EDLCs having improved energy density. At May 2011, Ruoff's group (Zhu et al., 2011) at university of Texas-Austin claimed they have developed graphene-based EDLCs through an industrial viable method having energy density of 75 $Wh.kg^{-1}$ which is more than one order magnitude higher than conventional supercapacitors. It is foreseen by 2020, half of graphene's market (~ $675 million) belongs to supercapacitors which clearly illustrate the impact of graphene-based materials on this field (http://www.bccresearch.com/report/AVM075A.html). It should be noted that hybrid systems of batteries and supercapacitors are indentified as the most effective and reliable solution for applications where lifecycle and reliability are vital including cars (http://www.grapheneenergy.net/applications.html).

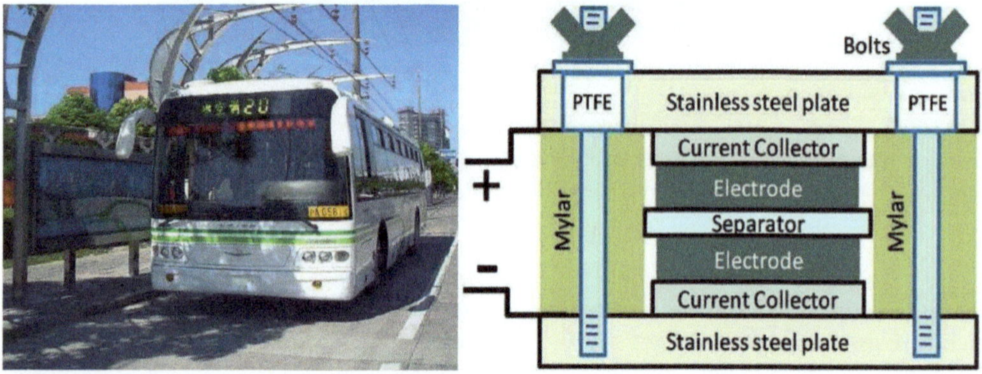

Figure 29. Left side: Sinautec's Ultracapacitor Bus, an urban bus powered by on-board supercapacitors and batteries which charged at bus stop. Right side: a conventional set-up for supercapacitors in which forming a double layer of electrolyte ions on the surface of high surface area conductive electrodes store energy (Stoller et al., 2008).

At June 2010, researchers at MIT introduced new concept of energy storage device, having both high density (comparable to Li-ion batteries) and power density (even higher than supercapacitors) (Lee et al., 2010). Again nanocarbons did excellent as electrode. The whole idea was to exchange Li ions between the surfaces of two nanostructured carbon electrode having functional groups (Jang et al., 2011). In fact, charging of Li-ion batteries is timely because Li ions must intercalate into cathode and anode which takes time. This strategy to store energy may be implemented in next generation of automobiles but further investigation are required to clarify the exact chemistry governing the device and also confirm feasibility and other issues.

3.8. Nanotechnology in solar cell in automobile

Solar cells are used to produce electricity from sunlight. This system has been gradually developed in different industries as it is an environmentally benign method of producing electricity and helps industries to reduce fuel consumption. In recent years, research to find new sources of energy in automobiles are being carried out (Levitsky et al., 2010, Ong et al., 2010).

It has been found that solar cells can be used as an additional source of energy supporting some of electronic devices used in an automobile. Silica-based solar system is a conventional system for producing electricity from sunlight. However, production and large scale use of this system is complicated and costly. Therefore, attempts have been carried out to produce new solar system usable for automobile economically. In silicon based solar cells, the electron needed will be supplied by silicon after exposure to sunlight. The produced electron can be transferred to semiconductors from electrodes. In solar cells, an organic material like chlorophyll can be used as substrate. A large surface area layer based on nanoporous titanium oxide is used for the transmission to the electrodes. This system as shown in Fig. 30 consists of two glass plates each of them have a transparent electrode (Levitsky et al., 2010, Ong et al., 2010).

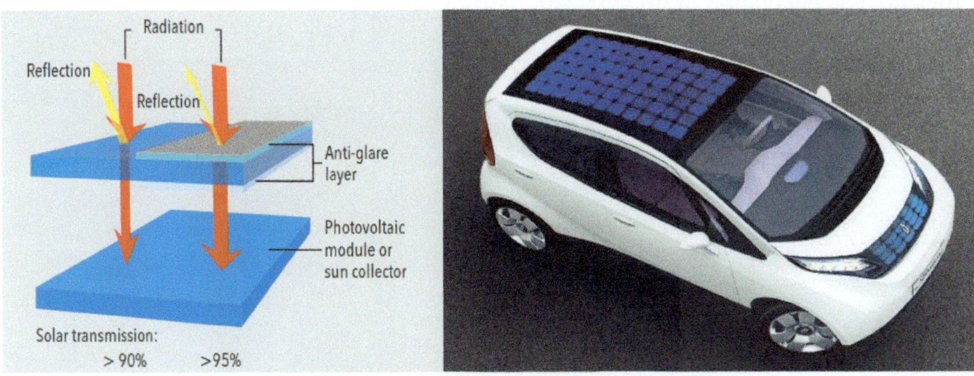

Figure 30. Solar cells based on nanotechnology for modern cars.

As it can be seen in Fig. 30, one of the plates covered with layer of dye-sensitive titanium oxide and another one is coated with platinum as catalyst. However, in conventional silicon-based solar cells, the efficiency is low due to light reflection at the solar glass pan. The reduction is approximately 10 percent for even high efficiency solar cells. This problem has been solved in new generation of solar cells using sol-gel method. Using this technique, a coating layer is applied over the glass pans. This coating could reduce light reflection from glass pans resulting in an increase in solar cell efficiency up to 6 percent. The sol used for this purpose includes a mixture of silica balls at two different sizes. To obtain best anti-reflective properties from coating, mixture of particles with diameter of 10 nm and 30 nm should be used. In order to apply coating layer over the cells, the glass pans should be immersed in the tank containing nano SiO_2 sol. The optimum antireflective properties can be obtained at 120 nm thickness. The sol becomes dry using gel and nanoporous layer can be obtained after hardening glass coated pans at 600-700 °C. This novel coating layer has very low refractive index (of only 1.25) resulting in light reflection at 400-2500 nm. In this way, solar transmission will be increased from 90 percent for conventional one to 95 percent for this solar cell (Levitsky et al., 2010, Ong et al., 2010).

3.9. Nanotechnology based catalyst for reduction of exhaust emission

Todays, environmental regulations forced car producers to reduce exhaust emission of automobiles. Using new generations of fuels can be one way achieving this target. However, at this time, technology needed for large scale production of fuel has not been developed maturely. Use of catalysts is a conventional approach reducing exhaust emission. These catalysts are made of high-grade steel housings containing catalytically active materials. These active materials are able to convert exhaust pollutants to nitrogen, steam, and carbon dioxide. Three most important polluting elements that exhaust included carbon monoxide, hydrocarbons and nitric oxides. To eliminate or reduce these pollutants emission from exhaust gas, three kinds of catalysts are needed. Nanotechnology can play an important role in converting toxic pollutants to non-toxic gases. It is well known that increasing surface area of catalyst enhance its catalytic activity. Designing catalytic materials to absorb nitric oxides from exhausted gas has become a big challenge for car manufacturers. To solve this problem, new generation of catalyst with high capability of NOx-absorbing are developed. The mechanism by which this system works is presented in Fig. 31 (Hvolbk et al., 2007, Kim et al., 2006, Nilsson et al., 2005, Zhou et al., 2010).

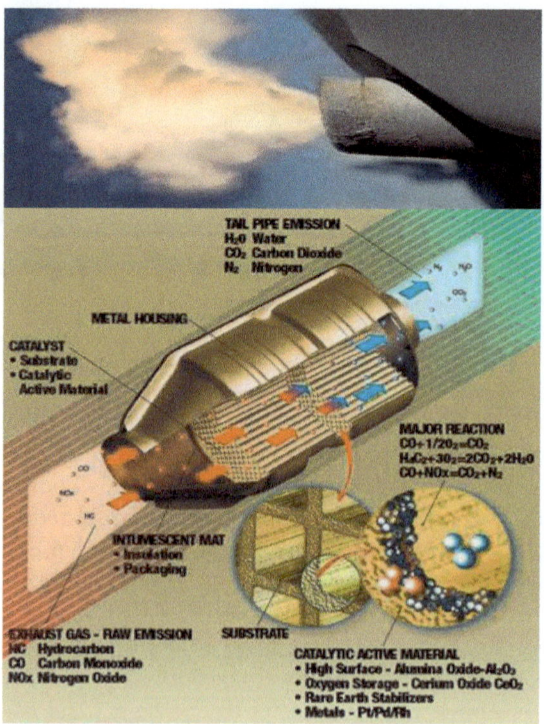

Figure 31. Three-way nano-structured catalyst for cleaning exhaust from pollutants (Hvolbk et al., 2005).

Recent researches revealed that Au nanoparticles having sizes lower than 5 nm are very effective catalysts. There are many different mechanisms indicating catalytic activity of nano sized Au particles. The most important of them are the quantum size effects, charge transfer to or from the support or support induced strain, oxygen spill-over to or from the support, oxidation state of Au and the role of very low-coordinated Au atoms in nanoparticles. The metal nobility depends on the metal surface ability to oxidize or chemisorption of oxygen. From periodic table of elements, Au is the only metal with endothermic chemisorption energy. Because of this behavior, Au is a metal which could not bind with oxygen at all. Au is a very good catalyst for oxidation of carbon monoxide (CO) presented in exhausted gas in automobile. The activity of Au depends on particle size as the best activity can be seen at particle sizes < 5 nm (Hvolbk et al., 2005).

3.10. Ultra-reflecting layer for automobile mirror

3.10.1. Mirrors with high optical and self-cleaning properties

New generation of mirrors and headlights used in cars are based on glass and polymer components with high optical quality and efficiency. Nanotechnology is employed to achieve these unique properties. To this end, ultra-reflecting thin layer (with thickness lower than 100 nm) based on aluminum oxide can be applied over the surface of mirrors or headlights. Applying ultra-thin layers over the mirrors can also help us to equip surfaces with fat, dirt water and repellent features. Using chemical vapor deposition (CVD) technique, nanometric hydrophobic and oelophobic layers can be applied over the surface of mirrors. It has been found that fluoro-organic materials are able to improve hydrophobicity and oelophobicity of the surface at thicknesses of 5-10 nm. This nanometric layer could also produce smooth surface which impurities like water drop, dirt, oil and fingerprints can be easily cleaned. This ultra-thin layer has high resistance against friction and makes it applicable at longer times. As it can be seen in Fig. 32, the layer could chemically bonded to the surface of mirror from the side consisting anchor groups. The chemical groups at other side of the layer produce hydrophobic surface.

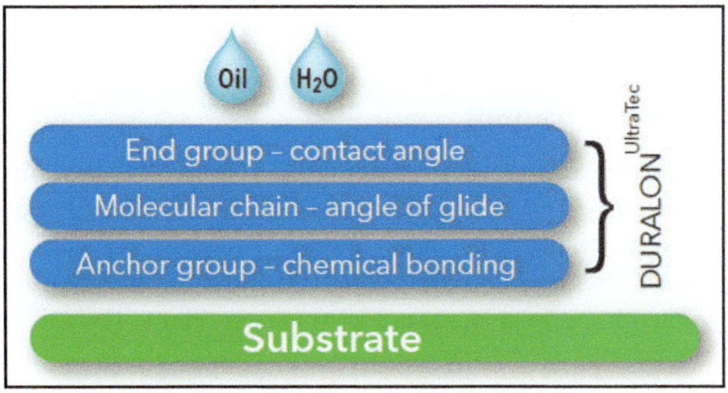

Figure 32. The composition of ultra-reflecting layers used on modern mirrors.

3.10.2. Anti-glare rear mirrors

Nowadays, safety regulations for car drivers lead to produce rear view mirrors to obtain an appropriate view at dawn and dusk. This can be done by equipping glasses with functional layer composite with electrochromic properties. For the glasses equipped with this technology, the optical properties will be changed by applying a certain voltage which moves the charges to intermediate layer. The incoming light will be absorbed by color centers produced by ions at the electrodes. As a result, small quantity of light will be reflected (Fig. 33).

Figure 33. The conventional mirrors (a) and modern anti-glare rear mirrors for cars (b).

Like charging and discharging of a car battery, the glass can get back to its original properties as the pole changes. This glass equipped by a rear sensor which could measure and control the glaring light of following vehicles. As soon as the glaring light disappears, the mirror gets back to original state.

3.11. Nano-filters for air cleaning

In addition to safety improvement and fuel consumption reduction, the enhancing comfort for customers is of importance. The air quality inside the car is an important factor affecting customers comfort. The air inside cars contains particles and gaseous pollutions which need to be filtered. Achieving this target needs equipping cars with high quality interior air filters with high efficiency of pollens, spores and industrial dusts filtration. This filter could be used at variable temperature (-40 °C to 100 °C) and humid conditions. Nanofibers are utilized to produce novel filters with superior properties compared with conventional filters (Fig. 34).

For the fibers in nanometric range size the classical fluid dynamic laws is not true anymore. Because of the lower air resistance of nanofibers compared with micron-sized fibers, air can be transported through filter easier with lower air pressure loss. This shows that the new filter work at lower level of energy.

This technology is also applied for soot filters. Using nanofibers in soot filters, the emission of pollutants in passengers and utility cars will be reduced. Using nanofilters, the dirt particles could not adhere to the foam materials used at roof of car and prevent polluting it.

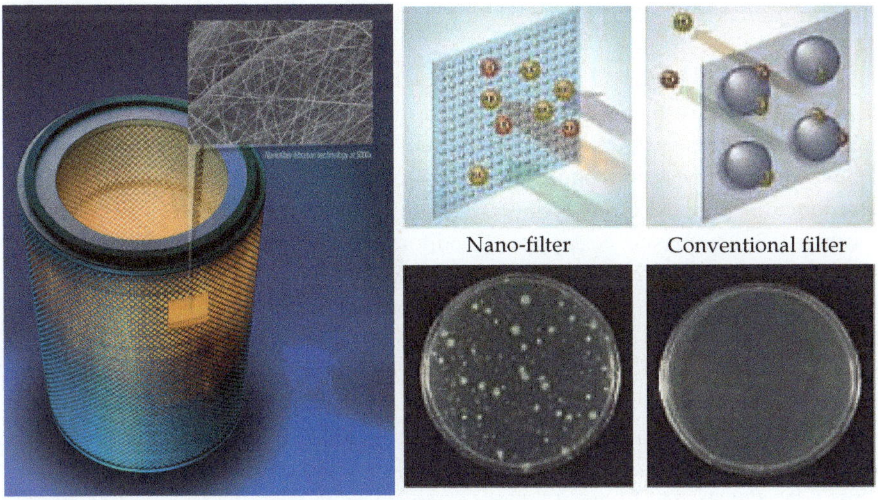

Figure 34. Nano-filters for air cleaning in car interior.

Author details

Mohsen Mohseni, Bahram Ramezanzadeh, Hossein Yari and Mohsen Moazzami Gudarzi
Amirkabir University of Technology, Iran

4. References

Amerio, E.; Fabbri, P.; Malucelli, G.; Messori, M.; Sangermano, M.; Taurino, R. (2008) Scratch resistance nano-silica reinforced acrylic coatings. Prog. Org. Coat. 62., 129.

Arico, A. S.; Bruce, P.; Scrosati, B.; Tarascon, J.-M. & van Schalkwijk, W. (2005). Nanostructured Materials for Advanced Energy Conversion And Storage Devices. Nat. Mater. 4, 5, 366-377

Armand, M. & Tarascon, J. M. (2008). Building Better Batteries. nature 451, 7179, 652-657

Banerjee, A.N. (2011) The design, fabrication, and photocatalytic, utility of nanostructured semiconductors: focus on TiO2-based nanostructures, Nanotechnology, Science and Applications, 4, 35-65.

Bashyam, R. & Zelenay, P. (2006). A Class of Non-Precious Metal Composite Catalysts for Fuel Cells. nature 443, 7107, 63-66

Barletta, M.; Bellisario, D.; Rubino, G.; Ucciardello, N. (2010). Scratch and wear resistance of transparent topcoats on carbon laminates. Prog. Org. Coat. 67., 209.

Bautista, Y.; Gómez , M.P.; Ribes, C.; Sanz, V. (2011). Correlation between the wear resistance and the scratch resistance for nanocomposite coatings. Prog. Org. Coat, 70., 4.,178-185.

Béguin, F. & Frackowiak E. (2006). Nanotextured Carbons for Electrochemical Energy Storage, In: Carbon Nanomaterials, Gogotsi Y, (1st ed.), 295-319, CRC Press, 0-8493-9386-8, Florida

Bellayer, S.; Bourbigot, S.; Flambard, X.; Rochery, M.; Gilman, J.W. & Devaux, E. Proceedings of the 4th AUTEX conference. Roubaix: ENSAIT; 2004. 3.

Bourbigot, S.; Le Bras, M.; Flambard, X.; Rochery, M.; Devaux, E. & Lichtenhan, J.D. Polyhedral oligomeric silsesquioxanes: applications to flame retardant textiles. In: Fire retardancy of polymers: new applications of mineral fillers. Le Bras, M.; Wilkie, C.A.; Bourbigot, S.; Duquesne, S. & Jama, C. editors. p. 189-201, Royal Society of Chemistry; 2005, London:. Ce´te, A. P.; Benin, A. I.; Ockwig, N. W.; O'Keeffe, M.; Matzger, A. J. & Yaghi, O. M. (2005). Porous, Crystalline, Covalent Organic Frameworks. Science 310, 5751, 1166-1170

Brinker, CJ., Scherer, GW. (1990). Sol-gel science: the physics and chemistry of sol-gel processing. Academic Press, San Diego.

Brooman, E.W. (2002). Modifying Organic Coatings to Provide Corrosion Resistance: Part II-Inorganic Additives and Inhibitors. Met. Finish. 100 (2002) 42.

Brooman, E.W. (2002). Modifying Organic Coatings to Provide Corrosion Resistance--Part II1: Organic Additives and Conducting Polymers. Met. Finish. 100 (2002) 104.

Burress, J. W.; Gadipelli, S.; Ford, J.; Simmons, J. M.; Zhou, W. & Yildirim, T. (2010). Graphene Oxide Framework Materials: Theoretical Predictions and Experimental Results. Angew. Chem. Int. Ed. 49, 47, 8902-8904

Chan, C. C. (2007). The State of the Art of Electric, Hybrid, and Fuel Cell Vehicles. Proceedings of the IEEE 95, 4, 704-718,

Chan, C. K.; Peng, H.; Liu, G.; McIlwrath, K.; Zhang, X. F.; Huggins, R. A. & Cui, Y. (2008). High-Performance Lithium Battery Anodes Using Silicon Nanowires. Nat Nano 3, 1, 31-35

Chen, S; Liu, W.; Yu, L. (1998). Preparation of DDP-coated PbS nanoparticles and investigation of the antiwear ability of the prepared nanoparticles as additive in liquid paraffin. Wear. 218., 153–158.

Chen, P.; Wu, X.; Lin, J. & Tan, K. L. (1999). High H2 Uptake by Alkali-Doped Carbon Nanotubes under Ambient Pressure and Moderate Temperatures. Science 285, 5424, 91-93

Chen, P.; Xiao, T.-Y.; Li, H.-H.; Yang, J.-J.; Wang, Z.; Yao, H.-B. & Yu, S.-H. (2012). Nitrogen-Doped Graphene/ZnSe Nanocomposites: Hydrothermal Synthesis and Their Enhanced Electrochemical and Photocatalytic Activities. ACS Nano 6, 1, 712-719

Cheng, K.C.; Chen, F.R. & Kai, J.J. (2007) Electrochromic property of nano-composite Prussian Blue based thin film. Electrochimica Acta, 52, 2007, 3330–3335.

Chinas-Castillo. F.; Spikes, H.A. (2003). Mechanism of action of colloidal solid dispersions, Trans. ASME. 125., 552–557.

Chmiola, J.; Yushin, G.; Gogotsi, Y.; Portet, C.; Simon, P. & Taberna, P. L. (2006). Anomalous Increase in Carbon Capacitance at Pore Size Less Than 1 Nanometer. Science 312, 5794, 1760-1763

Cho,S.I.; Kwon,W.J.; Choi, S.J.; Kim, P.; Park, S.A.; Kim,J.; Son, S.J.; Xiao, R.; Kim, S.-H.; & Lee, S. B. (2005) Nanotube-Based Ultrafast Electrochromic Display, Advanced Materials, 17 (2), 171-175.

Chung, S.-Y.; Bloking, J. T. & Chiang, Y.-M. (2002). Electronically Conductive Phospho-Olivines as Lithium Storage Electrodes. Nat Mater 1, 2, 123-128

Cui, L.-F.; Yang, Y.; Hsu, C.-M. & Cui, Y. (2009). Carbon-Silicon Core-Shell Nanowires as High Capacity Electrode for Lithium Ion Batteries. Nano Lett. 9, 9, 3370-3374

Cole, G.S.; Bin, F. (1993) Proceed. ASM, Mater. Cong. Pittsburg, PA, October 17–21 ASM International Materials Park, OH 13.

Courter, J.L. (1997). Mar resistance of automotive clearcoat: I. Relationship to coating mechanical properties. J. Coat. Technol. 69., 866., 57.

Cummins, D.; Boschloo, G.; Ryan, M.; Corr, S.D.; Rap, N. & Fitzmaurice, D. (2000). Journal of Physical Chemistry B, 104, 2000, 11449.

Dahotre, N.B.; Nayak, S. (2005). Nanocoatings for engine application. Surface & Coatings Technology. 194., 58– 67.

Das, A.; et al. (2008). Modified and unmodified multiwalled carbon nanotubes in high performance solution-styrene-butadiene and butadiene rubber blends. Polymer. 49., 24., 5276.

Dastjerdi, R. & Montazer, M. (2010). A review on the application of inorganic nano-structured materials in the modification of textiles: Focus on anti-microbial properties. Colloids and Surfaces B: Biointerfaces, 79 (1), August 2010, 5-18.

Demirdeven, N. & Deutch, J. (2004). Hybrid Cars Now, Fuel Cell Cars Later. Science 305, 5686, 974-976

Dillon, A. C.; Jones, K. M.; Bekkedahl, T. A.; Kiang, C. H.; Bethune, D. S. & Heben, M. J. (1997). Storage of Hydrogen in Single-Walled Carbon Nanotubes. nature 386, 6623, 377-379

Dhoke, S.K.; Khanna, A.S. (2009). Electrochemical behavior of nano-iron oxide modified alkyd based waterborne coatings. Mater. Chem. Phys. 117., 550-556.

Dimitrakakis, G. K.; Tylianakis, E. & Froudakis, G. E. (2008). Pillared Graphene: A New 3-D Network Nanostructure for Enhanced Hydrogen Storage. Nano Lett. 8, 10, 3166-3170

Ebisawa, M.; Hara, T.; Hayashi, T.; Ushiko, M. (1991). Production Process of Metal Matrix Composite (MMC) Engine Block. SAE Paper No.-910835.

Frackowiak, E.; Metenier, K. & Bertanga, V. (2000). Supercapacitor Electrode from Multiwalled Carbon Nanotubes. Appl. Phys. Lett. 77, 15, 2421-2423

Fujishima, A.; Zhang, X. & Tryk. D.A. TiO2 photocatalysis and related surface phenomena, Surface Science Report 63 (2008) 515-582.

Garcia-Heras, M.; Jimenez-Morales, A.; Casal, B.; Galvan, J. C.; Radzki, S.; Villegas, M. A. (2004). J. Alloys Compd. 219., 219.

Gong, K.; Du, F.; Xia, Z.; Durstock, M. & Dai, L. (2009). Nitrogen-Doped Carbon Nanotube Arrays with High Electrocatalytic Activity for Oxygen Reduction. Science 323, 5915, 760-764

Greeley, J.; Stephens, I. E. L.; Bondarenko, A. S.; Johansson, T. P.; Hansen, H. A.; Jaramillo, T. F.; RossmeislJ; ChorkendorffI & Nerskov, J. K. (2009). Alloys of Platinum and Early Transition Metals as Oxygen Reduction Electrocatalysts. Nat. Chem. 1, 7, 552-556

Groenewolt, M. (2008). Highly scratch resistant coatings for automotive applications., Prog. Org. Coat. 61.,106.

Hara, Y.; Mori, T.; Fujitani, T. (2000). Relationship between viscoelasticity and scratch morphology of coating films. Prog. Org. Coat. 40., 39.

Hernandez-Padron, G.; Rojas, F.; Garcia-Garduno, M.; Canseco, M.A.; Castano, V.M. (2003). Development of hybrid materials consisting of SiO2 microparticles embedded in phenolic-formaldehydic resin polymer matrices. Mater Sci Eng A. 355., 338.

http://en.wikipedia.org/wiki/Platinum

http://media.daimler.com

http://www.bccresearch.com/report/AVM075A.html

http://www.economist.com/sciencetechnology/tq/PrinterFriendly.cfm?story_id=10789409 (In Search of the Perfect Battery. The Economist. 2008.)

http://www.grapheneenergy.net/applications.html

Holzapfel-Metallveredelung, Application areas of nanotechnological corrosion protection.

Hong, J. W.; Kang, S. W.; Choi, B.-S.; Kim, D.; Lee, S. B. & Han, S. W. (2012). Controlled Synthesis of Pd-Pt Alloy Hollow Nanostructures with Enhanced Catalytic Activities for Oxygen Reduction. ACS Nano 6, 3, 2410-2419

Horrocks, A.R. (2011). Flame retardant challenges for textiles and fibres: New chemistry versus innovatory solutions. Polymer Degradation and Stability, 96, 2011, 377-392.

Hu, C.-C.; Chang, K.-H.; Lin, M.-C. & Wu, Y.-T. (2006). Design and Tailoring of the Nanotubular Arrayed Architecture of Hydrous RuO2 for Next Generation Supercapacitors. Nano Lett. 6, 12, 2690-2695

Huang, P.Y.; Chao, Y.C. & Liao, Y.T. (2004). Preparation of fluoroacrylate nanocopolymer by miniemulsion polymerization used in textile finishing. Journal of Applied Polymer Science, 94(4), 2004, 1466-1472.

Hvolbk, B.; Janssens, T. V. W.; Clausen, B.S.; Falsig, H.; Christensen, C.H.; Nórskov, J.K. (2007). Catalytic activity of Au nanoparticles. Nanotoday. 2., 4., 14.

Iijima, S. (1991). Helical Microtubules of Graphitic Carbon. nature 354, 6348, 56-58

Iijima, S. & Ichihashi, T. (1993). Single-Shell Carbon Nanotubes of 1-nm Diameter. nature 363, 6430, 603-605

Jacobson, M. Z.; Colella, W. G. & Golden, D. M. (2005). Cleaning the Air and Improving Health with Hydrogen Fuel-Cell Vehicles. Science 308, 5730, 1901-1905

Jang, B. Z.; Liu, C.; Neff, D.; Yu, Z.; Wang, M. C.; Xiong, W. & Zhamu, A. (2011). Graphene Surface-Enabled Lithium Ion-Exchanging Cells: Next-Generation High-Power Energy Storage Devices. Nano Lett. 11, 3785–3791

Jardret, V.; Lucas, B.N., Oliver, W. (2000). Scratch Durability of Automotive Clear Coatings: A Quantitative, Reliable and Robust Methodology. J. Coat. Technol. 72., 907., 79.

Jasinski, R. (1964). A New Fuel Cell Cathode Catalyst. nature 201, 4925, 1212-1213

Ji, L.; Rao, M.; Zheng, H.; Zhang, L.; Li, Y.; Duan, W.; Guo, J.; Cairns, E. J. & Zhang, Y. (2011). Graphene Oxide as a Sulfur Immobilizer in High Performance Lithium/Sulfur Cells. J. Am. Chem. Soc. 133, 46, 18522-18525

Ji, X.; Lee, K. T. & Nazar, L. F. (2009). A Highly Ordered Nanostructured Carbon-Sulphur Cathode for Lithium-Sulphur Batteries. Nat Mater 8, 6, 500-506

Kabacoff, L.T. (2002). Nanoceramic coatings exhibit much higher toughness and wear resistance than conventional coatings. Materials Science and Technology, The AMPTIAC Newsletter. 6., 1., 37.

Kaempgen, M.; Chan, C. K.; Ma, J.; Cui, Y. & Gruner, G. (2009). Printable Thin Film Supercapacitors Using Single-Walled Carbon Nanotubes. Nano Lett. 9, 5, 1872-1876

Kakac, S.; Pramuanjaroenkij, A. (2009) Review of convective heat transfer enhancement with nanofluids. International Journal of Heat and Mass Transfer. 52., 13-14.,3187–3196.

Kang, B. & Ceder, G. (2009). Battery Materials for Ultrafast Charging and Discharging. nature 458, 7235, 190-193

Kang, K.; Meng, Y. S.; Breger, J.; Grey, C. P. & Ceder, G. (2006). Electrodes with High Power and High Capacity for Rechargeable Lithium Batteries. Science 311, 5763, 977-980

Kasten, L. S.; Grant, J. T.; Grebasch, N.; Voevodin, N.; Arnold, F. E.; Donley, M. S. (2001). Surf. Coat. Technol. 140., 11.

Kiliaris, P. & Papaspyrides, C.D. (2010). Polymer/layered silicate (clay) nanocomposites: An overview of flame retardancy, Progress in Polymer Science, 35, 902–958.

Kim, J., et al. (2006). Oxygen adsorption and oxidation reactions on Au(2 1 1) surfaces: Exposures using O2 at high pressures and ozone (O3) in UHV. Surf. Sci. 600 (2006) 4622.

Kimura, G. & Yamada, K. (2009). Electrochromism of poly(3,4-ethylenedioxythiophene) films on Au nano-brush electrode. Synthetic Metals, 159 (9-10), 914-918

Koenigsmann, C.; Santulli, A. C.; Gong, K.; Vukmirovic, M. B.; Zhou, W.-p.; Sutter, E.; Wong, S. S. & Adzic, R. R. (2011). Enhanced Electrocatalytic Performance of Processed, Ultrathin, Supported Pd-Pt Core-Shell Nanowire Catalysts for the Oxygen Reduction Reaction. J. Am. Chem. Soc. 133, 25, 9783-9795

Koenigsmann, C.; Sutter, E.; Chiesa, T. A.; Adzic, R. R. & Wong, S. S. (2012). Highly Enhanced Electrocatalytic Oxygen Reduction Performance Observed in Bimetallic Palladium-Based Nanowires Prepared under Ambient, Surfactant less Conditions. Nano Lett. 12, 4, 2013-2020

Kole, M.; Dey, T. K. (2010). Thermal conductivity and viscosity of Al2O3 nanofluid based on car engine coolant. Journal of Physics D. 43., 31., ID 315501.

Kötz, R.; Müller, S.; Bärtschi, M.; Schnyder, B.; Dietrich, P.; Büchi, F. N.; Tsukada, A.; Scherer, G. G.; Rodatz, P.; Garcia, O.; Barrade, P.; Hermann, V. & Gallay, R. (2001). Supercapacitors for Peak-Power Demand in Fuel-Cell-Driven Cars. Electrochemical Society Proceeding 21, 564-575

Lee, S.H.; Deshpande, R.; Parilla, P.A.; Jones, K.M.; To, B.; Mahan, A.H. & Dillon, A.C. (2006) CrystallineWO3 Nanoparticles for Highly Improved Electrochromic Applications. Advanced Materials, 18, 2006, 763–766.

Lee, J. K.; Smith, K. B.; Hayner, C. M. & Kung, H. H. (2010). Silicon Nanoparticles-Graphene Paper Composites for Li Ion Battery Anodes. Chem. Commun. 46, 12, 2025-2027

Lee, S. W.; Yabuuchi, N.; Gallant, B. M.; Chen, S.; Kim, B.-S.; Hammond, P. T. & Shao-Horn, Y. (2010). High-power lithium batteries from functionalized carbon-nanotube electrodes. Nat. Nanotech. 5, 7, 531-537

Lefevre, M.; Proietti, E.; Jaouen, F. d. r. & Dodelet, J.P. (2009). Iron-Based Catalysts with Improved Oxygen Reduction Activity in Polymer Electrolyte Fuel Cells. Science 324, 5923, 71-74

Levitsky, I.A. (2010). Hybrid Solar Cells Based on Carbon Nanotubes and Nanoporous Silicon, IEEE Nanotechnology Magazine. doi. 10.1109/MNANO.2010.938654.

Li, Y.; Zhou, J.; Tung, S.; Schneider, E.; Xi, S. (2009). A review on development of nanofluid preparation and characterization. Powder Technology. 196., 2., 89.

Li, Y.; Zhao, Y.; Cheng, H.; Hu, Y.; Shi, G.; Dai, L. & Qu, L. (2011). Nitrogen-Doped Graphene Quantum Dots with Oxygen-Rich Functional Groups. J. Am. Chem. Soc. 134, 1, 15-18

Lim, B.; Jiang, M.; Camargo, P. H. C.; Cho, E. C.; Tao, J.; Lu, X.; Zhu, Y. & Xia, Y. (2009). Pd-Pt Bimetallic Nanodendrites with High Activity for Oxygen Reduction. Science 324, 5932, 1302-1305

Lin, S.S.; Patterson, D.J. (1993). Piston-Ring Assembly Friction Modeling by Similarity Analysis. SAE Paper No.-930794., 129.

Liu, C.; Fan, Y. Y.; Liu, M.; Cong, H. T.; Cheng, H. M. & Dresselhaus, M. S. (1999). Hydrogen Storage in Single-Walled Carbon Nanotubes at Room Temperature. Science 286, 5442, 1127-1129

Liu, Y.Y.; Tang, J.; Wang, R.; Lu, H.; Li, L.; Kong, Y.; Qi, K. & Xin, J.H. (2007). Artificial lotus leaf structures from assembling carbon nanotubes and their applications in hydrophobic textiles. Journal of Materials Chemistry, 17(11), 2007, 1071-1078.

Liu, R.; Wu, D.; Feng, X. & Müllen, K. (2010). Nitrogen-Doped Ordered Mesoporous Graphitic Arrays with High Electrocatalytic Activity for Oxygen Reduction. Angew. Chem. Int. Ed. 49, 14, 2565-2569

Lowry, M.S.; Hubble, D.R.; Wressell, A.L.; Vratsanos, M.S.; Pepe, F.R. & Hegedus, C.R. (2008). Assessment of UV-permeability in nano-ZnO filled coatings via high throughput experimentation. Journal of Coating Technology and Research, 5 (2), 233–239.

MacLean, H.L.; Lave, Lester B. (2003). Evaluating automobile fuel/propulsion system technologies. Energy Combust. Sci. 29., 1.,1-69.

Magasinski, A.; Dixon, P.; Hertzberg, B.; Kvit, A.; Ayala, J. & Yushin, G. (2010). High-Performance Lithium-Ion Anodes Using a Hierarchical Bottom-Up Approach. Nat Mater 9, 4, 353-358

Mayrhofer, K. J. J.; Juhart, V.; Hartl, K.; Hanzlik, M. & Arenz, M. (2009). Adsorbate-Induced Surface Segregation for Core–Shell Nanocatalysts. Angew. Chem. Int. Ed. 48, 19, 3529-3531

Miller, J. R. & Simon, P. (2008). Electrochemical Capacitors for Energy Management. Science 321, 5889, 651-652

Mirabedini, S.M.; Sabzi, M.; Zohuriaan-Mehr, J.; Atai, M. & Behzadnasab, M. (2011). Weathering performance of the polyurethane nanocomposite coatings containing silane treated TiO2 nanoparticles. Applied Surface Science, 257, 2011, 4196–4203.

Monk, P. M. S.; Mortimer, R. J. & Rosseinsky, D.R. (2007). Electrochromism and Electrochromic Devices. Cambridge University Press, ISBN:9780521822695, London.

Montazer, M.; Behzadnia, A.; Pakdel, E.; Rahimi, M.K. & Bameni Moghadam, M. (2012a). Durable antibacterial and cross-linking cotton with colloidal silver nanoparticles and butane tetracarboxylic acid without yellowing, Colloids and Surfaces B: Biointerfaces, 89, 196-202.

Montazer, M.; Alimohammadi, F.; Shamei, A. & Rahimi, M.K. (2012b). synthesis of nano-silver/polypropylene plastics for antibacterial application. Current Applied Physics, In Press, Corrected Proof, Available online 9 February 2012

Nilsson, A., et al. (2005). Catal. Lett. The electronic structure effect in heterogeneous catalysis. 100 (2005) 111.

Nobel, M.L.; Picken, S.J.; Mendes, E. (2007). Waterborne nanocomposite resins for automotive coating applications. Progress in Organic Coatings. 58., 96–104.

Ong, P-L.; Euler, W. B.; Levitsky, I. A. (2010). Hybrid solar cells based on single-walled carbon nanotubes/Si heterojunction. Nanotechnology. 21., 10., 105203.

Osborne, J. H.; Blohowiak, K. Y.; Taylor, S. R.; Hunter, C.; Bierwagen, G.; Carlson, B.; Bernard, D.; Donley, M. S. (2001). Prog. Org. Coat. 41, 217.

Padhi, A. K.; Nanjundaswamy, K. S. & Goodenough, J. B. (1997). Phospho-olivines as Positive-Electrode Materials for Rechargeable Lithium Batteries. J. Electrochem. Soc. 144, 4, 1188-1194

Park, N.; Hong, S.; Kim, G. & Jhi, S.-H. (2007). Computational Study of Hydrogen Storage Characteristics of Covalent-Bonded Graphenes. J. Am. Chem. Soc. 129, 29, 8999-9003

Pawlicka, A. (2009). Development of Electrochromic Devices. Recent Patents on Nanotechnology, 3, 177-181.

Peramunage, D. & Licht, S. (1993). A Solid Sulfur Cathode for Aqueous Batteries. Science 261, 5124, 1029-1032

Presting, H.; Ko¨nig, U. (2003). Future nanotechnology developments for automotive applications. Mater Sci Eng C. 23., 6–8., 737.

Proietti, E.; Jaouen, F. d. r.; Lefevre, M.; Larouche, N.; Tian, J.; Herranz, J. & Dodelet, J.P. (2011). Iron-Based Cathode Catalyst with Enhanced Power Density in Polymer Electrolyte Membrane Fuel Cells. Nat. Commun. 2, 416

Qu, L.; Liu, Y.; Baek, J.-B. & Dai, L. (2010). Nitrogen-Doped Graphene as Efficient Metal-Free Electrocatalyst for Oxygen Reduction in Fuel Cells. ACS Nano 4, 3, 1321-1326

Ramezanzadeh, B.; Mohseni, M.; Yari, H.; Sabbaghian, S. (2009). An evaluation of an automotive clear coat performance exposed to bird droppings under different testing approaches. Progress in Organic Coatings, 66., 149-160

Ramezanzadeh, B.; Mohseni, M.; Yari, H.; Sabbaghian, S. (2010a). A study of thermal mechanical properties of an automotive coating exposed to natural and simulated bird droppings. Journal thermal analysis calorimetry, 102, 13-21

Ramezanzadeh, B.; Mohseni, M.; Yari H. (2010b)The role of basecoat pigmentation on the biological resistance of an automotive clear coat. Journal Coating Technology Research, DOI 10.1007/s11998-010-9254-5

Ramezanzadeh, B.; Mohseni, M.; Yari, H. (2010c). The Effect of Natural Tree Gum and Environmental Condition on the Degradation of a Typical Automotive Clear Coat. Journal of Polymers and the Environment, 18 (4) , 545-55

Ramezanzadeh, B.; Moradian, S.; Khosravi, A.; Tahmassebi, N. (2010d). Effect of polysiloxane additives on the scratch resistance of an acrylic melamine automotive clearcoat "J. Coat. Technol. Res. DOI 10.1007/s11998-010-9239-4.

Ramezanzadeh, B. & Attar, M.M. (2011a) Characterization of the fracture behavior and viscoelastic properties of epoxy-polyamide coating reinforced with nanometer and micrometer sized ZnO particles. Progress in Organic Coatings, 71, 3, 242–249

Ramezanzadeh, B.; Mohseni, M.; Yari H. (2011b). On the electrochemical and structural behavior of biologically degraded automotive coatings; Part 1: Effect of natural and simulated bird droppings. Progress in Organic Coatings 71 (1) , 19-31.

Ramezanzadeh, B.; Mohseni, M.; Yari H. (2011c). Studying the effects of the chemical structure of an automotive clearcoat on its biological degradation caused by tree gums. Journal of Coatings Technology Research 8 (3) , 375-387

Ramezanzadeh, B.; Moradian, S.; Tahmasebi, N.; Khosravi, A. (2011d). Studying the role of polysiloxane additives and nano-SiO2 on the mechanical properties of a typical acrylic/melamine clearcoat. Prog. Org. Coat. 72., 4., 621-631.

Ramezanzadeh, B.; Moradian, S.; Khosravi, A.; Tahmasebi, N. (2011e). A new approach to investigate scratch morphology and appearance of an automotive coating containing nano-SiO2 and polysiloxane additives. Prog. Org. Coat. 72., 3., 541-552.

Ramezanzadeh. B.; Moradian, S.; Yari, H.; Kashani, A.; Niknahad, M. (2011f). The effect of basecoat pigmentation on the scratch resistance and weathering performance of an acrylic–melamine basecoat/clearcoat automotive finish.Tribology International. doi:10.1016/j.triboint.2011.10.016.

Ramezanzadeh, B. Mohseni;, M.; Karbasi, A. (2011g). Preparation of sol–gel-based nanostructured hybrid coatings; part 1: morphological and mechanical studies. J Mater Sci, DOI 10.1007/s10853-011-5819-7.

Ramezanzadeh, B.; Moradian, S.; Yari, H.; Kashani, A. & Niknahad, M. (2012a). The effect of basecoat pigmentation on the scratch resistance and weathering performance of an acrylicmelamine basecoat/clearcoat automotive finish. Tribology International, 47 , 77-89

Ramezanzadeh, B.; Mohseni, M.; Mohammad Rabea, A. & Yari, H. (2012b) . Attributing the resistance against simulated tree gum of an acrylic/melamine film loaded with an active silicone additive to its surface free energy. International Journal of Adhesion and Adhesives, 31 (7) , 775-783

Rao, V.D.N.; Kabat, D.M.; Rose, R.; Yeager, D.; Leong, D.Y. (1997). Material Systems for Cylinder Bore Applications—Plasma Spray Technology. SAE Paper No.-970023.

Rapoport, L.; Feldman, Y.; Homyonfer, M.; Cohen, H.; Sloan, J.; Hutchison, J.L. (1999). Inorganic fullerene-like material as additives to lubricants: structure–function relationship. Wear. 225–229., 975–982.

Rashvand, M.; Ranjbar, Z. & Rastegar, S. (2011). Nano zinc oxide as a UV-stabilizer for aromatic polyurethane coatings. Progress in Organic Coating, 71, 362–368.

Rastegar, F.; Richardson, D.E. (1997). Alternative to chrome: HVOF cermet coatings for high horse power diesel engines, Surf. Coat. Technol. 90., 1–2., 156-163.

Sanchez, E.; Bannier, E.; Cantavella, V.; Salvador, M.D.; Klyatskina, E.; Morgiel, J.; Grzonka, J.; Boccaccini, A.R. (2007). Deposition of Al2O3-TiO2 Nanostructured Powders by Atmospheric Plasma Spraying. Journal of Thermal Spray Technology. DOI: 10.1007/s11666-008-9181-5.

Sangermano, M.; Messori, M. (2010). Scratch Resistance Enhancement of Polymer Coatings. Macromol. Mater. Eng. 295., 603.

Schlapbach, L. & Zuttel, A. (2001). Hydrogen-Storage Materials for Mobile Applications. nature 414, 6861, 353-358

Schulz, U.; V. Wachtendorf, T. Klimmasch, P. Alers. "The influence of weathering on scratches and on scratch and mar resistance of automotive coatings" Prog. Org. Coat, 42 (2001) 38.

Serrano, A. K.; Jackson, C. L. & Scott, K. (2003). The Promise of Fuel Cell-Based Automobiles. Bull. Mater. Sci. 26, 2, 207-214

Serrano, E.; Rus, G. & Garcea-Martenez, J. (2009). Nanotechnology for Sustainable Energy. Renewable and Sustainable Energy Reviews 13, 9, 2373-2384

Shchukin, D.G.; Mçhwald, H. (2007). Self-Repairing Coatings Containing Active Nanoreservoirs, DOI: 10.1002/smll.200700064.

Sheffer, M.; Groysman, A.; Starosvetsky, D.; Savchenko, N.; Mandler Mandler, D. (2004). Corros. Sci. 46., 2975.

Shen, W.; Jiang, B.; Scholten, A.; Schwenke, R.; Mi, L.; Seal, C.; Wang, P. (2004). A Quantitative Index for Mar and Scratch Resistance of Materials for Automotive Glazing Applications and Quantitative Evaluation of Damages by Different Scratching Modes. Tribogy Letters. 17., 3., 637-644.

Simon, P. & Gogotsi, Y. (2008). Materials for Electrochemical Capacitors. Nat. Mater. 7, 11, 845-854

Steele, B. C. H. & Heinzel, A. (2001). Materials for Fuel-Cell Technologies. nature 414, 6861, 345-352

Stoller, M. D.; Park, S.; Yanwu, Z.; An, J. & Ruoff, R. S. (2008). Graphene-Based Ultracapacitors. Nano Lett. 8, 10, 3498-3502

Subrahmanyam, K. S.; Kumar, P.; Maitra, U.; Govindaraj, A.; Hembram, K. P. S. S.; Waghmare, U. V. & Rao, C. N. R. (2011). Chemical Storage of Hydrogen in Few-Layer Graphene. Proceedings of the National Academy of Sciences 108, 7, 2674-2677

Tahmassebi, N.; Moradian, S.;, Ramezanzadeh, B.; Khosravi, A.; Behdad, S. (2010) Effect of addition of hydrophobic nano silica on viscoelastic properties and scratch resistance of an acrylic/melamine automotive clearcoat. Tribo Intern. 43., 685.

Tarascon, J. M. & Armand, M. (2001). Issues and Challenges Facing Rechargeable Lithium Batteries. nature 414, 6861, 359-367

Thorstenson, T.A.; Huang, J.B.; Urban, M.W. (1994). Mobility and distribution of silicone additives in coatings; a spectroscopic study. Prog. Org. Coat. 24., 1994., 341.

Tollefson, J. (2010). Hydrogen: Fuel of the Future? nature 464, 1262-1264

Trisaksri, V.; Wongwises, S. (2007). Critical review of heat transfer characteristics of nanofluids. Renewable and Sustainable Energy Reviews. 11., 3., 512.

Tzeng, S. C.; Lin, C. W.; Huang, K. D. (2005). Heat transfer enhancement of nanofluids in rotary blade coupling of fourwheel- drive vehicles. Acta Mechanica. 179., 1-2., 11.

Turner, J. A. (2004). Sustainable Hydrogen Production. Science 305, 5686, 972-974

Venkataraman, R.; Sundararajan, G. (1996). The sliding wear behaviour of Al·SiC particulate composites—I. Macrobehaviour. Acta Mater. 40., 2., 451-460.

Vreugdenhil, A. J.; Woods, M. E. (2005). Prog. Org. Coat. 53., 119.

Wang, X. Q.; Mujumdar, A. S. (2007). Heat transfer characteristics of nanofluids: a review. International Journal of Thermal Sciences. 46., 1., 1.

Wang, X. Q.; Mujumdar, A. S. (2008). A review on nanofluids—part I: theoretical and numerical investigations. Brazilian Journal of Chemical Engineering. 25., 4., 613.

Wang, S.; Yu, D.; Dai, L.; Chang, D. W. & Baek, J.-B. (2011). Polyelectrolyte-Functionalized Graphene as Metal-Free Electrocatalysts for Oxygen Reduction. ACS Nano 5, 8, 6202-6209

Wang, X.; Ding, B.; Yu, J. & Wang, M. (2011). Engineering biomimetic superhydrophobic surfaces of electrospun nanomaterials. Nano Today, 6, 510-.

Winter, M. & Brodd, R. J. (2004). What Are Batteries, Fuel Cells, and Supercapacitors? Chem. Rev. 104, 10, 4245-4270

Wu, Y.Y.; Tsui, W.C.; Liu, T.C. (2007). Experimental analysis of tribological properties of lubricating oils with nanoparticle additives. Wear. 262., 819–825.

Wu, G.; More, K. L.; Johnston, C. M. & Zelenay, P. (2011). High-Performance Electrocatalysts for Oxygen Reduction Derived from Polyaniline, Iron, and Cobalt. Science 332, 6028, 443-447

Wu, H.; Chan, G.; Choi, J. W.; Ryu, I.; Yao, Y.; McDowell, M. T.; Lee, S. W.; Jackson, A.; Yang, Y.; Hu, L. & Cui, Y. (2012). Stable Cycling Of Double-Walled Silicon Nanotube Battery Anodes through Solid-Electrolyte Interphase Control. Nat. Nanotech. advance online publication

Wuest, G.; Barbetaz, G.; Keller, S. (1997). The Key Advantages of the Plasma-Powder Spray Process for the Thermal Spray Coating of Cylinder Bores in Automotive Industry. SAE Paper No.-970016.

Xu, J,; Pang, W.; Shi, W. (2006). Synthesis of UV-curable organic–inorganic hybrid urethane acrylates and properties of cured films. Thin Solid Films. 514., 69.

Xie, H.; Chen, L. (2009). Adjustable thermal conductivity in carbon nanotube nanofluids. Physics Letters Section A. 373., 21., 1861.

Xie, H.; Yu, W.; Li, Y. (2009). Thermal performance enhancement in nanofluids containing diamond nanoparticles. Journal of Physics D. 42., 9.

Yari, H.; Moradian, S.; Ramezanzadeh, B.; Kashani, A.; Tahmasebi, N. (2009a). The effect of basecoat pigmentation on mechanical properties of an automotive basecoat/clear coat system during weathering. Polymer Degradation Stability, 94., 1281-1289

Yari, H.; Mohseni, M.; Ramezanzadeh, B. (2009b). Comparisons of Weathering Performance of Two Automotive Refinish Coatings: A Case Study. Journal Applied Polymer Science, Vol 111, Issue 6., 2946-2956

Yari, H.; Mohseni, M.; Ramazanzade, B.; Naderi, N. (2009c). Use of analytical techniques to reveal the influence of chemical structure of clear coat on its biological degradation caused by bird-droppings. Progress in Organic Coatings, 66., 281-290

Yari, H.; Mohseni, M.; Ramazanzade, B. (2011). A mechanistic study of degradation of a typical automotive clear coat caused by bird droppings. Journal of coatings and technology Research, DOI 10.1007/s11998-010-9273-2

Yari, H.; Moradian, S.; Tahmasebi, N. & Arefmanesh, M. (2012a). The Effect of Weathering on Tribological Properties of an Acrylic Melamine Automotive Nanocomposite, Tribology Letters, 46:123–130.

Yari, H.; Mohseni, M.; Ramezanzadeh, B. & Mohammad Rabea, A. (2012b). Investigating the degradation resistance of silicone-acrylate containing automotive clearcoats exposed to bird droppings. Progress in Organic Coatings, Accepted for publication

Yahyaei, H.; Mohseni, M.; Bastani, S. (2011). Using Taguchi experimental design to reveal the impact of parameters affecting the abrasion resistance of sol–gel based UV curable nanocomposite films on polycarbonate. J Sol-Gel Sci Technol. 59., 95.

Yeo, S.Y.; Lee, H.J. & Jeong, S.H. (2003). Preparation of nanocomposite fibers for permanent antibacterial effect. Journal of Materials Science, 38(10), 2143-2147.

Yin, L.; Wang, J.; Lin, F.; Yang, J. & Nuli, Y. (2012). Polyacrylonitrile/Graphene Composite as a Precursor to a Sulfur-Based Cathode Material for High-Rate Rechargeable Li-S Batteries. Energy & Environmental Science 5, DOI: 10.1039/c2ee03495f

Yoshino T.; Kobayashi, K.; Araki, S.; Nakamura K. & Kobayashi, N. (2012). Electrochromic properties of electrochemically fabricated nanostructure nickel oxide and manganese oxide films. Solar Energy Materials & Solar Cells, 99, 43–49.

Yu, M., Gu, G.; Meng, W.D.& Qing, F.L. (2007). Superhydrophobic cotton fabric coating based on a complex layer of silica nanoparticles and perfluorooctylated quaterny ammonium silane coupling agent. Applied surface science, 253, 2007, 3669-3673.

Yu, W.; France, D. M.; Choi, S. U. S.; Routbort, J. L. (2007). Review and assessment of nanofluid technology for transportation and other applications. Tech. Rep. 78, ANL/ESD/07-9, Argonne National Laboratory.

Yu, W.; Xie, H.; Chen, L.; Li, Y. (2009). Investigation of thermal conductivity and viscosity of ethylene glycol based ZnO nanofluid. Thermochimica Acta. 491., 1-2., 92.

Yu, W.; Xie, H. (2011). A Review on Nanofluids: Preparation, Stability Mechanisms, and Applications, Journal of Nanomaterials. doi:10.1155/2012/435873.

Yuranova, T. , Rincon, A.G.; Pulgarin, C.; Laub, D.; Xantopoulos, N.; Mathieu, H. J.& Kiwi, J. (2006). Performance and characterization of Ag-cotton and Ag/TiO2 loaded textiles during the abatement of E-coli. Journal of Photochemistry and Photobiology a-Chemistry, 181(2-3), 363-369.

Zhang, K.; Zhang, L. L.; Zhao, X. S. & Wu, J. (2010). Graphene/Polyaniline Nanofiber Composites as Supercapacitor Electrodes. Chem. Mater. 22, 4, 1392-1401

Zhang, M. & Dai, L. (2012). Carbon Nanomaterials as Metal-Free Catalysts in Next Generation Fuel Cells. Nano Energy 1, advance online publication,

Zheng, G.; Yang, Y.; Cha, J. l.; Hong, S. S. & Cui, Y. (2011). Hollow Carbon Nanofiber-Encapsulated Sulfur Cathode for High Specific Capacity Rechargable Lithium Batteries. Nano Lett. 11, 10, 4462-4467

Zhou, J.; Yang, J.; Zhang, Z.; Liu, W.; Xue, Q. (1999). Study on the structure and tribological properties of surface-modified Cu nanoparticles. Mater. Res. Bull. 34., 9., 1361–1367.

Zhou, W.; Wu, H.; Hartman, M. R. & Yildirim, T. (2007). Hydrogen and Methane Adsorption in Metal Organic Frameworks: A High-Pressure Volumetric Study. The Journal of Physical Chemistry C 111, 44, 16131-16137

Zhou, X., et al. (2010). New Fabrication and Mechanical Properties of Styrene-Butadiene Rubber/Carbon Nanotubes Nanocomposite. J Mater Sci Tech. 26., 12., 1127.

Zhou, X.; et al. (2010). New Fabrication and Mechanical Properties of Styrene-Butadiene Rubber/Carbon Nanotubes Nanocomposite. Journal of Materials Science & Technology. 26., 12., 1127-1132.

Zhu, Y.; Murali, S.; Stoller, M. D.; Ganesh, K. J.; Cai, W.; Ferreira, P. J.; Pirkle, A.; Wallace, R. M.; Cychosz, K. A.; Thommes, M.; Su, D.; Stach, E. A. & Ruoff, R. S. (2011). Carbon-Based Supercapacitors Produced by Activation of Graphene. Science 332, 6037, 1537-1541

Chapter 4

Biolubricants and Triboreactive Materials for Automotive Applications

Amaya Igartua, Xana Fdez-Pérez, Iñaki Illarramendi,
Rolf Luther, Jürgen Rausch and Mathias Woydt

Additional information is available at the end of the chapter

1. Introduction

Replacing hydrocarbon-based oils with biodegradable products is one of the ways to reduce adverse effects on the ecosystem caused by the use of lubricants. The application of low or no sulphur, ash and phosphorous (lowSAP) ester- or polyglycol based oils are intended for passenger car engine lubricants as substitutes for hydrocarbon-based oils. The study is focused on passenger car motor oils (PCMO) with reduced metal-organic additives. This is necessary in order to reduce the ash build-up in the after treatment system and therefore improve its efficiency and lifetime. High fuel efficiency and long drain intervals are requested, as well.

In a modern diesel or Gasoline engine, the engine oils has to fulfill quite a number of different functions, such as lubricating and cooling the system, wear protection, soot and particle handling with less deposit tendency and so on.

2. Engine oils – Developed of new passenger car motor bio oils

2.1. Relevant aspects of influence

Generally speaking, engine oils consist of three major parts. The biggest part (per volume) is taken by the base fluid. It is responsible for the wear free hydrodynamic regime in lubrication and provides the basic lubricity. Besides, it serves as solvent for the additives. The additives can be subdivided into the group of surface active and lubricant active additives.

The first group consists of EP/AW-additives to reduce wear at heavily loaded tribo- contacts and so called friction modifiers (FM) that are used to reduce the coefficient of friction. These additives are polar and tend to adhere to the metallic surface of the engine components. Conventionally, they contain sulphur and phosphorous usually as metal-organic compounds. The other additive group is unpolar or less polar and consists of antioxidants (AO) to prolong the oil service life, dispersancy and detergency (D&D) to keep the engine clean and soot in dispersion and of corrosion inhibitors (Ci). Metal-organic compounds are widely used in that field, the most prominent being zinc-dialkyl-dithiophosphate (ZnDTP). These three components are not independent of each other; they all contribute to overall lubricant performance. The complex interplay of the components is shown schematically in figure 1 before.

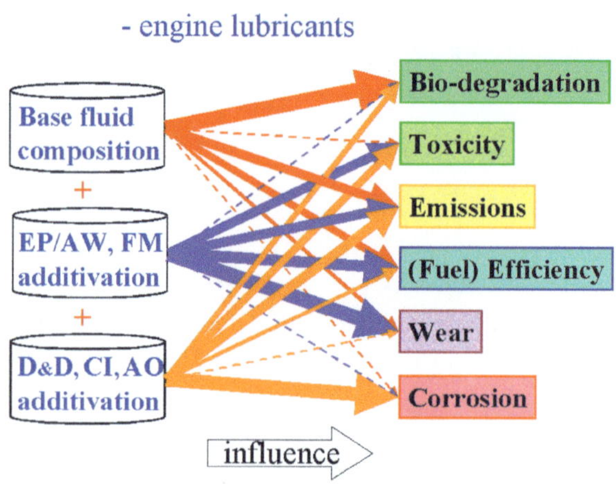

*The width of the arrows indicate the relative importance of the individual components

Figure 1. Importance of engine oil component.

2.2. Emissions – Laboratory procedure

One focus was the development of lubricants which are compatible with latest exhaust gas after-treatment systems. This is the reason for the desired reduction in phosphorous, sulphur and metal-organic content (expressed as sulphated ash). These elements are suspected to contribute towards catalyst poisoning and deposit formation in particle traps. However, one has to keep in mind that there are two different ways how engine oil constituents can influence the performance of after-treatment systems.

Figure 3 shows the result from the field test with a car equipped with a particle trap ('filtre à particule', FAP) and conventional engine oil. After the test the FAP was dismantled and the found deposits were analysed. Most residues were fuel related (FbA=fuel based ash), but the engine oil showed an influence, too. Interesting is the comparison between metal ions found as deposits and those present in the ACEA A3/B3 engine oil. Apparently, zinc is four times more critical towards after treatment deposits than calcium. Therefore, it developed engine oils without any Zinc; even so this is the most common metal organic compound in engine oils today.

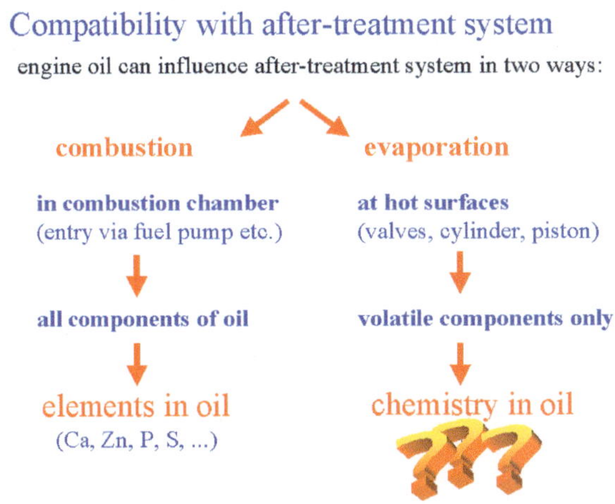

Figure 2. Engine oil composition and influence on after-treatment system.

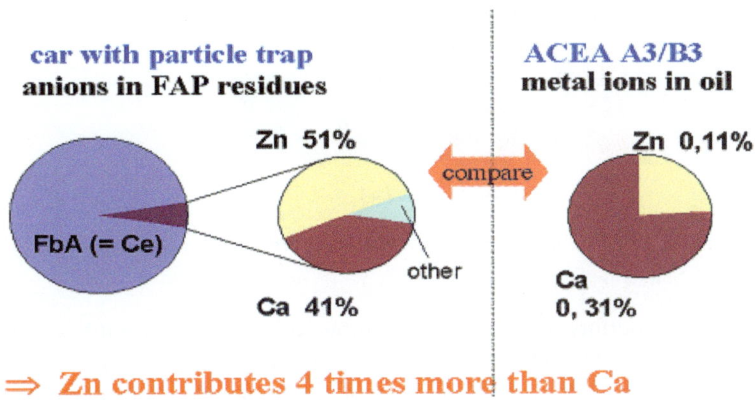

Figure 3. Analysis of residues found in particle filter (result of a field test). Zinc seems to give over-proportional residues in the after-treatment system.

Some engine oil is inevitably entering the fuel path, e.g. via the fuel pumps. The fuel/engine oil mixture is then burned in the combustion chamber and all chemical elements of the engine oil are likely to reach the after treatment system. Thus, phosphorous, sulphur and sulphated ash are appropriate parameters to characterise the oil compatibility with the exhaust gas systems.

The second way of contact between engine oil and after-treatment system arises from evaporation. Hot lubricated surfaces in the engine, e.g. cylinder walls, give rise to the evaporation of volatile substances. These are not necessarily present in the fresh lubricant but may be created by thermal cracking of higher molecular weight compounds at these elevated temperatures.

From the past, it is well known that the type of base oil has a significant influence on raw emissions. An example is the correlation between Noack evaporation loss and particle emissions (see figure 4). This correlation, together with concerns about the oil consumption led to the limitation of the Noack evaporation loss to 15% respectively 13% according to the ACEA regulation.

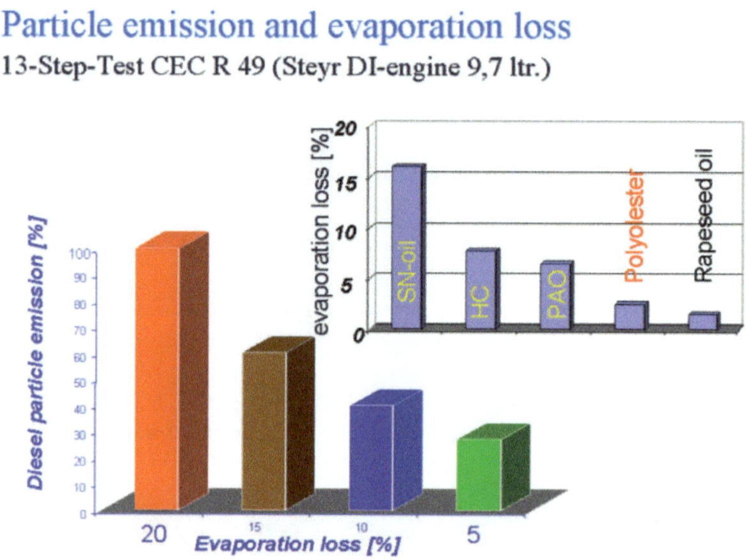

Figure 4. Correlation between Noack Evaporation loss and particle emission in a direct injection HD Diesel engine

Besides the base oil volatility, the volatility of some additives becomes interest. The content of chemical elements in the oil and their abundance in the exhaust gas need not to be proportional. For example, it is well-known that different phosphorous compounds with identical phosphorous content will show great differences in thermal stability and volatility.

In order to evaluate the element specific evaporation of fully formulated engine oils, we established a new test procedure. The new test procedure uses the Noack evaporation test equipment (DIN 51581, ASTM D5800) and the test condition 1h @ 250°C. However, not only the total mass loss but also the evaporation loss of additive elements is determined. A comparison of the two reference engine oils and the new candidate fluids in this test is shown in figure 5. To display hydrocarbon, sulphur and phosphorous emissions in one diagram, the test results have been normalised to a typical reference oil (=100% emissions).

Obviously, both developed engine oils have reduced evaporation losses compared to today's conventional oils. The hydrocarbon loss, dominated by the base oil, was cut to less than half of the original value. This should reduce the particle emission. On the additive side, sulphur and phosphorous volatility have been significantly reduced.

Under the described conditions, volatile metal-organic compounds in significant quantities have not been found in any of the tested oils.

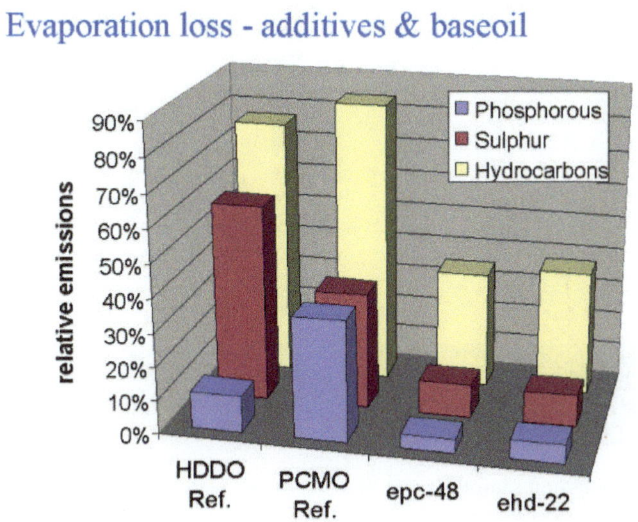

Figure 5. Element specific Noack evaporation losses (relative two reference engine oils).

2.3. Engine cleanliness & low ash

The main challenge in formulating engine oils with low ash and sulphur is the question of engine cleanliness. Metal soaps and similar components (commonly called detergents) are responsible for preventing deposits in the engine, predominantly at the piston rings and groves. Besides, their alkalinity is necessary to neutralise acidic combustion products. Today, the most prominent substance used is calcium sulphonate–accounting for approximately 80% of lubricant ash and sulphur. Current engine oils have typical values of 0.3% to 0.4% Calcium, corresponding to sulphated ash contents of about 1.3% to 1.5%. Thus, a reduction to 0.5% as expected by passenger car manufactures cannot be achieved by just reducing the concentration of well known additives, but needs a new additive technology.

To evaluate the capabilities of new D&D-additives, it made intense use of the so-called Wolf-test. This test is a modification of the well known panel-coker idea. The engine oil to be tested is dropping on a hot metal plate. This steel panel is mounted at a fixed angle to allow the oil to flow down the panel slowly. At the lower end it is captured in a glass vessel to be recirculated to the panel again. After a fixed period of time the test is stopped, the panels are washed and the deposit formation is observed. Weighting and a visual rating of the panels by an expert allows judging the deposit formation tendency.

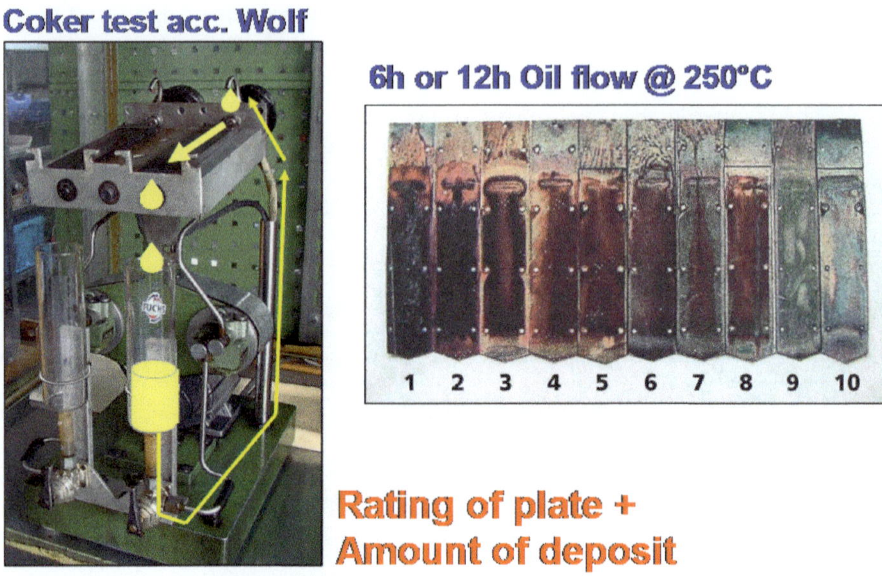

Figure 6. Wolf test apparatus and sample panels showing the rating scheme.

The results obtained in the WOLF test give a good indication for engine cleanliness in real engines. Figure 7 shows the correlation between the Wolf test and the VW TDI test according to CEC L-78-T-99 as established with new oils in the past. The CEC L-78 test belongs to the ACEA sequence of engine tests and is used to rate piston cleanliness. Results greater or equal to 65 points are supposed to be good in this engine test. Thus deposits in our Wolf screening test should be below approximately 8 to 10 mg. However, only a test in a real fired engine can give reliable information about the actual performance.

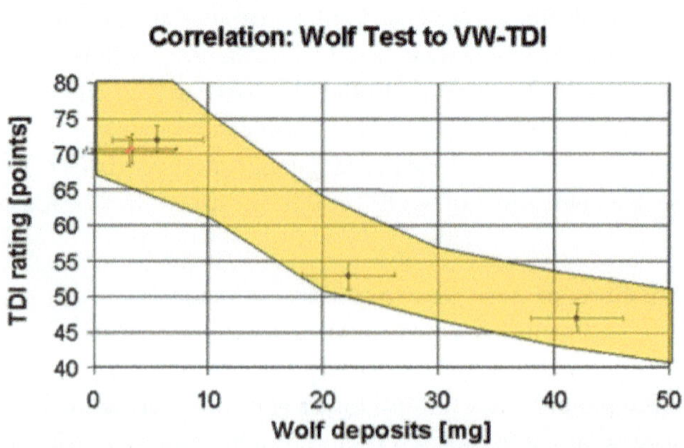

Figure 7. Correlation between screening test and actual engine test with VW TDI according to CEC L-78-99.

As explained earlier, the major challenge in formulation the new oils is the reduction in crucial additive elements. Sulphur and phosphorous are the elements found in most anti-wear agents, while the metal-organic compounds constitute important anti-oxidants and engine cleanliness additives. All three additive groups can have negative influence on the after-treatment system. Therefore so called lowSAP oils are required. The understanding of lowSAP is the reduction of sulphur ash (as a measure for metal-organic compounds) and phosphorous. The figure 8 demonstrates the comparison between the passenger motor oil reference lubricant and the new developed oils.

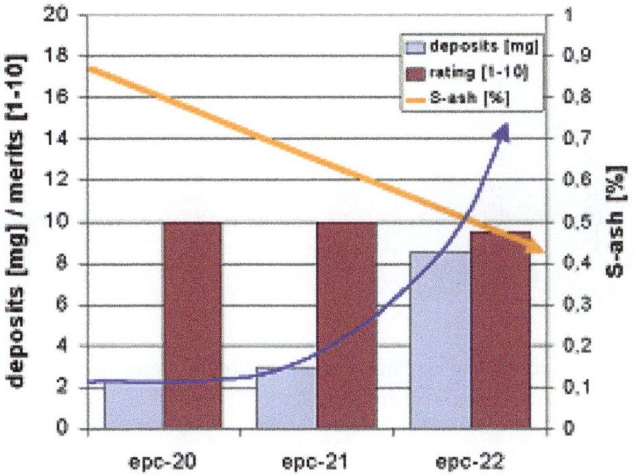

Figure 8. Results of Wolf test showing the dependency of deposit formation from the sulphated ash content of the engine oil, even with most modern technology

The (sulphur) ash content is reduced to one third, phosphorous and sulphur to less than 50% of today's product. Despite this strong decrease in metal-organic compounds, the deposit formation tendency – as screened with the panel coker test according to Wolf – remains as low as the reference oil (figure 9.b). This target was reached by using a new ash less EP/AW additive technology. However, the influence of the concentration of metal-organic compounds remains visible even with newest technologies.

2.4. PCMO - EP/AW properties

The second concern while using lowSAP oils is wear protection. To check this performance, screening tests have been performed. Next figure 10 displays the results obtained with the SRV test rig. On the left hand side, wear scars according to DIN 51834-2 are displayed. This is a 2 hour test at constant load. Here, both new engine oils show slightly smaller wear scars. On the right hand side, results according to ASTM 5706 are displayed. In this SRV test, the load is increased in fixed intervals until seizure occurs. Higher load corresponds to a better protection at highest load spikes. The reference oil and both new candidate oils easily supersede the values typically found in engines.

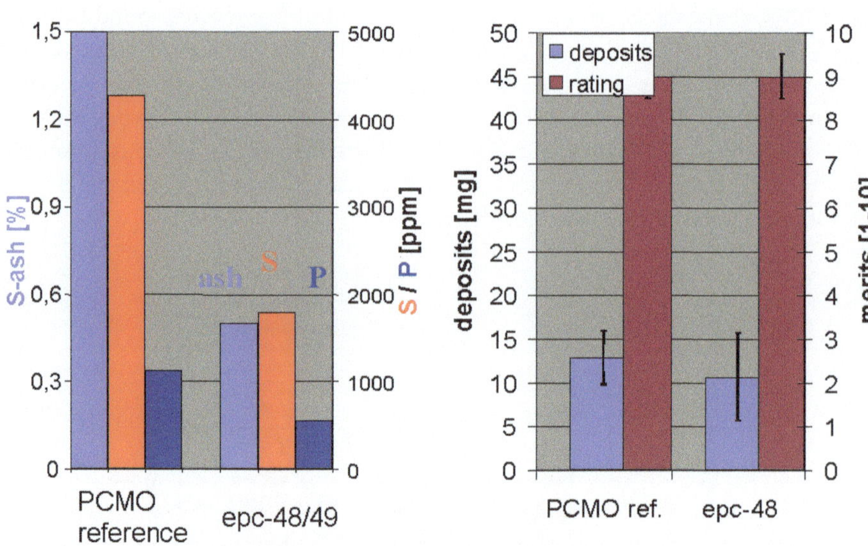

Figure 9. a) Comparison of additive elements of reference oil and epc-48 and 49
b) Wolf comparison between reference oil and epc-48

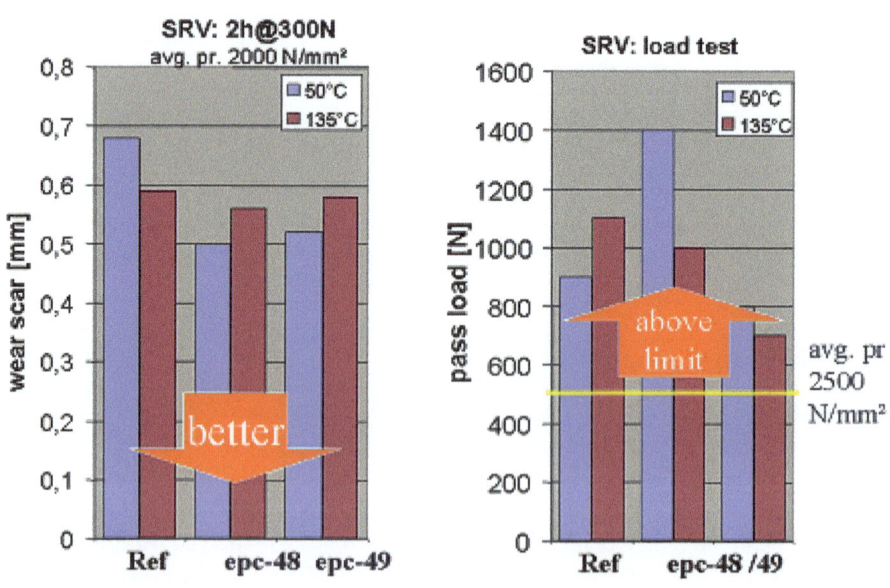

Figure 10. SRV test results for the reference oil and the new epc-48 and epc 49

2.5. Fuel efficiency test

The fuel efficiency potential of epc 50 was tested in an M111-FE engine test bench which is part of different ACEA specification. The matter of this test is to evaluate the fuel consumption under defined conditions. The fuel saving compared to the 15W40 CEC reference oil RL 191 is calculated in %. The formidable result of 3,9% fuel benefit demonstrate the potential of epc 50.

2.6. Passenger car motor oil candidates

At least two candidate oils named epc-48 and epc-49 passes the laboratory optimization process. A comparison of the passenger car specifications and the candidate oil data is given in next table 1.

Epc 48 and epc 49 are based on the same base fluid composition but differ in additivation. Due to their high viscosity index, the desired high temperature, high shear viscosity (HTHSV) at 150°C of 2.9 mPas is reached at significantly lower viscosities at 100°C, 40°C and at -25°C. This offers the potential for improved fuel efficiency.

To meet the passenger car oil specifications, the additives and also their treat rates have to be balanced well.

		PCMO-Specifications	epc-48	epc-49
SAE grade		0W-30 or 5W-30	0W-20	
Viscosity	kin.V. 40°C	< 55mm²/s	44mm²/s	
	kin.V. 100°C	> 9.3mm²/s	8.8mm²/s	
	CCS -25°C	< 2900mPas	1660mPas	
	HTHS 150º C	2,9 - 3,1mPas	2.9mPas	
Shear stability (Bosch)		stay in grade	Excellent	
Pour point		< -35°C	-45°C	
Elements	P	< 800ppm	560ppm	560ppm
	S	< 2000ppm	1800ppm	1800ppm
	Zn	< 500ppm	free	Free
sulphated ash		< 0,5%	0,5%	0,5%
Noack evaporation loss		< 13%	5,2 %	
ICOT oxidation stability		140 h pass	O.K.	

Table 1. Candidate engine oils for the PCMO specification

Table 2 illustrates the developing and optimising process for passenger car motor oils regarding to oxidation stability and deposit formation tendency.

	epc-0	epc-1	epc-2	epc-3	epc-4	epc-8
ICOT	fail	pass	fail	pass	pass	pass
Wolf Test	pass	pass	pass	pass	n.d	n.d
	epc-9	**epc-10**	**epc-11**	**epc-12**	**epc-13**	**epc-14**
ICOT	fail	n.d	fail	n.d	n.d	n.d
Wolf Test	n.d	n.d	n.d	pass	pass	pass
	epc-15	**epc-17**	**epc-18**	**epc-19**	**epc-20**	**epc-21**
ICOT	n.d	fail	n.d	n.d	fail	n.d
Wolf Test	pass	n.d	n.d	n.d	pass	pass
	epc-22	**epc-23**	**epc-24**	**epc-25**	**epc-26**	**epc-27**
ICOT	n.d	fail	n.d	pass	pass	fail
Wolf Test	pass	n.d	n.d	n.d	n.d	pass
	epc-28	**epc-29-36**	**epc-37**	**epc-38**	**epc-39**	**epc-40**
ICOT	Pass	n.d	pass	fail	pass	pass
Wolf Test	n.d	n.d	pass	pass	pass	pass
	epc-41-47	**epc-48**	**epc-49**	**epc-50**	**epc-51**	**epc-52**
ICOT	n.d	pass	pass	pass	n.d	fail
Wolf Test	n.d	pass	pass	pass	n.d	n.d
	epc-53	**epc-54**	**epc-55**	**epc-56**	**epc-57**	**epc-58-61**
ICOT	fail	n.d	fail	fail	fail	n.d
Wolf Test	pass	n.d	pass	pass	pass	n.d

Table 2. Developing PCMO candidates – Optimization in oxidation stability and deposit formation tendency

One important aspect of the engine oil specification is to meet all viscosity properties according to SAE J300. During the study a change in the in SAE J300 requirements occurred. The changes relates to the low temperature cranking viscosity (CCS viscosity) and low temperature pumping viscosity (MRV). The defined measuring temperature switched to a lower value whereas the limit value was adjusted to the lower temperature.

Due to this epc 48 and epc 49 was developed in accordance to the old SAE J300 viscosity specification. Nevertheless epc 48 and 49 fulfil also the new SAE J300 low temperature requirements.

To follow a line in a consequent way, these oils have to be biodegradable and non-toxic to the aqueous environment according to the directive EC/1999/45, coherent with other international standards.

Biodegradability

Table 3 and figure 11 show results obtained with new biolubricants developed. The reference substance is an internal reference for OECD 301F biodegradation test. The passenger car motor oil (PCMO) reference was not readily biodegradable. The new PCMO developed pass the limit of biodegradation (60% in 28 days).

Time	% Degradation				
Day	Ref. Sust.	EPC48 (Test a)	EPC48 (Test b)	EPC 49 (Test a)	EPC 49 (Test b)
0	0	0	0	0	0
7	90.96	45.11	37.46	40.76	41.72
14	90.38	70.34	55.05	57.65	61.63
21	89.50	81.42	66.13	66.10	74.06
28 days	89.50	87.54	78.36	76.04	84.00

Table 3. Results of biodegradability tests of EPC-48 and EPC-49 samples. Fully formulated motor oils.

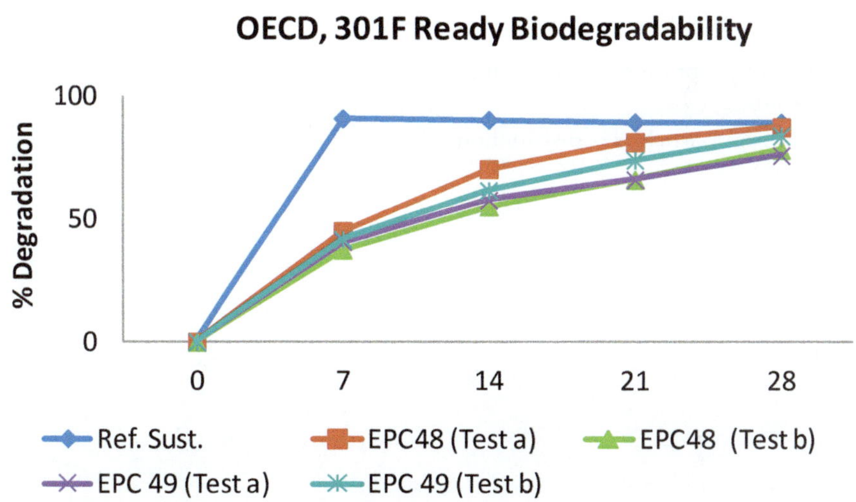

Figure 11. Results of biodegradability of EPC-48 and EPC-49 samples. Fully formulated motor oils.

Aquatic toxicity

The following basic level test are proposed and included in the EU hazard assessment of substances: Daphnia acute immobilization test (OECD 202; Table 4).

Lubricant	*EC50 (mg/l)	Classification
EPC-48	EL50>100	"Not harmful" to aquatic organisms
EPC-49	EL50>100	"Not harmful" to aquatic organisms

*EC50: Effective Concentration. The EC50 value means the mean effective concentration of a substance that produces a particular, previously defined behaviour in 50% of the organisms of a test series.

Table 4. OECD 202, Daphnia Magna Toxicity Test Results. Fully formulated motor oils.

3. Definition of Tested Prototypes - Triboreactive coating for the tribo-system piston ring and cylinder liner outside of engines

3.1. Plasma powder characterization

Three types of triboreactive powders were used for the production of ring prototypes. Piston rings were produced with the reference coating, CIE-TARABUSI-PL72, and with the new coatings E1, E2 and E3 with successful characteristics in terms of hardness, porosity, adherence and machine ability. In all cases an intermediate layer is applied to obtain good adherence with the substrate.

Production of plasma powder with Magnéli phases and deposition of plasma sprayed coatings containing Magnéli phases or triboactive suboxides.

Plasma powder and coating were characterized with:

- Metallography, optical microscopy and image analysis
- Profilometry
- Mastersizer for particle size distribution
- SEM + EDX
- X-ray diffraction
- Raman spectroscopy

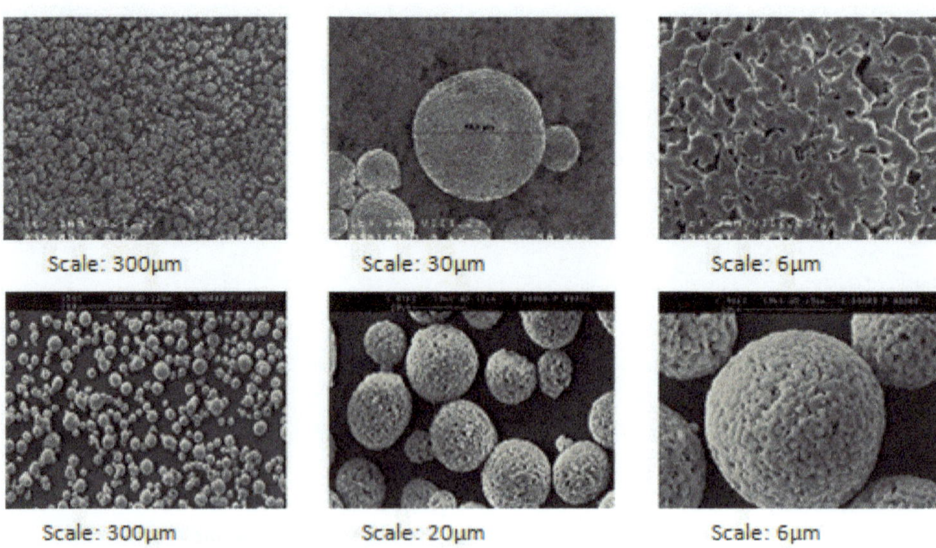

Figure 12. SEM morphology of sintered and agglomerated Ti_nO_{2n-1} powder and (Ti,Mo)(C,N)+Ni+Mo plasma powder

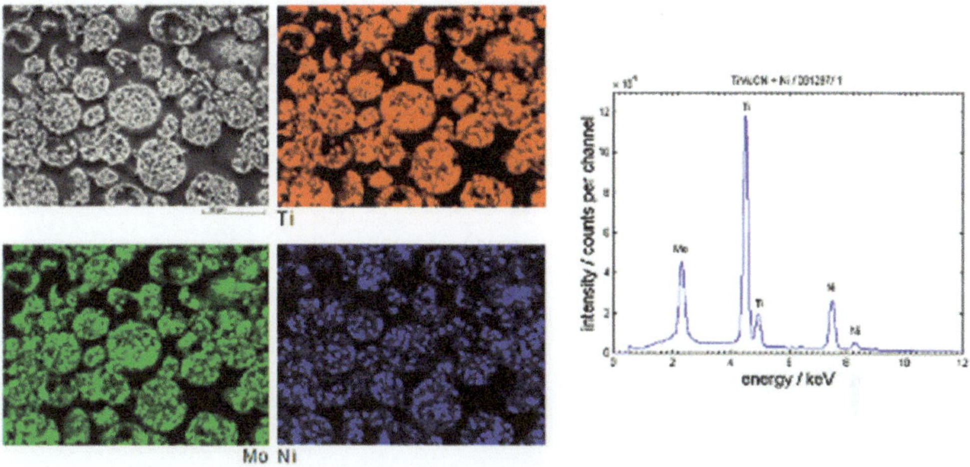

Figure 13. Cross section of (Ti,Mo)(C,N)+Ni+Mo plasma powder before thermal spraying with element distribution and EDX spectrum

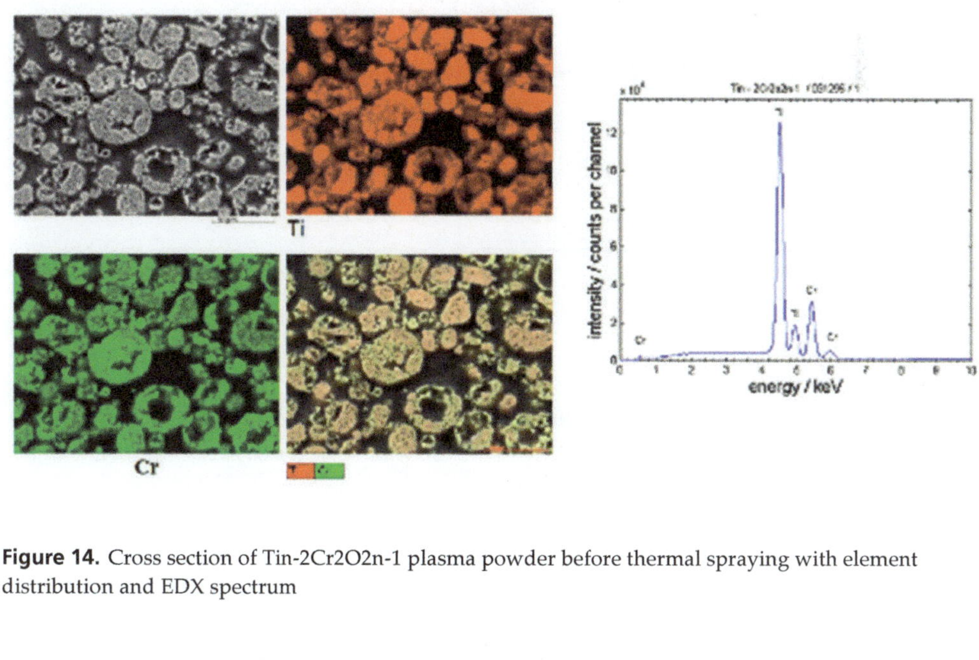

Figure 14. Cross section of Tin-2Cr2O2n-1 plasma powder before thermal spraying with element distribution and EDX spectrum

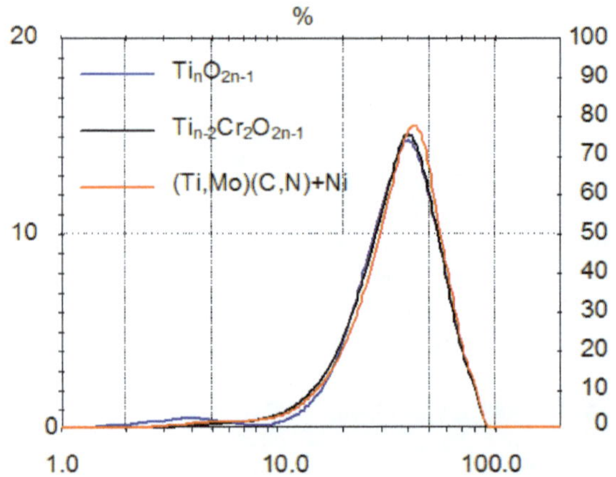

Figure 15. Particle size distribution of three plasma powders (TinO2n-1,Tin-2Cr2O2n-1 and (Ti,Mo)(C,N)+Ni) without US-treatment.

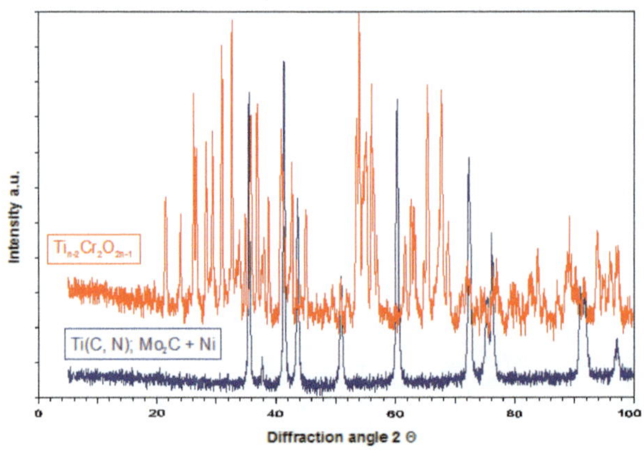

Figure 16. XDR spectra of Ti(C,N)Mo2C+Ni and Tin-2Cr2O2n-1 plasma powders

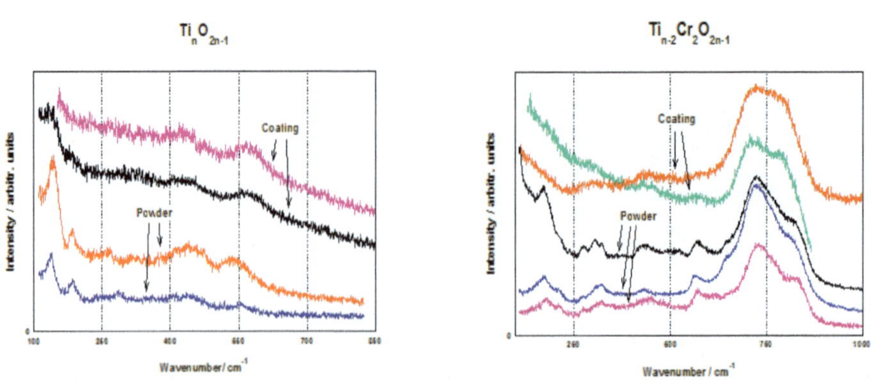

Figure 17. Raman spectra of TinO2n-1 and Tin-2Cr2O2n-1 plasma powders and plasma coatings

Powder	Structure of particles (Metallography; SEM)	Grain size fraction (FhG Dresden) [μm]	Particle size after US-bath d50 [μm]	LRS	XRD	Composition data sheet (weight %)
TinO2n-1 (n = 4…6) IKTS	**Agglomerated and sintered**	**20-63 μm**	**25,01**	**TinO2n-1 (n = 4 …6)**	**Ti407 Ti509 Ti6011**	
(Ti,Mo)(C,N) + Ni + Mo	**Agglomerated and sintered**	**15-45 μm**	**22,33**	**TiCN/ TiMoCN/ TiC1? Ni (not directly measurable with LRS, but thermally induced formation of NiTiO3 and NiMoO4)**	**TiC TiC0,7N0,3 [(TiMo)(CN)] Ni4Ti3 probably: Mo2N and Ni3Mo3N Mo2C (very low content)**	**Ti 41,5-42 soll, O 0,20; N 3,43; C 10,09; Ni 20 soll; Mo 21,5-22 soll (not determined)**
Tin-2Cr2O2n-1 $6 \leq n \leq 9$, solid solution of 21-23 mol% Cr2O3 in TiO2 HC Starck	**Agglomerated and sintered**	**15-63 μm**	**21,84**	**Magnéli phases (TinO2n-1) TiO44 - and CrO44 - complexes – Coating thermal stable against laser radiation**	Cr0.15Ti0.85O1.925 Cr0.46Ti0.54O1.77 **Cr2Ti6O15 Cr2TiO5 or CrTi2O5 Ti8015**	Ti 38,2; Cr 21,8; O about 40; N < 0,1; C < 0,01; Ti/Cr ≈ 1,75 Phases: Cr0.46Ti0.54O1.77 Cr0.15Ti0.85O1.925 CrTi2O5; Cr2Ti6O15 Ti8O15

* determined with thermogravimetry

Table 5. Characterisation of thermal spray powders

1: TiC, (TiMo)(CN) and TiCN of cubic structure exhibit only broadened, second order Raman peaks; the detected Raman peaks can be attributed to these phases, but attribution is not unambiguous.

- TinO2n-1 powder (n = 4 – 6) is not stable against thermal stress.
- Atmospheric plasma spraying as well as laser radiation cause oxidation with increasing formation of rutile. With APS process Magnéli phases cannot be fully transferred from the TinO2n-1 spray powder to the APS coatings.
- Magnéli phases are stabilized against thermal stresses, laser radiation and oxidation, if chromium oxide was added to TinO2n-1.

- Both, Tin-2Cr2O2n-1 powder and thermal sprayed coating contain high amount of stable H-Magnéli phases. Tin-2Cr2O2n-1 can be sprayed with APS to produce triboactive coatings.
- Vacuum plasma spraying (VPS) in reducing gas atmospheres using hydrogen addition to plasma gas allows to produce coatings with Magnéli phases. With VPS process Magnéli phases of TinO2n-1 (n = 4 – 6; IKTS) powder can be transferred to the coatings. Additionally, other Magnéli phases were detected with XRD.
- The substoichiometry and electric conductivity of TiO2-x can be effected by the amount of hydrogen in plasma gas. Substoichiometry and electric conductivity increase with the amount of hydrogen in plasma gas as reducing medium.
- Magnéli phases were found with LRS and XRD.
- Owing to the stability of the Tin-2Cr2O2n-1 powder, thermal sprayed coatings with stable Magnéli phases can be produced with APS process - using Ar/He process gas mixture with Ar = 35 SLPM and H2 = 8 SLPM.
- Thermal sprayed (Ti,Mo)(C,N)+Ni+Mo coatings have been produced with HVOF and plasma spraying process.
- Thermal sprayed coatings (Ti- and TiCr-suboxides with crystallographic planar defect structure (Magnéli phases)) on flat disks have been produced for prescreening tribotests and evaluation of interactions of different lubricant / coating combinations.

3.2. Piston ring prototypes

Part Name	Dimensions		Material	Coating
	Nominal Ø	Height		
Cylinder Liner	96	-	Perlitic cast iron AT182	-
Top piston Ring	96	2	Nodular graphite cast iron AT126	PL72
				PL E1
				PL E2
				PL E3

Table 6. System/Component Definition

	PL E3 Ti_nO_{2n-1}	PL E1 $Ti_{n-2}Cr_2O_{2n-1}$	PL E2 (Ti, Mo)(C,N)+23NiMo 1.batch	PL E2 (Ti,Mo)(C,N)+23NiMo 2.batch
Argon pressure (psi)	50	50	50	50
Hydrogen pressure (psi)	25	25	25	25
Electric intensity (A)	575	575	575	575
Distance to part (mm)	110	110	110	110
Carrying gas pressure (rpm)	20	20	20	20
Part rotating speed (rpm)	300	300	300	300
Gun feed (mm/s)	10	10	10	10
Powder flow (g/m)	20	20	20	11,5

Table 7. Process parameters of atmospheric spraying of triboactive coatings on piston rings.

The characteristics of the coatings, as porosity, hardness adherence and light tightness, are collected in table below.

Feature	PL72	PL E3	PL E2 (Batch 2)	PL E1
	Reference	Ti_nO_{2n-1}	(Ti, Mo) (C,N)+23NiMo	$Ti_{n-2}Cr_2O_{2n-2}$
Process	Casting Rough machining APSpraying OS grinding Sides grinding Face honing Gap-milling			
Adherence to substrate	By means of intermediate layer			
Incidents by manufacturing	Normal behaviour	The coating wears off very easily by honing. The coating breaks by gap milling	Highly porous layer. Barrel shape hard to recognize due to high roughness. Easy to machine	Normal behaviour
Porosity of coatings	5%	2%	10%	5%
Hardness (HV0.5)	450-480	631-670	640-660	522-554
Roughness Rpk (μm)	0.2	0.15-0.32	0.37-1.09	0.19-0.25
Thickness (μm)	~250	> 50	~200	~250
Adherence	Good	Good	Fair	Good
Light tightness	Good			

Table 8. Summary of properties of atmospheric plasma sprayed triboactive coatings on piston rings.

Reference Components:

These components were designed and manufactured by CIE-TARABUSI with own knowledge and technology.

Piston: Design for Turbo Diesel application with oil cooling gallery and steel struts for expansion control, in aluminum alloy AT12 with 12% silicon content and iron insert for the top groove improved wear resistance. The piston head is coated by hard anodizing to avoid thermal cracks. The piston skirt has a graphite coating to improve lubrication during running-in. Basic dimensions: Diameter 96 mm, Total height 87.2 mm, Compression height 50.45 mm.

1st ring: Plasma sprayed PL-72 rings in spheroidal cast iron material AT126, heat treated martensitic, ISO 6621/3MC 53 type, of diameter 96 mm and height of 2 mm. Pl-72 ring is high wear and scuffing resistant coating made up by Molybdenum, Molybdenum carbide and Nickel-Chromium compounds widely used in powered Diesel applications, where Chromium coatings cannot be used due to the elevated temperature of the engine.

2nd ring: Chromium plated taper ring with conical periphery and positive torsion internal step, manufactured in STD material AT110, non heat treated gray cast iron, ISO 6621/3 MC 11 type, of diameter 96 mm and 2 mm height

3rd ring: Chromium plated helicoidal spring loaded oil control ring with symmetric bevels, produced in STD material AT110 non heat treated gray cast iron, ISO 6621/3 MC 11 type. The helicoidal spring is made of oil hardened spring steel, of 96 mm of diameter and 3 mm of height.

Pin: DIN 17Cr3 carburized, hardened and tempered steel pin, with 30 mm of outer diameter, 15 mm of inner diameter and 80 mm length.

E1: Top ring coated with titanium chromium oxide

Top ring: As the reference ring but coated by atmospheric plasma with E1 powder developed. The composition of the powder is given below:

E1: Tin-2Cr2O2n-1, $6 \leq n \leq 9$, 21-23 mol % Cr2O3, Magnéli phases sintered and agglomerated, 15-63μm:

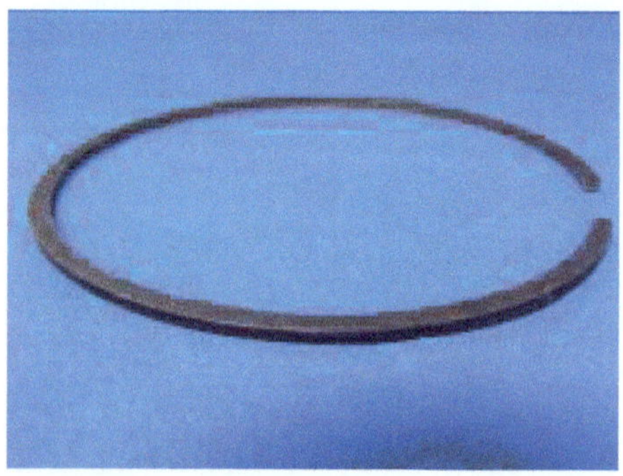

Figure 18. Prototype E1: Tin-2Cr2O2n-1,

E2: Top ring coated with titanium-molybdenum carbide-nitrides

Top ring: As the reference ring but coated by atmospheric plasma with E2 powder developed. The composition of the powder is given below:

E2: 77 (Ti,Mo) (C,N) – 20 Ni – 3 Mo, Ti/Mo ratio ≈ 2:1, sintered and agglomerated:

Figure 19. Prototype E2: 77 (Ti,Mo) (C,N)–20Ni–3 Mo

E3: Top ring coated with titanium oxide

Top ring: As the reference ring but coated by atmospheric plasma with E1 powder developed. The composition of the powder is given below:

E3: TinO2n-1 (Ti_4O_7, Ti_5O_9, Ti_6O_{11}) sintered and agglomerated, 20-63μm:

Figure 20. Prototype E3: TinO2n-1

4. Friction and wear simulation tests. Combining bio oils with triboreactive coatings

4.1. Piston ring cylinder liner simulation

To study these new oils have been tested different piston rings coated with triboreactive powders deposited by Plasma spray against cast iron cylinder liner, using the PCMO oil as reference. In the next figure 21, the configuration to carry out the simulation tests developed by TEKNIKER is shown.

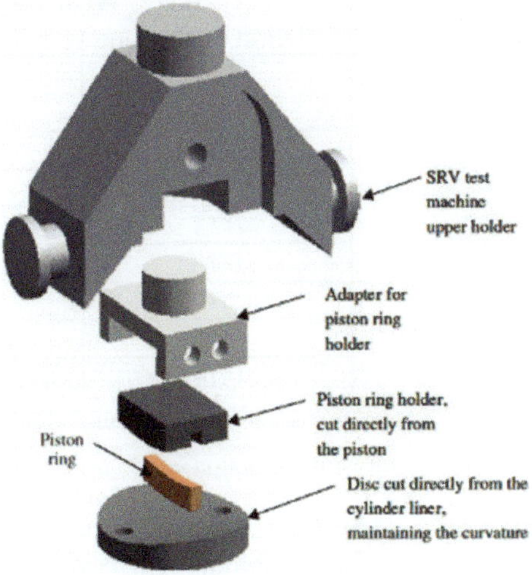

Figure 21. Cylinder liner–Piston ring configuration test.

The test conditions were:

Load and time:

- 41 min, load: 50/2000N in scuffing tests. The load is increased in fixed intervals until seizure occurs. Higher load corresponds to a better protection at highest load spikes.
- 90minutes and 300N in Wear Test,

The rest of the conditions are similar in both tests: frequency: 50 Hz, stroke: 3mm, temperature: 200ºC.

Piston rings	Oil	Extreme Pressure Conditions		
		Time at which failure occurs (minutes)	The highest friction load at which no failure occurs (N)	Type of failure*
Chrome plated	Ref.	28	1300	Stroke<0,3 mm
E3:Ti_nO_{2n-1}		36	1800	Stroke<0,3 mm
E1:$Ti_{n-2}Cr_2O_{2n-1}$		29	1400	Stroke<0,3 mm
Chrome plated	EPC-48	30	1400	Stroke<0,3 mm
E3:Ti_nO_{2n-1}		32	1600	Stroke<0,3 mm
E1:$Ti_{n-2}Cr_2O_{2n-1}$		41	2000	No failure
Chrome plated	EPC-49	32	1500	Stroke<0,3 mm
		31	1500	Stroke<0,3 mm
E3:Ti_nO_{2n-1}		36	1800	Stroke<0,3 mm
		38	1900	Stroke<0,3 mm
E1:$Ti_{n-2}Cr_2O_{2n-1}$		41	2000	No failure

*The failure is considered when the first micro-welding appears.

Table 9. Cylinder liner-Piston ring Test Results.- Scuffing Results

Piston rings	Oil	Wear Conditions	
		Friction coefficient	Cylinder liner mass lost (mg)
Chrome plated	Ref.	0.11	0.4
E3:Ti_nO_{2n-1}		0.12	0.2
E1:$Ti_{n-2}Cr_2O_{2n-1}$		0.11	0.4
Chrome plated	EPC-48	0.13	0.3
E3:Ti_nO_{2n-1}		0.14	0.0
E1:$Ti_{n-2}Cr_2O_{2n-1}$		0.13	0.3
Chrome plated	EPC-49	0.13	0.6
E3:Ti_nO_{2n-1}		0.14	0.2
E1:$Ti_{n-2}Cr_2O_{2n-1}$		0.14	1.5

*The roughness of E2 77(Ti,Mo) (C,N)-20Ni-3Mo are very high, so these piston rings will not be tested in laboratory simulation.

Table 10. Cylinder liner-Piston ring Test Results.- Friction and Wear Results

Also, the new oils developed (EPC-48 and EPC-49) have similar or better behaviour than the PCMO reference oils. It seems that the combination triboreactive coating/polar lubricant increases the load carrying capacity and the extreme pressure properties.

4.2. Specific wear energy

Specific wear energy is a criterion, which takes both into account: it is the ratio of the friction work spent in the interface divided by the mass loss due to the wear.

$$E_w = \frac{E}{\Delta m} = \frac{v_m F_N \int_{ti}^{tf} \mu_{fr}(t)dt}{\Delta m}$$

The specific wear energy Ew is the amount of energy needed to wear a certain mass of matter. Consequently, the higher the value of the Ew is, the more difficult it is to wear the material.

The Figures 22 and 23 show the cylinder liner wear specific energy.

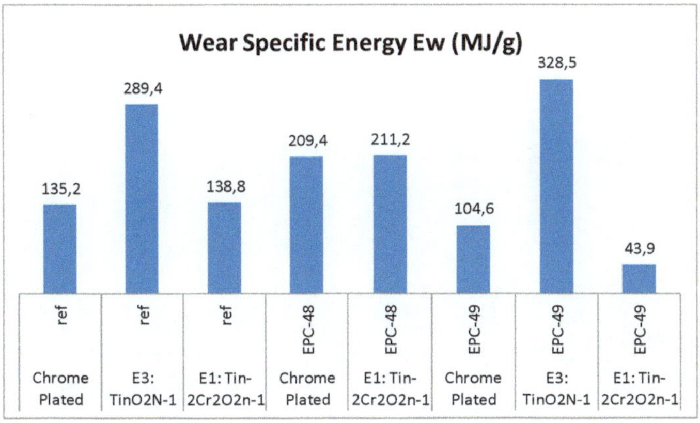

Figure 22. Cylinder liner wear specific energy. Friction and Wear Simulation Tests.

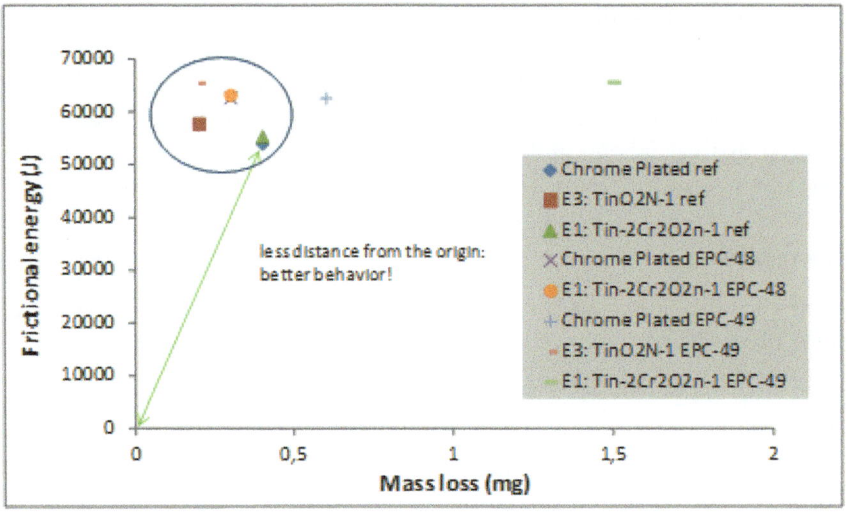

Figure 23. Frictional energy/mass lost. Friction and Wear Simulation Tests.

5. Engine tests. Pass/No pass results for biodegradable fluids in combination with triboreactive materials. Evaluation of friction, wear, life, oil consume

The piston ring component engine tests have been performed by CIE TARABUSI in a 2.7 L turbo-charged 4-stroke diesel engine, using as reference the Mo-based plasma coating PL72 (reference used by CIE-Tarabusi).

Figure 24. Testing of piston rings in a turbo diesel engine arrangement

The test conditions and parameters were fixed for three different type of tests: (a) Scuff test to determine the scuffing resistance of the coating at high temperature and pressure with minimum component clearance, (b) Hot test, long run at high power and high speed to set the reference values for wear and (c) Cycle test between low and high thermal conditions to determine the thermal fatigue and adherence resistance of the developed coatings. The components were measured before and after the test.

5.1. Tests results

The pistons and rings are inspected visually after the scuff, life tests and cyclic tests. The appearance observations of the tested components are presented comparatively to the reference components. We summarized the aspect of the prototypes as follows.

Reference components

The piston has good appearance for the time of running and type of tests run. The wear pattern of the piston skirt is smooth and wide. The rings have normal periphery and side wear. The pin also has a typical appearance with normal signs of friction effects.

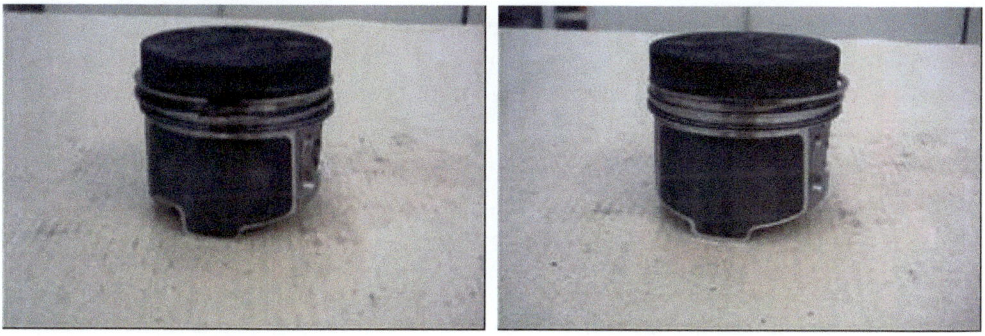

Thrust Side Anti-thrust Side

Figure 25. Testing of Piston Rings with reference coating and reference oil

E1, E2 and E3 coated top Rings

E2 and E3 coatings have good appearance as it is the case for the reference rings. Coating E1, on the contrary, breaks off during the tests. The picture below shows the periphery of the coated rings.

E3: TinO2n-1 E2: (Ti, Mo)(C,N)+Ni+Mo E1: Tin-2Cr2O2n-1

Figure 26. Engine Tests.- Testing of piston rings with triboreactive coatings, aspect of parts after test with standard oil

The engine test results are as follows:

1) Reference engine test: The studied tribological pair is the cylinder liner of 96mm internal diameter and made of uncoated pearlitic cast iron AT182 and top piston rings of 96mm outside diameter, height 2mm and made of nodular graphite cast iron AT126 coated with PL72 standard atmospheric plasma layer. The results are the reference for wear, scuff resistance and adherence. Engine oil consumption and blow-by values are also considered for comparison.

These tests were performed with the reference materials and using the oil SAE 5W30. The scuffing test (2h30m) was done with 100% load and speed engine conditions. Water and oil temperatures were measured. Neither piston nor piston rings show scuff marks after testing. Concerning hot test (100h), the engine conditions were 105% load, 100% speed, water temperature 110ºC and oil temperature 130ºC. Results were an oil consumption of 48,3 g/h and blow by 55,7 l/min. The cold/warm test consists of (100h) with cycle time of 14 min.

2) Engine testing of piston rings with new coating E2 (1st batch) and reference oil SAE 5W30. After 14 hour of hot test running the engine is stopped due to high blow-by. In the engine disassembly it is observed that the E2 coated rings show scuffing and the liners are severely deteriorated.

3) Engine testing of piston rings with new coating E3 and reference oil SAE 5W30. After an initial test failure due to reasons not caused by the new rings, the tested ring set goes successfully under the hot test running (100 h) and the cold/warm cycles (100 h). The E3 coated rings show no sign of scuffing and a periphery wear 30% more than the reference coating. On the contrary the induced liner wear is lower more than 50%.

4) Engine testing of piston rings with new coating E1 and reference oil SAE 5W30 was successfully concluded after hot test running (100h) and cold/warm cycles (100h). The ring periphery wear of E1 is equal to the reference coating and the induced wear in the cylinder is importantly reduced in around 70%. Unfortunately it is observed after the engine disassembly that the top ring of piston no. 4 shows peripheral erosion and breakage.

5) Engine testing of piston rings with new coating E2 (2nd batch) and reference oil SAE 5W30 was successfully concluded after hot test running (100h) and cold/warm cycles (100h). The ring periphery wear of E2 is 80% higher than the reference coating and the induced wear in the cylinder is reduced in around 40%.

6) Reference engine test with EPC-49 oil: The studied tribological pair is the cylinder liner of 96mm internal diameter and made of uncoated pearlitic cast iron AT182 and top piston rings of 96mm outside diameter, height 2mm and made of nodular graphite cast iron AT126 coated with PL72 standard atmospheric plasma layer. The results are the reference for wear, scuff resistance and adherence with special triboreactive epc-49 oil.

These tests were performed with the reference materials and using the oil EPC-49. The scuffing test (2h30m) was done with 100% load and speed engine conditions. Neither piston nor piston rings show scuff marks after testing. Concerning hot test (100h), the engine conditions were 105% load, 100% speed, water temperature 110ºC and oil temperature 130ºC.

7) Engine testing of piston rings with new coating E3 and EPC-49 oil. The tested ring set goes successfully under scuff test (2.5 h), the hot test running (100 h) and the cold/warm cycles (100 h). The E3 coated rings show no sign of scuffing and a periphery wear and cylinder bore induced wear performance similar to the result obtained with the reference components.

TEST	TIME (h)	OIL	DWG	COATING	COAT COMPOSITION	TEST TYPE	BB (l/min)	OC (gr/h)	RADIAL CYL WEAR (μ/100 h)	RADIAL RING WEAR (μ/100 h)	PASS/NO PASS	NOTES
1	102.5 100	STD	E11	PL72	Mo+Mo2C+NiCr	SCUFF/DURATION CYCLES	55.7	48.3	2.5	10.1	PASS	REFERENCE TEST
2	15.75	STD	E23	PL-E2 (1ST BATCH)	(Ti,Mo)(C,N)+ Ni+Mo	SCUFF/DURATION	-	-	-	-	NO-PASS	TEST STOPPED DOR HIGH BLOW-BY (>75 l/min). SCUFFING OF TOP RINGS AND CYLINDER DAMAGE
3	80.1	STD	E24	PL-E3	TinO2n-1	SCUFF/DURATION	54.4	-	-	-	NO-PASS	SEIZURE OF PISTON OF CYLINDER NO. 1
3 BIS	102.5 100	STD	E24	PL-E3	TinO2n-1	SCUFF/DURATION CYCLES	45.6	42.2	0	13	PASS	OK
4	102.5 100	STD	E22	PL-E1	Tin-2 Cr2O2n-1	SCUFF/DURATION CYCLES	60.2	43.3	0.5	10.4	NO-PASS	TOP RING OF PISTON NO. 4 SHOWS PERIPHERAL EROSION AND BREAKAGE
5	102.5 100	STD	E23	PL-E2 (2ND BATCH)	(Ti,Mo)(C,N)+ Ni+Mo	SCUFF/DURATION CYCLES	66.4	48.3	1.4	17.8	PASS	OK
6	102.5 100	EPC-49	E11	PL72	Mo+Mo2C+NiCr	SCUFF/DURATION CYCLES	49.4	20.9	2.1	14.7	PASS	OK
7	102.5 100	EPC-49	E24	PL-E3	TinO2n-1	SCUFF/DURATION CYCLES	50.6	20.5	2.2	14.7	PASS	OK

Table 11. Main Test Results

In figure 27 the difference between measurements are presented. These values are a combination of the wear and deformation that the components have suffered during the test.

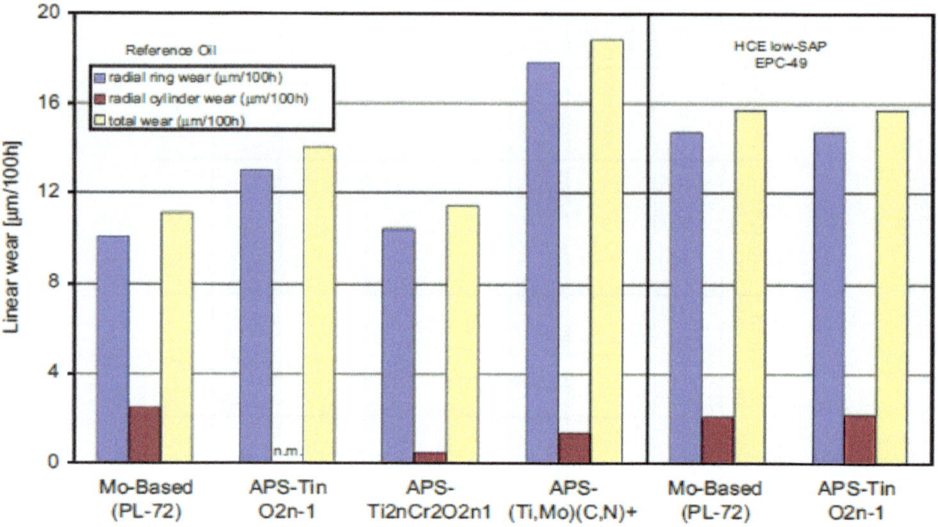

Figure 27. Testing of piston rings in a turbo diesel engine testing arrangement.

6. Conclusions

The main conclusions of the tests are summarized as follows:

- Good wear and anti-scuffing properties of Titanium Oxide followed by the (Ti, Mo)(C,N)+23NiMo coatings.
- Wear resistance of TinO2n_1 coating is similar to standard Mo-based coating for both standard and triboreactive oil.

The best extreme pressure properties (in tribological tests) were found for the Titanium Chromium Oxide coating but it was detached from the iron substrate during the engine tests.

Wear results with triboreactive oil EPC-49 are similar to the standard oil, and oil consumption results were reduced by 45%.

Author details

Amaya Igartua and Xana Fdez-Pérez
Ik4 Tekniker, Spain

Iñaki Illarramendi
CIE- Tarabusi, Spain

Rolf Luther and Jürgen Rausch
FUCHS, Germany

Mathias Woydt
Bam, Germany

Acknowledgement

The authors express their gratitude to the European Union for the financial support given to this work through the GROWTH Project no. GRD2-2001-50119, Contract no. G3RD-CT-2002-00796-EREBIO and the Spanish Minister of Science and Technology reference MAT2002-11137-E. The authors also would like to thank all the remaining partners from EREBIO project FALEX, CRSA, GUASCOR, RENAULT, FERESPE, BRITO, and INEGI for giving their approval for this publication.

Also the authors give their gratitude to the European Union for the financial support through the seventh framework, Theme 7, Sustainable Surface Transport, POWERtrain for Future Light-duty vehicles, Grant agreement no.: 234032

7. References

Igartua, A.; Barriga, J., Aranzabe, A.; (2005) *Biodegradable Lubricants.* Virtual Tribology Institute Edition, ISBN 83-70204-418-X..

Schmidt, R.; Klingenberg, G., Woydt, M., (2006) *Tribological, thermophysical and viscosinetric properties of lubricants interacting with triboactive materials.* BAM Research Report 277.

Fitamen, E.; Tiquet, L., Woydt, M., (2006) *In: ASTM D02 Symposium on Automotive Lubricants—Testing and Additive Development*; 3–5 December 2006.

Desplanches, G.; Criqui, B., Linneman, T., Woydt, M., (2006) *Plenary paper al 15th International Colloquium Tribology*, TAE Esslingen; 17–19 January 2006.

Rakopoulos, CD.; Antonopoulos, KA., Rakopoulos, DC., (2006) *Energy conversion and Management* 2006;47(11–12):1550–73.

Ramadhas, AS. ; Muraleedharan, C., Jayaraj, S., (2003) *Renewable Energy 2003*; 30(12): 1789–800.

Miura, M.; (2004) *Journal of Japanese Society of Tribologists* 2004; 49(10):793–8.

Ulosoy, Y., Tekin, Y., Centinkaya, M., (2004) *Energy Sources* 2004; 26(10):927–32.

Megahed, OA.; Abadı´a RI, Nabil, D. (2004) *Energy Sources* 2004; 26(2):119–26.

Tanabe, H.; (2003) *Journal of Japanese Society of Tribologists* 2003; 48(6):442–6.

Hashimoto, T., (2003) *Journal of Japanese Society of Tribologists* 2003; 48(4):278–82.

Chen, CI.; (2003) *Tribology Letters* 2003; 14(2):83–90.

Gmble, RJ.; Priest M., Taylor CM., (2003) *Tribology letters* 2003; 14(2):147–56.

Section 2

Electronics

Chapter 5

Batteries Charging Systems for Electric and Plug-In Hybrid Electric Vehicles

Vítor Monteiro, Henrique Gonçalves, João C. Ferreira and João L. Afonso

Additional information is available at the end of the chapter

1. Introduction

Nowadays, energy efficiency is a top priority, boosted by a major concern with climatic changes and by the soaring oil prices in countries that have a large dependency on imported fossil fuels. A great part of the oil consumption is currently allocated to the transportation sector and a large portion of that is used by road vehicles. According to the international energy outlook report, the transportation sector is going to increase its share in world's total oil consumption by up to 55% by 2030 [1]. Aiming an improvement of energy efficiency, a revolution in the transportation sector is being done. The bet is in the electric mobility, mostly supported by the technological developments in different areas, as power electronics, mechanics, and information systems.

Different types of Electric Vehicles (EVs) are being developed nowadays as alternative to the Internal Combustion Engines (ICE) vehicles [2][3], namely, Battery Electric Vehicles (BEV), Plug-in Hybrid Electric Vehicles (PHEV), in its different configurations [3], and Fuel-Cell Electric Vehicles (FCEV). This chapter presents batteries charging systems for Electric and Plug-in Hybrid Electric Vehicles. To simplify the reading and to contribute to a simple understanding, from now on, in this chapter, it will be used the terminology of Electric Vehicle (EV) to define these two types of vehicles.

EVs are increasingly popular, as demonstrated by the numerous vehicles recently made available in the market by almost all automakers. The main energy storage systems of these vehicles are the electrochemical batteries, the ultracapacitors and the full-cells. However, taking into account nowadays limits of energy storage of those technologies, the vehicles have limited range autonomy. Different energy storage systems configurations can be implemented [3][4][5], however, the electrochemical batteries still are the most used technology to store energy. Nevertheless, they are usually used in conjunction with ultracapacitors to store energy during transient moments, as during the vehicle regenerative braking. Actually, the ultracapacitors are used in this way to receive a significant amount of energy in a short time, and to provide this energy to the next acceleration, or to help charging the batteries.

The electrical power grids were not designed for this new type of load, which corresponds to the batteries charging systems of EVs, therefore the impact caused by the proliferation of EVs cannot be neglected [6]. The challenge is to rebuild the electrical power grids, as early as possible, as "smarter" as possible, and the most environmentally friendly as possible. To achieve these targets arise the Smart Grids, which are not characterized as a single technology or device, but rather as a vision of a distributed electrical system, supported by reference technologies, as integrated communications, Power Electronics devices, Energy Storage Systems (ESS), and Advanced Metering Infrastructures (AMI). The Smart Grids intend to reduce the energy costs, and simultaneously to achieve a sustainable balance between production and consumption, increasing the reliability of the power grids and the power quality of the electrical energy delivered to the loads.

As aforementioned, EVs represent a new type of load that introduces new problems, but that also brings new possibilities of actuation. The problems arise from the possibility of occurring simultaneous charging of a large number of vehicles, which can overload the power grid, and from the effects of non-sinusoidal current consumption of the batteries charging systems. Among the opportunities, stands the fact that these vehicles have enormous potential to regulate the consumption profile from the power grid, by smoothing the natural intermittency of the renewable energy sources, and ensuring the power grid stability in terms of voltage and frequency, if they allow collaboration with the electrical power grid to store and deliver energy of the batteries in parked vehicles. One factor which suggests that such benefits may exist relates to the fact that private vehicles are parked on average 93-96% of their lifetime, during which time each vehicle represents an idle asset [7]. So, the energy stored in EVs' batteries may be suitable for providing regulation services, spinning reserves and peak power demand. This interactivity between the vehicles and the power grid is expected to be one of the key technologies in the future of the Smart Grids and batteries charging systems, and is called Vehicle-to-Grid (V2G). The new paradigms of Smart Grids and V2G bring a profound change to the present systems. In [8] is presented one approach to the future of the power grids focusing the impact of PHEVs in Smart Grids, and in [9] is presented a concrete case of a technology to integrate EVs with Smart Grids.

To solve the aforementioned problems different approaches can be taken into account, among them can be implemented a coordinated charging of the EVs, or a regulation of the required power of the vehicles according to the power grid capabilities. Different authors present studies about how the charging systems affect the distribution power grid, and how they contribute to the degradation of the power quality. In [10] is presented a comparative study of the performance of two types of batteries charging systems, and in [11] is studied the effect of EVs batteries charging systems on a substation transformer that supplies commercial, residential, and industrial loads during a peak of consumption in a summer day. A report published by the California Energy Commission [12] presents a study about the impact of residential EVs batteries charging systems. It shows that, for the use of the GM EV1 vehicle, the Total Harmonic Distortion (THD) of the current presents a variation from 3% (at the beginning of charging, with a unitary power factor) to 28.11% (at the end of the charging, with a power factor equal to 0.96). So, it is clear that the simultaneous use of a great number of EVs batteries charging systems connected to the electrical distribution grid can cause a significant degradation of the electrical power quality. In the particular case of Portugal, in [13] is analyzed the impact of PHEVs in the electric utility system, where it is approached a Portuguese consumption profile. These studies have only considered the Grid-to-Vehicle operation (G2V) since it is expected that this is the first scenario that will be found. The impacts of the batteries charging rates of EVs on the Smart Grid distribution systems are approached in [14]. For this goal were compared different charging rates, during different charging periods along a day. It was also considered the existing system load profiles, and it was evaluated the overall performance of the electrical distribution system. The integration of EVs in the power grid is approached in several papers in the literature (e.g., [15]), as well as the integration of PHEVs in different parts around the world, like China, [16]. In [17] is analyzed the impact of EVs in an isolated electrical power grid.

The power quality is an important issue in order to ensure the proper functioning of the power grid system and the loads connected to it. These requirements should be a characteristic of both parts of the system: the energy supplied by the power grid, as well as the energy consumed by the equipment connected to the grid [18]. Besides the harmonics, other power quality problems, as inter-harmonics, noise (electromagnetic interference), momentary interruptions, sags, swells, flicker, notches, and transients can also occur [19][20]. But in what regards the integration of the electric mobility in a Smart Grid, the degradation of the power quality is mainly caused by the non-linear current consumption of the batteries charging systems. This is reflected in the THD of the consumed current and also in the voltage THD, due to the line impedance [21]. As a way to mitigate these problems, the EVs' batteries charging systems should have sinusoidal current consumption and unitary power factor. This is true for both home chargers and public charging stations.

2. Smart Grid: Electric mobility integration

It is predictable that in the near future, in a real full scale Smart Grid scenario, the power grid should meet the increasing demand of energy in a reliable and efficient way, maintaining the required stability and interfacing renewable energy resources, as a large network of microgrids.

Figure 1 shows a draft of a scenario for a micro Smart Grid with: a microgeneration power station with solar photovoltaic panels and micro wind turbines (which produce energy); some EVs with G2V and V2G capabilities (which can receive or provide electrical energy);

and Energy Storage Systems (which, like the EVs, can receive or provide electrical energy). Beyond the flow of energy between the parts, there is also the sharing of information, controlled by a Collaborative Broker [22]. In this figure are also shown the blocks of the Maximum Power Point Trackers (MPPTs) (for the micro solar photovoltaic panels and micro wind turbines), and the blocks of the AC-DC and DC-AC converters to adjust the levels of the voltages and the currents between both sides [23].

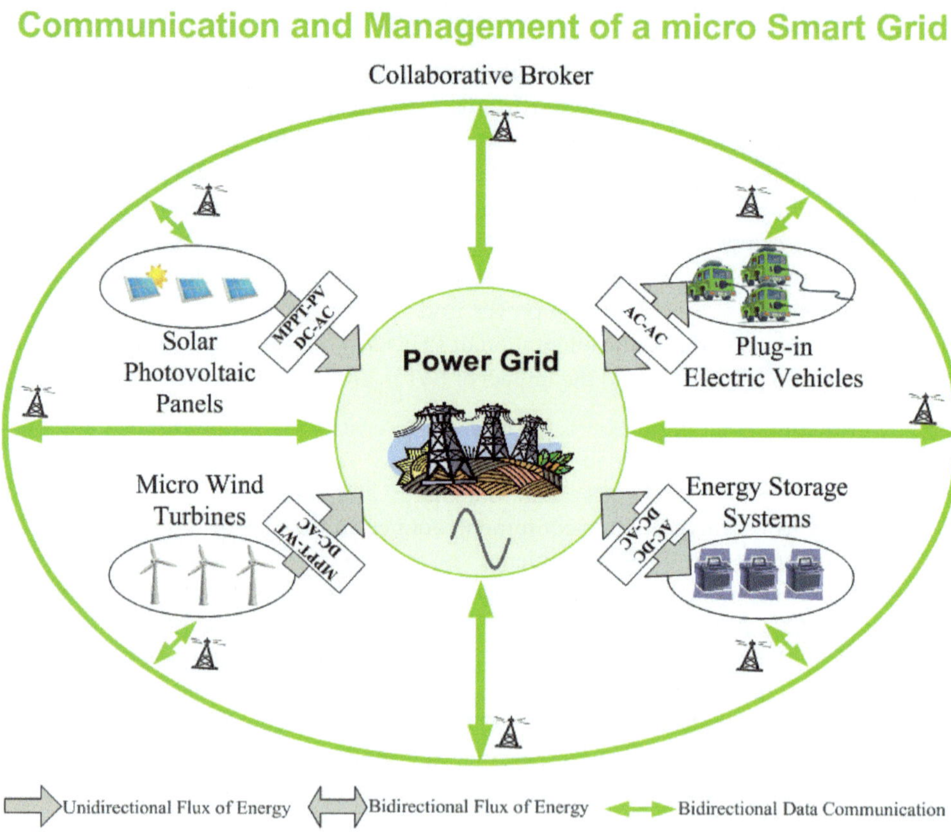

Figure 1. Scenario of a micro Smart Grid.

Such power grid scenario, with the EVs smart charging systems, will allow the communication of the vehicles with the local utilities to ensure that the batteries are charged when the electricity is cheapest and the impact of the charging systems on the grid is smallest. The use of computerized charging stations which constantly monitor the EV charging process, in order to optimize the charging rate, will be of extremely important to preserve the batteries lifespan. In Figure 2 is shown in detail the integration of EVs (in a typical charging park) with microgeneration renewable energy sources (solar photovoltaic panels and micro wind turbines), and Energy Storage Systems (ESS), in a Smart Grid context.

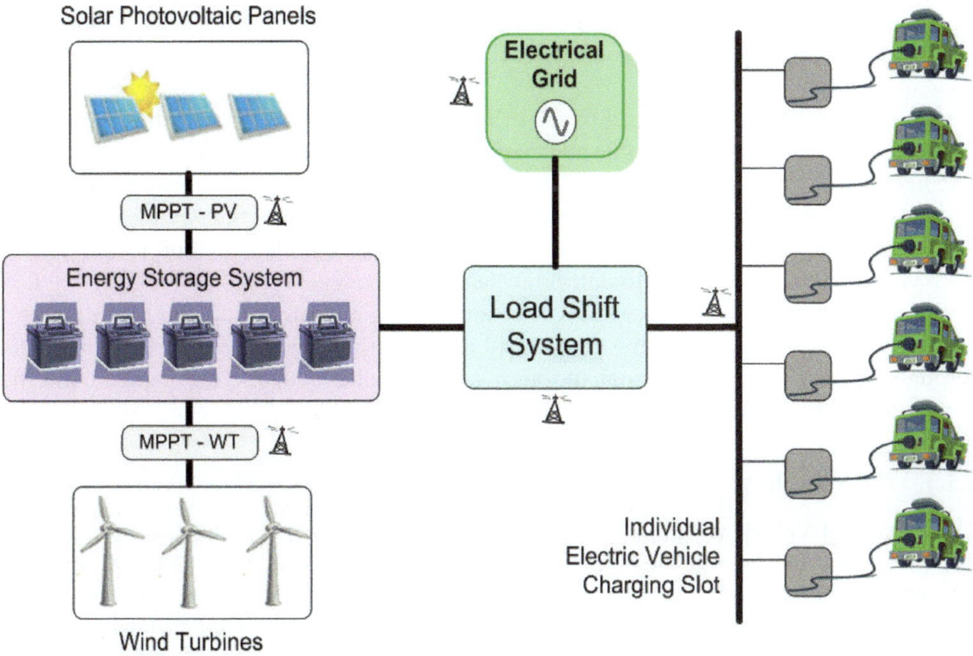

Figure 2. Integration of Electric Vehicles, renewable energy sources (solar photovoltaic panels and micro wind turbines), and Energy Storage Systems (ESS), in a Smart Grid context.

2.1. Interaction modes with Electric Vehicles

As mentioned before, enormous advantages arise from the interaction of EVs with the power grid. Focusing the interaction of EVs with the different systems where they can be connected, several concepts can be defined.

2.1.1. Grid-to-Vehicle

The concept of Grid-to-Vehicle (G2V) is the simplest process of integration of the EVs batteries charging system with the power grid. It is not required any communication between both systems and only exists energy flow from the power grid to the EVs. Nowadays, this is the most common (and almost unique) batteries charging process for EVs, and it will be the first approach to the massive integration of these vehicles.

2.1.2. Vehicle-to-Grid

As defined by Mid-Atlantic Grid Interactive Cars Consortium (MAGICC) [24], Vehicle-to-Grid (V2G) technology utilizes the stored energy in the EVs batteries to contribute with electricity back to the electrical power grid, when the grid operators request it. This way the

EVs can receive energy during the excess of production and deliver it back to the power grid during the periods of great demand, balancing the energy production and consumption, and also stabilizing the intermittency production from renewable energy sources, improving their integration into the power grid. Thus, V2G is described as a system where EVs can communicate with the power grid to sell or to buy energy, in order to establish the power demand [25]. To make it possible the batteries charging system needs assistance of an intelligent and collaborative system to control both processes, and also to add extra functionalities, as finding the periods with cheaper prices to charge the batteries, to identify available charging slots in public areas, and to provide useful information to the drivers. Each vehicle should have two fundamental elements: a connection for the bidirectional flow of the electrical energy; and a logical connection for the communication and control in both sides. The control signal from the power grid operator to the vehicles can be implemented through a broadcast radio signal, through a cell phone network, through an internet connection, or through Power Line Communication (PLC) [26].

2.1.3. Vehicle-to-Home

The concept of Vehicle-to-Home (V2H) is similar to the V2G concept; however it can avoid the grid infrastructure and the electricity tariff problems associated with V2G, because the bidirectional flux of energy is between the vehicle and the house. Thereby, V2H can be used to manage and regulate the profile of electricity demand in a house, controlling the use of the loads and the stored energy available in the vehicle. It also can be used out of the power grid, in isolated electrical systems, and in conjunction with renewable energy sources, increasing their effectiveness.

2.1.4. Vehicle-to-Building

A specific version of V2G, denominated Vehicle-to-Building (V2B), is a concept that consists in using the stored energy in the batteries of EVs as an energy source of back-up to compensate the energy consumption profile in a commercial scale (e.g., in companies and shopping center parks).

2.2. Batteries charging process

As previously commented, and as demonstrated in [27], the load profile of the EVs batteries charging systems has high importance to the power grid management. Depending on the design of the electrical grid and the type of charging and discharging processes, EVs can be a problem or a benefit to the power grid. The batteries charging and discharging processes can be realized in two different ways: controlled and uncontrolled.

In the controlled way, the batteries are charged or discharged in accordance to the capability and the needs of the power grid, and in accordance with the conveniences of the vehicle´s owner, which is the main figure in this process. In this scenario of real time control, several parameters should be taken into account, namely, energy price (to sell or to buy) and batteries State-of-Charge (SoC) and State-of-Health (SoH - reflects the batteries degradation along their lifetime). In [28] is proposed an autonomous distributed V2G control scheme, and in [29] is presented a coordinated charging scheme for multiple PHEVs in a residential distribution grid. Some technical solutions for the integration of EVs in the electrical power grid can be found in [30].

In the uncontrolled way, as the name suggests, there is no control over the charging system. The only control that exists from the moment in which the vehicle is plugged to the end of the process, is the decision if the process of charging or discharging can start immediately or after a fixed time delay (controlled by the vehicle´s owner in accordance with his convenience). Since there is no control over the charging systems, peaks of power consumption can occur in the electrical power grid, during some periods in which exist a large number of EVs simultaneously charging their batteries. These power peaks can bring overload problems to the power grid. Similar problems can occur when the energy stored in the batteries is delivered back to the electrical power grid. Despite these problems, nowadays, the uncontrolled way is the most common charging procedure. As predicted in [31], in Portugal, the smart charging (in controlled way) will be a necessity in midterm, in order to prevent a large demand of energy peaks over the power grid.

2.3. Types of access

The batteries charging systems can be of two types: public chargers and residential chargers. Public chargers are an optimal solution to charge the batteries of the vehicles using energy from several sources of energy (as wind or sun) and can be deployed at strategic places around a town or city, like for example, at companies, public buildings and shopping centers parking lots. On the other hand, residential chargers are designed to deliver low power, in an efficient way, since in general they are used to make a complete charge of the batteries during long periods of time (slow charging). The main benefits of these two types of chargers are the comfort for the user, and the freedom of the user to charge the batteries when he wants, according with the best prices of energy. The main disadvantage is that, since each charging process is independent, the limit of overload of the electrical power grid can be easily reached.

2.4. Charger specifications

Currently, the majority of EVs are designed with on-board unidirectional batteries charging systems. Besides the on-board batteries charging systems, some vehicles allow the charge of their batteries with off-board chargers. An on-board batteries charging system refers to a charger implemented inside the vehicle. The user only has access to the input of the charging system. This type of charger is connected to the AC electrical grid voltage and is used to slowly charge the batteries – it is denominated as "slow charging". On the other hand, an off-board batteries charging system is implemented outside the vehicle. It is given access to the DC voltage of the batteries and is used to charge the batteries as fast as possible – it is denominated as "fast charging".

Regarding the way that the charger can be connected to the vehicle, there are two different approaches: conductive or inductive. The conductive batteries charging system is made through a physical contact between the vehicle and the power grid. In counterpart, with the inductive batteries charging system there is no physical contact between the vehicle and the power grid. Independently of the charger type, the interaction between the EVs and the power grid should comply with regulatory standards, as the International Electrotechnical Commission (IEC) norms (IEC 62196 and IEC 61851).

3. Electric vehicles batteries charging systems

As aforementioned some vehicles allow the charge of their batteries with off-board batteries charging systems, like public charging stations, but almost every vehicle is equipped with its own on-board batteries charging systems. This charging system is an AC-DC power circuit that must be controlled in order to respect the vehicles' batteries nominal characteristics to preserve their lifespan. Additionally, it should monitor the batteries during their operation to prevent damages during the charging or discharging processes. The AC-DC power circuit can be implemented with different topologies according to the characteristics desired for the system. Figure 3 illustrates graphically the main categories in which can be divided the different topologies of the batteries charging systems for EVs.

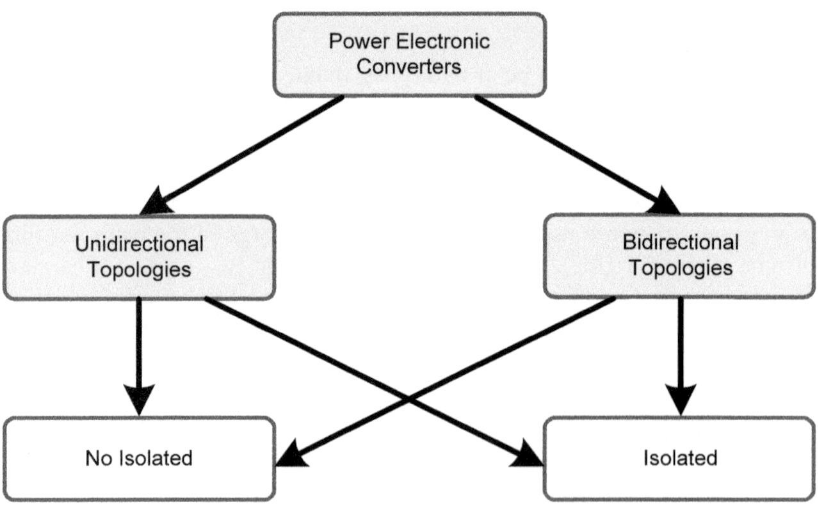

Figure 3. Main categories of AC-DC power converters topologies used in EVs batteries charging systems.

3.1. Power electronics circuits topologies

Typically, the power electronics circuit topology of batteries charging systems is formed by two power electronics converters: an AC-DC converter followed by a DC-DC converter. Both power converters can have different topologies and can be arranged together in different ways, with and without isolation between them. The AC-DC converter is used to rectify the AC voltage from de power grid to a DC voltage. The DC-DC converter is used to adapt the rectified voltage to a level of voltage compatible with the batteries' voltage and also to control the batteries charging process. In Figure 4 is shown the main AC-DC power converters topologies used to rectify the power grid AC voltage, and in Figure 5 is shown the main DC-DC power converters topologies used to control the batteries charging and discharging processes.

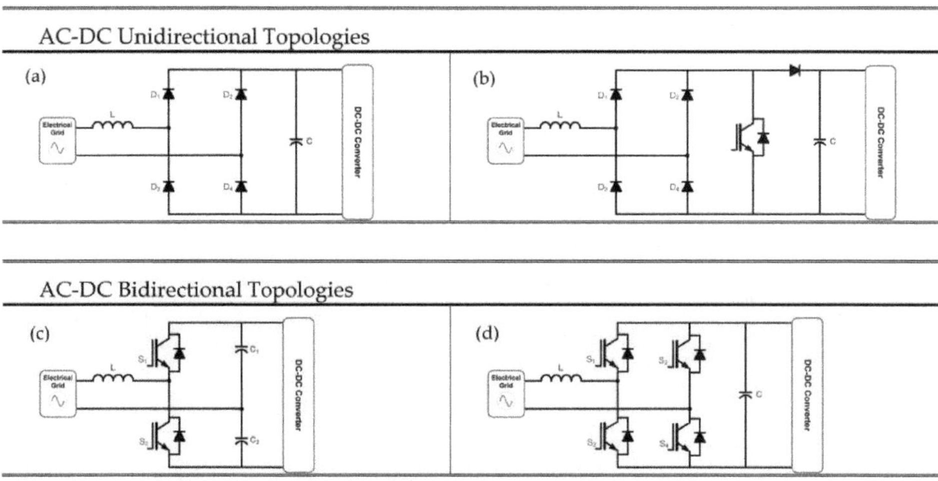

Figure 4. Main AC-DC power converters topologies used in EVs batteries charging systems: (a) unidirectional full-bridge non-controlled converter; (b) unidirectional full-bridge non-controlled converter followed by a boost converter; (c) bidirectional half-bridge full-controlled converter; (d) bidirectional full-bridge full-controlled converter.

The simplest AC-DC power converter uses diodes as rectifying components, as illustrated in Figure 4 (a). It is easy to implement, cheap to construct, and less susceptible to damages. However, the output voltage, and the consumed current are not controlled and consequently, the waveform of the consumed current is not sinusoidal, and so this type of converter contributes to the degradation of the electrical grid power quality. On the other hand, using power switching semiconductors is possible to control the waveform of the consumed current, as well as the output voltage. When compared with the AC-DC power converter with diodes, it has the disadvantages of having a more complex power electronics circuit and control system (which can be digital or analogue, and higher implementation and maintenance costs. Nevertheless, it has the advantage of the sinusoidal current consumption that does not degrade the power grid quality. The AC-DC power converter presented in Figure 4 (b) adds to the diodes rectifier a DC-DC boost converter that will make the Power Factor Correction (PFC). With this topology, it is possible to control the waveform and the power factor of the consumed current. Despite being a good alternative to control the waveform of the consumed current, it has the disadvantage of only operating in unidirectional mode (G2V). To make possible a bidirectional energy flow it is necessary a topology like the one presented in Figure 4 (c) and (d). The difference between these two converters is the number of power semiconductors and capacitors, and the resulting level of the output voltage. The level of the output voltage of the converter presented in Figure 4 (c) is the double of the converter presented in Figure 4 (d).

The most basic topology of DC-DC power converter used in batteries charging systems is the DC-DC buck converter. This topology, presented in Figure 5 (a), is not isolated and only allows the unidirectional operation (G2V). Combining the structure of this topology with the structure of a DC-DC boost converter it is possible to obtain a bidirectional topology, as shown in Figure 5 (c). During the batteries charging process, the energy flows from the power grid to the batteries (G2V), and the converter operates as a DC-DC buck converter. When the energy flows from the batteries to the power grid (V2G), the converter operates as a DC-DC boost converter. If it is wanted galvanic isolation it should be used high frequency transformers. In Figure 5 (b) is presented an isolated unidirectional topology of DC-DC converter, and in Figure 5 (d) is presented an isolated bidirectional topology of DC-DC converter. As it is noticeable, when compared with the non-isolated topologies, these isolated topologies use a greater number of power semiconductors, and besides, their control systems are more complex.

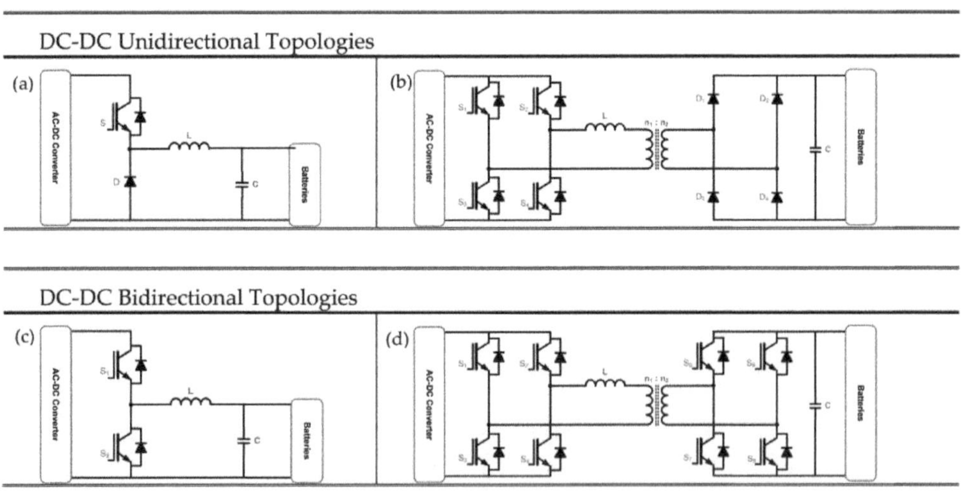

Figure 5. Main DC-DC power converters topologies used in EVs batteries charging systems: (a) unidirectional buck converter; (b) unidirectional isolated converter; (c) bidirectional buck boost converter; (d) bidirectional isolated converter.

The typical structure of an EV batteries charging system results from the combination of the AC-DC and DC-DC converters with the respective digital control system. For an appropriate control it should be measured the voltage and current in the power grid side, the DC link voltage, and the voltage and the current in the batteries. In Figure 6 is presented a block diagram with both converters, the digital control system and the points of measurement.

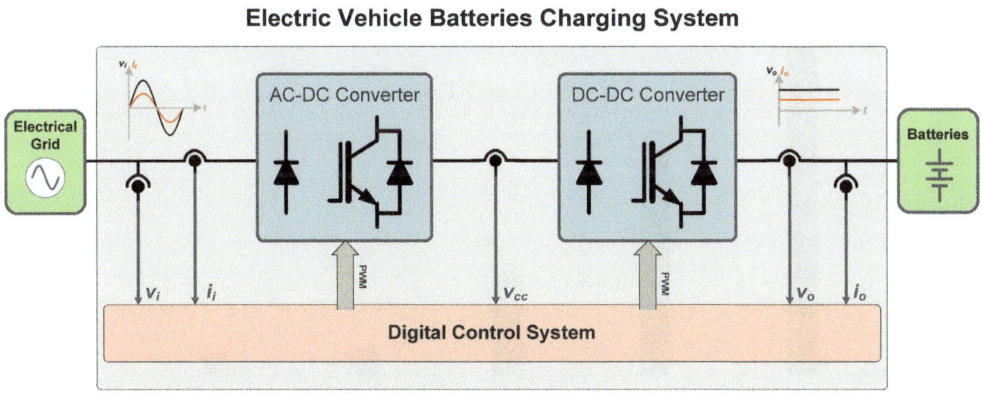

Figure 6. Structure of an Electric Vehicle batteries charging system.

3.2. Typical waveforms

For better understanding the influence of the AC-DC power converters topologies in the power grid quality, it should be analyzed their typical waveforms, with and without sinusoidal current consumption. The typical waveform of the current consumed by a conventional AC-DC power converter, without sinusoidal current consumption, presented in Figure 4 (a), is shown in Figure 7. It is visible that the current is not sinusoidal. Due to the line impedance the power grid voltage waveform is also affected. The harmonic spectrum of this current is shown in Figure 8. The Total Harmonic Distortion (THD) of the consumed current is 77.9%.

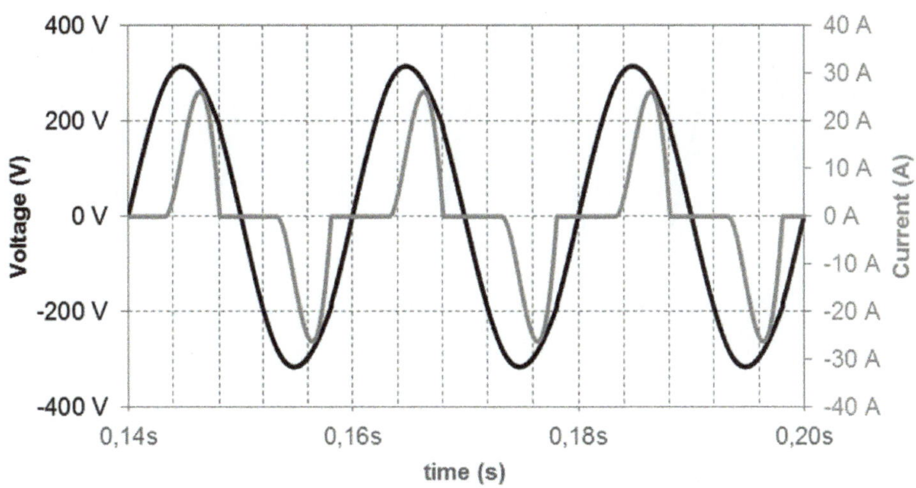

Figure 7. Typical waveforms of a conventional AC-DC power converter.

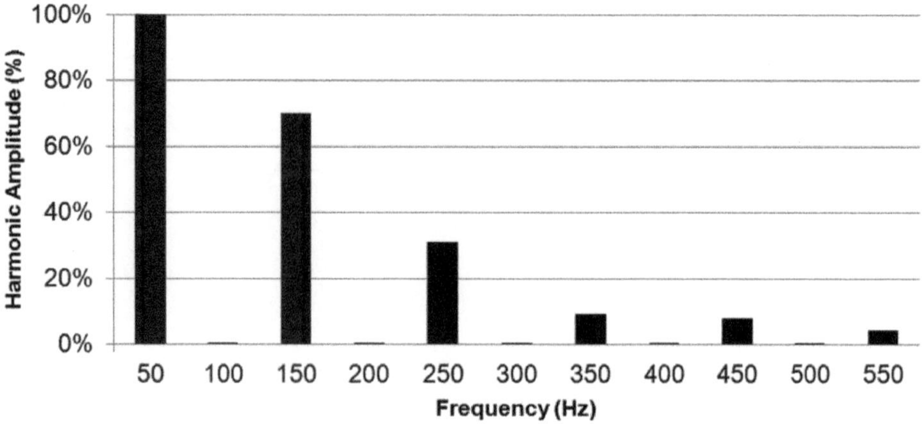

Figure 8. Harmonic spectrum of the current waveform of a conventional AC-DC power converter.

The typical waveform of the current consumed by AC-DC power converters with sinusoidal current consumption, presented in Figure 4 (b), (c), and (d), is shown in Figure 9. As illustrated, the current waveform is sinusoidal and with unitary power factor. The harmonic spectrum of this current is shown in Figure 10. The THD% of this consumed current is 0.4%.

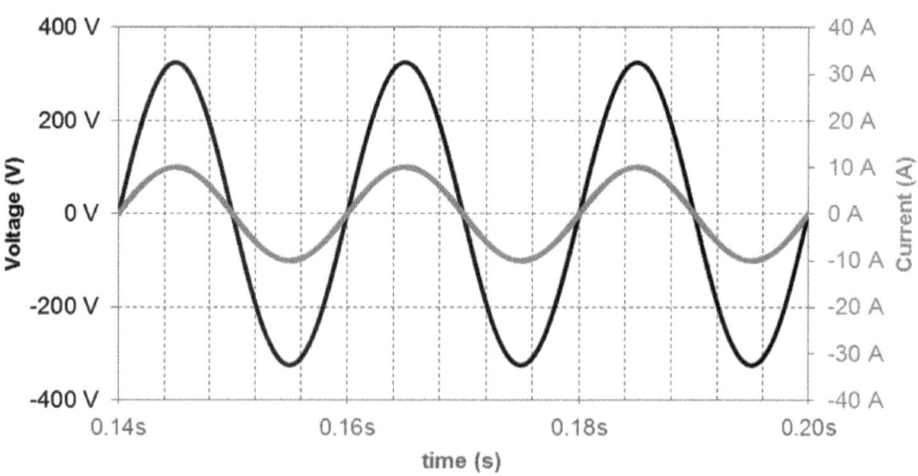

Figure 9. Typical current waveform of an AC-DC power converter with sinusoidal current consumption.

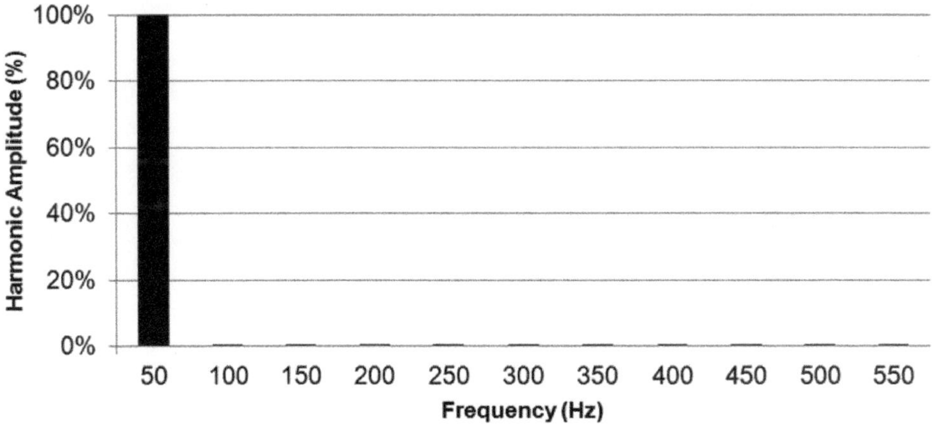

Figure 10. Harmonic spectrum of the current waveform of an AC-DC power converter with sinusoidal current consumption.

Taking into account the predictable increase in the utilization of EVs, the AC-DC power converters with sinusoidal current consumption are more appropriate to be implemented in the batteries charging systems of these vehicles. During the V2G process, when the energy stored in the batteries is delivered back to the power grid, it is also important to have control over the current. The typical current waveform during this process is presented in Figure 11. The harmonic spectrum of this current is shown in Figure 12. The THD% is 0.7%. As illustrated, the current is in phase opposition with the voltage of the power grid.

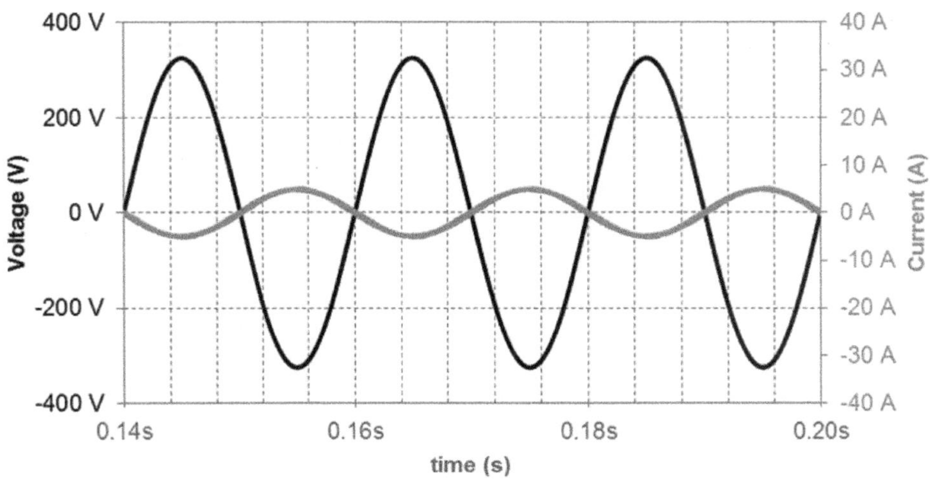

Figure 11. Typical current waveform of an AC-DC power elecronic converter with sinusoidal current, delivering energy back to the power grid.

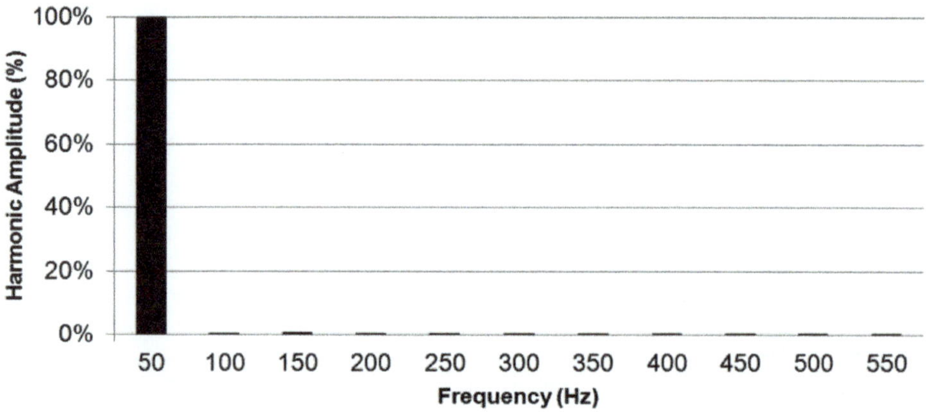

Figure 12. Harmonic spectrum of the typical current waveform of an AC-DC power converter with sinusoidal current, delivering energy back to the power grid.

4. Developed electric vehicle batteries charging system

It was developed a laboratory prototype of a 3 kW batteries charging system that works with sinusoidal current consumption and unitary power factor, and that allows the charging of the batteries with different algorithms: constant-voltage, constant-current, and constant-current followed by constant-voltage; in accordance with the State-of-Charge (SoC) level of the batteries and with their technology, namely, lithium, nickel, and lead-acid. Figure 13 shows the schematic of the developed batteries charging system.

This batteries charging system also allows bidirectional flow of energy between the power grid and the batteries, operating in both modes with sinusoidal current, and therefore, it can

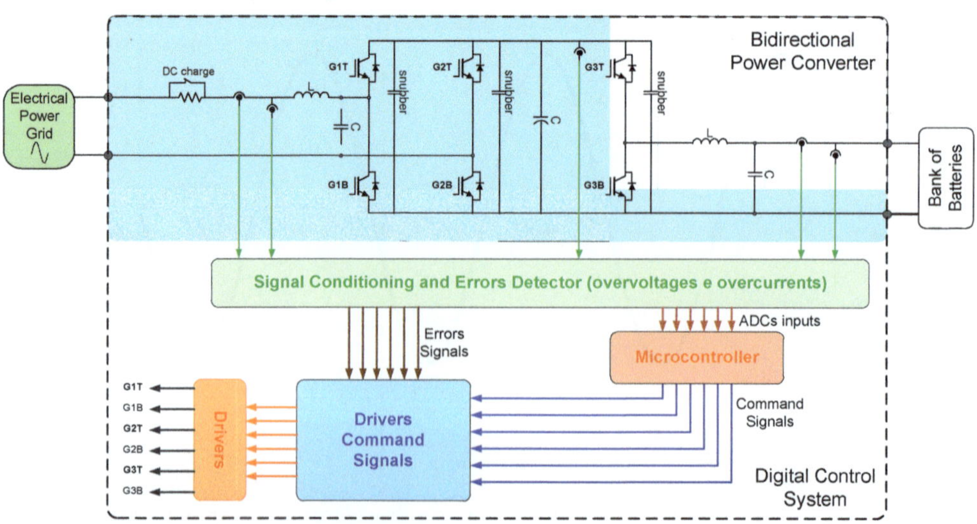

Figure 13. Schematic of the developed Electric Vehicle batteries charging system.

be considered as a smart charger. As illustrated in Figure 13, it is constituted by two main parts: the bidirectional power converter, which uses inductances, capacitors and IGBTs as switching power semiconductors; and the control system, that is constituted by the microcontroller, the signal conditioning circuit, the command drivers and the drivers.

4.1. Laboratory prototype

The laboratory prototype of the implemented smart EV batteries charging system is shown in Figure 14. In this figure can be seen the different parts of the developed smart charging system,

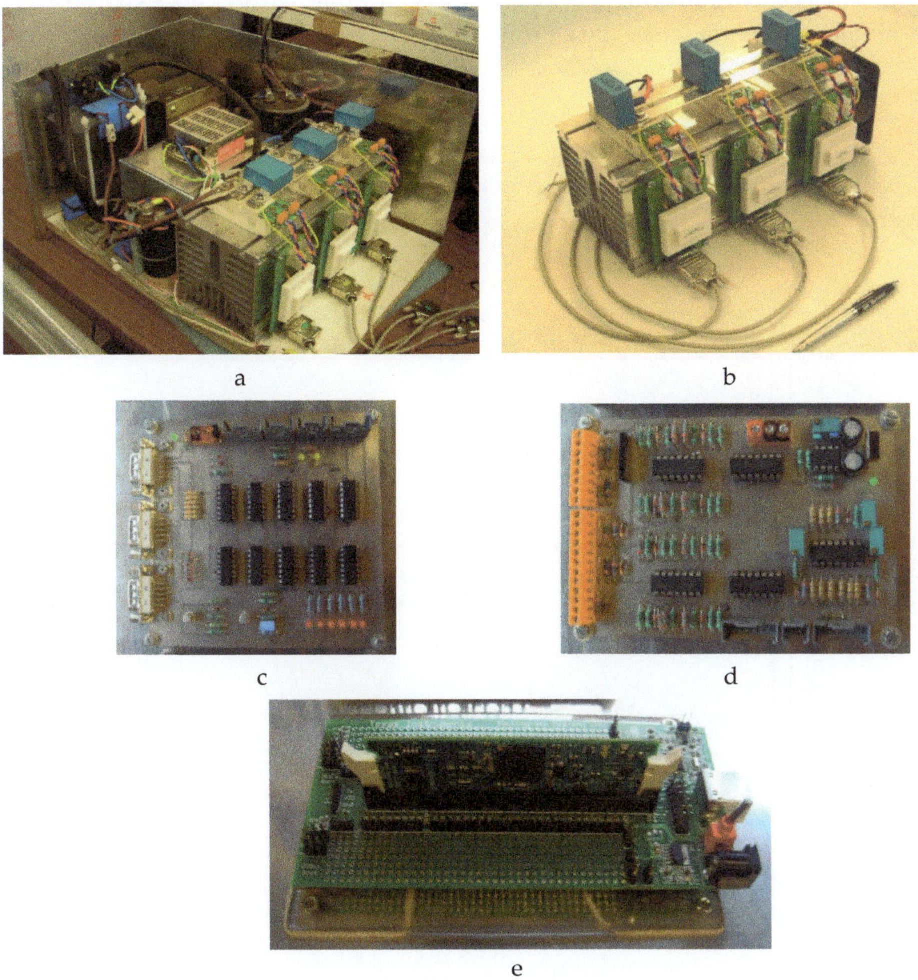

Figure 14. Developed smart charging system for Electric Vehicle batteries: (a) Charging system overview; (b) AC-DC and DC-DC Bidirectional power converters; (c) Command drivers board; (d) Signal conditioning and errors detector board; (e) Microcontroller DSP TMS320F28335 board.

namely: the bidirectional power converters, the IGBTs drivers, the command drivers board, the conditioning and errors detector board, and the microcontroller DSP TMS320F28335.

4.2. Experimental results

This laboratory prototype was tested for different conditions of operation. In Figure 15 are presented the power grid voltage and the current consumed in steady state by the batteries charging system. As it can be seen the consumed current is sinusoidal and in phase with the power grid voltage. Figure 16 shows the waveforms of the power grid voltage and current during the functioning as V2G. Since the vehicle is providing energy back to the power grid the current is in phase opposition to the voltage. The power quality specifications are fulfilled, and the current is sinusoidal.

Figure 17 presents the results obtained with the same batteries charging system, but operating as a conventional charger (without sinusoidal current consumption). It is possible to see that a conventional batteries charging system consumes highly distorted currents.

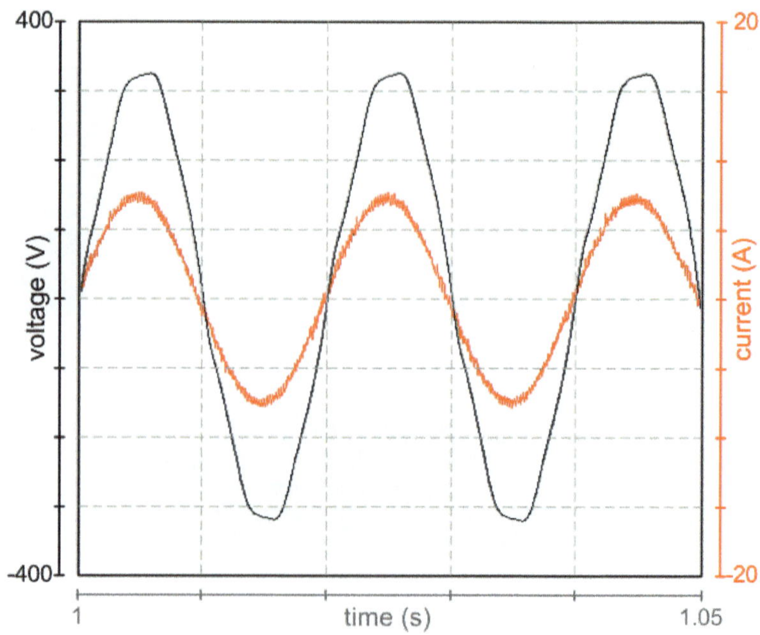

Figure 15. Experimental results: electrical grid current and voltage (in phase) during batteries charging - operation as G2V.

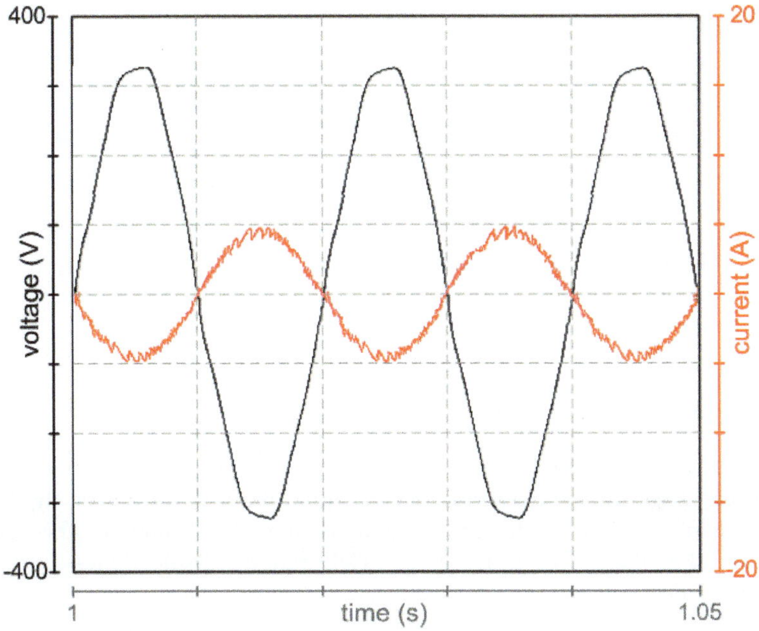

Figure 16. Experimental results: electrical grid current and voltage (in phase opposition) during batteries discharging - operation as V2G.

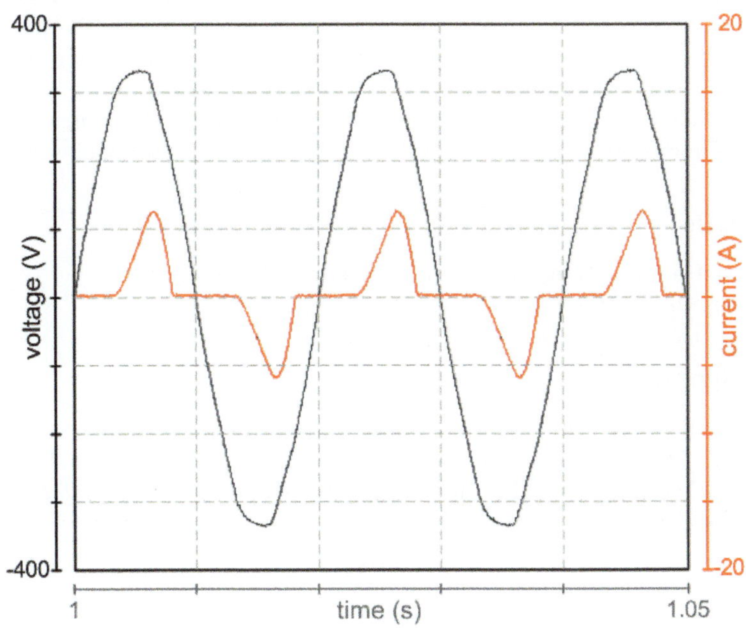

Figure 17. Experimental results: electrical grid current and voltage with a conventional charger during batteries charging - operation as G2V.

5. Conclusion

Economic and environmental reasons are making EVs a reality of nowadays. The main energy storage elements used in these vehicles are batteries of different technologies. Batteries need to be charged and the more common source for that is the power grid. However, the spread use of EVs will bring consequences to the power grid, mainly in terms of load management and electric power quality, which are associated to the batteries charging systems. Many developments in the area of the batteries charging systems are being made, with the development of new topologies and control strategies. In parallel, the development of Smart Grids and the spread of micro renewable energy production systems have created a new paradigm in power grids.

This chapter assessed the electric mobility integration in the Smart Grid context, focusing different approaches to the operation of EVs batteries charging systems, their different topologies and features, modes of operation, typical waveforms, and impact in the power grid in terms of power quality. It became apparent that if the vehicles charging systems consume current with sinusoidal waveform and with unitary power factor, the THD of the currents in the power grid is drastically reduced, the RMS values of the currents are minimized, and thus the power grid voltage's THD and amplitude are less affected by the operation of the EVs batteries charging systems.

Also, in this chapter it was presented a laboratory prototype of a bidirectional EV batteries charging system, and shown some experimental results, which allows mitigating the power quality degradation of the power grid. During the batteries charging process, the voltage and the current in the batteries are controlled in order to maximize their lifespan, and at the same time consuming from the power grid sinusoidal current with unitary power factor. This batteries charging system also permits to deliver back to the power grid part of the energy stored in the batteries, which can be, in the near future, an interesting solution during short periods of time, when occur peaks of energy demand in the power grid.

Author details

Vítor Monteiro, Henrique Gonçalves, João C. Ferreira and João L. Afonso
Centro Algoritmi, University of Minho,Guimarães, Portugal

Acknowledgement

This work is financed by FEDER Funds, through the Operational Program for Competitiveness Factors – COMPETE, and by National Funds through FCT – Foundation for Science and Technology, under the projects FCOMP-01-0124-FEDER-022674 and MIT-PT/EDAM-SMS/0030/2008.

6. References

[1] "International Energy Outlook 2009," Energy Information Administration Office of Integrated Analysis and Forecasting U.S. Department of Energy Washington DC, May 2009.

[2] A. Khaligh, Z. Li, "Battery, Ultracapacitor, Fuel Cell and Hybrid Energy Storage Systems for Electric, Hybrid Electric, Fuel Cell, and Plug-In Hybrid Electric Vehicles: State-of-the-Art," IEEE Transactions on Vehicular Technology, vol.59, issue:6, pp.2806-2814, 2010.
[3] C.C. Chan, A. Bouscayrol, K. Chen, "Electric, Hybrid, and Fuel-Cell Vehicles: Architectures and Modeling," IEEE Transactions on Vehicular Technology, vol.59, issue:2, pp.589-598, 2010.
[4] A.I. Antoniou, J. Komyathy, J. Bench, A. Emadi, "Modeling and Simulation of Various Hybrid-Electric Configurations of the High-Mobility Multipurpose Wheeled Vehicle (HMMWV)," IEEE Transactions on Vehicular Technology, vol.56, issue:2, pp.459-465, 2010.
[5] S. Onoda, S. Lukic, A. NAsiri, and A. Emadi, A PSIM-based modeling tool for conventional, electric, and hybrid electric vehicles studies, IEEE 56th Vehicular Technology Conference Proceedings. Year 2002.
[6] João C. Ferreira, P. Trigo, A.R. Silva, H. Coelho, João L. Afonso, "Simulation of Electrical Distributed Energy Resources for Electrical Vehicles Charging Process Strategy," Published by IEEE Computer Magazine - dedicated to the Second Brazilian Workshop on Social Simulation (BWSS 2010), pp. 82-89, São Paulo – Brazil, 24-25 October, 2010.
[7] H. Turton, F. Moura, "Vehicle-to-Grid Systems for Sustainable Development: An Integrated Energy Analysis, " Technological Forecasting and Social Change, vol.75, issue:8, pp.1091-1108, 2008.
[8] A. Ipakchi, F. Albuyeh, "Grid of the Future," IEEE Power and Energy Magazine, vol.7, issue:2, pp.52-62, 2009.
[9] S.S. Hernandez, P.P. Galindo, A.Q. Lopez, "EPV PROJECT. Technology to integrate EV inside smart grids," 7th IEEE International Conference on the European Energy Market (EEM), pp.1-6, 2010.
[10] M.M. Morcos, N.G. Dillman, C.R. Mersman, "Battery Chargers for Electric Vehicles," IEEE Power Engineering Review, vol.20, issue:11, pp.8-11, 2000.
[11] P.T. Staats, W.M. Grady, A. Arapostathis, R.S. Thallam, "A Procedure for Derating a Substation Transformer in the Presence of Widespread Electric Vehicle Battery Charging," IEEE Transactions on Power Delivery, vol.12, issue:4, pp.1562-1568, 1997.
[12] Frank Lambert, "Secondary Distribution Impacts of Residential Electric Vehicle Charging," California Energy Commission, May 2000.
[13] C. Camus, C.M. Silva, T.L. Farias, J. Esteves, "Impact of Plug-in Hybrid Electric Vehicles in the Portuguese Electric Utility System," IEEE POWERENG International Conference on Power Engineering, Energy and Electrical Drives, pp.285-290, 2009.
[14] A.S. Masoum, S. Deilami, P.S. Moses, A. Abu-Siada, "Impacts of Battery Charging Rates of Plug-in Electric Vehicle on Smart Grid Distribution Systems," IEEE PES Innovative Smart Grid Technologies Conference Europe, pp.1-6, 2010.
[15] J.A.P. Lopes, F.J. Soares, P.M.R. Almeida, "Integration of Electric Vehicles in the Electric Power System," Proceedings of the IEEE, vol.99, issue:1, pp.168-183, 2011.
[16] Y. Song, X. Yang, Z. u, "Integration of Plug-in Hybrid and Electric Vehicles Experience from China," IEEE PES Power and Energy Society General Meeting, pp.1-6, 2010.
[17] P. Kadurek, C. Ioakimidis, P. Ferrao, "Electric Vehicles and their Impact to the Electric Grid in isolated systems," IEEE POWERENG Power Engineering, Energy and Electrical Drives, pp.49-54, 2009.

[18] M. Bollen, J. Zhong, F. Zavoda, J. Meyer, A. McEachern, F. Lopez, "Power Quality aspects of Smart Grids," ICREPQ'10 International Conference on Renewables Energies and Power Quality, pp.1-6, 2010.

[19] R. Alves, D. Gonçalves, J.G. Pinto, J.Batista, João L. Afonso, "Development of an Electrical Power Quality Monitor Based on a PC", IECON 2009 - The 35th Annual Conference of the IEEE Industrial Electronics Society, Porto, Portugal, 3 5 November, 2009.

[20] João L. Afonso, Carlos Couto, Júlio Martins, "Active Filters with Control Based on the p-q Theory", IEEE Industrial Electronics Society Newsletter, vol. 47, issue:3, pp.5 10, September, 2000.

[21] V. Monteiro, H. Gonçalves, João L. Afonso, "Impact of Electric Vehicles on Power Quality in a Smart Grid Context", 11th IEEE International Conference on Electrical Power Quality and Utilization (EPQU), Lisbon-Portugal, 17-19 October, 2011.

[22] João C. Ferreira, Alberto R. Silva, V. Monteiro, João L. Afonso, "Collaborative Broker for Distributed Energy Resources", International Symposium on Computational Intelligence for Engineering Systems (ISCIES), ISEC, Coimbra-Portugal, 16-18 November, 2011.

[23] L.G.B. Rolim, A. Ortiz, M. Aredes, R. Pregitzer, J.G. Pinto, J.L. Afonso, "Custom Power Interfaces for Renewable Energy Sources", Proceedings of ISIE 2007- 2007 IEEE International Symposium on Industrial Electronics, Vigo, Spain, 4-7 June, 2007.

[24] MAGICC - Mid-Atlantic Grid Interactive Cars Consortium, "MAGICC Brochure" [Online]. Available: http://www.magicconsortium.org

[25] V. Monteiro, João C. Ferreira, J.G. Pinto, D. Pedrosa, João L. Afonso, "iV2G Charging Platform" 13th International IEEE Conference on Intelligent Transportation Systems (ITSC), pp. 409-414, Madeira-Portugal, 19-22 September, 2010.

[26] W. Kempton, J. Tomic, "Vehicle-to-Grid Power Fundamentals: Calculating Capacity and net Revenue," Journal of Power Sources, vol.144, issue:1, pp.268-279, 2005.

[27] S. Shahidinejad, S. Filizadeh, E. Bibeau, "Profile of Charging Load on the Grid Due to Plug-in Vehicles," IEEE Transactions on Smart Grid, vol.PP, issue:99, pp.1, 2011.

[28] Y. Ota, H. Taniguchi, T. Nakajima, K.M. Liyanage, J. Baba, A. Yokoyama, "Autonomous Distributed V2G (Vehicle-to-Grid) considering Charging Request and Battery Condition," IEEE PES Innovative Smart Grid Technologies Conference Europe (ISGT Europe), pp.1-6, 2010.

[29] K. Clement, E. Haesen, J. Driesen, "Coordinated Charging of Multiple PHEV in Residential Distribution Grids," IEEE PES Power Systems Conference and Exposition, pp.1-7, 2009.

[30] R.J. Rei, F.J. Soares, P.M.R. Almeida, J.A. Pecas Lopes, "Grid Interactive Charging Control for Plug-in Electric Vehicles," 13th International IEEE Conference on Intelligent Transportation Systems, pp.386-391, 2010.

[31] R. Freire, J. Delgado, J.M. Santos, A.T. Almeida, "Integration of Renewable Energy Generation with EV Charging Strategies to Optimize Grid Load Balancing," 13th International IEEE Conference on Intelligent Transportation Systems, pp. 392-396, 2010.

Chapter 6

Antennas for Automobiles

Niels Koch

Additional information is available at the end of the chapter

1. Introduction

In recent years, we spent more and more time in our cars. So it became obvious to implement Car Entertainment Systems into the car for comfort and driver information. Car entertainment began with AM-reception on short wave bands. FM-tuners on VHF bands followed soon, with stereo sound, cassette players and CD-players to entertain passengers. Today we know a number of different analog- and digital broadcasting systems, such as DAB, DMB, DRM, DVB, as well as satellite broadcasting systems such as SDARS.

For driver information, modern navigation systems not only help to find the most efficient route, but also give an overview of traffic situation. Car-to-Car Communication and Car-to-Infrastructure Communication is currently one of the most popular field of researches for efficient car information systems.

All of these systems have in common that they are wireless systems, hence a number of antennas must be used satisfying all different services in different frequency bands. In modern cars we find up to 24 antennas placed on the vehicle and inside the vehicle. In the next years the number will even rise.

This Chapter is structured into different subsections. At first we set the requirements for vehicular antennas. Then we search for locations where to place the antenna for optimal reception and for which service and wireless system the best antenna technology is selected. All this is fundamentally supported by best-practice examples, including how to beat fading effects.

2. Requirements for vehicular antennas

In order to have a good reception in a vehicle, some prerequisites must be fulfilled.

First of all, and this seems very obvious, the antenna must receive from any direction around the car. If this requirement cannot be satisfied with one antenna only, then an antenna array (two or more antennas) shall be considered.

In general, an antenna shall be as high over ground on the vehicle as possible. The higher the antenna is placed over ground, the better is can receive and transmit.

Then the antenna must be integrated into the car easily. The distance to the receiver shall be not too far so that received signals are not extra attenuated before they are used. The surrounding of the antenna may influence the antenna performance severely. Hence the materials and distances around the antenna must be considered. As a rule of thumb, a box of 3 times antenna size around the antenna shall be unobstructed. That means for VHF-antennas with 75 cm length, this requirement can never be fulfilled on regular cars, but can be easily achieved with Telephone or GPS-antennas above 1 GHz.

The engine generates spurious noise which can disturb reception, therefore the antenna shall be placed as far away from the engine as possible, but taking all other requirements mentioned before into account. So in total there will be a trade-off between height over ground, omnidirectional reception and reducing spurious emissions influences.

Sometimes the polarization of the antenna is of some importance, as some signals are transmitted specially polarized, e.g. SDARS is left hand circular polarized.

Summary of requirements for ideal antenna placements

- Antennas must be high above ground and receive from all azimuth directions
- Antennas must be unobstructed, >3*size
- Minimum coupling with other metallic structures or antennas
- Distance to receiver (cable-length) short
- Distance to spurious emissions as large as possible
- Some antenna types require large ground plane, either galvanic connected or coupled
- Polarization of the transmitted service and antenna shall be considered

3. Overview on vehicular antenna placements

Watching different cars on the road in terms on antenna placements, it seems that there are a number of placements to be found.

Figure 1 displays the best practices to place antennas to vehicles.

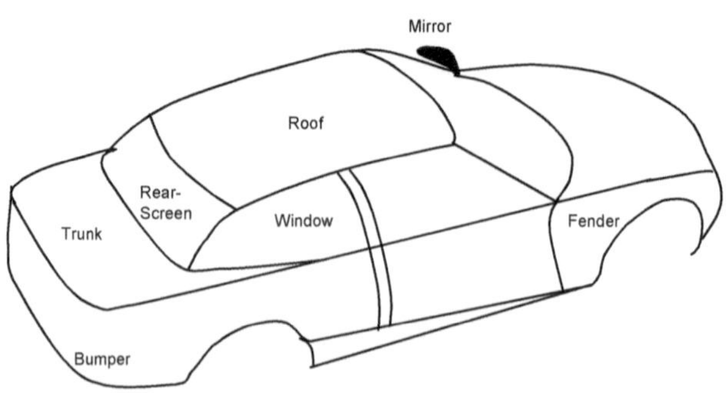

Figure 1. Suitable antenna placement on a regular passenger car

3.1. Roof

Most car manufacturers use the roof to place an antenna. This has an obvious reason, because the roof of a car is the high above ground and unobstructed. This provides a good reception into every direction. Mostly, omni directional reception is required anyway, so placing the antenna on the roof is one of the best options for most of the vehicles, except convertibles.

3.2. Spoiler

Some sportive cars exhibit a spoiler for better down force on higher speeds. When the spoiler is made of plastic, it can be used to place antennas inside. Racing cars use this technique for telemetry communication.

Regular hatchback cars can have a little spoiler, in which a number of antennas can be implemented. This method works exceptionally well and is the second preferred place, whenever a spoiler is present.

3.3. Screens and windows

Placing antenna structures into windscreens, side windows or rear-windows has become very popular for premium car manufacturers since 1980. As most cars have glass windows, to which an antenna structure can be applied, it is a cheap but effective method. Here the antenna structure can be either on the glass or along the frame. The glass itself is usually big enough to inherit large antenna structures or many different antennas. On-glass antennas require a larger engineering effort but can be manufactured with low costs once the structure is developed. Especially when the design of a car is of importance and roof antennas would not suite aesthetic aspects, then on-glass antennas is the way forward.

3.4. Fender and bumper

Some of the fenders or bumpers are made of plastic, which suite for placing the antennas behind as they can offer enough free room. However, special care must be taken for easy repair, as bumper and fenders can crash. Also take into consideration that engine noise and low height above ground might degrade antenna performance.

3.5. Trunk cover

Alternatively to the roof, the trunk cover is a suitable position to contain a number of antennas. Especially for convertible vehicles, where the roof and screens can be hidden, the trunk cover is advised to place antennas into. However, the trunk cover must be made of plastic or a double-layer structure with metal frame and plastic cover on top.

3.6. Mirrors

Light trucks and sport utility vehicles (SUV) offer huge side mirrors in comparison to normal cars. The shell of the mirror is mostly plastic. Inside these mirrors a number of higher frequency antennas can be placed. Some truck side mirrors are large enough to inherit a combination of FM-receiving antenna at VHF-band, Telephoning antenna from GSM and CDMA systems, a GPS navigation antenna and on top a SDARS satellite broadcasting antenna. Regular vehicle side mirrors can offer some space for higher frequency services above 1 GHz, e.g. Car-2-Car communication at 5,9 GHz.

3.7. Summary of ideal antenna placements

There are many positions possible to place antennas on a vehicle, but not all are good for each type of car. For good reception of wireless systems, there are a number of factors to be considered.

As stated in the chapter 2 on general requirements, the antenna must be as high as possible above ground for long path transmission. The antenna must be unobstructed to communicate into all directions well.

Combining both prerequisites translates into the rule of thumb, that antennas shall be operate as freely as possible, which means, the more an antenna packed into a structure, the less efficient it is.

Some antenna structures are easier to handle even in tight environment and some antenna technologies are very fragile in terms of antenna characteristics in tight metallic surroundings. Selecting the appropriate antenna structures helps finding a compromise between antenna position and performance.

4. Overview on possible antenna technologies

Antennas can be placed on the vehicle on many places, but only a few positions are ideal. To define which position is ideal, this must be considered for the service and frequency band as well as in conjunction with selected antenna technology. Here we give a short overview about the most popular antenna technologies

4.1. Monopoles

Monopole antennas are the most popular antenna type, as they are very easy to handle, easy to manufacture, easy to implement to the vehicle and very cheap.

A monopole antenna consists of an antenna foot and the rod of some length. Placing the antenna on the roof of a car requires just a hole in the metallic structure for the antenna foot and the cable to be attached.

The rod can be a stiff metallic stick of a quarter wavelength ($0,25*\lambda$) or longer. There are rods which are telescopic or even flexible. Monopoles provide a broad range of applications in every frequency band, from VHF sound broadcasting up to car-2-car communication at 5.9 GHz. The monopole can be placed in the roof center, on an edge of the roof or on the frame, or even on a bumper or fender. This antenna technology is quickly implemented to nearly anywhere on the vehicle. In conjunction with a very competitive pricing due to quick development and easy manufacturing, monopoles are the first choice in the automotive industry.

The monopole antenna requires a direct connection to ground, meaning the metallic structure of the vehicle. The monopole will not work efficiently when the metallic ground is small in comparison to the wavelength λ. Minimum 0,25*λ is required for ground plane to neglect effects.

However, aesthetically speaking, the rod antenna on a roof or fender does not satisfy modern design emotions, so car designers try to avoid simple rod antennas on their cars. Especially for premium car manufacturers, rod antennas are meant to be avoided in modern car antenna system concepts.

4.2. Patch antenna

In principal, a patch antenna is a flatted monopole. Figure 2 shows the principal structure of a patch antenna. A metallic plate of about half wavelength (0,5*λ) is placed over a metallic ground plane, which are separated either by air or a low loss dielectric material. The RF is ideally fed directly to the radiating patch.

This antenna type is becoming more and more popular in the automotive industry. At first just for high frequency applications beyond 1 GHz, such as for GPS reception at 1,5 GHz or SDARS satellite broadcasting radio reception at 2,3 GHz, WLAN and Car2Car Communication at 5-6 GHz, but nowadays increasingly seen for telephoning antennas for GSM and CDMA at 800/900 MHz.

As this antenna type is unobtrusively flat and can be implemented into the vehicles structure, e.g. behind a fender or bumper or in a plastic trunk cover. Patch Antennas require a large metallic ground structure, preferably flat of minimum 1*λ around the antenna. If the vehicle cannot fulfil this requirement at the position with its surrounding metallic structure, then the patch antenna shall provide its own ground plane in the antenna structure.

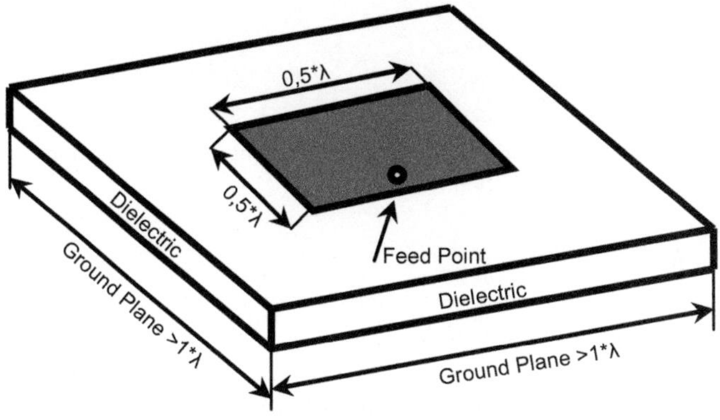

Figure 2. Principal structure of a squared patch antenna

4.3. On-glass antenna

This antenna type became popular in the premium car industry in the 1980, when design topics became increasingly important. With on-glass antennas it becomes possible to hide a number of antennas for the regular user's eye. As most vehicles have glass windows, an antenna structure can be implemented there.

The most straight forward antenna technique for on-glass follows a slot antenna principle. Here the metallic car structure becomes part of the antenna and the glass (windscreen, side window or rear window for example) is used as the slot in the metallic structure. Coupling this slot with a thin conductor, the slot resonates and can be used as antenna. The feeding position is of some importance. With a clever selection of the feeding position, different modes can be activated, which can help to excite certain polarities or radiation behaviour.

Figure 3 shows the slot antenna principle which explains on-glass antenna function.

As defrosting conductor elements are already placed to the rear window, these conductors can be applied for such on-glass antenna structures. Here, a simple filter is used to separate the DC-current for defrosting from the radio frequency (RF).

The development of such on-glass antennas can take some time, as car structure and glass window holes in the metallic structure are both parts of the antenna jointly, which influence each other. So changing the shape of the car will change the on-glass antenna characteristics. But once a suitable structure is found, the manufacturing process is very cheap.

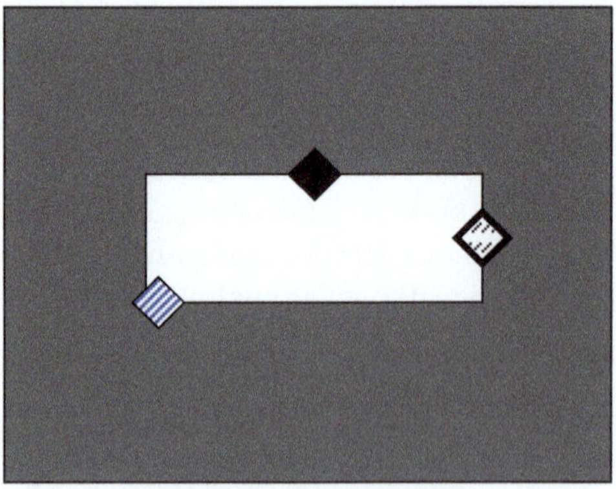

Figure 3. Slot antenna principle for on-glass antennas. Different feed point options are indicated which excite different modes for special radiation characteristics

4.4. Glued foil antenna

Similar to slot antennas and patch antennas, these structures can be placed on metallic foil, which is glued to a plastic element.

So very flat antenna structures can be produced, which can be placed into rear-view-mirrors, side-mirrors, spoilers, bumper and fenders or any non-metallic parts of the vehicle.

However there are some disadvantages to be mentioned, which are the very narrow usable bandwidth and the problem to connect coaxial cables to the flat foil antenna. In addition, moisture and spray water shall be avoided.

The connection problem is often solved by coupling the signal to the structure, but losses must be taken into account. The structure shape is virtually unlimited so that a huge variety can be found in books and journals.

Figure 4 shows one example of foil antenna structure.

4.5. Fractal antenna

A very modern field of antenna technology is summarized as fractal antennas. It has been found that not only some structures repeat their resonance with n*λ, but also when repeating the structure in itself. This fractal breakdown of a wire structure or a patch antenna structure leads to multiple resonances of the antenna. When tuning the structure that many frequency ranges can be used, only one antenna could be used to serve most of the required services.

As the benefit is that only a few antennas need to be placed, which is good for small cars, the disadvantage is the very limited usable bandwidth where the structure resonates.

Figure 5 shows an example of a fractal antenna principle.

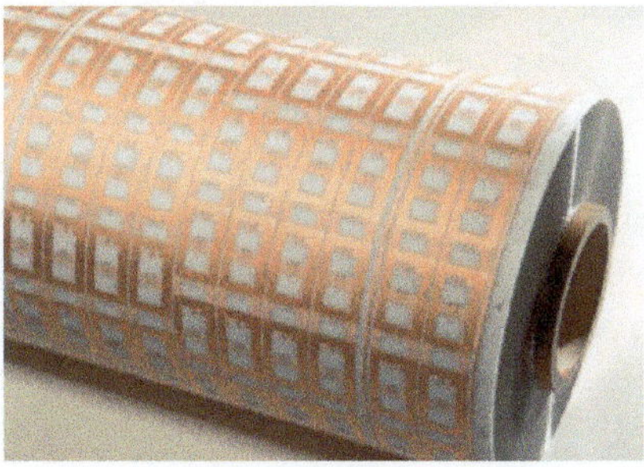

Figure 4. Thin foil antenna structures on a roll

Figure 5. Sierpinski fractal antenna structure

5. Selecting antennas according to services

Broadcasting systems reign a long history so backward compability is a major requirement also for modern cars. Although the number of mobile shortwave listeners is very limited, each car entertainment system offers shortwave-reception.

Table 1 summarizes the existing broadcasting systems.

5.1. Short- and medium wave bands

Short wave (SW) and medium wave (MW) band provide a worldwide coverage due to ground wave propagation and wave reflection at ionosphere layers in the atmosphere. As these layers are approximately 80 km to 300 km high, a large distance can be bridged. For large-area countries such as Australia, Brazil or Canada just to name a few, a nation-wide broadcasting coverage can easily be done on MW-bands. From history we adopted AM-modulated signals with relatively poor sound quality, compared to today's quality expectations. Vehicular SW-antennas are difficult to implement, as the car is always too small even for quarter wavelength antennas. So an electrically shorted monopole must be used then, which is applied as rod antenna. In some occasions the rear window or a plastic spoiler can be used.

5.2. Digital Radio Mondiale (DRM)

In order to overcome poor sound problems in SW/MW-bands, transmission became digital. Digital Radio Mondiale (DRM) decodes sound information to a digital data stream which is modulated with OFDM and broadcasted on SW/MW-bands. The receiver can decode this data stream and correct transmission errors. The sound quality is intuitively better than analog broadcasting but coverage is found to be reduced. That is mainly due to the fact that digital wireless systems have an abrupt go-nogo border, while analog systems degrade gradually until signals disappear in noise [1].

5.3. VHF-band and UHF-band

Around 1950, VHF-frequency spectrum was utilized in Europe. Radio propagation in 100-MHz-band can reach approximately 30% beyond the optical horizon, so called radio horizon. This leads to a limited coverage of sound or TV broadcasting stations. Depending on transmitter location and radiated transmit power the typical coverage area is about 100 km^2. For a nation wide broadcasting network, a high number of transmitters must be installed. Especially in hilly terrain uncovered areas occur due to radio shadows. When reception is needed there, filling transmitters are needed. On VHF-band (76-109 MHz) sound broadcasting is transmitted while on UHF-band (400-800 MHz) TV-broadcasting is located. Typically, a dual layer broadcasting network is rolled out in VHF- and UHF-bands. The first layer consists of transmitters with high power on prominent locations, covering a large area. The second layer fills coverage gaps with low-power transmitters.

As each transmitter requires a separate frequency in order not to interfere with each other, the dedicated frequency spectrum is quickly used up. Clever frequency allocation and frequency reuse is needed.

For VHF and UHF a number of antenna techniques can be used. The cheapest version is using a monopole, but on-glass slot antenna structures offer a very efficient way as well. If the vehicle has a plastic spoiler or a plastic fender/bumber, it can inherit VHF/UHF antennas.

5.4. Digital broadcasting systems

In 1987 development of digital broadcasting systems began and a number of digital transmission standards derived since then. Digital Audio Broadcasting (DAB), former Eureka-147 Project, was the start into digital data processing and digital broadcasting transmission. DAB decodes sound information to a digital data stream which is modulated with 4-PSK-OFDM and broadcasted on VHF-band. The receiver can decode this data stream and correct transmission errors. The sound quality is intuitively better than analog broadcasting. Similar to all other digital wireless systems, DAB reception is experienced having an abrupt go-nogo border, which is often misunderstood that coverage is smaller than analog systems [1].

One of the benefits of digital broadcasting system is that more programs can be transmitted with one frequency allocation. While in analog broadcasting one radio station used one frequency, now up to 64 radio stations can be received on one single frequency. This offered new broadcasting capacity for more divert channels covering more genres and clientele. Another benefit of digital transmission is the higher sound quality.

With better coding capabilities using MP3 and MP4 codecs a more efficient transmission could be implemented. Digital Multimedia Broadcasting (DMB) is using MP4 AAC+ codec in contrast to DAB, where MUSICAM codec is implemented. For video transmission, Digital Video Broadcasting (DVB) is used. DVB is similar to DAB and DMB, except that video signal compression H.264 is added to sound compression algorithms and higher intrinsic modulation schemes are used, often 16QAM-OFDM. DVB is successively replacing analog TV-transmitters in Europe.

The placement for DAB antennas shall be properly selected, keeping the spurious emission noise in mind.

5.5. IBOC system

In USA, another method of digitalization is used. Beside the analog modulated signal, an additional digital modulated signal with same content is transmitted, so called in-band on channel (IBOC) signal. Figure 6 displays the spectrum of such an IBOC signal.

Having good reception, the tuner decodes the digital information and provides high-quality sound. Reaching the go-nogo-border of digital transmission, the tuner switches to traditional analog broadcast. The listener may observe a small degradation in sound quality but is still able to follow the program.

In Europe, IBOC was tested in Switzerland in a range limited test environment [2]. It is unlikely that IBOC will be introduced in Europe, as VHF-band channel raster of 200 kHz interferes with US-IBOC channel bandwidth requirements of 400 kHz. To fit into the existing channel raster, one of the redundant digital sidebands can be removed. This method is known as FMeXtra and is currently under test in a testbed [3].

For IBOC receiving systems, the selection of suitable antenna structures are identical as VHF/UHF systems, but due to accompanied digital sidebands, the spurious emission noise must be considered when placing the antenna, similar to DAB systems.

5.6. Satellite Digital Audio Radio System (SDARS)

For large area countries such as Australia, Brazil, Canada, Russia or USA, just to name a few, traditional nationwide terrestrial broadcasting systems are very expensive to install. Using low orbit satellites offers an efficient nationwide coverage. In USA, Satellite Digital Audio Radio Systems (SDARS) in 2.3-GHz-band was implemented and succeeded with a sheer overwhelming number of different programs for all sorts of listeners. Even a monthly fee could not stop listeners to attend SDARS in USA.

Above 1 GHz there are patch antennas first choice. As GPS and SDARS is transmitted from satellites high above ground, the antenna shall receive from sky best, which limits the placements to be 360° unobstructed to sky. Hence the roof and exposed spoilers are suitable positions, in some occasions mirror housings as well. Convertible cars can utilize the trunk cover or the upper edges of the fender.

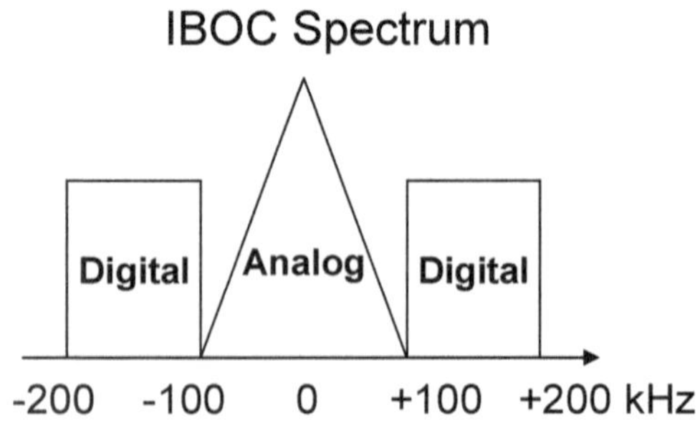

Figure 6. Spectrum of IBOC system with analog and digital modulation containing identical information

System Name	Abbreviation	Frequency Band	Modulation
Long-Short-Medium Wave	LW/MW/SW	172 kHz - 30 MHz	Amplitude Modulation AM
Ultra short Wave	VHF	76 MHz - 109 MHz	Frequency Modulation FM
In-Band On-Channel	IBOC	76 MHz - 109 MHz	Quadratur Phase Shift Keying, QPSK
Digital Broadcasting System	DAB	174 MHz - 210 MHz	Orthogonal Frequency Division Multiplex, OFDM
Digital Broadcasting System with enhanced codec	DAB+	174 MHz - 210 MHz	Orthogonal Frequency Division Multiplex, OFDM
Digital Multimedia Broadcasting	DMB	174 MHz - 210 MHz	Orthogonal Frequency Division Multiplex, OFDM
terrestrial Digital Video Broadcasting	DVB-T	480 MHz - 860 MHz	Orthogonal Frequency Division Multiplex, OFDM
terrestrial Digital Video Broadcasting with enhanced codec	DVB-T2	480 MHz - 860 MHz	Orthogonal Frequency Division Multiplex, OFDM
Satellite Digital Audio Radio System	SDARS	2,30 GHz - 2,33 GHz	Orthogonal Frequency Division Multiplex, OFDM

Table 1. Overview of broadcasting systems

5.7. Best practices - Where to place antennas for services

When looking at various vehicles on the market, there will be seen a number of antenna solutions on the vehicle structure. The most popular antenna placement is a monopole rod antenna on the roof of the car, as this method fulfils the most important basic requirements, high above ground and unobstructed.

The radiation characteristic is quasi omni directional, when the metallic roof is large enough.

For frequencies >100 MHz, which refers to FM-broadcasting reception, TV-reception and mobile telephoning systems, monopole antennas on vehicle roof are perfectly suited.

In addition, they are very easy to implement and with about 5 EUR or less per piece very cheap. These are the reason why the majority of the vehicle manufacturers apply monopole rod antennas.

Sometimes the car designer does not like antennas on the structure. Then hidden antenna concepts must be used. A very common way is to place the antennas into a plastic spoiler. As the spoiler is usually high above ground and unobstructed for good airflow, it fulfils basic requirements for good reception. Racing cars use the spoiler structure for telemetry communications. Sometimes regular hatchback cars contain a spoiler, into which VHF-antennas for sound and TV broadcasting, GPS-antennas and nowadays Car-2-Car Communication antennas can be easily integrated. Of cause it is also possible to apply telephoning and satellite broadcasting SDARS antennas into spoiler structures.

When there is no sportive spoiler but the design aspects require invisible antennas, the antennas can be placed into the screens. Here the slot antenna concept is used, which requires some development but once the structure is found, it is easy and cheap to manufacture. Usually the rearwindow is used when the engine is in front, offering the maximum distance from spurious emission noise. In rare occasions where the engine is in the backside of the car, e.g. Porsche 911 model, then the windscreen is used. However, the antenna structure shall not effect visibility then.

Alternatively sidewindows can be used for antenna structures but often they are too small for multiple antennas and VHF-antennas.

As foil- and fractal antennas became popular for mobile phone antennas, this concept has also been applied to the car industry. Today we find fractal foil glued antenna structures in rearview mirrors, e.g. garage door opener, car entry systems and Bluetooth- / WLAN antennas, just to name a few.

Light trucks and SUVs comprise some large side mirrors, in which even low frequency antennas can be placed, such as for FM-reception on VHF bands and mobile telephoning systems.

The only disadvantage is that in a case of an accident with damaging the side mirror, the wireless service can be damaged as well or reception becomes poor.

In order to avoid a complete loss of service, either both side mirrors can be equipped or a combination with other antenna locations can be used.

The side effect of this method is that multiple antennas can provide better reception. This method is known as diversity reception.

6. Diversity - Combating fading

In vehicular receiving environments fading of signals occur due to multipath reception. When a radio signal is transmitted, it reaches the receiver on a direct path as well as on reflected paths from buildings, landscape and obstructions, see Figure 7. The reflected paths reach the receiver at different times than the direct path. This leads to superposition of multiple signals of the same content. Signals can add or subtract each other at the receiving zone, leading to a varying loudness impression in amplitude modulated signals and signal dropouts in frequency modulated signal. In digital systems, bit errors can occur.

A number of methods were invented to reduce the audible effect of multipath fading.

One method is to apply a number of receiving antennas on different positions of the car. Different antennas at different locations are assumed to receive different signal components. This effect is known as spatial diversity. A switch selects the best receiving antenna according to the signal strength and noise level. Figure 8 shows the principle of switching spatial diversity for VHF-FM reception. Another method is to use both tuners in a dual-tuner concept which are connected to individual antennas. Due to the different location of the antennas phase differences occur which are corrected by phase shifters. Then, both phase corrected receiving signals can be added in-phase and can prevent fading dropouts. This method is known as phase diversity.

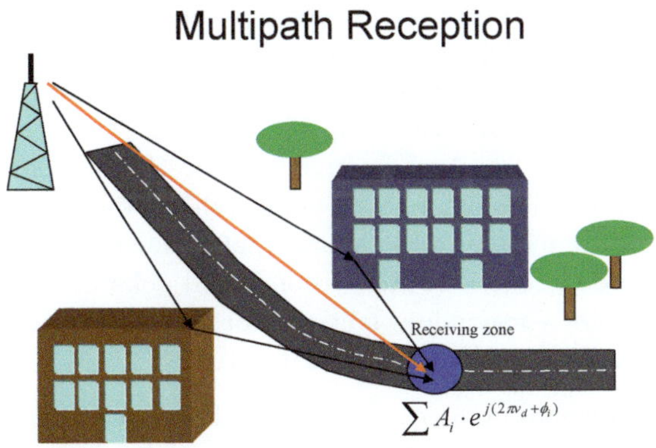

Figure 7. Principle of multipath environment where transmitter signal is reflected, diffracted and attenuated by environment. At the receiving zone signal paths are summed.

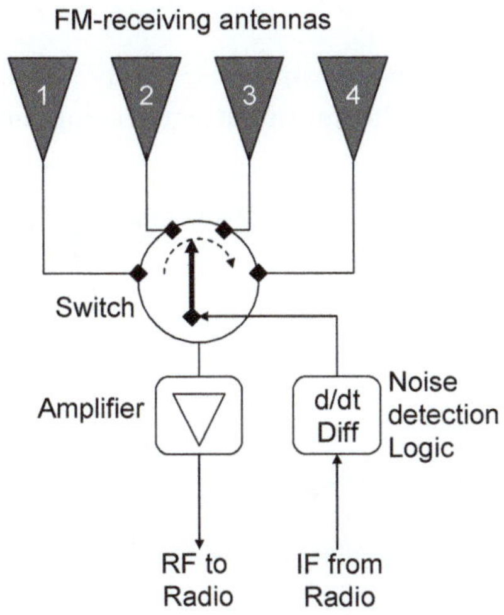

Figure 8. Principle of switched diversity in modern tuners [1]

In digital broadcasting systems, bit streams can be combined to correct transmission errors. Here, two receivers operate individually but synchronized by internal clocks. The digital data streams are compared with each other and bit errors corrected when necessary. The combined data stream provides a more consistent data rate which results in a better quality of service for the listener. Figure 9 shows principle bit stream diversity for DAB.

6.1. Best practices - Where to place diversity antennas

When placing diversity antennas, there must be some requirements to be fulfilled.

The main requirement is that antennas used for diversity reception must be mutually decoupled, meaning that one antenna shall not receive the identical signal than the other, better to receive a different phase or a different component of the signal. Only then the fragments can be assembled to a better signal. To achieve this, different antennas shall be placed at least 3 wavelength apart, so that signal can be received at the different time and different phase. This method is known as "spatial diversity". If there is not enough space for a large gap between antennas, an alternative is to receive different phase components, which is known as "phase diversity". In a premium vehicle, diversity reception for FM or TV is achieved with a combination of phase and spatial diversity. For instance, placing 2 antennas in the rear-window, one is vertically dominated polarized, the other horizontally dominated polarized. Both rear-window antennas provide good reception, as nearly all phase components of the multipath signal can be received.

However, having both antennas in the rear of the vehicle, the reception to the front side can be shadowed. This can be overcome with a third antenna, either a monopole on rooftop or a structure in the spoiler (if the vehicle has one) or using a structure in the front fender. For higher frequency ranges, the mirrors can be used, e.g. fractal foil structure glued to the plastic housing.

Combining rear-window and side-mirror antennas, a fully decoupled but omni directional reception is achieved where most of the multipath signal components can be processed.

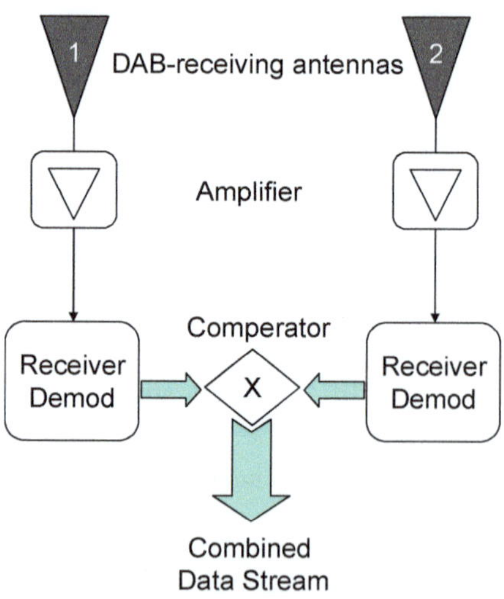

Figure 9. Principle of bit stream combining method for digital broadcasting systems [1]

7. Outlook and future trends

From the performance point of view, a lot can be optimized in the entertainment system in the future. Especially broadcasting reception is deemed to be improved. Historically, the tuner was installed in the center console, while the receiving antenna was on the fender. The cable length was comparably short. A disadvantage of this solution is that engine spurious emission noise is easily picked up by the antenna and disturbs reception.

Modern cars however offer a number of receiving antennas for diversity reception in the rear-window, side-window, bumper and fender for instance, but long cabling ways attenuate RF signals.

The wide range of broadcasting standards requires multiple tuners buried in the car. Integration and size reduction is a major playground in R&D departments. Transceivers of modern mobile phones are approximately 30x30 mm^2 or less and 3 mm thick, offering multi-frequency and multi-standard operation already. With SDR-tuners it will become possible in near future to provide compact multi-standard broadcasting receivers exploiting diversity gain by MIMO concepts. This allows integrating such receivers into - or at least close to - the antennas. Reception performance will improve drastically unless EMC problems occur.

Another mega trend of this decade is a permanent internet connection. With UMTS and WLAN it is already possible to connect laptops and mobile phones to the internet while riding in car. In near future, the vehicle itself gets connected to the internet. The mobile phone standard Long-Term-Evolution (LTE) will support this trend. The merge of internet services and vehicular entertainment functionality will provide efficiency and convenience to the passengers. The sheer endless list of new service ideas for the drivers and passengers is overwhelming and becoming unique selling points for car manufactures. They will offer new services to drivers, from intelligent traffic routing, parking aid to firmware updates inside the car. Passengers will be able to stream music and videos as well as communicate while surfing the internet.

In conjunction with passenger entertainment, Car-2-Car Communication systems offer new information sources for the driver to plan a trip and offers new safety features during the ride.

One of many applications is that traffic flow is constantly monitored by ego-speed history and current data recognition, which will be broadcasted to other vehicles on the road nearby. Comparing own speed information and vehicles in front of own position allows a rolling "look-ahead" traffic situation estimation of the upcoming path. This will help to adjust driving speed to enable a better traffic flow as well as avoid traffic jams.

Another safety aspect of C2C information exchange is on crossroads. Here a central station receives signals from vehicles approaching and can guide them through the cross-traffic. With this, right-of-way accidents can be reduced as well as traffic flow improved.

Author details

Niels Koch
Altran GmbH & Co.KG, Germany

8. References

[1] Koch, N. (2008). Diversity for DAB – Worth the Effort ?, *9th Workshop Digital Broadcasting,* Fraunhofer Institute IIS Erlangen, Germany

[2] Henk, C.M, Hamelink, S.G (2008). FMeXtra – the Principle and its Application, *9th Workshop Digital Broadcasting*, Fraunhofer Institute IIS Erlangen, Germany
[3] Ruoss, M. (2008). The Digitalization of the FM-Band in Europe, *9th Workshop Digital Broadcasting*, Fraunhofer Institute IIS Erlangen, Germany

Chapter 7

Power Electronic Solutions to Improve the Performance of Lundell Automotive Alternators

Ruben Ivankovic, Jérôme Cros, Mehdi Taghizadeh Kakhki,
Carlos A. Martins and Philippe Viarouge

Additional information is available at the end of the chapter

1. Introduction

Until in the early 1960s, automobiles used a DC generator called dynamo. The availability of affordable power diodes in the beginning of 1960s paved the way for the widespread use of three-phase claw-pole alternators (or Lundell alternators) for the generation of electric power in motor vehicles. After more than 50 years, this system is still the most economic choice in today's vehicles due to its low manufacturing cost. However, the efficiency and output power of the Lundell alternators are limited. This is a major drawback for its use in modern vehicles requiring an increase in electrical power. Many alternatives are being considered to replace the Lundell alternator such as the salient pole machines, however they require large investments in manufacturing infrastructure. In this context, this chapter focuses on the improvements of Lundell alternators that could represent the best solutions for the short term. First, we present the conventional automotive generating system, its performance and the limitation of modelling methods. We also discuss various solutions to increase the output power of Lundell alternators without any geometry modification. These include the use of reconfigurable windings and replacement of the diode rectifier by different electronic converters such as a synchronous rectifier or an interleaved PWM rectifier will be considered.

2. Generating system with a Lundell alternator

Today, the great majority of electrical generator systems installed on combustion powered vehicles are based on a three-phase wound-field synchronous machine. The conventional automotive generator has a claw pole rotor with a single excitation coil wound axially. It is often named "Lundell" alternator (Fig. 1). The excitation coil is surrounded by two solid iron pole pieces, or claw poles, and is fed via a pair of slip rings and two carbon brushes.

The stator is composed of a slotted laminated iron core and a three-phase overlapped winding, wye or delta connected. It is a wave-winding in most cases. Fig. 1 shows a dismantled Lundell alternator (Delcotron 22SI) and its main components. The number of rotor poles for passenger vehicles is generally fixed at 12 poles. The stator has often a single slot per pole and per phase (36 total slots). High power alternators for some applications like buses, trucks or other special vehicles have higher pole numbers (between 14 and 18). Increasing the number of poles reduces the inductance of the stator winding, and as a result, increases the short circuit current. Unfortunately the magnetic losses will also increase due to the higher electrical frequency. So the choice of pole number is actually based on a compromise between the magnetic losses and alternator power requirements. The alternator is coupled to the combustion motor through a belt. In passenger cars the maximum alternator speed (typically 8000 RPM) is about two times more than that of the engine crankshaft (a pulley ratio of 1:2). The Lundell alternator is generally characterized by its form factor (a relatively large diameter compared to its length) which facilitates thermal dissipation.

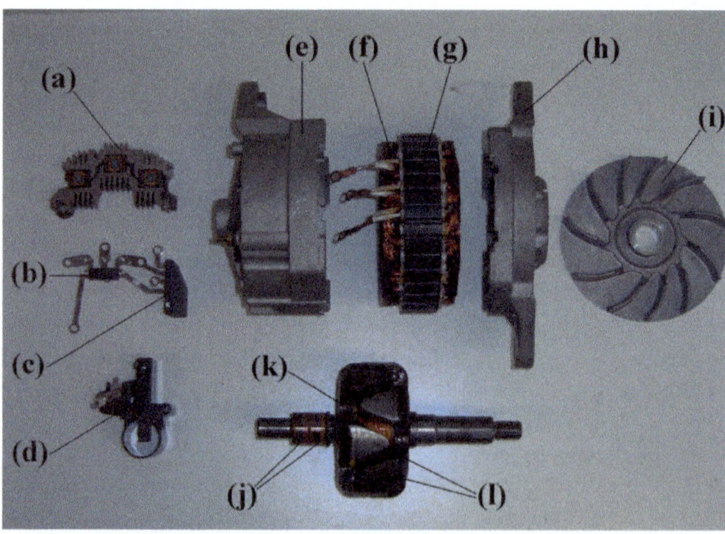

Figure 1. Dismantled alternator parts: (a) 6 diode full- bridge power rectifier, (b) excitation rectifier, (c) regulator, (d) brush assembly, (e) aluminum rear housing, (f) stator winding, (g) stator laminated core, (h) aluminum front housing, (i) aluminum fan, (j) slip rings, (k) excitation winding, (l) claw-shaped pole pieces.

2.1. Electrical circuit

Fig. 2 shows the simplified schema of a vehicle generation system. Usually six diodes in a full-bridge configuration are used to rectify the output current. The rectifier is divided in two sets of three diodes. The metal casing of the first set is typically pressed into a heat sink (or welded to the heat sink) for better thermal dissipation.

The output power is controlled by regulating the field current. The regulator maintains a constant output voltage on the battery despite the varying alternator speed and variable load conditions. This voltage depends highly on the ambient temperature and the chemical characteristics of the battery and necessitates a temperature compensation by the regulator. The regulator-excitation circuit is often supplied by an additional half- bridge rectifier instead of the battery (Exciter rectifier).

In the alternators with wye connected windings, two or more other auxiliary diodes are connected between the neutral point and the main rectifier output terminals. In this way it is possible to rectify the induced third harmonic voltage and increase the output current at high speeds.

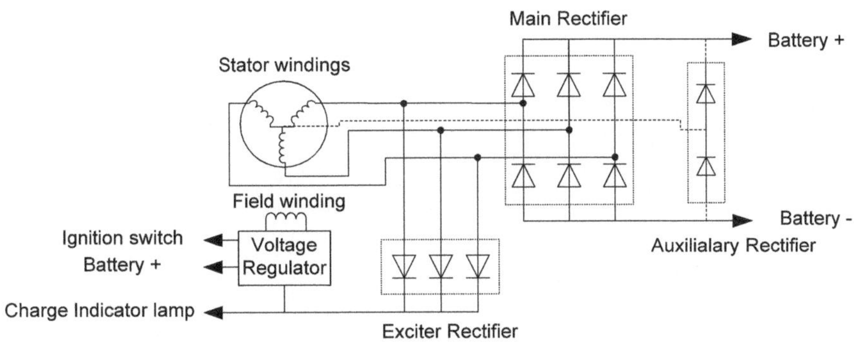

Figure 2. Simplified schema of conventional vehicle generating system with a Lundell alternator.

2.2. Alternator performance curves

Performance curves of an alternator are used to show its performance across the whole speed range. Fig. 3 shows the performance curves of a 100A commercial alternator obtained for a battery voltage of 14 V, an ambient temperature of 25° C, and for the maximum excitation current (near 6A).

The output current curve is characterized essentially by three operation points. The first one is the generation starting speed (Ω_{DG}) or 0-Ampere speed at which the alternator reaches its rated voltage without delivering power. The second one is the maximum output current at the highest speed, which corresponds approximately to the DC short-circuit current of the alternator (I_{SC}). Also, a special attention is paid to the output power requirement at engine idle speed (Ω_R) at which the alternator must deliver at least the power needed for long-term consumers. No output power is required below the idle speed. Another operating point often mentioned is the speed at rated current (Ω_N).

2.3. Alternator efficiency

Generally, Lundell alternators are characterized by low efficiency due to important mechanical, copper, and magnetic losses. The efficiency of the alternator varies widely depending on load conditions, alternator speed and alternator size. As shown in fig. 3, the efficiency is only 53% at idle speed (losses: 675W) and about 42% at 6000 RPM (losses: 2100W). Larger and heavier alternators may be more efficient for the same speed and load conditions, however, this advantage can be compensated by the increase in fuel consumption due to the higher weight (0.1 L/100km fuel consumption for each additional 10 kg for a medium sized vehicle)(Beretta, 2008). Table 1 shows the distribution of losses in an alternator (Beretta, 2008) for two different speeds (full excitation field).

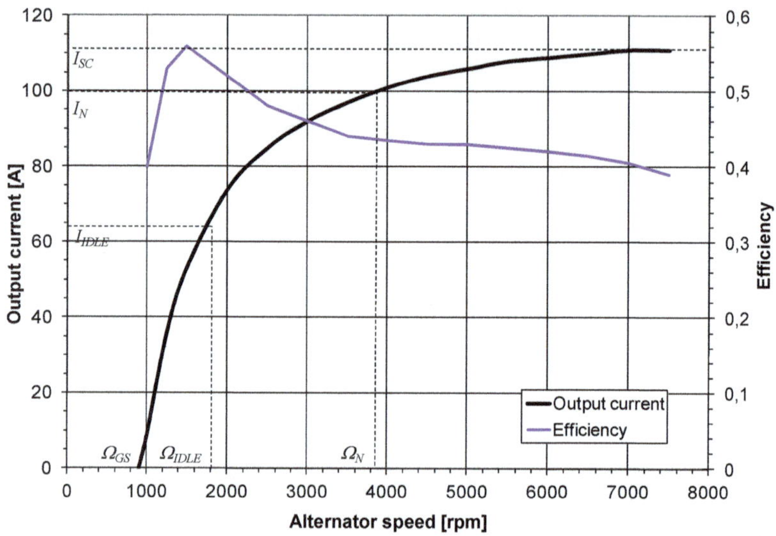

Figure 3. Characteristic of the output maximum current and efficiency as a function of speed for an alternator (Delcotron 22SI type12V-100A) at a battery voltage of 14V, ambient temperature of 25° C and maximum excitation current.

	1800 RPM	6000 RPM
Mechanical losses	2%	6%
Excitation losses (rotor)	7%	3%
Magnetic losses (stator)	21%	20%
Copper losses (stator)	49%	57%
power rectifier losses	21%	14%

Table 1. Distribution of losses in an alternator for two different speeds (full excitation field).

2.3.1. *Mechanical losses*

Mechanical losses are generated by brush and bearing friction and losses created by the claw pole rotor and fan (windage losses). The mechanical losses increase considerably at higher speeds (Bosch, 2003). In the current air cooled alternators, one or two fans are used for convection cooling. These fans add to the alternator (aerodynamic) losses and are also responsible for an important part of the alternator audible noise at higher speeds.

2.3.2. *Copper losses*

As shown in Table 1, the main losses in the alternator are the stator copper losses. Increasing the filling factor will reduce copper losses, however, this is limited by the industrial production constraints (Beretta, 2008).

Other sources of ohmic losses include the losses in the rectifier diodes, regulator losses and and losses occurring due to the contact resistance between the slip rings and brushes.

2.3.3. *Magnetic losses*

At nominal excitation current and at lower speeds, the output current is low and the stator is fully saturated due to the weak magnetic reaction. This will result in important magnetic losses in the stator core. Increasing the speed and output current while maintaining the same excitation reduces the flux density in the stator. With this demagnetisation effect the magnetic losses become proportional to speed itself rather than the square of speed (as expected in no-load conditions). Reducing the lamination thickness can lead to a significant reduction of the magnetic losses (0.5 mm or 0.35 mm instead of 1 mm).

Since the claw pole parts are made of solid forged iron, the eddy currents can easily circulate. This will also add to magnetic losses at very low speeds and low loads where the flux density in the air-gap (generated mainly by excitation field) is modulated by the stator slot openings. At higher speeds, the stator magneto-motive force (MMF) produces space harmonic fields in the air-gap which are augmented by the slot openings and again produce claw-pole eddy currents (Boldea, 2006). To reduce magnetic losses in the rotor, it is possible to use a laminated material but the assembly process is more complex (Bretta, 2008).

2.3.4. *Cooling of the alternator*

An efficient cooling method is necessary to limit the temperature in the motor beyond the permissible limits (a junction temperature of 200°C for the diodes and the stator winding).

The thermal dissipation in an alternator is mainly by convection through one or two cooling fans. The small amount of heat generated in the rotor can be dissipated by conduction through alternator bearings.

As an alternative to air cooling, it is possible to cool the alternator more efficiently with a circulation of the engine coolant in the alternator housing. It also reduces the alternator's noise level by omitting the cooling fan(s).

3. Alternator modeling for performance analysis

In order to evaluate the performance of the alternator under variable load conditions, both the machine and rectifier systems shall be modeled with appropriate magnetic and electrical models.

3.1. Electrical modeling of the Lundell alternator with an equivalent circuit

3.1.1. Electrical modeling of the machine

A 3-phase synchronous machine can be modeled by its equivalent circuit as shown in fig. 4. The parameters of the electrical alternator model are derived from the no-load EMF versus rotor field current characteristic $E_V(j_{rotor})$ and the short circuit characteristic $I_{CC}(j_{rotor})$. The stator cyclic inductance ($L(j_{rotor})$) is derived from both tests by applying Eq. 1 where ω is the electrical frequency.

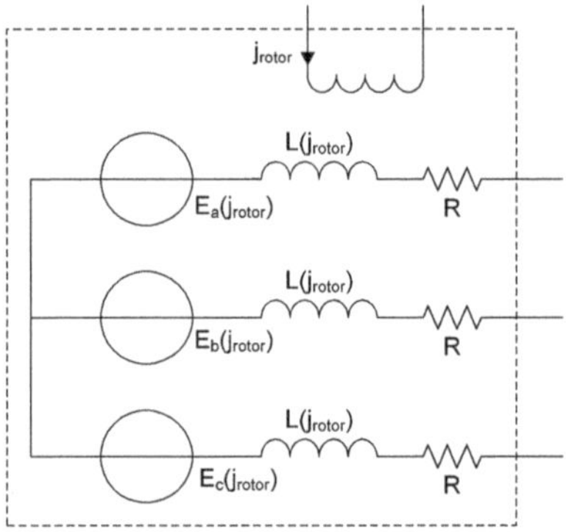

Figure 4. Equivalent circuit model of a 3-phase synchronous machine

A high rotor field current produces magnetic saturation and the stator inductance value decreases. The accuracy of such a model highly depends on the parameters identification or computation method. The armature phase resistance is measured at the rated temperature rise.

$$L(j_{rotor}) = \frac{E_V\left(j_{rotor}\right)}{\omega \cdot I_{CC}\left(j_{rotor}\right)} \tag{1}$$

3.1.2. Modeling of the alternator- rectifier

It is possible to evaluate analytically the steady state performance of a 3-phase alternator - rectifier system connected to a voltage source (fig. 5) as shown in (Figueroa et al., 2005). This type of modeling uses the electrical equivalent circuit of the machine to directly determine the steady state copper losses, current harmonic contents and DC bus current.

It is faster than a step by step simulation method as employed by simulators for power electronics (PSIM, Saber) and it is well adapted for use within an optimization loop for system evaluation purposes. However, it uses simplifications which compromise the accuracy of the results including rotor saliency, and magnetic losses in the machine.

To evaluate the simulation method, an experimental set-up is used to compare the output performance of the alternator with various rectifier topologies (fig. 6). Torque is measured by a rotating torque sensor and a strain gauge. The alternator is driven by a 5 kW DC motor via a belt and pulley system. A position encoder is also available. The rectifier has been taken out from the machine and the voltage regulator has been removed so that the excitation current could be imposed using an external current source. The same experimental set-up will allow the comparison of various converter topologies and winding configurations for alternators. It has been used for all the experimental work presented in this chapter.

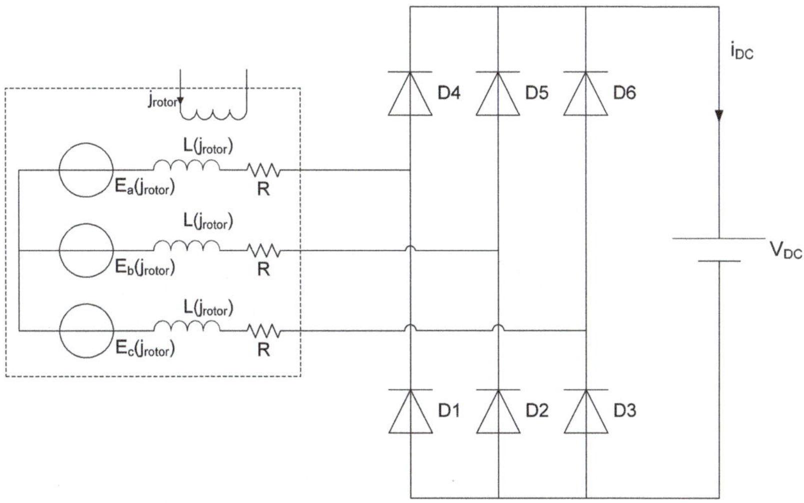

Figure 5. Circuit model of a generation system with a diode rectifier

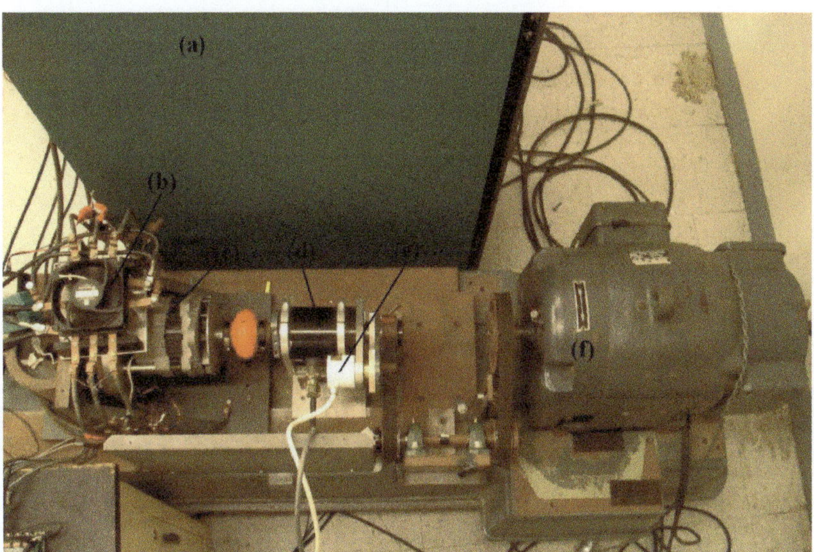

Figure 6. Test bench, a) converter (b) conventional rectifier, (c) alternator, (d) torque sensor, (e) position sensor, (f) DC motor

Fig. 7 shows the characteristics of the output maximum current and efficiency versus speed provided by the manufacturer along with those obtained by simulation. The simulation slightly overestimates the output current particularly in the central part of the curve. This could be explained by the absence of magnetic losses in the simulation, assumption of a sinusoidal EMF and inaccuracies in parameter identification. Despite these drawbacks, it can be seen that the proposed model is able to provide a current versus speed curve of satisfactory accuracy. In the same figure, it can be seen that the absence of the magnetic and mechanical losses leads to a significant overestimation of the efficiency.

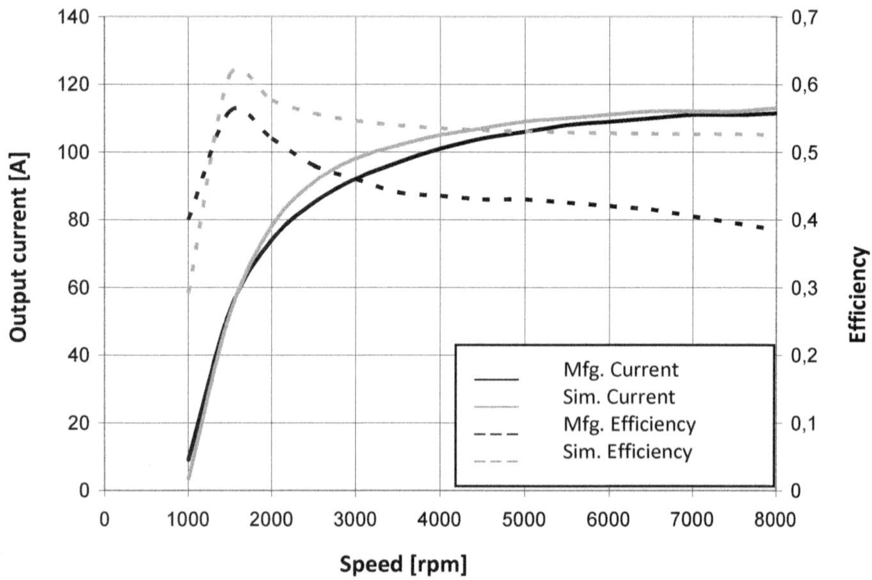

Figure 7. Comparative analysis of simulation and experimental results (from manufacturer) of the alternator output characteristics

3.2. Simulation limitations

The use of a PWM controlled rectifier (fig. 8) instead of a diode rectifier allows for the following main benefits: boosting operation for increasing the output power at low speed and power factor correction in the machine for maximization of output power.

In Lundell alternators, the magnetic circuit is saturated for the rated field current. With a Conventional Diode Rectifier (CDR), the armature reaction always has a demagnetizing effect since the voltage is in phase with the current. In case of controlled rectifiers, the armature reaction can have a magnetizing effect at low speeds depending on the power angle. Therefore, the power increase due to power angle control is less significant than what one could expect when magnetic saturation is increased by the armature reaction. As shown in fig. 7, the simplified electrical model provides a good current estimation in the case of the diode rectifier while the magnetic saturation of armature field is neglected. This is not the case with a PWM rectifier and the output current is often over estimated by the simulation as shown in fig. 9.

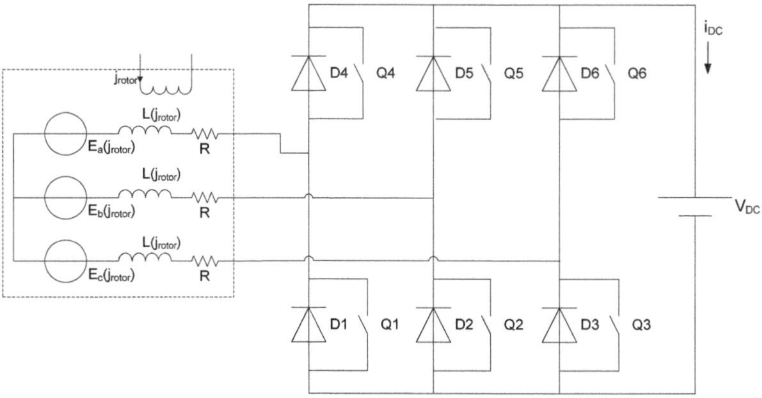

Figure 8. Circuit model of a generation system with a PWM controlled rectifier

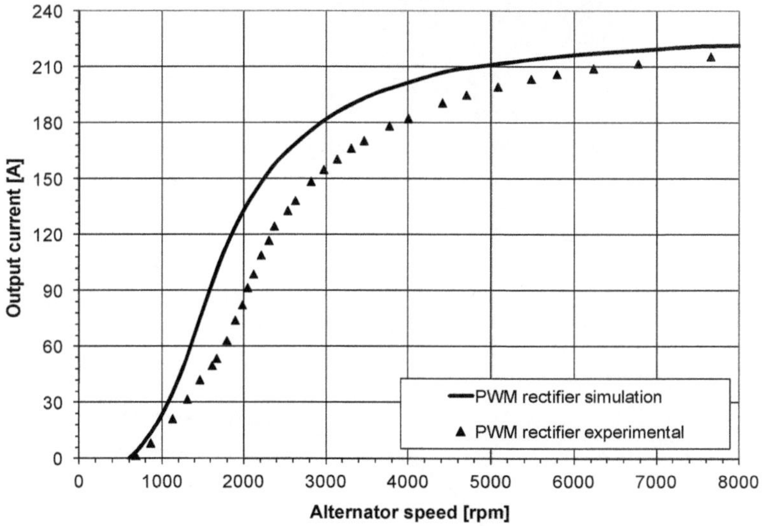

Figure 9. The output current obtained by simulation for a diode rectifier and a PWM full-bridge rectifier compared to experimental results

3.3. Magnetic modeling

Different magnetic models can be used to compute the parameters of the machine's equivalent circuit, provided that they take account of magnetic saturation.

A first one is based on a magnetic reluctance network that takes into account the machine geometry and the magnetic material B(H) characteristic (Ostovich et al., 1999). The analytical method based on the reluctance network is fast due to its simplifying assumptions.

A second method is based on 3 Dimensional Finite Element (3D FE) modeling (Küppers & Henneberger, 1997) which is particularly interesting for the analysis and evaluation of saturation and magnetic losses. Finite element methods are the most accurate but are time consuming for a variable speed machine and under variable load conditions.

Fig. 10 shows the results of 3D FE simulation of a claw pole alternator with 36 slots and 12 poles. With the 3D FE modeling, the electrical parameters of the alternator can be directly obtained by applying two methods: a method using the scalar potential and another one using the vector potential (Cros et al., 2008). Fig. 11 shows the output current versus speed characteristics obtained with the electrical parameters of scalar potential and the vector potential models. The right characteristic is between the two curves (Henneron et al., 2004).

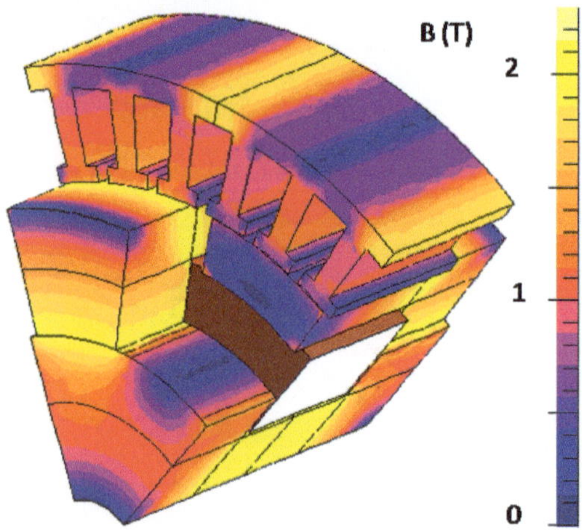

Figure 10. The results of 3D FE simulation of a claw pole alternator

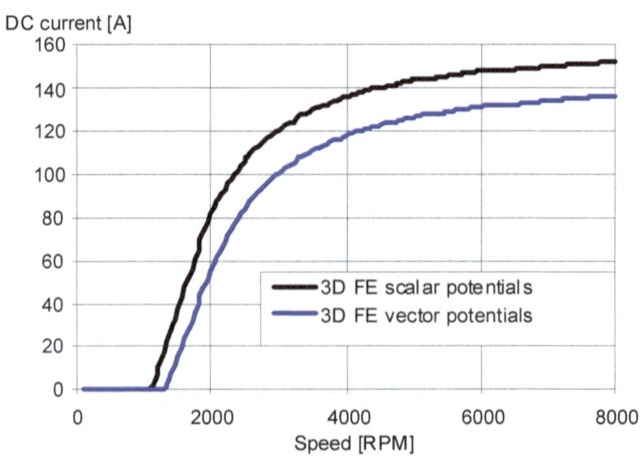

Figure 11. Simulated DC current output vs speed characteristics with two 3D FE methods

4. Improving the alternator's performance

This section focuses on various possible solutions to increase the output power of Lundell alternators without any geometry modification. This can be achieved by a winding reconfiguration to modify the number of turns per phase. A low number of turns increases power at high speed and a high number of turns improves the idle current. Another way is to rewind the alternator with a lower number of turns and to replace the conventional diode rectifier by an active PWM rectifier. The active rectifier boosts the alternator voltage in order to attain acceptable performances during low speed operation.

4.1. Influence of the number of turns on the performance

The number of turns in the stator winding has a significant effect on the output performances of an alternator connected to a Conventional Diode Rectifier (CDR) and a battery. The output current and the efficiency during high-speed operation can be easily increased by reducing the number of conductors per phase. However, the reduction of the number of turns presents also an unacceptable drawback which is a severe reduction of the output power during idle operation. On the other hand, one should increase the number of turns to improve performance at low speeds.

If the total copper cross section is not modified, the maximum current density is always the same and so the copper losses are not increased. Equation (2) shows the electrical parameters variation of the stator equivalent model (resistance R, cyclical inductance L and no-load flux Φ) according to the number of conductors per slot N when the total copper cross section is kept constant.

$$L = N^2 \cdot L_0 \quad , \quad R = N^2 \cdot R_0 \quad , \quad \phi = N \cdot \phi_0 \tag{2}$$

Fig. 12 shows a comparison of output characteristics with a number of turns per phase divided by two. The stator winding with the highest number of turns halves the speed of the generation starting point but it produces twice less output current during high-speed operation. One can notice that the winding with twice less turns provides better performance (higher current with lower copper losses) as soon as the speed exceeds 2700 rpm.

4.2. Winding reconfiguration

Winding reconfiguration is an interesting approach to improve output power, efficiency and satisfy idle current specifications.

4.2.1. Delta-wye and the series-parallel reconfiguration

One solution is to divide each phase winding in several coil groups and to modify the winding configuration by using several switches. It is then possible to obtain different connections of the coil groups (parallel, series, delta, wye) as a function of the speed. Fig. 13 shows the delta-wye and the series-parallel reconfiguration (Liang et al., 1999).

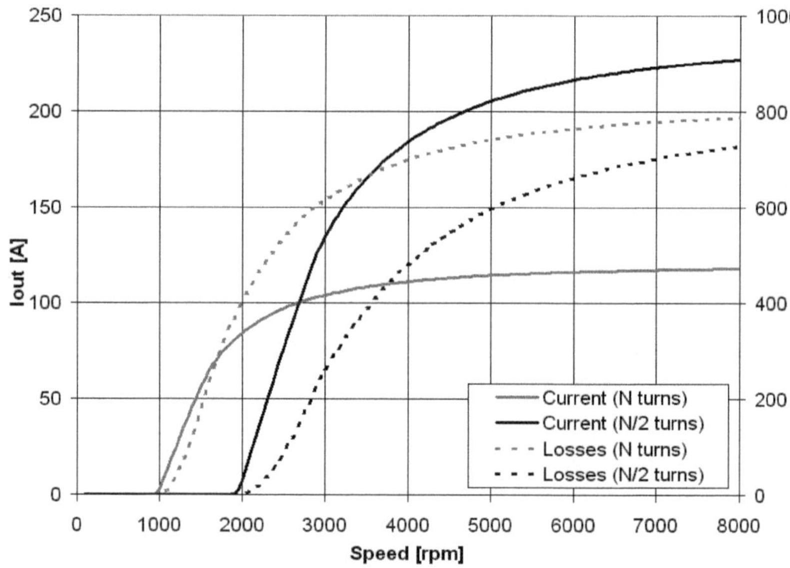

Figure 12. Influence of winding number of turns on the output performance with a diode rectifier

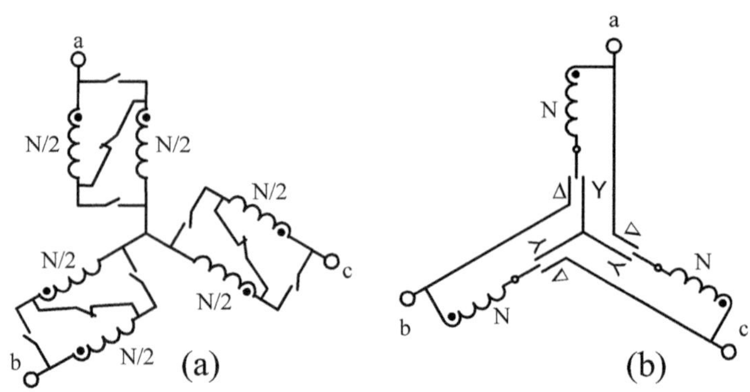

Figure 13. a) Series- Parallel reconfiguration, b) Wye- Delta winding reconfiguration

4.2.2. Multi-winding stator with series/parallel diode rectifier

One can also use a multi-winding stator with different number of turns to adapt the output characteristics (output current and efficiency) with a single voltage output. In this case, the different windings are connected to different diode rectifiers and must be magnetically decoupled (Cros et al., 2003). In order to compare the relative performance, we consider a given stator with the electrical parameters of a single-winding configuration (L_0, R_0 and Φ_0). Equation (3) shows the new electrical parameters of a m-winding configuration in the same stator.

$$L_i = N^2 \cdot \frac{L_0}{m} \quad , \quad R_i = N^2 \cdot \frac{R_0}{m} \quad , \quad \phi_i = N \cdot \frac{\phi_0}{m} \tag{3}$$

If we consider m=2, it is possible to optimize the first winding system for the lower speeds and the second one for the higher speeds by choosing the right number of turns in each winding while maintaining the same copper volume per slot. For example, one may use N conductors for the first winding and N/2 conductors for the second one, as shown in fig. 14.

Fig. 15. shows a comparison of output characteristics between a single-winding configuration with N conductors per slot and the double-winding configuration of fig. 14. Note that the double-winding configuration provides higher output current when the speed is greater than 2700 rpm but the current is reduced at low-speed.

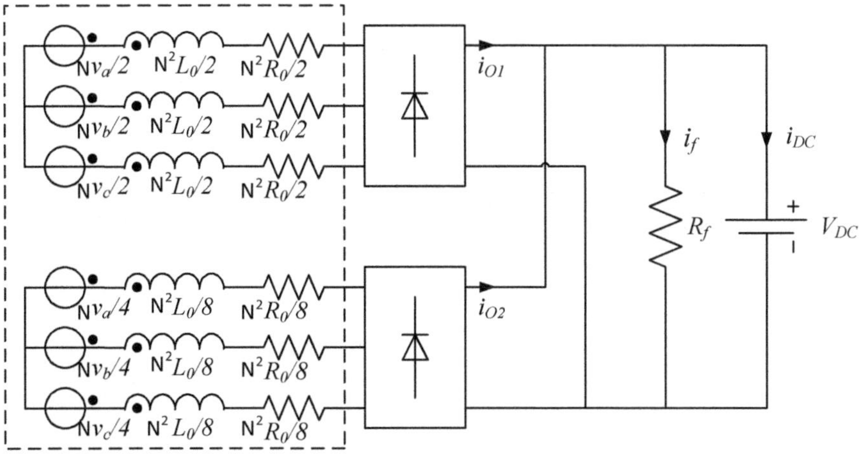

Figure 14. Multi- winding alternator with a number of turns equal to N for the first winding and N/2 for the second one.

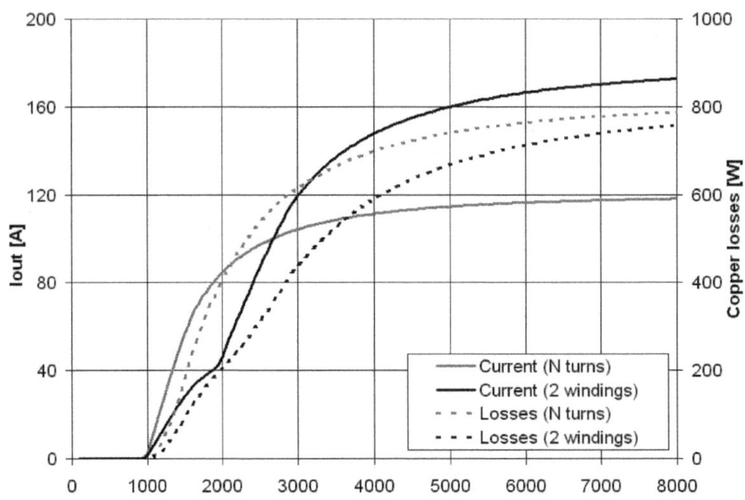

Figure 15. Performances comparison with a multi-winding stator configuration

To improve the performance over the whole speed range, it is more interesting to use the same number of turns in the two winding systems and make parallel/series reconfigurations at the rectifier outputs as shown in fig. 16. This method is easier to implement than the AC phase winding reconfiguration. It uses only one unidirectional switch and two additional diodes. Fig. 17 shows the performances of a double-winding configuration using the same number of conductors per slot as the original single-winding. A series connection provides the same output current as the original alternator, during low-speed operation when diode voltage drop is neglected. Once the speed exceeds 2700 rpm, a parallel connection is used to obtain higher output current and lower copper losses.

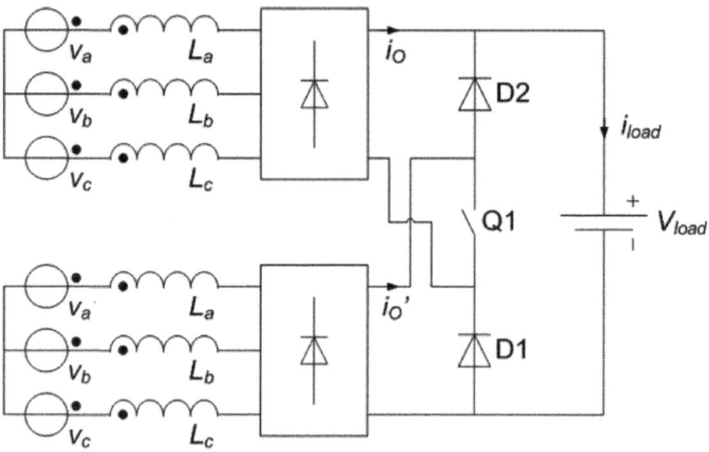

Figure 16. Reconfigurable parallel/series diode rectifiers

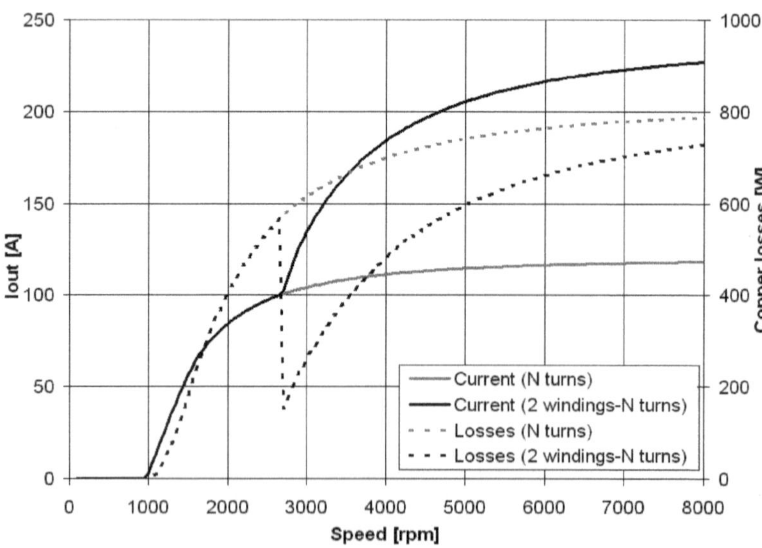

Figure 17. Performance of reconfigurable parallel/series diode rectifiers

4.2.3. Determination of parameters for rewound and multi-winding alternators

The winding scheme of a reconfigured alternator compared to the original one (Delcotron 22SI Type 12V-100A) is depicted in Fig. 18. In order to minimize the magnetic coupling, the first 3-phase winding is wound using half stator (18 slots) and the second 3-phase winding using the other half. The new lap windings have the same coil pitch as the original alternator and 12 conductors per slot. The alternator parameters are given in Table 2. The magnetic coupling between the two 3-phase winding systems has been measured for different rotor positions and the maximal measured mutual inductance between them reaches 4% of the measured self-inductance.

Assuming that magnetic coupling is negligible, the electrical parameters of a multiple three phase winding system can be computed from the original alternator parameters (L_0, R_0, Φ_{NL0} with N_0 conductors per slot). When the same stator is rewound with m three-phase windings and N conductors per slot, the new electrical parameters can be expressed as:

$$L = \frac{N^2}{m{N_0}^2} \cdot L_0 \quad , \quad R = \frac{N^2}{m{N_0}^2} \cdot R_0 \quad , \quad \phi_{NL} = \frac{N}{mN_0} \cdot \phi_{NL0} \tag{4}$$

The analysis of these equations and the experimental results (Table 2) confirms that both winding systems are magnetically decoupled.

	Original	Rewound	Double-Winding
Armature connection	Delta		
Pairs of poles	6		
Stator slots number	36		
Nominal field current	6 A		
Turns per slot	11	6	12
Wire cross section	1.75 mm^2	3.30 mm^2	1.65 mm^2
Resistance at 25°C	0.1 Ω	0.035 Ω	0.07 Ω
Cyclic Inductance (I_f = 6A)	390 μH	115 μH	225 μH
No-load flux (I_f = 6A)	28.6 mWb	15.8 mWb	16.0 mWb
Generation-starting speed (I_f = 6A)	910 rpm	1670 rpm	1650 rpm
Output current at 8000rpm (I_f =6A)	116 A	215 A	218 A
Efficiency at 8000rpm (I_f = 6A)	38 %	53 %	53.5 %

Table 2. Alternator parameters (Delcotron 22si type 12v-100a)

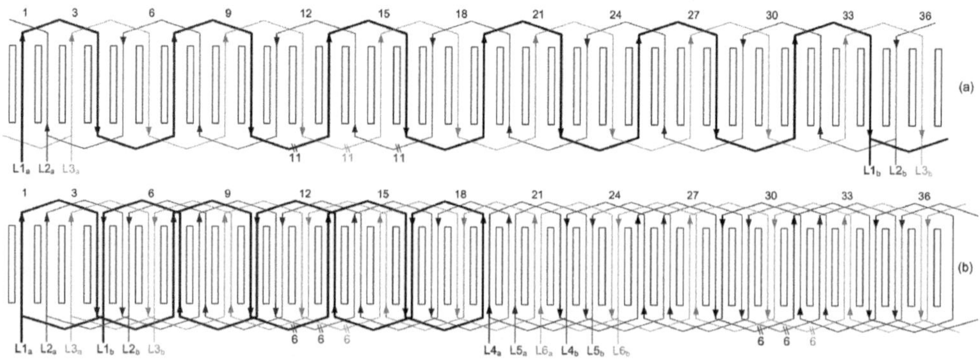

Figure 18. Winding scheme: (a) original 3-phase Delcotron alternator, (b) double-winding Delcotron alternator

4.3. Alternative converter solutions

4.3.1. Synchronous rectifier

A synchronous rectifier is an interesting alternative to conventional diode rectifiers. The main advantage of such converters is the reduced rectifier losses particularly at higher speeds if MOSFETs with low on-resistance (e.g. 4 mΩ (Beretta, 2008)) are used. Besides, it may be also employed in applications where bidirectional power transfer is required such as stop-start system (Beretta, 2008). In a start-stop system developed by Citroën, the claw-pole machine performs the functions of both starter and alternator. In the starter mode, the phase current and EMF in each phase are synchronised for maximum torque using a position sensor. In the alternator mode, the converter operates as a synchronous rectifier. The voltage drop may be as low as 0.2 V for an output current of 120 A in contrast to 0.8 V to 1.1 V for a diode. The losses can be 60 % lower compared to a diode rectifier (Beretta, 2008). In both modes of operation the switching losses will be much lower than PWM converters due to lower switching frequency. In the alternator mode the switching frequency may be as high as 2800 Hz (at 4000 rpm), however the drain currents are kept close to zero during MOSFET switching (Beretta, 2008).

4.3.2. PWM controlled full-bridge rectifier (PFBR)

In order to exploit the alternator at its maximum capabilities, it is interesting to set an optimum power angle as proposed in (Liang et al.,1999) and (Liang et al.,1996). This can be achieved with a PWM full-bridge rectifier (PFBR) as already shown in fig. 8. For each point of operation, the modulation index k and the angle θ between phase voltage and back-EMF are adjusted to maximize the output power with sinusoidal PWM control.

A PFBR is a quite expensive and complex solution; it counts for several active switches and requires rotor position sensing or complex sensorless algorithms (Boldea, 2006). However, like a synchronous rectifier, it offers bidirectional power flow control.

Fig. 19 shows the maximum output current and efficiency obtained by using a sinusoidal PWM and a control technique maximizing output power. It also shows the output current curves obtained with original and rewound alternators connected to a conventional diode rectifier. If compared to the rewound alternator with CDR (RACDR), the PFBR connected to the same alternator increases the output power for speeds below 4000 rpm. (For winding parameters see Table 1). During high speed operation, it is preferable to operate in synchronous rectification mode.

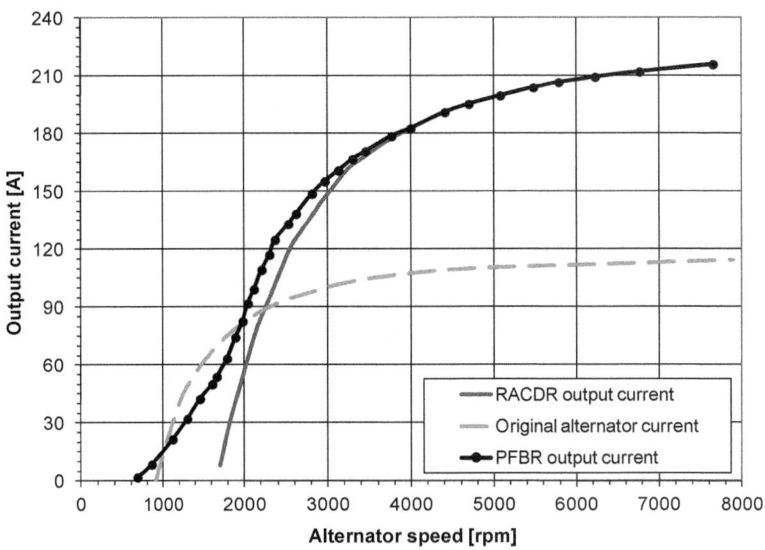

Figure 19. Experimental output current and efficiency vs. speed using a PFBR

The comparison of the output current characteristics with the original alternator shows that the current generated with the PFBR is lower in the range of 1000 to 2000 rpm (Fig. 19). Note that idle power requirement is not satisfied with the PFBR. This is partly due to the magnetic saturation and significant voltage drops across active switches.

4.3.3. *Other PWM rectifiers*

If bidirectional power flow is not required, the three single-phase BSBR structure shown in fig. 20 is a simpler solution. It has twice less active switches and all of them are referenced to the ground. Active switches can be reduced to only one using a Boost Switched-Mode Rectifier (BSMR), shown in fig. 21. With this topology, it is not necessary to use a rotor position sensor but the power angle can't be controlled.

4.3.4. *High frequency ripple reduction of the output current using interleaving*

All presented PWM rectifier topologies deliver an output DC current with abrupt current steps and high *di/dt's*. These high-amplitude fast-moving current transitions generate RF noise that flows to the battery and then pollutes other loads connected to the battery. Attenuation of these current variations to an acceptable level requires a very large output capacitor. This can lead to considerable efforts to comply with EMC standards (Maxim, 2001). Battery current ripple results in battery heating and a corresponding rise in temperature.

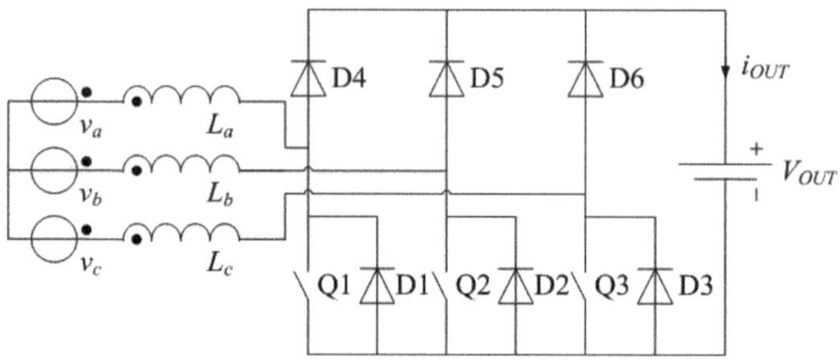

Figure 20. Boost semi-bridge rectifier (BSBR)

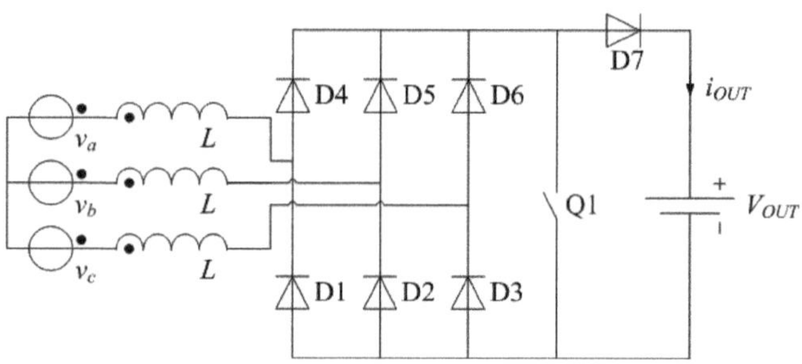

Figure 21. Boost switched-mode rectifier (BSMR)

Benefits of interleaved converters on current ripple, components stress and EMI reduction are well known for several different applications (Consoli et al., 2004), (Crebier et al., 2005). The main constraint of interleaved structures is the magnetic coupling between the different windings. Both converters must be connected to different three-phase windings that are not magnetically coupled to avoid a decrease of performances. Fig. 22 shows an example of two interleaved BSMRs with two identical three-phase winding systems having a same back-EMF. Q1 and Q2 are driven by two signals having the same duty-cycle and the same frequency but the phase is shifted by 180 degrees.

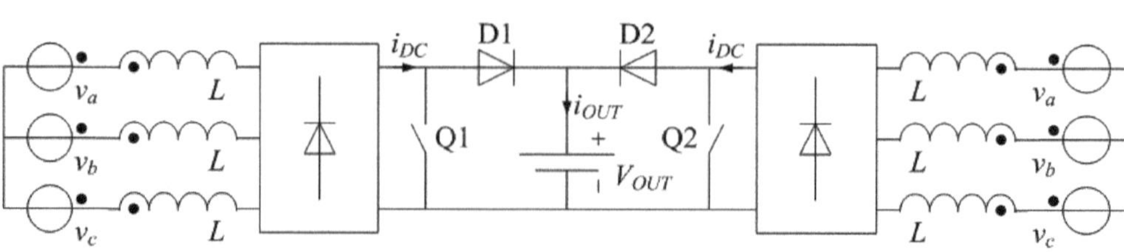

Figure 22. Two interleaved BSMRs connected to two identical three-phase windings

Interleaving decreases the current ripple for any value of duty-cycle and it allows for total ripple cancellation when $D = 0.5$. Equation (5) gives the optimal duty-cycle for a number m of interleaved rectifiers where k_i is an integer.

$$D_{opt} = \frac{1}{m}\left(m - k_i\right) \quad with \quad 0 < k_i < m \tag{5}$$

4.3.5. PWM BSMR with high frequency ripple cancellation

The structure of fig. 23 could be also implemented using the double-winding stator system presented above and two BSMRs operated with fixed optimal duty-cycle ($D = 0.5$). The rectifier control is thus extremely simple and no position sensing is required. Experimental current waveforms are shown in fig. 23 (left). When comparing the output current ripple with the one obtained with the non-interleaved converter presented in fig 23 (right), one can appreciate the considerable output current ripple reduction due to interleaving. (from 66A rms to 6.1A rms).

4.3.6. PWM BSBR with high frequency ripple cancellation

It is also interesting to operate two interleaved BSBRs with a fixed duty-cycle of 0.5 to obtain an ideal ripple cancellation. Hence the simplest control mode consists in driving all the switches of the first converter with the same gate signal and all the switches of the second converter with the same complementary signal. This method doesn't require additional rotor position sensing. Since the duty cycle is fixed at 0.5, no more control of the power angle is achievable.

The BSBR has a voltage drop per device less than the BSMR. This leads to better performance over the whole speed range

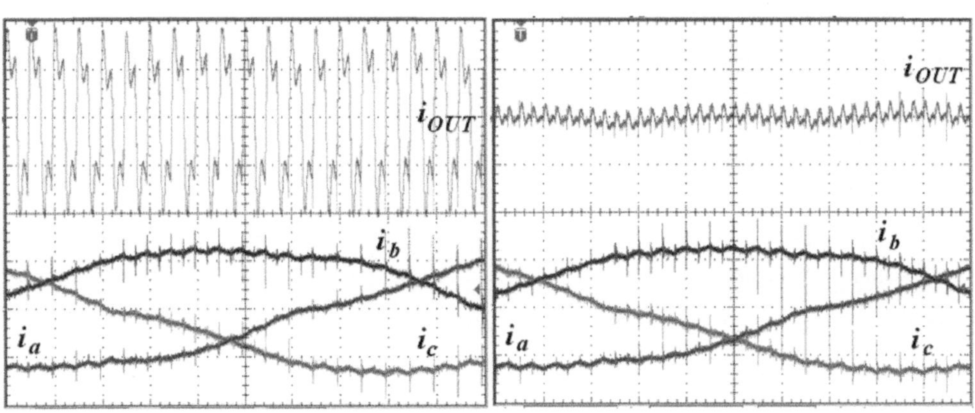

Figure 23. Experimental current waveforms for: left) interleaved BSMRs; right) non-interleaved BSMRs, at 2000rpm with If = 6A. (Vertical scale: 60 A/div, horizontal scale:200 μs/div)

4.4. General comparison

To compare the power improvement provided by the different rectifier topologies, the average power output $P_{avg.}$ and efficiency $\eta_{avg.}$ are estimated over the same drive cycle (a combination of the two vehicle speed cycles EPA UDDS and EPA US06) using the steady state performance curves with the rated field current. The results are given in Table 2.

Over the same driving cycle, the average power increase with respect to the original alternator ranges from 62% to 67% depending on the topology. The average efficiency has been improved by 10.6 to 11.5 percentage points.

The average output power and the efficiency are very similar for all the structures connected to a rewound machine since all structures regain the conventional diode rectifying mode during high-speed operation. In fact, the controlled rectifier is essentially used at idle speed.

Remarkable differences are although noticeable on the idle mode output power. Note that none of the topologies regains the original idle power. The closest idle power is obtained with the CSPR (-13%). Furthermore, the BSMR and BSBR deliver surprisingly more output power when interleaving is used. This can be partially explained by lower ESR losses in the output capacitor. However, the fact that two different machines are used (single-winding alternator and double-winding alternator) can have an influence too.

The approximate cost estimation for each solution can be derived from the number and ratings of the semiconductors and from the output filter size. It is assumed that the machine cost is not affected by rewinding and the filter size is proportional to the AC ripple component of the output current.

For semiconductors sizing, only rms current ratings are taken in account. In fact, in all topologies, each switch has to withstand the same reverse voltage which corresponds to V_{OUT} (voltage transients neglected). Current ratings are normalized with respect to I_{SC}. Table 3 shows the parts count and their ratings for each topology.

Topology	P_{idle}		$P_{avg.}$		$\eta_{avg.}$
Original Alternator	1059W	1.0pu	1454W	1.0pu	42.4%
Rewound alternator with PFBR	899W	0.85pu	2430W	1.67pu	53.6%
Rewound alternator with BSBR	831W	0.78pu	2413W	1.66pu	53.9%
Rewound alternator with BSMR	779W	0.74pu	2350W	1.62pu	53.4%
Double-winding alternator with interleaved BSBR	869W	0.82pu	2419W	1.66pu	53.8%
Double-winding alternator with interleaved BSMR	799W	0.75pu	2353W	1.62pu	53.0%
Double-winding alternator with CSPR	918W	0.87pu	2389W	1.64pu	53.5%

Table 3. Idle power, average output power and efficiency over the same driving cycle for each topology

In the case of interleaved structures, semiconductor parts are multiplied by two. However, the active silicon area remains the same and distribution of switching power can eventually even be advantageous. Also note that for non-interleaved structures using PWM, the rms value of the ripple component of the output current is about 10 times more significant than with other structures. The resulting filtering requirements for EMC standards compliance can easily lead to bulky and expensive solutions.

Topology	Number of Slow diodes	Slow diode rms current (I_{SD})	Number of Fast diodes	Fast diode rms current (I_{FD})	Number of Active switches	Active switch rms current (I_Q)	Output current AC ripple (I_{AC})
Original Alternator	6	0.52pu	-	-	-	-	0.04pu
PFBR	-	-	6	0.52pu	6	0.37pu	0.48pu
BSBR	3	0.52pu	3	0.52pu	3	0.37pu	0.5pu
BSMR	6	0.52pu	1	1.0pu	1	1.0pu	0.5pu
Interleaved BSBR	6	0.26pu	6	0.26pu	6	0.18pu	0.04*pu
Interleaved BSMR	12	0.26pu	2	0.5pu	2	0.35pu	0.04*pu
CSPR	12	0.26pu	2	0.5pu	1	0.5pu	0.04pu

* neglecting HF ripple through inductances

Table 4. Semiconductor parts count and normalized current ratings for each topology

5. Conclusion

The low efficiency and the limitation of the output power are major drawbacks of the Lundell alternator. Replacing these alternators with other types of machine is not an economic choice because of their low manufacturing cost. However, improving the performance of existing machines is still the best way for the short term future. This chapter discusses the performance of the conventional automotive alternators and various modeling methods for the simulation of the alternator-rectifiers. Some improvements without geometry modification have been proposed and validated using standard frame of automotive alternator. The results show that the modification of the number of turns and replacing the diode rectifier with other electronic converters could significantly increase the output current and the efficiency. However, there are many other possibilities for enhancing the performance which have not been considered. For example, the transition from 14V to a 42V system, will allow to increase the efficiency of the power electronics rectifier by reducing the conduction losses. The optimization of the claw-pole machine, the use of hybrid structure with permanent magnets, laminations of higher quality, and liquid cooling are alternative methods to enhance the performance.

Author details

Ruben Ivankovic, Jérôme Cros, Mehdi Taghizadeh Kakhki,
Carlos A. Martins and Philippe Viarouge
Laval University, Canada

6. References

Beretta, J. (2008). *Electronique, électricite et mecatronique automobile,* Lavoisier, ISBN:978-2-7462-1245-9, Paris

Boldea, I. (2006). Automotive Claw-Pole-Rotor Generator Systems, In: *Variable speed generators,* Taylor & Francis Group, ISBN 0-8493-5715-2

Bosch (2003). Alternators, In: *Automotive electrics and electronics,* Bauer, H., pp. 112-163, SAE International, ISBN 0-7680-0508-6, Stuttgart

Consoli, A.; Cacciato, M.; Scarcely, G. & Testa, A. (2004). Compact, reliable efficiency. *Industry Applications Magazine, IEEE,* vol.10, no.6, pp. 35- 42, Nov.-Dec. 2004

Crebier J.C.; Revol, B. & Ferrieux J.P. (2005). Boost chopper derived PFC rectifier: Interest and Reality. *IEEE Transactions on Industrial Electronics,* Vol.52, No.1, February 2005, pp 36-45

Cros, J.; Paynot, C.; Figuerora, J. & Viarouge, P. (2003). Multi-Star PM brushless DC motor for traction application, *EPE'2003,* Toulouse, 2-4 sept. 2003.

Cros, J; Radaorozandry, L; Figueroa, J & Viarouge, P. (2008). Influence of the magnetic model accuracy on the optimal design of a car alternator, in *Int. Journal for Computation and Mathematics in Electrical and Electronic Engineering -COMPEL,* Vol 27, Issue 1, pp. 196-204, 2008.

Figueroa, J.; Cros, J. & Viarouge, P. (2005). Analytical model of a 3-phase rectifier for the design of a car alternator, *Proceedings of Electrimacs conference,* Tunisia, Apr. 2005

Henneron T., Clénet S., Cros J. & Viarouge P. (2004), "Evaluation of 3D Finite Element Method to Study and Design a Soft Magnetic Composite Machine", *Magnetics, IEEE Transactions on,* Vol.40 , No2, pp786 – 789.

Kuppers, S.; Henneberger, G. (1997), "Numerical procedures for the calculation and design of automotive alternators," *Magnetics, IEEE Transactions on,* vol.33, no.2, pp.2022-2025, Mar 1997

Liang, F.; Miller, J.M. & Xu, X. (1999). A vehicle electric power generation system with improved output power and efficiency. *Industry Applications, IEEE Transactions on,* vol.35, no.6, pp.1341-1346, Nov/Dec 1999

Liang, F.; Miller, J.M. & Zarei, S. (1996). A control scheme to maximize output power of a synchronous alternator in a vehicle power generation system, in *Conf. Rec. IEEE-IAS Annual Meeting,* Oct. 1996, pp. 830-835

Maxim/ Dallas semiconductors (2001), Circuit tradeoffs minimize noise in battery-input power supplies, In: *Application note 653,* 8 pages, 22 January. 2001. Available from: http://www.maxim-ic.com/an653

Nipp, E. & Norberg, E. (1998) , "On the feasibility of switched stator windings in permanent magnet motors for traction drives," *Railroad Conference, 1998. Proceedings of the 1998 ASME/IEEE Joint* , vol., no., pp.33-39, 15-16 Apr 1998

Ostovic, V.; Miller, J.M.; Garg, V.K.; Schultz, R.D. & Swales, S.H. (1999), "A magnetic-equivalent-circuit-based performance computation of a Lundell alternator," *Industry Applications, IEEE Transactions on* , vol.35, no.4, pp.825-830, Jul/Aug 1999

Perrault, D. & Caliskan, V. (2004). Automotive Power Generation and Control. *IEEE Transactions on Power Electronics,* Vol. 19 No. 3, May 2004, pp 618-630.

Chapter 8

A Road Ice Sensor

Amedeo Troiano, Eros Pasero and Luca Mesin

Additional information is available at the end of the chapter

1. Introduction

The investigation of ice formation found important applications in many different fields. For instance, the accretion of ice is a common occurrence on the aircraft, due to the high velocity and humidity, and low temperature at upper air. The detection of ice formation on aircrafts is a critical issue since not only the aerodynamics change, but there is also the possibility of ice liberation from the aircraft during the flight, and falling ice could strike and possibly damage the engine [23]. Another applications of the investigation of ice formation are on the seas, in order to prevent iceberg formation and possible accident due to the crash with boats [39]. An adequate assessment of the environmental conditions of road surfaces may significantly contribute to enhance traffic safety, since corresponding decision made by administrators should be based on this information [18]. An important application of the investigation of ice formation is on the runway of airports, in order to improve safety during take off and landing of the aircrafts. Moreover, a correct estimation of the environmental condition of road and runway surfaces can be useful to reduce costs due to de-icing systems and ice control [27]. Finally, an indication of the environmental condition of the pavements could be adopted on the walkways, to prevent people falls.

2. State of the art

The problem of ice detection is not a new one: over the time, several solutions have been proposed depending on the application [9]. The proposed solutions can be divided into two families: detection of ice based on optical methods and using physical sensors.

Optical methods are indirect techniques that detect icy conditions applying image-processing algorithms to the images acquired by different systems. Wide-area ice detection can be obtained processing images by weather satellites [17, 26, 37], which are used to monitor the climate of the Earth. Weather satellites can be either polar orbiting, seeing the same swath of the Earth, or geostationary, hovering over the same spot on Earth by orbiting over the equator while moving at the speed of the Earth's rotation. One of the most used optical method is based on Time-Domain Reflectometry (TDR) technique [4, 32]. TDR is a measurement technique commonly used to determine characteristics of electrical lines by observing reflected waveforms. The time-domain reflectometry technique can also be used for detection of ice and water on the road, based on the principle that the propagation velocity of an electromagnetic wave in a transmission line depends on the relative permittivity of the material; and the relative permittivity, in turn, depends on the water or ice content of the medium. Ice detection can be also obtained using microwave systems [6]. A MicroWave Radiometer (MWR) is a system that measures energy emitted at sub-millimetre to centimetre wavelengths (at frequencies of 1-1000GHz), known as microwaves. Studying the physical processes associated with the energy emission at these wavelengths, it is possible to estimate a variety of surface and atmospheric parameters, including air and sea surface temperature, salinity of sea ice, precipitations, and sea ice. River and sea ice can be detected applying specific algorithms to the data collected by acoustic Doppler current meters [11]. These instruments are designed for monitoring water level, velocity, and flow of rivers and seas. Since the velocity distribution of water changes significantly when ice is present, specific algorithms can be developed to determine the presence of ice in rivers or seas. Ocean ice can be detected processing the images obtained by MODerate resolution Imaging Spectroradiometer (MODIS) [22, 39]. MODISs are instruments launched into Earth orbit by NASA in 1999 on board of satellites. MODIS captures images of the entire Earth using spectral bands ranging in wavelength from $0.4\mu m$ to $14.4\mu m$, at different spatial resolutions. The last optical method used for the detection of ice is based on ultrasonic waveguides [3, 7, 16], that are structures which guide waves, such as electromagnetic or sound. Ultrasonic guided waves can recognizes water or ice since the amplitude of the A0 mode Lamb wave is influenced by water loading and ice formation on the surface of the aluminum plate of the waveguide.

In order to detect ice directly, physical sensors are used. An ice detection system based on a physical sensor can be obtained using micro-fabricated diaphragms as sensing elements [29]. Accumulation of ice on a diaphragm leads to an increase in its effective stiffness. The sensor is composed by a diaphragm, which is the sensing element and is electrically isolated, and an electrode. When a voltage is applied between the diaphragm and the electrode, an attractive electrostatic force will cause the diaphragm to deform toward the electrode, increasing the capacitance between the electrode and the diaphragm. For a given actuation voltage, it was shown that an ice-covered diaphragm exhibits a smaller deflection than a corresponding ice-free diaphragm. Another kind of ice detection system designed for aircrafts is obtained using fiber optics [12, 13, 20, 21]. When there is no ice accretion on the probe surface, the light from the emitter propagates through the source fiber bundle into the air, and no signal can be detected by the destination fiber bundle. When there is ice accreting on the probe, light entering the volume of ice is in part transmitted and in part reflected into the destination fiber bundle, which couples with the detector that determines the thickness of ice depending on the light intensity distribution. Physical sensor for ice detection system can also be based on resonant piezoelectrics [30]. Resonant structures are commonly used in sensor systems due to their high sensitivities to small variations in applied loads. Ice detection systems based on resonant sensors are based on the principle that accretion of ice on a cylindrical probe leads to a decrease in the measured resonant frequency, due to the increase in effective mass of the resonator. In such mass-loaded resonant ice detection sensor, variations in probe design are usually restricted to favor ice accretion and minimize resonant frequency shifts due to accumulation of water and other fluids.

3. The ice detection system

The previously described methods do not fit the requirements for a direct detection of ice in the road or in the runway surfaces. For example, optical methods cannot be used for this application because they are indirect techniques. Moreover, resonant probes are too power consuming and optical fibers are too expensive for the development of this sensor. Since the sensor should be embedded in the road or in the runway, it has to resist on the pressure of trucks or airplanes, or also possible chemical substances present on the road or on the runway, so diaphragms cannot be used for the presented application. For these reasons, an innovative sensor for water and ice detection was designed.

The ice detection system consists of a relative permittivity measurement [25, 33–36]. In general, permittivity is defined as a measure of how an electric field affects and is affected by a dielectric medium, and is determined by the ability of a material to polarize in response to the field, and thereby to reduce the total electric field inside the material. The relative permittivity (also called dielectric constant) is the ratio of the amount of stored electrical energy when a potential is applied, relative to the permittivity of the vacuum. The relative permittivity depends on temperature and measurement frequency. The relation between the relative permittivities of water and ice, and the temperature and measurement frequency, is shown in Figure 1 [5, 31, 38] (note the different scales of the frequency axes). It can be seen that the relative permittivity of ice at approximately -1°C is substantially constant within a range from DC to about 1kHz, and decreases in the range of approximately 2kHz to several hundred kHz. On the other hand, the relative permittivity of water at approximately 1°C is substantially constant up to approximately 10^9Hz and decreases within the range from 10^9Hz to 10^{10}Hz. The relative permittivity of air can be assumed to be equal to 1 and constant for each frequency and temperature. Therefore, at low frequencies the relative permittivities of water and ice are similar, while the relative permittivity of air is different from the others. Instead, at high frequencies the relative permittivities of air and ice are similar, while the relative permittivity of water is different from the others [25, 33–36]. So, estimating the relative permittivity, it is not possible to distinguish reliably water and ice at low frequencies (lower than 1kHz), but only air can be identified; on the other hand, ice and air cannot be distinguished at high frequencies (between 10kHz and 1GHz), but it is possible to identify only the presence of water. Thus, for an exact classification of the material placed over the sensor, distinguishing among ice, water, or air, it is necessary to estimate the relative permittivity of the material on the surface of the sensor at two frequencies (low and high frequency).

The relative permittivity can be sensed by a capacitance measurement. In general, the capacitance value of an electrode assembly depends on the geometrical configuration and dimensions of the surfaces of the electrodes, and on the permittivity and thickness of the material placed between the electrodes. The device consists in a multi-frequency capacitance measurement. As low measurement frequency 200Hz was considered; while 20MHz was chosen as high measurement frequency. The sensor includes a pair of concentric conductive electrodes (with geometry and dimensions shown in Figure 2). Such electrode geometry was chosen since the capacitance measurement is invariant to the rotations of the electrode assembly and to the position of raindrops or pieces of ice with respect to the electrodes. The pair of concentric ring conductive electrodes, which are the sensing device, was directly implemented on a PCB, as showed in Figure 2. The external maximum ring dimension was imposed by the box which includes the sensor, while the external minimum ring and internal circle dimensions were choosen in order to explore a largest area between the two electrodes,

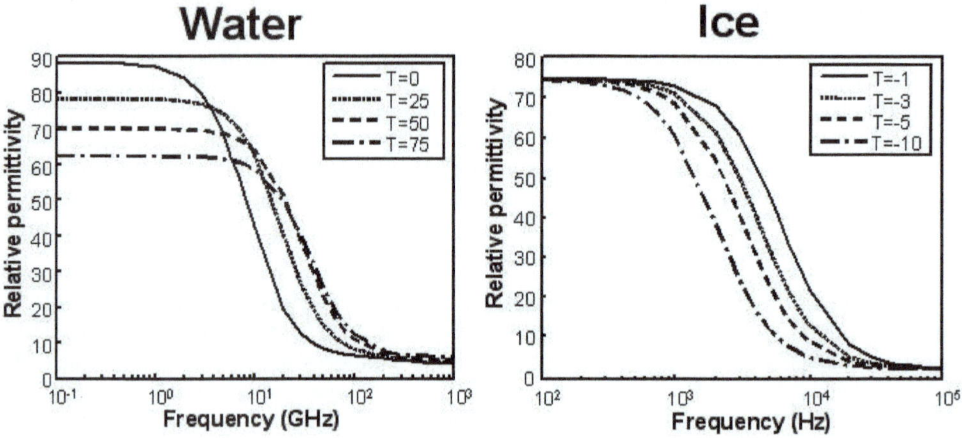

Figure 1. Representation of the relative permittivity of water and ice at different frequencies and temperatures.

but still large enough to get a measurable value of capacitance. The capacitance value of the pair of concentric ring conductive electrodes was measured using a high resolution RCL meter (Fluke - PM 6306). The capacitance value was in the order of 0.3pF.

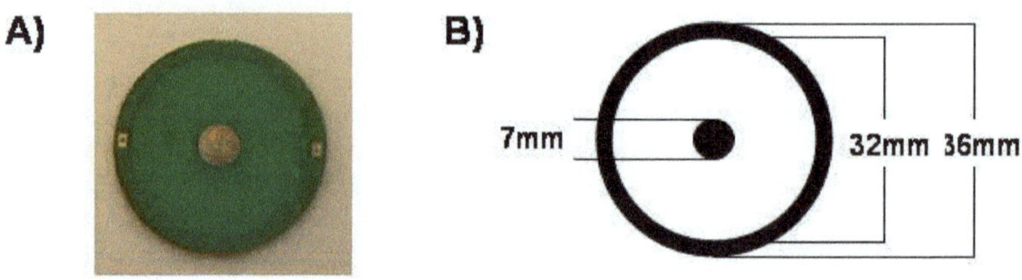

Figure 2. Representation of the pair of concentric ring conductive electrodes. A) Top view of the electrode assembly. B) Geometry of the electrode assembly.

The sensor electrodes of the ice detector are directly connected to a capacitance measurement system, which is a transfer charge circuit. In general, the transfer charge circuit includes a sensing capacitor, a frequency generator, and a charge detector. The schematic layout of the capacitive measurement circuit is shown in Figure 3. The sensing capacitor is obtained by the pair of concentric conductive electrodes and the material placed over the electrodes, and it is indicated in the Figure as C_X. The frequency generator is obtained by a reference voltage source V_R and a controllable switch S_1 to provide different frequencies. In the basic measurement circuit, the charge detector comprises only a reference capacitance C_S that is connected to C_X by closing the switch S_2. Note that S_2 opens when S_1 is closed, and viceversa. The charge Q_X stored in the electrode assembly is:

$$Q_X = C_X \cdot V_R \tag{1}$$

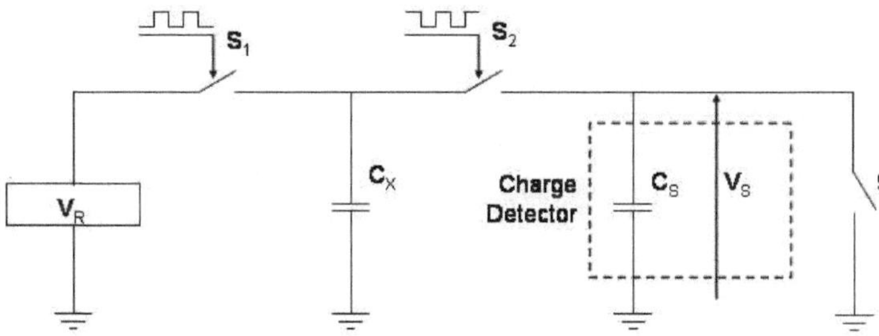

Figure 3. Schematic layout of the transfer charge circuit

When the switch S_2 is closed, the charge stored in the electrode assembly is transferred to the detection capacitance:

$$Q_X = (C_S + C_X) \cdot V_S \tag{2}$$

Since the value of the sensor electrode is several orders of magnitude lower than C_S, nearly all charge stored in C_X is transferred to the detection capacitance:

$$Q_X = C_X \cdot V_R = C_S \cdot V_S \tag{3}$$

and therefore, measuring the voltage level V_S reached by the detection capacitance, the capacitor value of the electrode assembly can be computed as:

$$C_X = C_S \cdot \frac{V_S}{V_R} \tag{4}$$

In order to increase accuracy in measuring the very small value of the capacitance C_X, the sensor was charged n times and its charge was transferred n times to the reference capacitor before taking a measurement of V_S. Therefore the value of the capacitance of the electrode assembly is given by:

$$C_X = \frac{C_S}{n} \cdot \frac{V_S}{V_R} \tag{5}$$

At the end of the measurement process, C_S is discharged by closing the switch S_3. In the currently available ice sensor, the reference voltage V_R was equal to 3V (tolerance of 0.2%), the value of the reference capacitance C_S was equal to 2.2nF (tolerance equal to 1%), and the number of times in which the reference capacitor was charged before taking a measurement (n) was 50. The voltage V_S was first amplified by a factor 150 by an analog circuit and then sampled by the analog to digital converter (10 bit resolution) of a microcontroller (8051 microcontroller from Silicon Laboratories Inc). The microcontroller is also used to drive the switches S_1, S_2, and S_3. These operations are performed for both measurement frequency in sequence, low and high frequencies, using a sampling frequency of 0.2Hz. The transfer charge circuit is implemented on a PCB using commercially available low power and low cost components. The PCBs of the electrodes arrangement and of the transfer charge circuit are directly connected together.

As it stated above, the value of capacitance of the electrode assembly (without water or ice placed over it) is about 0.3pF. However, parasitic capacitances of wires and electronic circuits are, in general, on the order of several pF. Therefore, the value of parasitic capacitances is bigger than that of the unknown capacitance, and consequently it is not possible to estimate the capacitance of the electrode assembly using the previously described circuit. In order to prevent error in the data due to parasitic capacitances, a differential capacitance measurement circuit was used. Parasitic capacitances are first estimated in a preliminary calibration phase, performed in laboratory, measuring the capacitance of the sensor without placing water or ice over the electrode assembly, and the measured value is stored in the flash memory of the microcontroller. This value results from the combination of the capacitance of the dry electrode assembly and the parasitic capacitances. Then, during the measurement, the obtained value of V_S is subtracted by the calibration value in order to remove the contribution of parasitic capacitances (also the value of the dry electrode assembly is removed). The calibration value stored in the flash memory of the microcontroller, which is subtracted to the measured value, is obtained by the digital to analog converter (10 bit resolution) of the microcontroller. A schematic layout of the differential capacitance measurement circuit is shown in Figure 4.

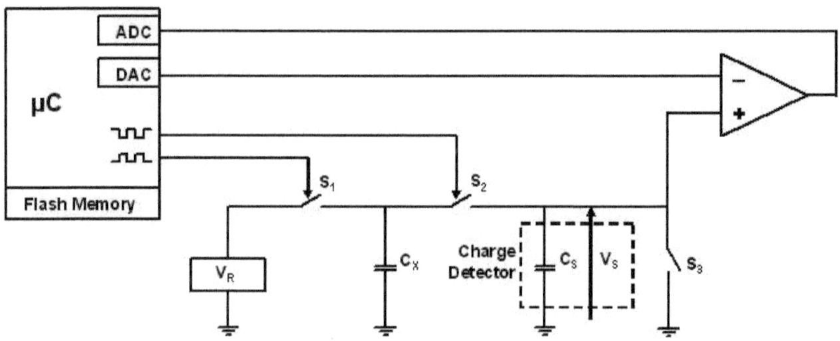

Figure 4. Schematic layout of the differential capacitance measurement circuit.

The device also comprises a temperature sensor, to account for the variation of the relative permittivity with temperature. The temperature obtained by the sensor is sampled by the analog to digital converter (10 bit resolution) of the microcontroller, using a sampling frequency of 0.2Hz.

The ice detection system arrangement (circuit and electrode assembly) was included in a metallic box filled with resin, which protects the circuitry from infiltration of water or chemical agents. The only exposed parts are the Arnite covering the sensor (on the top) and the connector for the power supply and the RS 485 communication protocol (on the bottom). Arnite (Polyethylene Terephthalate) was chosen for protection purposes since its relative permittivity is nearly constant within the range of temperatures and measurement frequencies in which the sensor is used. Moreover, it has a high value of hardness to steel ball (180 $\frac{N}{mm^2}$) and resistance to traction (255 $\frac{N}{mm^2}$). A layer 3 mm thick of Arnite was mounted over the electrode assembly. The final prototype of the ice detection system is shown in Figure 5.

4. Mathematical model of the sensor

A mathematical model of the sensor was designed. The sensor was described using a multilayer electrostatic model [15]. Indeed, at the maximum measurement frequency the wave length of the electromagnetic field is more than two orders of magnitude greater than the dimensions of the sensor. Three layers were included: dielectric (the Arnite protecting the

Figure 5. Picture of the final prototype of the ice detection system.

sensor) covering the electrodes (z_1), air or water or ice placed over the sensor (z_2), and air (z_3), as showed in Figure 6 (z_i indicates the thickness of each layer). Each layer was assumed to be homogeneous, isotropic, of constant thickness, and with a constant value of relative permittivity (ε_i). These three layers were positioned over the concentric circular electrode arrangement constituted by a circular electrode of radius 3.5mm (r_0) and a ring electrode with external radius of 18mm (r_{EXT}) and an internal one of 16mm (r_{INT}).

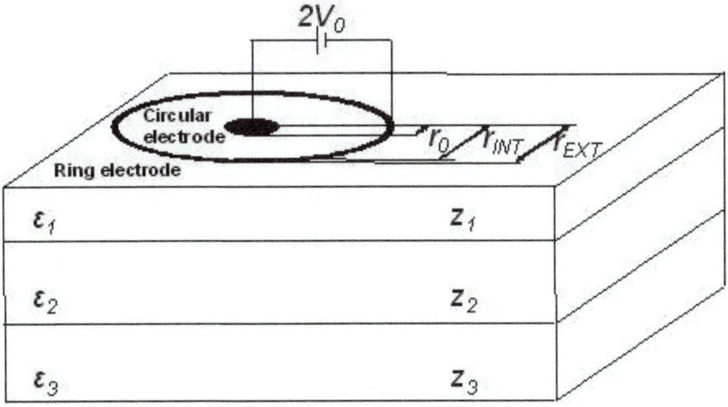

Figure 6. Representation of the three layers model of the sensor.

For the first layer, the Arnite, a constant value of relative permittivity ε_1 was assumed, equal to 3.5. For the second layer, the relative permittivity of water, ice, or air was modelled. The relative permittivity of water ε_2 was modelled as [24]:

$$\varepsilon_2 = \varepsilon_\infty + \frac{\varepsilon_s - \varepsilon_i}{1 + j\frac{f}{f_1}} + \frac{\varepsilon_i - \varepsilon_\infty}{1 + j\frac{f}{f_2}} \tag{6}$$

where all parameters (static permittivity ε_s, high frequency permittivity ε_∞, intermediate frequency permittivity ε_i, and relaxation frequencies f_1 and f_2) are functions of temperature.

The relative permittivity of ice (again indicated with ε_2) was modelled as a Debye model [8]:

$$\varepsilon_2 = \varepsilon_\infty + \frac{\varepsilon_s - \varepsilon_\infty}{1 + j2\pi\tau f} \tag{7}$$

in which the static permittivity ε_s was 75, the high frequency permittivity ε_∞ was 3.2, and the relaxation time τ was dependent on temperature, assuming a linear variation between the values $1.4 \cdot 10^{-4}$ at -20°C and $2.5 \cdot 10^{-5}$ at 0°C [28]. The relative permittivity ε_3 of the third layer, the air, was assumed constant and equal to 1.

The mathematical model of the sensor is an electrostatic equation for dielectrics. Gauss's law for dielectrics [15], is considered:

$$-\nabla \cdot (\varepsilon \nabla \phi) = 0 \tag{8}$$

where ϕ is the potential (V) and ε the relative permittivity (F/m). Since the mathematical model is constituted by three layers, the electrostatic equation for dielectrics was studied in different sub-domains in which Laplace equation $\Delta\phi = 0$ was solved (with $\Delta = \nabla^2$). Each sub-domain indicates a layer. Such sub-domains were coupled by the interface conditions between two layers requiring that the potential and the dielectric displacement $\vec{D} = \varepsilon\nabla\phi$ were continuous crossing the interface.

As the electrodes are circular (see Figure 6), the problem was considered symmetrical in cylindrical coordinates (ρ, θ, z). A mixed boundary value problem was studied, imposing opposite value of potential (indicated as V_0 in Figure 6) on the two electrodes and vanishing dielectric displacement on the other part of the boundary. Due to the cylindrical symmetry of the problem, the potential does not depend on the angle θ, but only on the radius ρ and on the depth within the layers z. The problem can be stated as follows:

$$\begin{cases} \Delta\phi_1 = \frac{\partial^2\phi_1}{\partial\rho^2} + \frac{1}{\rho}\frac{\partial\phi_1}{\partial\rho} + \frac{\partial^2\phi_1}{\partial z^2} = 0 & \rho > 0,\ 0 < z < z_1 \\ \Delta\phi_2 = \frac{\partial^2\phi_2}{\partial\rho^2} + \frac{1}{\rho}\frac{\partial\phi_2}{\partial\rho} + \frac{\partial^2\phi_2}{\partial z^2} = 0 & \rho > 0,\ z_1 < z < z_1 + z_2 \\ \Delta\phi_3 = \frac{\partial^2\phi_3}{\partial\rho^2} + \frac{1}{\rho}\frac{\partial\phi_3}{\partial\rho} + \frac{\partial^2\phi_3}{\partial z^2} = 0 & \rho > 0,\ z_1 + z_2 < z < z_1 + z_2 + z_3 \\ \phi_1 = \phi_2 & \rho > 0,\ z = z_1 \\ \phi_2 = \phi_3 & \rho > 0,\ z = z_1 + z_2 \\ \varepsilon_1\frac{\partial\phi_1}{\partial z} = \varepsilon_2\frac{\partial\phi_2}{\partial z} & \rho > 0,\ z = z_1 \\ \varepsilon_2\frac{\partial\phi_2}{\partial z} = \varepsilon_3\frac{\partial\phi_3}{\partial z} & \rho > 0,\ z = z_1 + z_2 \\ \phi = V_0 e^{j2\pi ft} & \rho \le \rho_0,\ z = 0 \\ \phi = -V_0 e^{j2\pi ft} & \rho_{INT} \le \rho \le \rho_{EXT},\ z = 0 \\ \frac{\partial\phi}{\partial\rho} = 0 & \rho = 0 \\ \frac{\partial\phi}{\partial z} = 0 & \rho_0 < \rho < \rho_{INT} \cup \rho > \rho_{EXT},\ z = 0 \\ \frac{\partial\phi}{\partial z} = 0 & z = z_1 + z_2 + z_3 \end{cases} \tag{9}$$

where ρ_0 is the radius of the internal electrode, ρ_{INT} and ρ_{EXT} are the internal and external radiuses defining the ring shape electrode, and z_1, z_2 and z_3 are the thicknesses of the layers. The estimated capacitance is a monotonic increasing function of the thickness of the third layer (air), but saturates within a few simulated mm of air; thus, the simulated thickness of air was chosen to be 25mm, assuring that the capacitance curve saturates.

In order to solve the mathematical problem, the domain should be limited. The domain was limited imposing a maximum radius of 30mm where homogeneous Neumann boundary conditions $\frac{\partial \phi}{\partial \rho} = 0$ were assumed. The mathematical problem was solved using the finite difference method, which is a numerical method for approximating the solutions of differential equations using finite difference approximation of derivatives [19]. A non-uniform discretization of the domain was used, with increasing resolution close to the electrodes and to the interfaces. Specifically, the sampling step of the radius $\Delta\rho$ was 1% of the maximum radius (30mm) close to the electrodes and 3% of the maximum radius otherwise. The discretization step of the depth variable Δz was 1% of the sum of the thickness of the three layers $z_1 + z_2 + z_3$ close to the electrode surface $z = 0$ and to the interfaces $z = z_1$ and $z = z_2$, and 3% of the before indicated sum otherwise.

First and second order derivatives were discretized, using the finite difference method, with a second order approximation, both within the domain and on the boundary:

$$\begin{aligned} &\left.\frac{\partial u}{\partial x}\right|_{x_0} \cong au(x_0) + bu(x_0 + h_1) + cu(x_0 + h_2); \\ &\text{where } \mathrm{a} = -\frac{\mathrm{h_1+h_2}}{\mathrm{h_1 h_2}},\ \mathrm{b} = -\frac{\mathrm{h_2}}{\mathrm{h_1(h_1-h_2)}},\ \mathrm{c} = \frac{\mathrm{h_1}}{\mathrm{h_2(h_1-h_2)}} \\ &\left.\frac{\partial^2 u}{\partial x^2}\right|_{x_0} \cong au(x_0) + bu(x_0 + h_1) + cu(x_0 + h_2); \\ &\text{where } \mathrm{a} = \frac{2}{\mathrm{h_1 h_2}},\ \mathrm{b} = \frac{2}{\mathrm{h_1(h_1-h_2)}},\ \mathrm{c} = -\frac{2}{\mathrm{h_2(h_1-h_2)}} \end{aligned} \tag{10}$$

where h_1 and h_2 have the same sign when the considered point x_0 is on the boundary and different signs when x_0 is within the domain. A linear system of algebraic equations was obtained after discretization. The potential was estimated resolving such a system using Gauss elimination method. Given the potential, the charge q over the internal electrode (which is the same except for the sign as that over the external ring electrode) was estimated, using the integral form of Gauss's law [15], as:

$$q = \Re\left[\int\limits_{\rho \le \rho_0, z=0} \vec{D} \cdot \hat{n} dS\right] = 2\pi\varepsilon_1 \int\limits_0^{\rho_0} r \frac{\partial \Re[\phi]}{\partial r} dr \tag{11}$$

Finally, the capacitance value was obtained as [15]:

$$C = \frac{q}{2V_0} \tag{12}$$

The mathematical model of the sensor was validated using both analytical and experimental data [36]. Validation based on analytical data was obtained changing the boundary conditions of the model, allowing the simulation of a capacitor with parallel planar plates with a layer of dielectric covering the plates. The relative error between the analytical and simulated solutions was lower than 1.5%. Validation based on experimental data was performed simulating the dry sensor, in which a capacitance value of about 0.3pF was obtained. This is the capacitance value of the dry electrode arrangement, which was measured using a high resolution RCL meter (Fluke - PM 6306), indicating a capacitance value of about 0.3pF. Moreover, simulated and experimental values of capacitance were compared for different temperatures and frequencies, and placing water or ice over the sensor. Results shown a relative error between the experimental and simulated solutions lower than 2%.

5. Results

5.1. Laboratory results

The reliability of the estimates provided by the sensor was investigated during laboratory experiments. Laboratory tests were performed applying the same environmental conditions to more sensors and evaluating the dispersion of the time instants in which phase changes of water were detected. Nine sensors were placed at the same time in a climatic chamber (Angelantoni - Challenge 250; temperature range for climatic test from -40°C to +180°C). At the beginning of the experiment, the climatic chamber was set to ambient conditions, with a temperature of 25°C and humidity of 50%, for approximately 10 minutes in order to adjust the parameters of the sensor. Then, 1 mm of tap water was placed over each sensor. Sensors were left in ambient conditions for 10 minutes. Then, the climatic chamber was arranged to reach -20°C with a temperature gradient of -1°C per minute. During this period, water placed over sensors frozes. The climatic chamber kept the temperature of -20°C for approximately 10 minutes and then it was arranged to reach 25°C with a temperature gradient of 1°C per minute. During this period, the ice formed over the sensors melts. The climatic chamber kept the temperature of 25°C for approximately 10 minutes. Then each sensor was dried. Data were acquired for additional 10 minutes, with a temperature of 25°C and humidity of 50%. The experiment was repeated in three different days, in order to investigate the repeatability of the capacitance values and the reliability of the estimates provided by the algorithm.

During the experiments, values of capacitance at different measurement frequencies (200Hz, 500Hz, and 20MHz) were continuously acquired. In Figure 7 the data acquired and processed from the sensor eight during the first experiment are shown. In detail, Figure 7A shows the capacitance values (raw data) at different measurement frequencies. Variations of the values of capacitance in different states of the sensors are visible. During the dry state, values of capacitance (at both low, medium, and high frequencies) were close to zero, due to the calibration procedure. During the wet state, the values of capacitance at high, medium, and low frequency were close to 0.3pF, so there is no distinction between different measurement frequencies (since the relative permittivity of water is constant at both low, medium, and high frequencies). However, during the icy state, the values of capacitance at high frequency are close to 0.15pF while at medium and low frequency are close to 0.3pF (since the relative permittivity of ice is different at both low and medium frequencies with respect to that at high frequency), so the value of capacitance can be easily distincted for different measurement frequencies. Moreover, there are no differences between values of capacitance obtained for low and medium frequencies. Low pass (cut-off frequency of 0.002Hz, 100th order causal FIR filter) filtered values of capacitance are shown in Figure 7B. It can be seen that high frequency variations and instrumentation noise are not present in the filtered capacitance values.

First-order derivative of the filtered values of capacitance are shown in Figure 7C (below). Peaks are visible near the variation of the state of the sensor. During the state transition from dry to wet, a concurrent positive jump at high and low (and medium) measurement frequencies is visible. A negative jump at high measurement frequency and no jump at low (and medium) measurement frequency indicate the state transition from wet to icy. During the state transition from icy to wet, a positive jump at high measurement frequency and no jump at low (and medium) measurement frequency are visible. Finally, a concurrent negative jump at high and low (and medium) measurement frequencies indicates the state transition from wet to dry. In order to discriminate between different states of the sensor surface (dry, wet, or icy), an algorithm was implemented based on the jumps on the first-order derivative

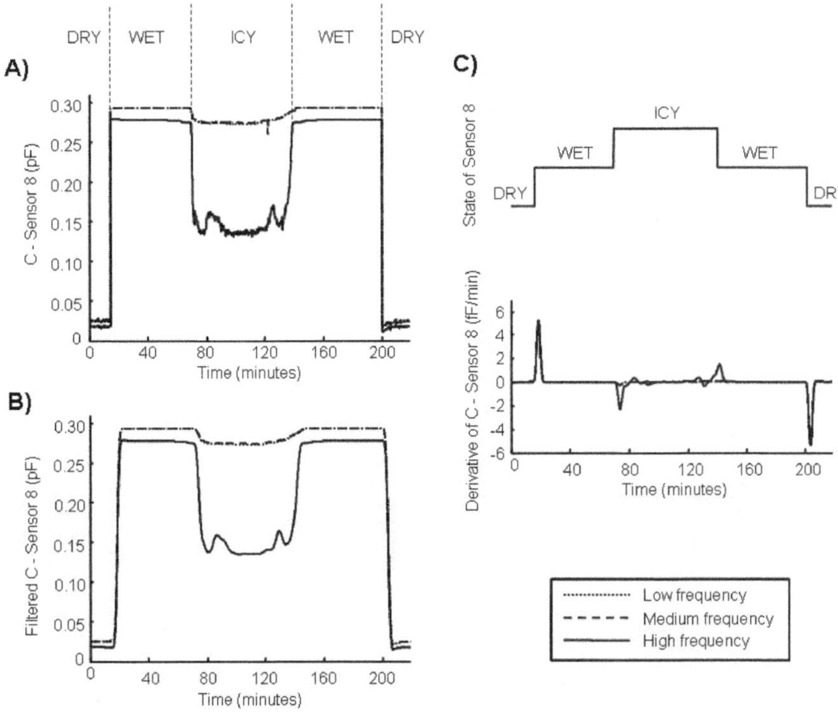

Figure 7. Signal processing algorithm applied to experimental data. A) Capacitance's values (raw data) obtained for the sensor eigth during the first experiment at different measurement frequencies. B) Low pass filtered values of capacitance obtained from the raw data. C) States of the sensor revealed by the algorithm (above). First-order derivative of capacitance values obtained from the low pass filtered data (below).

of the values of capacitance estimated at the low and high measurement frequencies. A jump was considered significant if the first-order derivative of the capacitance was higher than a threshold value estimated during the calibration. States of the sensor revealed by the algorithm are shown in Figure 7C (above). The estimated states agreed with the observed state of the sensor, showed in Figure 7A.

The reliability of the estimates provided by the sensor was investigated during laboratory experiments. Nine ice sensors were placed at the same time in a climatic chamber and the experiment descript above was performed, and the dispersion of the time instants in which phase changes of water were detected. Capacitance values (raw data) at two frequencies (low and high measurement frequencies) obtained for each sensor during the first experiment are shown in Figure 8A (medium measurement frequency is not reported since its capacitance values are very similar to that obtained at low measurement frequency). There are no difference between the capacitance values obtained at low and high measurement frequencies during the dry and wet states for different sensors; however, during the icy state there are a clear distinction. The capacitance values obtained in the icy state for different sensors are not the same due to manufacturing tolerances and low precision in controlling the thickness of water placed over the sensor. In Figure 8B, the internal temperature of the climatic chamber

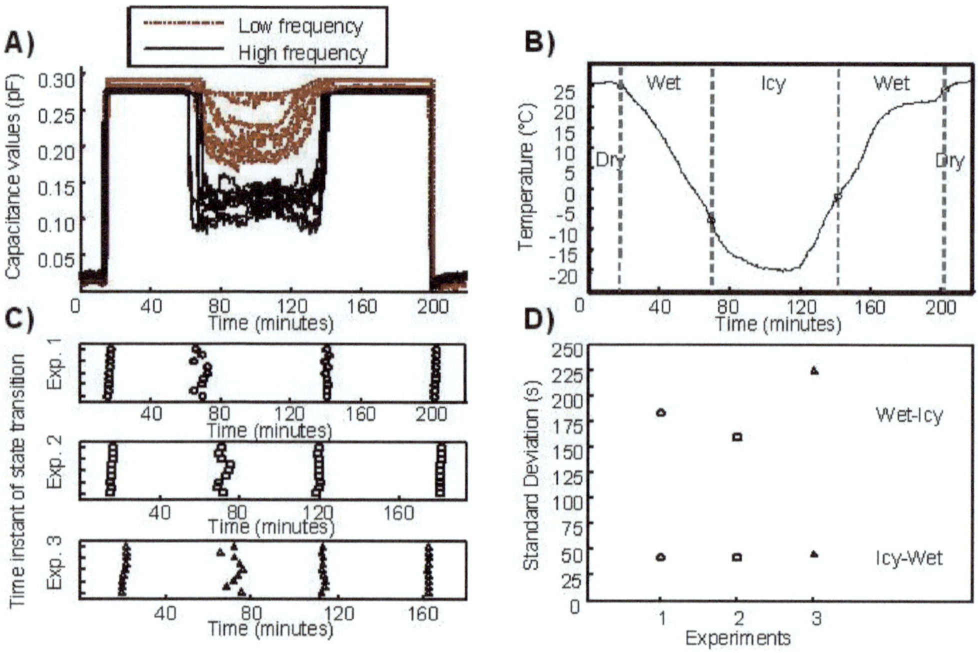

Figure 8. Statistical analysis of laboratory data. A) Values of capacitance (raw data) obtained for each sensor during the experiment. B) Internal temperature of the climatic chamber during the experiment. C) Time instants of state transitions estimated for each sensor during three experiments. D) Standard deviations of time instants of state transitions wet-icy and icy-wet estimated for each sensor during three experiments.

during the experiment is shown, and the state condition detected by one of the nine sensor is also indicated. The wet-icy transition is identified at about -7°C, whereas the icy-wet transition is identified when the internal temperature is about 0°C. So, the wet-icy transition was detected at lower temperatures than the icy-wet transition, probably due to the higher value of specific heat for water (about 4000 J kg^{-1} K^{-1}) with respect to ice (about 2000 J kg^{-1} K^{-1}), which determines a higher time to cool water than that needed to warm ice, keeping constant the magnitude of the temperature gradients imposed by the climatic chamber.

In order to study the repeatability of the estimates provided by the sensor, three experiments were performed. Time instants of state transitions estimated for each sensor during the three experiments are shown in Figure 8C. Differences of the estimated times of transition were in the order of a few minutes, which is related to the spatial heterogeneity of the icing and melting processes. The standard deviations of the time instants of the state transitions wet-icy and icy-wet estimated for each sensor are shown in Figure 8D. The time instants of the state transitions dry-wet and wet-dry were not considered, since the sensor were wetting and drying by the user. The standard deviation obtained for three experiments for the state transition wet-icy was larger than that obtained for the state transition icy-wet. Indeed, water started freezing from the surface proceeding downward, so a small difference in the thickness of the layer of water could determine a spread of the delays of different sensors in the identification of the presence of ice. On the other hand, the melting process started at the sensor surface (probably due to low energy dispersions from the device). Moreover, there is no relevant difference between the standard deviation values obtained in the same state transitions for three experiments.

5.2. Simulation results

The reliability of the capacitance estimates provided by the mathematical model of the sensor was investigated during a simulation tests, applying the same simulated conditions of the laboratory experiment. At the beginning of the test, the sensor has a simulated temperature of 25°C in the state dry. After 18 samples, the simulated state of the sensor became wet, and the temperature decrease up to -20°C with a temperature gradient of -1°C per sample. During this period, at 0°C, the water over the sensor frozes. When the simulated temperature reached -20°C, it increased up to 25°C with a temperature gradient of 1°C per sample. During this period, at 0°C, ice melts. When the simulated temperature reached 25°C, the state of the sensor became dry. The simulation test ends after 18 more samples.

Simulated capacitance values obtained using the mathematical model at different measurement frequencies and simulated temperature values are shown in Figure 9A. During the simulated dry and wet states of the sensor, the values of capacitance at high, medium, and low frequencies were close to 0.3pF and 0.6pF, respectively. During the simulated icy state of the sensor, the values of capacitance at high frequency were close to 0.45pF whereas at medium and low frequencies were close to 0.6pF. Thus, there were difference on the simulated capacitance values at different frequencies, reflecting the variation of the relative permittivity of ice with frequency. The obtained simulated capacitance values are very similar to those obtained during the laboratory test (see Figure 7), subtracting the calibration value, which is about 0.3pF. This demonstrates the validity of the mathematical model. Low pass filtered values of simulated capacitance are shown in Figure 9B. First-order derivative of the simulated filtered values of capacitance are shown in Figure 9C (below). Peaks are clearly visible near the variation of the state of the sensor. States of the sensor revealed by the algorithm, using simulated data, are shown in Figure 9C (above). The estimated states agreed with the simulated state of the sensor, showed in Figure 9A.

The mathematical model are used to characterize the sensor in different situations. Two simulation tests were performed, the first one oriented to the effect of the thickness of water or ice placed over the sensor, the second one oriented to the study of the capacitance values at different measuring frequencies when water or ice was placed over the sensor.

The values of capacitance obtained using the mathematical model at different thicknesses of water, from 0 (dry sensor) to 10mm, are shown in Figure 10A (temperature of 25°CC and measurement frequency of 20MHz, but equivalent results are obtained for lower frequencies). Capacitance values rose slightly increasing the thickness of water over the sensor. Simulated values of capacitance for different thicknesses of ice are also shown in Figure 10A (temperature of -10°CC and measurement frequency of 200Hz and 20MHz). Also increasing the thickness of ice formed over the sensor the capacitance values rise. Simulated values of capacitance obtained for different measurement frequencies of water and ice are shown in Figure 10B (thickness of 1mm, and temperature of 25°CC for water and -10°CC for ice). Increasing the measurement frequency, the values of capacitance decreased, following the same trend of the relative permittivity.

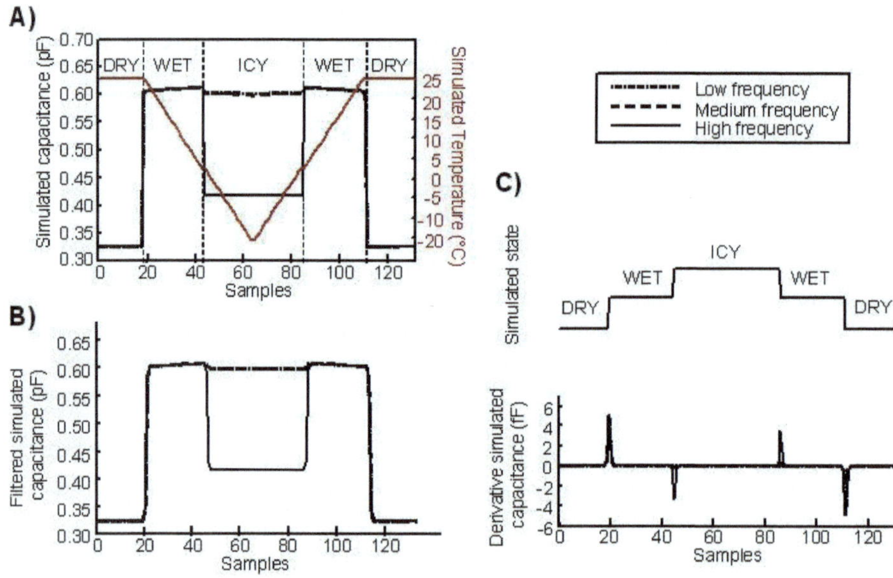

Figure 9. Signal processing algorithm applied to simulation data. A) Simulated capacitance values obtained during the simulation test at different measurement frequencies. B) Low pass filtered values of the simulated capacitance. C) States of the sensor revealed by the algorithm (above). First-order derivative of the low pass filtered simulated capacitance values (below).

5.3. Comparison of ice detection over the sensor and on the surface of a road

In order to detect ice formation on the surface of a road or a runway, embedding the sensors directly on the pavements is preferable. However, the surface of the sensor is flat and very different from that of the road, which is a rough surface due to the bitumen. Moreover, the sensor blocks percolation of water, which is very important in determining the road surface conditions. Thus, the icing and melting processes on the sensor and on the surface of a road may have a deviation, which is difficult to predict. In order to address this issue, some sensors were covered by bitumen [35, 36]. The bituminization process was performed in laboratory, and the experiment descript above was performed using bituminized sensors. The bitumen do not affect the measurement since the values obtained by bituminized sensors are the same of that obtained by sensors without bitumen. The reason is that bitumen has a very small value of relative permittivity ($\varepsilon_r = 3$, close to the value of relative permittivity of the Arnite) and this value does not change with the measurement frequencies [35, 36].

Nevertheless, the icing process over a bituminized sensor may be different from that of a road even if the two surfaces are the same, since percolation under them is different. For this reason the indications provided by the sensor were compared with the ice formation and melting over a road model identified by direct inspection. The road model was obtained using a road core with diameter of 30 cm and thickness of 20 cm. It is constituted by three layers, with asphalt and concrete with different granularity, allowing a good distribution of loads and a proper drain and filtration. In order to compare the indication of presence of ice provided by the sensor with the ice formation and melting over a road, experimental tests were performed applying the same environmental conditions to three sensors and to

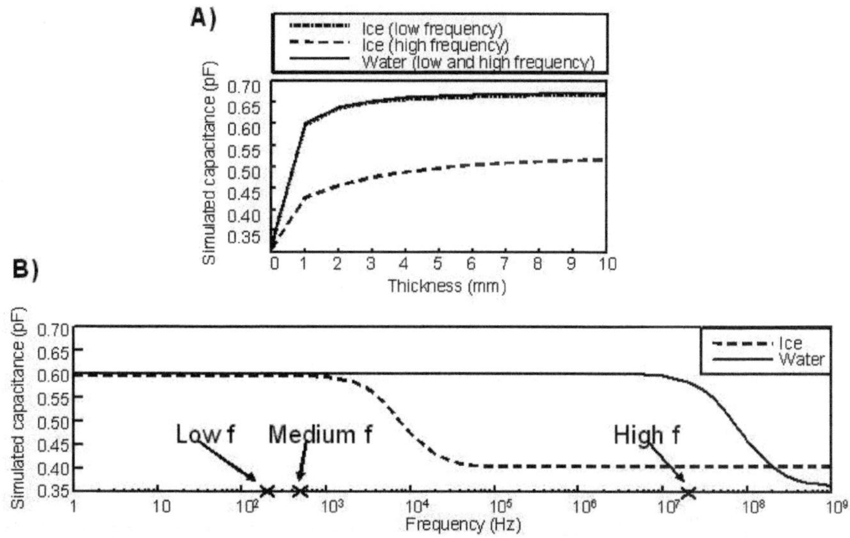

Figure 10. Simulation tests result. A) Simulated capacitance values at different thickness of water and ice (temperature of 25°CC for water and -10°CC for ice). B) Simulated capacitance values at different measurement frequencies when water and ice were placed over the sensor (thickness of 1 mm, and temperature of 25°CC for water and -10°CC for ice).

the road core, and evaluating the dispersion of the time instants in which phase changes were detected. Experimental tests were executed inserting the sensors and the road core in the climatic chamber. At the top of the climatic chamber, an USB webcam (Logitech - QuickCam Pro 9000; operative temperature range from -20°C to +60°C) was inserted to acquire images from the road core in order to detect the formation of the ice over it. A PC was used to store images from the USB webcam and sensory data from the sensors. Images and data were simultaneously acquired using a sampling frequency of 1 sample per minute.

Two different experimental tests were performed, the first imposing a linear gradient of temperature, the second leaving the climatic camber in static condition. Specifically, in the first test, the sensors and the road core were first introduced into the climatic chamber with a temperature of 25°C, for approximately 10 minutes in order to wait that the indications of the sensor became stationary. Then, 1 mm of tap water was placed over each sensor and at the center of the road core. Different temperature gradients were applied and the time instants in which the sensors identified ice formation and the road core surface froze were investigated. Specifically, the climatic chamber was arranged to reach -20°C with different temperature gradients equal to -0.25°C/min, -0.5°C/min, and -0.75°C/min. During this period, the water froze. Once reached the minimum temperature of -20°C, the climatic chamber kept stable conditions for approximately 10 minutes, and then it was arranged to reach 25°C with opposite temperature gradient. During this period, the ice melted. The climatic chamber kept the temperature of 25°C for approximately 10 minutes. Then, sensors and the road core were dried. In the second test, sensors and road core were first placed in the climatic chamber. Then, the chamber was arranged to reach -10°C with a temperature gradient of -1°C/min. Once reached the temperature of -10°C, the climatic chamber was arranged to keep stable conditions. One mm of tap water was placed over each sensor and at the center of the road core. After closing the chamber, the water placed over the sensors and the road core froze. Then, the climatic chamber was open at ambient conditions and the ice melted. Finally, sensors and the road core were dried. This experiment was repeated in three different days in order to investigate the repeatability of the data.

The results of the experiments are shown in Figure 11. The time intants of formation and melting of ice are shown for the sensors and compared to those in which the same happened over the road model. There was always a negative bias between the indication of the sensors and the conditions of the road. It is worth noticing that there is a deviation between the indications of the sensor and what happens over the road core, as the indications of the sensor anticipate the road conditions. Ice forms before over the sensor, due to the smooth surface of the sensor, which facilitates the formation of ice crystals on it than over the road. Moreover, ice melts before over the sensor than over the road core: this is probably due to the low power consumption of the electronics, which warms the surface of the sensor.

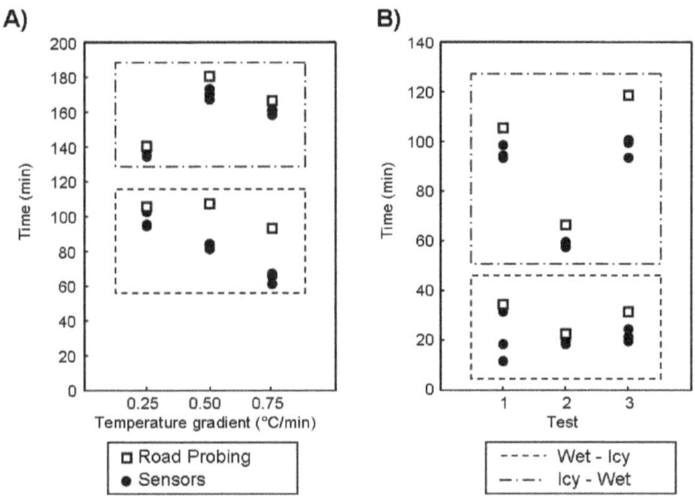

Figure 11. Time instants in which ice formed or melted over the road core compared to the indications of the sensors. A) First experiments, in which different gradients of temperature are applied. B) Second experiments, in which stable conditions at -10°C were maintained by the chamber till water froze and then it was switch off and the door was opened.

5.4. Effect of salt in the water

As one of the main problems in taking measures is the presence of salt spread over the road to avoid ice formation, the output of the sensor when covered by a ionic solution is investigated considering both a simulation model and experiments in the laboratory.

Simulations were performed using a mathematical model of the sensor which extends the one descript above. The model includes the electrode assembly and three planar layers over it, describing Arnite, a ionic solution (modelling a solution of water and salt), and air. Potential satisfies Laplace equation in the Arnite and in air (as no ion is present there) and Poisson equation in the ionic solution, where the ions contribute as a source term. Ions are present only in the ionic solution and satisfy continuity equation, which accounts for their diffusion and transport within the electric field generated by the external potential. The equations considered for each layer are the followings:

$$\begin{cases} \Delta\phi_1 = 0 \\ \Delta\phi_2 = -\frac{q}{\varepsilon}\cdot(n_+ - n_-) \\ \frac{\partial n_\pm}{\partial t} = D_\pm \cdot (\Delta n_\pm \pm \frac{q}{kT} n_\pm \Delta\phi_2) - D_\pm \cdot (\Delta n_\pm \pm \frac{q}{kT} N \Delta\phi_2) \\ \Delta\phi_3 = 0 \end{cases} \tag{13}$$

where q is the electrical charge of the ions, k is the Boltzman constant, T is the absolute temperature, $n_\pm$ indicates the positive or negative ion densities, $D_\pm$ is their diffusion coefficient (Na^+: $1.33 \cdot 10^{-9} m^2 s^{-1}$; Cl^-: $2.03 \cdot 10^{-9} m^2 s^{-1}$), and N is the density of the ions in thermodynamic equilibrium. The analysis is performed in the limit of small amplitude of the applied voltage (with respect to the thermal voltage), so that the response of the sample to the external signal is linear (the ion densities can be approximated with their value at equilibrium in the continuity equations) [2]. For the ions, the boundary condition was that the current density flowing across the boundary of the second layer (in which they are located) was zero. Experiments were also performed in the laboratory to assess the effect of dissolving different quantities of salt in the water covering the surface of the sensor. The values of capacitance at both high and low frequencies were acquired for one sensor covered by 2ml of water with different concentrations of salt, with salinity between 20% and 100%. The maximum value of salinity (100%) was obtained dissolving 36mg of salt per 100ml of water [10].

The impedance of the sensor at different concentrations of salt dissolved into the water is shown in Figure 12, for both simulations and experiments in the laboratory. Different percentages of salinity (from 20% to 100%) were considered. In the case of the experiments, for each salinity value, data were acquired for five minutes and mean and standard deviation of the measured capacitance were computed. Also data for pure water (without adding salt) are shown. Simulations indicated that the presence of ions in the water reduces the electrical resistance measured by the sensor monotonically with the increase of the concentration of the ions. Nevertheless, the imaginary part of the impedance is much larger than the real one for the two operating frequencies considered, so that the system is essentially insensitive to the presence of salt in the solution, as also indicated by the experiments.

5.5. Experiment in the field

Behavior and robustness of the sensors were studied in the field. Three sensors were embedded at the Turin-Caselle airport in order to detect formation of ice at the beginning, ending and in the middle of the runway, and increase safety during take off and landing of the aircrafts. Moreover, a complete weather station (PCE Group - PCE FWS-20) was placed close to the sensors to monitor meteorological variables, such as ambient humidity, temperature, pressure, velocity and direction of the wind, and quantity of rain falls. On field experiment lasted for a 36 days period (12/01/2010 - 17/02/2010).

Values of capacitance obtained by one of the three sensors (high and low measurement frequencies) during the in the field experiment are shown in Figure 13. In detail, capacitance

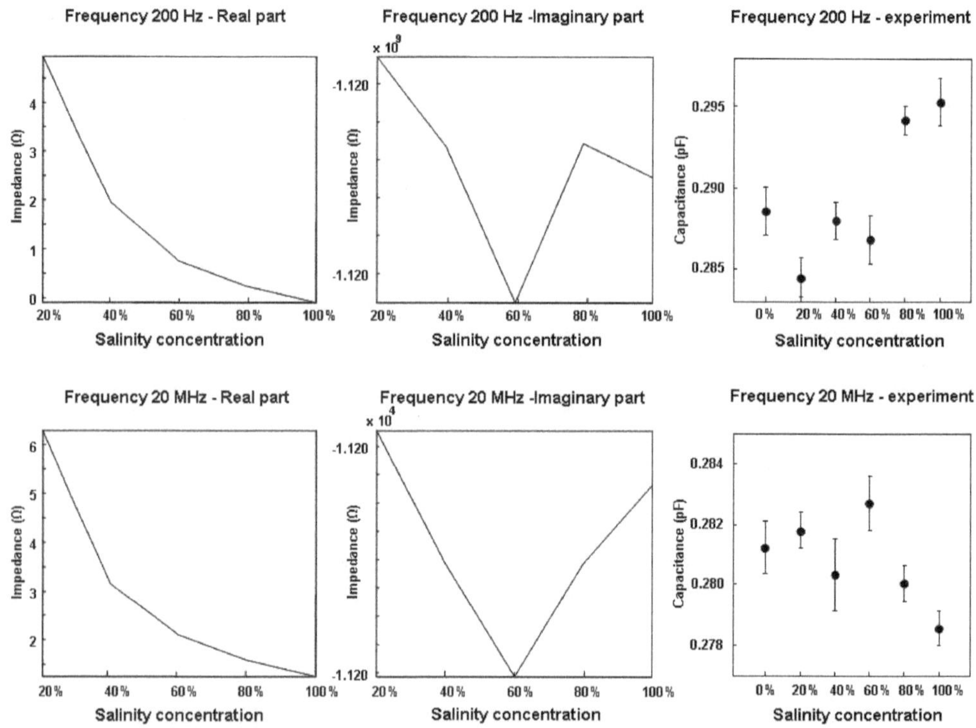

Figure 12. Impedance and capacitance of the system with different salinity, obtained in simulation (layer of water solution 5mm thick) and with experiments (2ml of solution were put over the sensor surface).

values (raw data) at low and high measurement frequencies are shown in Figure 13A. Variations of the values of capacitance over a 36 days period are clearly visible, since the state of the sensor (dry, wet, or icy) continuously changed. States of the standard sensor revealed by the algorithm are shown in Figure 13B. The algorithm was validated during the laboratory tests, so the states revealed for each sensor during the on field test were assumed correct. Environmental data obtained by the weather station are shown in Figure 13C (quantity of rain falls), Figure 13D (humidity of the air), and Figure 13E (temperature of the air). Environmental data are used to examine the correct estimation of the state of the sensor.

During the test period, the sensors revealed different road surface conditions (dry, wet, and icy). Wet condition of the surface of the runway was identified also in the case of fog, confirming that even a slow condensation of water over the sensor surface is associated to a jump in the value of capacitance that may be easily identified Transition to icy conditions were identified when the sensor was wet and temperature decreased to a negative value.

5.6. Relation between sensor output and weather conditions

Considerations about the correlation between weather data and state of the surface of the sensor are studied, with special attention to the weather data during ice formation and the relation between relative humidity and variation of the measured capacitance of the sensor.

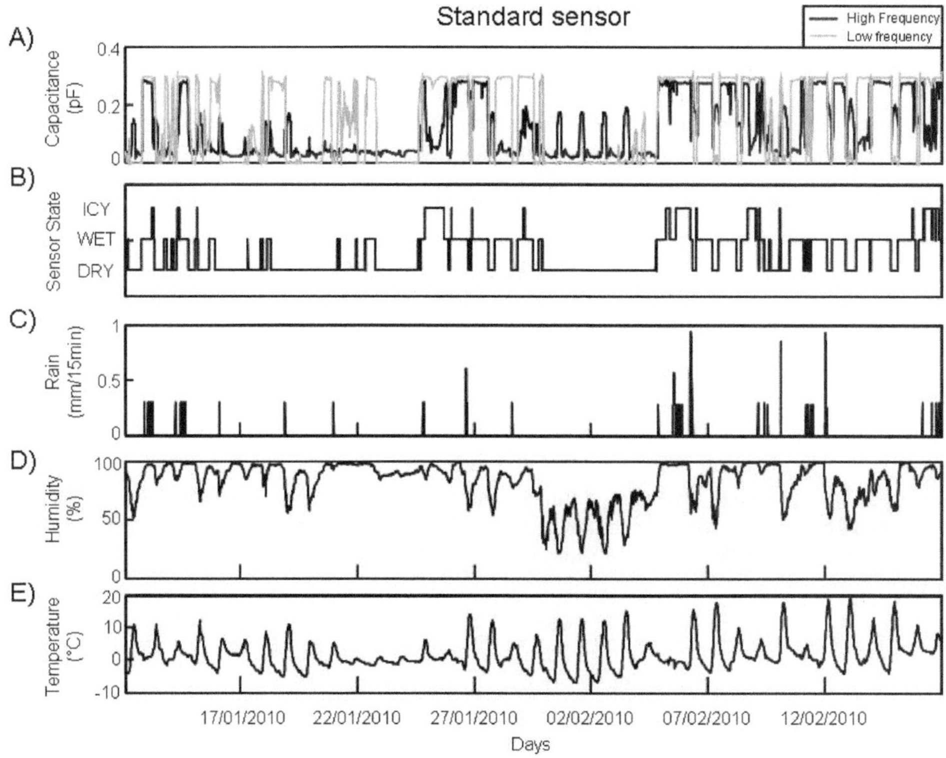

Figure 13. On field results obtained from a sensor placed at the Turin-Caselle airport over the period 12/01/2010 - 17/02/2010. A) Values of capacitance (raw data) obtained for the sensor. B) States of the sensors revealed by the algorithm. C) Quantity of rain falls. D) Ambient humidity. E) Ambient temperature.

Capacitance values from the sensors embedded at the airport were used. Weather conditions in airports are usually reported in the form of a METAR (METeorological Aerodrome Report), which is a format for the transmission of weather data (temperature, dew point, humidity, pressure, wind velocity and direction, quantity of rain falls, and solar radiation) together with considerations about the observed weather condition (sun, cloud cover, rain, snow, etc.). METAR data are on-line available for nearly every airport in the world. Reports are generated about twice an hour (sampling period is usually 30 minutes, but sometimes a pair of data is sampled with 10 minutes delay). The Turin airport is coded as LIMF in the METAR database. Data from January 2010 to September 2010 were used for the comparison with the indication provided by the sensors.

Data were first interpolated in order to be defined on the same time samples, with period of 1 minute. Data from the ice sensors and weather data were linearly interpolated. The observed weather conditions given by the METAR database were interpolated with a nearest neighbour approach. Correlation was estimated between the state provided by each sensor and the information of the observed weather condition provided by the METAR system. The observed weather conditions were grouped in 4 classes, indicating rain, fog, snowfall and none of the above. The first three observations were correlated with the wet state provided by the sensors. Since the road remains wet also after rain, fog or snowfall stopped, correlation was estimated only within periods in which the METAR indications were rain, fog, or snow, respectively. The icy state provided by the sensors were correlated with the following prediction algorithm: ice forms on the ground if the air temperature is lower than 0°C and the absolute difference between air temperature and the dew point is lower or equal to 1°C [1]. The dew point T_D, that is the temperature to which air must be cooled to trigger condensation of its water vapour, was estimated as [14]:

$$T_D = T - 31.25[2 - log(RH)] \tag{14}$$

where T is the air temperature in Celsius and RH the relative humidity. The values of air temperature and relative humidity were also provided by the METAR system. This algorithm indicates that the region of parameters T and RH allowing the formation of ice is the following:

$$\begin{cases} T < 0 \\ RH > 93\% \end{cases} \tag{15}$$

The correlation between the wet state estimated by the sensors and the rain indication provided by the present weather sensor of the METAR system is shown in Table 1. Correlation is always upper than 90%, it is higher during winter months with respect to the summer months.

	Sensor A	Sensor B	Sensor C
January	100.0%	100.0%	100.0%
February	96.0%	93.6%	100.0%
March	99.0%	97.9%	97.9%
April	100.0%	100.0%	98.1%
May	96.0%	95.6%	91.7%
June	94.6%	93.0%	92.6%
July	90.8%	93.4%	95.2%
August	94.7%	91.6%	93.1%

Table 1. Correlation between the wet state provided by the sensors and the rain indication provided by the METAR system.

The fog indication was obtained by the present weather sensor of the METAR system only during winter. Correlation between the wet state provided by the sensors and the fog indication of the METAR system is shown in Table 2. Correlation is higher than 51% in January and higher than 85% during February and March.

	Sensor A	Sensor B	Sensor C
January	58.9%	54.0%	51.3%
February	91.8%	94.7%	88.5%
March	100.0%	85.7%	96.5%

Table 2. Correlation between the wet state provided by the sensors and the fog indication provided by the METAR system.

Snow fell during the first three months of the year. Correlation between the wet state of the sensors and the snowfall indication provided by the present weather sensor of the METAR system is shown in Table 3. Correlation is higher than 92% in January and February, and decrease to about 80% in March.

	Sensor A	Sensor B	Sensor C
January	92.6%	100.0%	100.0%
February	99.6%	100.0%	100.0%
March	95.7%	79.2%	92.2%

Table 3. Correlation between the wet state provided by the sensors and the snow indication provided by the METAR system.

The ice indication was obtained by the sensor during the first three months of 2010. Correlation between the icy state provided by the sensors and the ice indication estimated by the prediction algorithm above described is shown in Table 4. Correlation is quite low for all the sensors.

	Sensor A	Sensor B	Sensor C
January	48.5%	64.6%	43.2%
February	78.1%	73.0%	62.2%
March	28.2%	36.1%	39.3%

Table 4. Correlation between the icy state provided by the sensors and approximation that ice can be found on the ground.

Correlation with rain and snowfall was higher than that with fog, which could wet the road only when it lasted a long time. Correlation between the indications of the sensors and of the prediction algorithm was low. This is due to many reasons. For example, in the case in which a thick layer of ice formed over the sensor and started to melt (as correctly predicted by the algorithm) due to the changed weather conditions, the sensor would still indicate that its surface is frozen, until the complete melting of the layer of ice. It should also be considered that the sensor surface could be wet after raining, even if the relative humidity measured by a weather station is lower than that required by the algorithm to predict the formation of ice: in such a condition, it is sufficient that the temperature lowers under 0°C to get ice over the sensor, regardless the value of the relative humidity. Moreover, the formation of ice depends on the concentration of the water covering the surface of the sensor, which could be dirty.

6. Conclusions

A low cost capacitive sensor for the estimation of road condition is presented. Performances of the sensor were investigated in simulations, laboratory and in the field experiments. Laboratory results showed that the sensor correctly identified the presence of air, water, or ice on its surface. Repeatability was satisfied, as different sensors provided the same indication with time delay of a few minutes during different laboratory experiments. Simulation results showed that the mathematical model correctly simulates the behavior of the sensor, allowing the simulation of different sensor conditions. Moreover, laboratory and simulation results showed that the sensor is not sensitive to salt present over the its surface. In the field results showed that the sensors placed at the Turin-Caselle airport worked properly, recognizing air, water, or ice over the runway. Finally, the device provided information consistent with the METAR message for the whole period of study.

The sensor may find applications in monitoring road conditions to support information systems assuring security and efficient maintenance of roads or airports during winter.

However, the descript sensor has a limitation. The measured value of capacitance depends on possible contaminations (e.g. dirt, or fuel) present in the water covering the sensor. Contaminants may change the relative permittivity of the water or ice, or even its variation with frequency, modifying the functionality of the sensor. For this reason, the detection of possible contaminations should be useful, in order to correct the wrong indication of the sensor.

7. Future works

Future works will be focused on the estimation of the thickness of the water or ice placed over the sensor. Since simulations showed that the capacitance values of the sensor increase with the increasing of the thickness of the water or ice placed over the sensor, the sensor can be also used for the estimation of this information. The estimation of the thickness of water can be useful to quantify, for example, the quantity of rain falls.

Nevertheless, the precision of the estimates of the sensor indicates its feasibility for different applications, such as ice forecasting using an artificial neural network, also using the local weather data available.

Finally, future works will be focused on the detection of the presence of snow or frost on the road and runway surfaces. At present, the sensor detects both snow and frost as water, but should be possible to distinguish between them. Moreover, the discrimination between different types of liquid solutions should be possible using the sensor.

Acknowledgment

This work was sponsored by the national projects AWIS (Airport Winter Information System) and ITACA (Innovazione Tecnologica, Automazione e nuovi Controlli Analitici per migliorare la qualità e la sicurezza dei prodotti alimentari piemontesi), both funded by Piedmont Authority, Italy.

Author details

Amedeo Troiano, Eros Pasero and Luca Mesin
Politecnico di Torino - Dipartimento di Elettronica,
Italy

8. References

[1] Baier, W., Shaw, R., Thompson, L. & Felch, R. [1976]. Agrometeorology of the maize (corn) crop, *Proceedings of Symposium on the Agrometeorology of the Maize (Corn) Crop*, Ames, Iowa, USA.

[2] Barbero, G. & Alexe-Ionescu, A. [2005]. Role of the diffuse layer of the ionic charge on the impedance spectroscopy of a cell of liquid, *Liquid Crystals* 32(7): 943Ű949.

[3] Bassey, C. & Simpson, G. [2006]. A comparison of the coplanar waveguide (cpw) and conductor-backed coplanar waveguide (cbcpw) for use as aircraft ice sensors, *Proceedings of Antennas and Propagation Society International Symposium*, Albuquerque, New Mexico.
[4] Bassey, C. & Simpson, G. [2007]. Aircraft ice detection using time domain reflectometry with coplanar sensors, *Proceedings of IEEE Aerospace Conference*, Big Sky, Montana.
[5] Buchner, R., Barthel, J. & Stauber, J. [1999]. The dielectric relaxation of water between 0°C and 35°C, *Chemical Physics Letters* 306: 57–63.
[6] Ezraty, R. [2003]. New-ice detection using microwave sensors, *Proceedings of International Geoscience and Remote Sensing Symposium (IGARSS)*, Toulouse, France.
[7] Gao, H. & Rose, J. [2009]. Ice detection and classification on an aircraft wing with ultrasonic shear horizontal guided waves, *IEEE Transactions on Ultrasonics, Ferroelectrics and Frequency Control* 56(2): 334–344.
[8] Herique, A. & Kofman, W. [1997]. Determination of the ice dielectric permittivity using the data of the test in antarctica of the ground - penetrating radar for mars '98 mission, *IEEE Transactions on Geoscience and Remote Sensing* 35(5): 1338–1349.
[9] Homola, M., Nicklasson, P. & Sundsbø, P. [2006]. Ice sensors for wind turbines, *Cold Regions Science and Technology* 46(2): 125–131.
[10] Howard, R. & Mendelssohn, I. [1999]. Salinity as a constraint on growth of oligohaline marsh macrophytes ii. salt pulses and recovery potential, *American Journal of Botany* 86(6): 795Ű806.
[11] Huhta, C. & Choquette, Y. [2005]. Ice detection and under-ice flow monitoring using a sontek argonaut-sw, *Proceedings of VIII Working Conference on Current Measurement Technology (IEEE/OES)*, Piscataway, New Jersey.
[12] Ikiades, A., Armstrong, D. & Howard, G. [2007]. Optical diffusion and polarization characteristics of ice accreting in dynamic conditions using a backscattering fibre optic technique, *Sensors and Actuators A: physical* 140(1): 43–50.
[13] Ikiades, A., Howard, G., Armstrong, D. J., Konstantaki, M. & Crossley, S. [2007]. Measurement of optical diffusion properties of ice for direct detection ice accretion sensors, *Sensors and Actuators A: physical* 140(1): 24–31.
[14] Iribarne, J. & Godson, W. [1981]. *Atmospheric thermodynamics*, Springer Science & Business.
[15] Jackson, J. [1998]. *Classical Electrodynamics*, Wiley.
[16] Jose, K., Sunil, G., Varadan, V. & Varadan, V. [2002]. Wireless idt ice sensor, *Proceedings of International Microwave Symposium*, Seattle, Washington.
[17] Kawano, K. & Kudoh, J. [2003]. Four-dimensional histogram method for sea ice detection using noaa avhrr images, *Proceedings of International Geoscience and Remote Sensing Symposium (IGARSS)*, Toulouse, France.
[18] Lee, N., Karimi, H. & Krakiwsky, E. [1989]. Road information systems: Impact of geographic information systems technology to automatic vehicle navigation and guidance, *Proceedings of Vehicle Navigation and Information Systems Conference*, Toronto, Canada.
[19] LeVeque, R. [2007]. *Finite Difference Methods for Ordinary and Partial Differential Equations*, Society for Industrial and Applied Mathematics (SIAM).
[20] Li, W., Zhang, J., Ye, L. & Zhang, H. [2009a]. Fiber-optic ice detection system for aeroplane application, *Proceedings of International Workshop on Intelligent Systems and Applications*, Wuhan, China.

[21] Li, W., Zhang, J., Ye, L. & Zhang, H. [2009b]. A fiber-optic solution to aircraft icing detection and measurement problem, *Proceedings of International Conference on Information Technology and Computer Science*, Kiev, Ukraine.

[22] Liu, Y., Keyb, J., Freya, R., Ackermana, S. & Menzel, W. [2004]. Nighttime polar cloud detection with MODIS, *Remote Sensing of Environment* 92(2): 181–194.

[23] Meindl, T., Moniaci, W., Pasero, E. & Riccardi, M. [2006]. An embedded hardware-software system to detect and foresee road ice formation, *Proceedings of IJCNN 2006*, Vancouver, Canada.

[24] Meissner, T. & Wentz, F. [2004]. The complex dielectric constant of pure and sea water from microwave satellite observations, *IEEE Transactions on Geoscience and Remote Sensing* 42(9): 1836–1849.

[25] Mesin, L., Troiano, A. & Pasero, E. [2010]. In field application of an innovative sensor for monitoring road and runway surfaces, *Proceedings of SENSORDEVICES 2010*, Venice, Italy.

[26] Muramoto, K., Saitoand, H., Matsuura, K. & Yamanouchi, T. [1997]. Cloud and ice detection using noaa/avhrr data, *Proceedings of International Geoscience and Remote Sensing Symposium (IGARSS)*, Singapore, Singapore.

[27] Pasero, E., Moniaci, W. & Raimondo, G. [2008]. Awis: an airport winter information system, *Proceedings of SIRWEC 2008*, Prague, Czech Republic.

[28] Petrenko, V. & Whitworth, R. [1999]. *Physics of ice*, Oxford University Press Inc.

[29] Roy, S., DeAnna, R., Izad, A. & Mehregany, M. [1998]. Miniature ice detection sensor systems for aerospace applications, *Proceedings of XI International Workshop on Micro Electro Mechanical Systems (MEMS)*, Heidelberg, Germany.

[30] Roy, S., Izad, A., DeAnna, R. & Mehregany, M. [1998]. Smart ice detection systems based on resonant piezoelectric transducers, *Sensors and Actuators A: phisical* 69: 243–250.

[31] Skolunov, A. [1997]. Frequency-temperature curve of the complex dielectric constant and refractive index of water, *Fibre Chemistry* 29(6): 367–373.

[32] Topp, G., Yanuka, M., Zebchuk, W. & Zegelin, S. [1988]. Determination of electrical conductivity using time domain reflectometry: Soil and water experiments in coaxial lines, *Water Resources Research* 24(7): 945–952.

[33] Troiano, A., Pasero, E. & Mesin, L. [2010]. An innovative water and ice detection system for road and runway surfaces monitoring, *Proceedings of PRIME 2010*, Berlin, Germany.

[34] Troiano, A., Pasero, E. & Mesin, L. [2011a]. Detection of ice formation over a road surface, *Proceedings of SENSORDEVICES 2011*, Nice, French.

[35] Troiano, A., Pasero, E. & Mesin, L. [2011b]. In the field application of a new sensor for monitoring road and runway surfaces, *Sensors & Transducers* 10: 71–83.

[36] Troiano, A., Pasero, E. & Mesin, L. [2011c]. New system for detecting road ice formation, *IEEE Transactions on Instrumentation and Measurement* 60(3): 1091–1101.

[37] Vivekanandam, J., Martner, B. & Politovich, M. [1998]. Aircraft icing detection using dual-wavelength and polarizationradar observations, *Proceedings of International Geoscience and Remote Sensing Symposium (IGARSS)*, Seattle, Washington.

[38] Von Hippel, A. [1988]. The dielectric relaxation spectra of water, ice, and aqueous solutions, and their interpretation. critical survey of the status-quo for water, *IEEE Transaction on Electrical Insulation* 23(5): 801–840.

[39] Wang, M. & Shi, W. [2009]. Detection of ice and mixed ice-water pixels for modis ocean color data processing, *IEEE Transactions on Geoscience and Remote Sensing* 47(8): 2510–2518.

Chapter 9

Automotive Networks Based Intra-Vehicular Communication Applications

Preeti Bajaj and Milind Khanapurkar

Additional information is available at the end of the chapter

1. Introduction

Intelligent Transportation System (ITS) includes eight wide areas of applications such as Advanced Traffic Information System, Advanced Traffic Management Systems, Automatic Vehicle Control System, Commercial Vehicle Operations, Advanced Public Transport Systems, Emergency, Infotainment Expectations, and Comfort to Drivers and Passengers, Safety, Electronic Payments etc. Safety, Mobility, Comfort are the three basic reasons for increasing the involvement of electronics in vehicles and in turn in Intelligent Transportation system.

Vehicular communication promises traffic management, efficient and easier maintenance, more fun and infotainment which ultimately lead to safer roads, enhanced comfort for driver and passengers (Safe Transportation). Categorically the Vehicular Communication can be classified as

- **Vehicle to infrastructure (V2I) and Infrastructure to Vehicle (I2V)**

In highway engineering, improving the safety of a roadway can enhance overall efficiency. Vehicle-Infrastructure communication targets improvements in both safety and efficiency. It is wireless communication catered by Vehicular Ad-Hoc Network, which provides communication between vehicles and nearby fixed equipments, usually described as roadside equipments or infrastructure.

- **Inter-vehicular (V2V)**

Inter Vehicle communication is governed by wireless communication and is catered by Mobile Ad-hoc Networks (MANETS) and Vehicular Ad-hoc Network (VANETS). A Vehicular Ad-Hoc Network, or VANET, is a form of Mobile ad-hoc network, to provide communications among nearby vehicles.

- **Intra-Vehicular**

Intra-vehicular communication describes as exchange of data within the ECUs of the vehicle, which are involved in vehicular applications. Major intra-vehicular communication is of wired type i.e. network based. There are some applications wherein wireless Intra-vehicular communication is accounted.

2. Automotive networks

One can visualize the complexities involved and the progressive advancements with around 8-10 ECUs in vehicle in early 90's through around 50 microcontrollers at the beginning of this decade i.e. 2000, Today, in high-end cars, it is common to have around 100 ECUs exchanging up to 2500 signals between them. Automotive networks Eliminates Unwieldy Wiring Harnesses, Increases Vehicle Safety and Reliability, Future expansion easily possible, Uniformity and Standardization in operations, Fast, Reliable and Efficient Inter-communication, Organizes the information stream, Assurance of faithful communication. Some of the Automotive Network Controllers are

- **Local Interconnect Network (LIN)**

The Local Interconnect network (LIN) is a serial communication protocol suited for networking sensors, actuators, and other nodes in real-time systems. It is 12 V Single wire, Universal asynchronous receiver-transmitter (UART), Single-Master, Multiple-Slave (up to 16 nodes), Up to 20kBits/s data rate automotive network protocol. LIN Bus Application Examples:

- Steering Wheel: Cruise control, Wiper, Turning light, Climate Control
- Seat: Seat Position Motors, Occupant Sensors, Seat Control Switch
- Door: Mirror Switch, Central ECU, Power Window Lift, Door Lock
- Safety and Investigation: Automotive Black Box.

- **Controller Area Network (CAN)**

The CAN bus is used in vehicles to connect engine control unit and transmission, or to connect the door locks, climate control, seat control, etc. It is Asynchronous Multi-master communication protocol serial bus having data rate of up to 1 Mbps. CAN Bus Application Examples:

- Safety Power Train: Electronic Parking Brake, Vacuum Leak Detection Automotive Black Box
- Chassis: Watchdog, Motor Control, Electronic Throttle Control Body Control: Low-end body controller (Lighting, Network Communication) Power Door, Power Sunroof, Power Lift Gate

2.1. Flex-Ray network

Flex-ray networking standards work on the principle of TDMA and have Dual-channel Architecture. It has Host processor which controls the communication process via communication controller and bus driver. Each flex ray node has two physical channels A and B facilitating data rate of up to 10 Mbps through per channel. It can be employed as network backbone with CAN and LIN. Flex-Ray Network Bus Application Examples:

- Data Backbone: for other buses (LIN, CAN, and MOST)
- Safety-critical Applications: X- by- wire, Automotive Black Box
- Distributed control system Applications: Power Train Applications,
- Châssis applications requiring computations across various ECUs

2.2. Media oriented system transport (MOST)

MOST is a high-speed multimedia network technology usually includes up to 64 MOST devices having plug n play operations. MOST can operate up to 15 uncompressed stereo audio channels or up to 15 MPEG1 channels for audio-video communication.

3. Steps to design intra-vehicular communication applications

- Depending upon the type of Intra-Vehicular Communication Applications network controller & host processor is to be decided.
- The intra-vehicular communication application can be homogenous automotive controller based or it can be of heterogeneous type
- Design and modification (after testing iterations) of interface Design prototype model (with suitable modeling tool / language like VHDL, LABVIEW, and MATLAB etc) of network controllers.
- Downloading the prototype file created with design tool in to suitable programmable platform (like FPGA, CPLD etc) and result Validation.
- Interfacing with electromechanical systems for simulation of real time data from sensors, actuators & experimenting for actions.
- Installation and Commissioning
- Testing the designed system in real time environment.

4. Intra-vehicular communication applications

In accordance with the steps mentioned above some of the intra vehicular communication applications governed categorically by various automotive controllers are listed as follows. LIN specifications implies that a LIN network can have single master and multiple (up to 16) Slaves. Figure 1 shows RTL view of LIN controller implemented on modelsim tool.

4.1. Master-slave local interconnect network for intra vehicular communication application

Figure 2 shows RTL view of the three node LIN network along with FSM and other components. The three transceivers are connected with the TXD and RXD of the respective three controllers. The transceivers have lin_bus signal forming LIN bus. The three lin_ bus signals are interconnected together to form a single wire LIN bus.

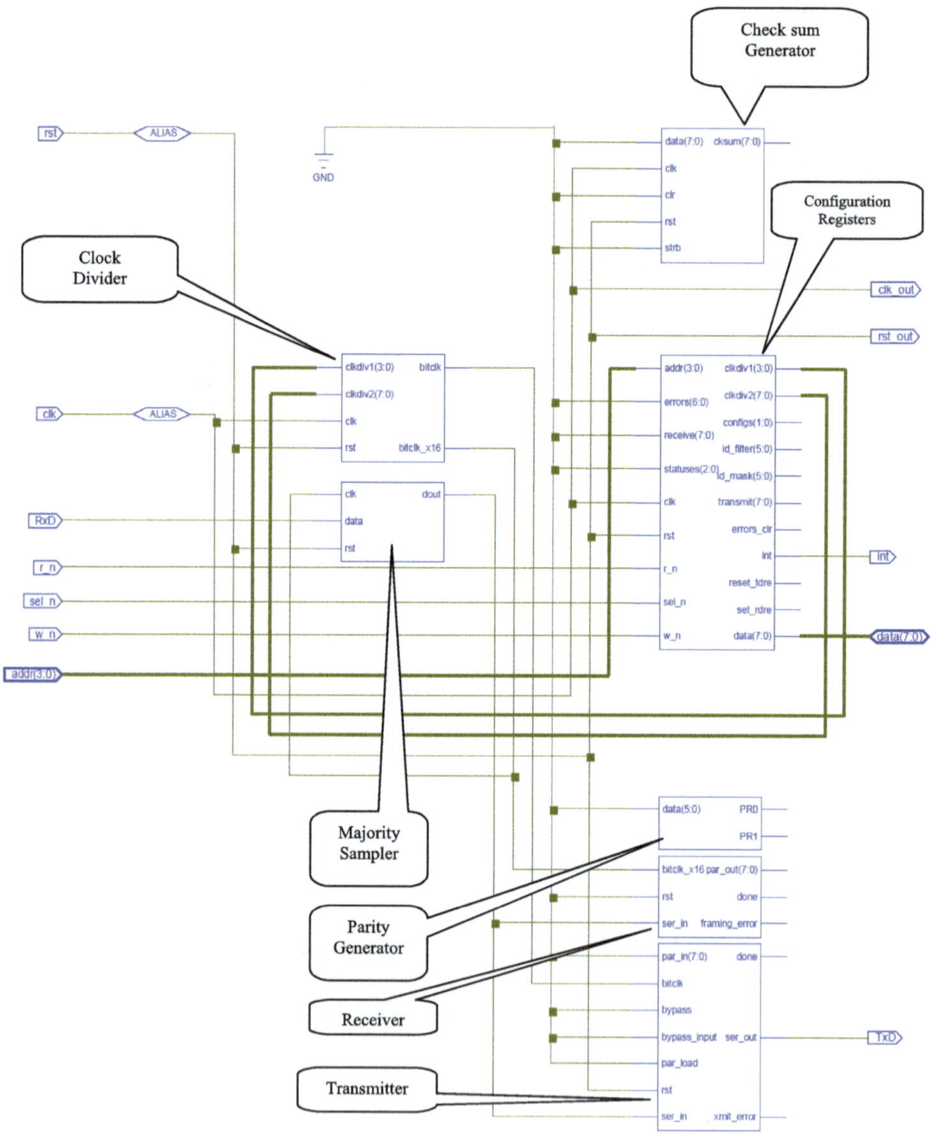

Figure 1. Modelsim RTL view of LIN Controller

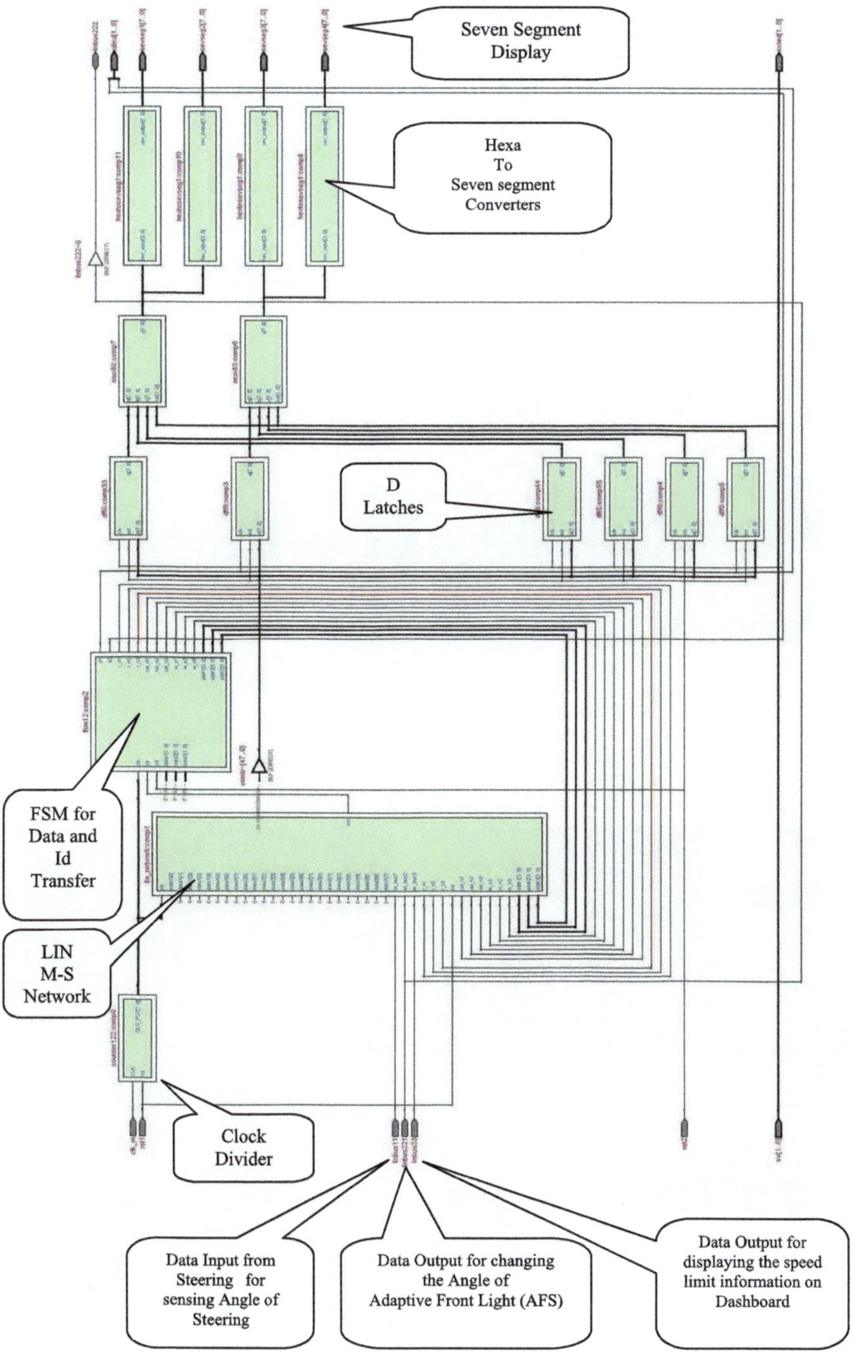

Figure 2. RTL view of the three node Local Interconnect Network

FPGA implemented three node Local interconnect network is used for sensing the steering wheel angle and controlling two dependent vehicle parameters viz: changing the angle of front light for better illumination and vision and also informing the driver safe speed limit on dash board. Figure 3 shows the implementation using FPGA DE1 board of ALTERA. The data transfer is displayed on display of DE1 board which has LED indication for indicating active slave node and the data along with id and masking bits are displayed on four seven segment displays.

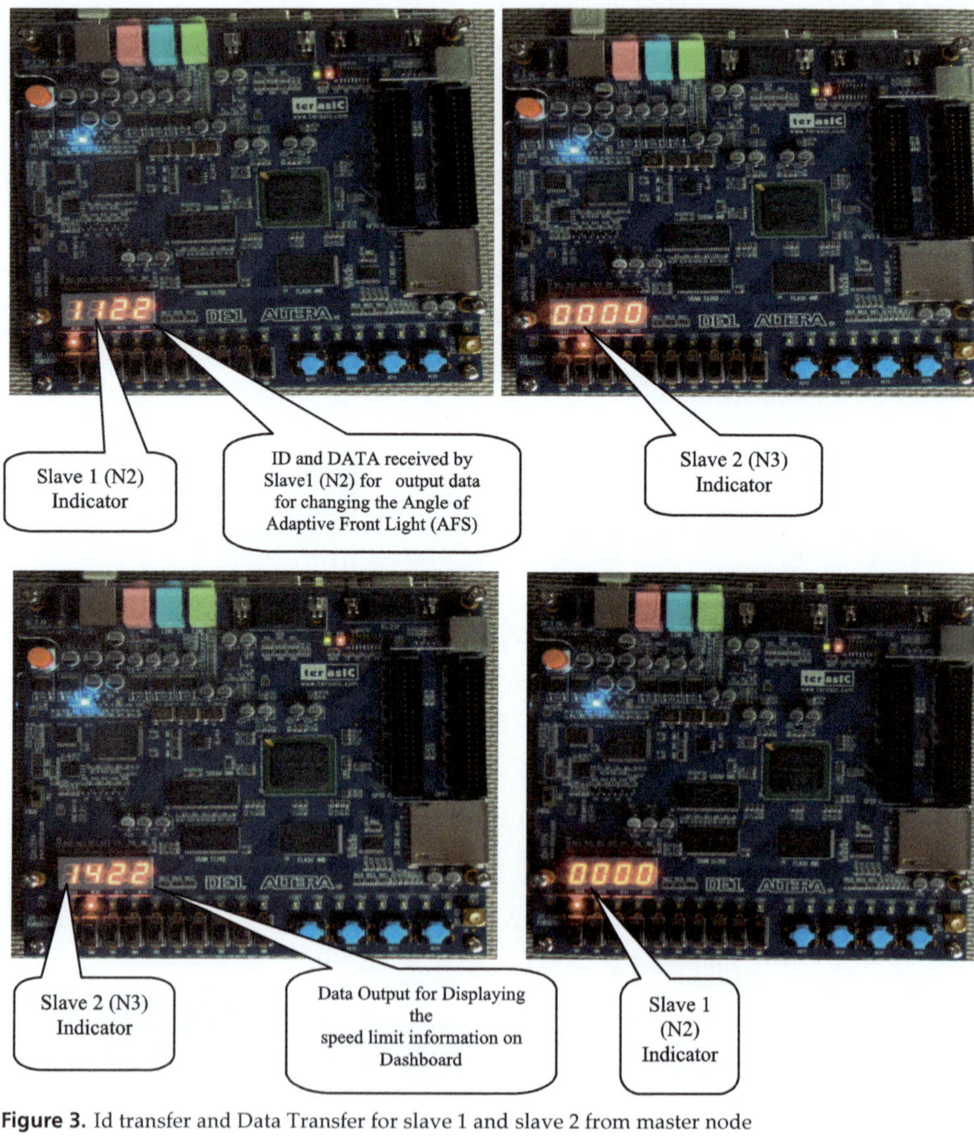

Figure 3. Id transfer and Data Transfer for slave 1 and slave 2 from master node

4.2. CAN for intra vehicular communication (black box) application

The design methodology for black box application using FPGA hosted CAN controller is shown in figure 4. The digital signal thus derived from output of ADC is given as input to CAN host FPGA which holds Data frame and directs it to Black box Memory. The black box will store the frame and will make available as and when required.

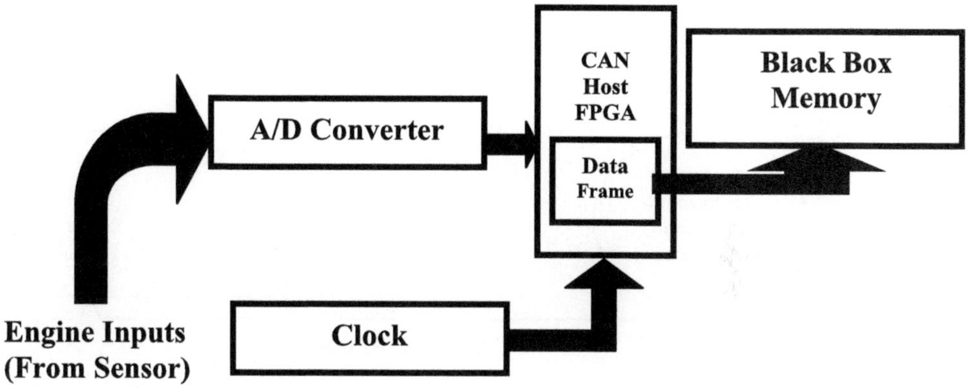

Figure 4. Block Diagram of Design Methodology

4.2.1. RTL – Schematic view of CAN host based black box

Prototype CAN is incorporated in designed Black Box application. Figure 5 shows the block diagram, schematic view of CAN controller used in FPGA based design of Black Box. The inputs and outputs are as explained above for RTL view and in addition to that Figure 5 also shows the details of schematic view of Data, Identifier, DLC, etc which constitute CAN frame. Input to system is output of ADC which is simulated signal of emergency situation in vehicle and the output is CAN Data frame containing data of parameter which gets stored in memory for revival and further referral analysis. The other inputs are clk (clock), rst (Reset) and tx_ctrl (transmission control).

The data frame that is to be transmitted by the CAN-Host will consist of the following fields: Identifier field, control field, data field and CRC field. The emergency situation in Vehicle i.e. output of sensor element from the Engine of a car can be simulated using ADC converter in which the Analog input is varied by potentiometer or the same can be simulated by operating DIP switches on XILINX FPGA Board. Then according to the input from the sensor the output of the ADC is needed to be interpreted or calibrated. This digital signal is given as input to CAN host FPGA which is employed to operate specific application in the vehicle. The instantaneous CAN Frame of that particular application is hold of and is stored in memory of intelligent vehicle black box.

The proposed design of black box as shown in figure 6 will record and store the multiple vehicle parameters through different automotive controllers and their networks hosting on different FPGAs. The diagram shows the black box memory and data retrieval system to retrieve data stored in system memory.

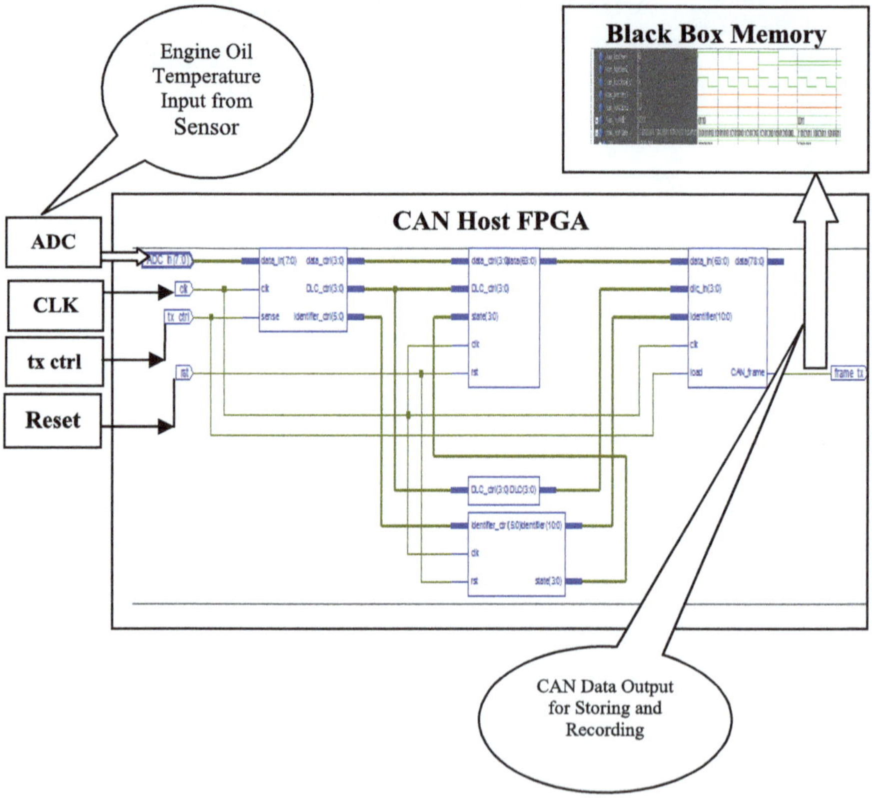

Figure 5. Block Diagram and Schematic View of CAN host FPGA based Black Box

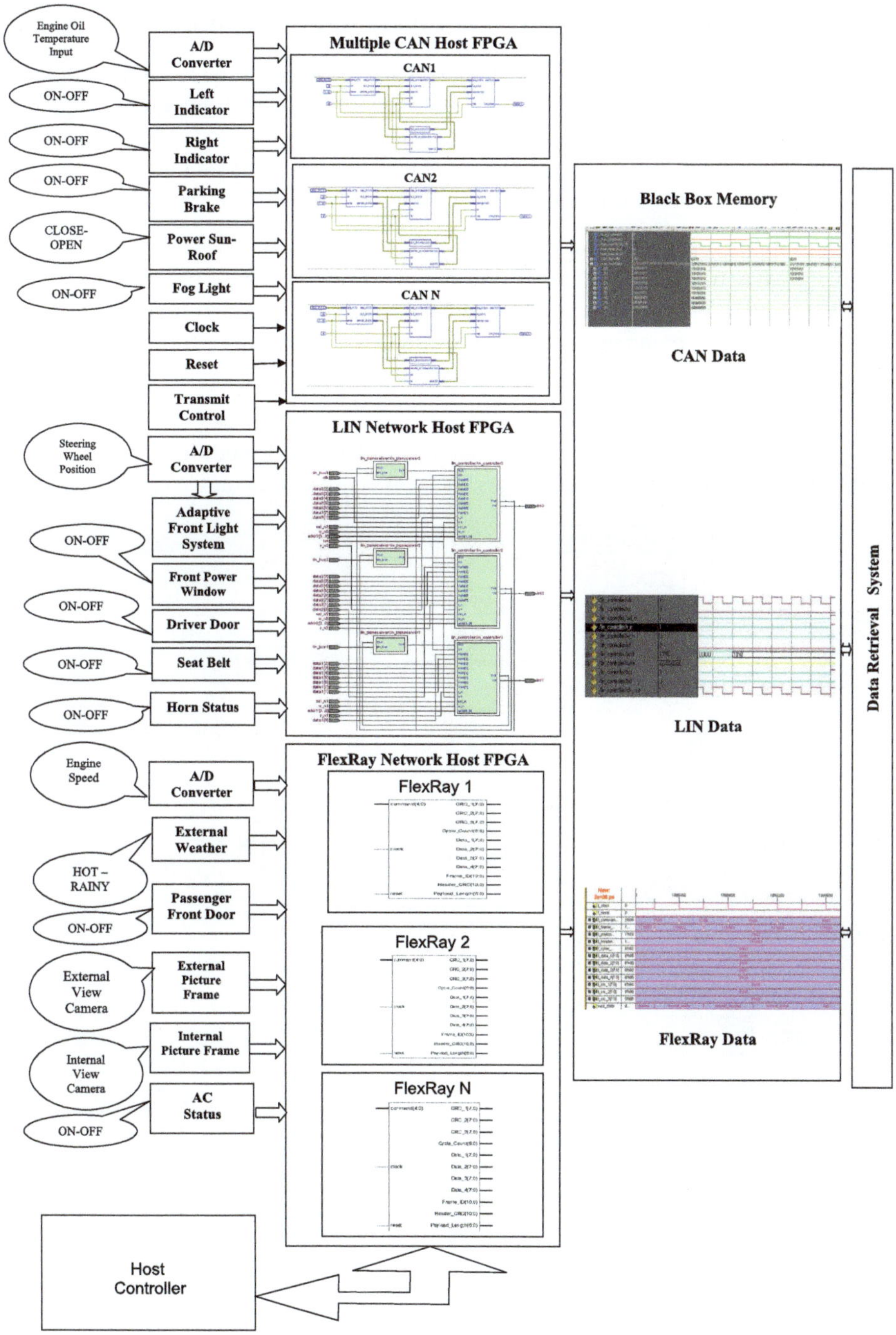

Figure 6. Multi Network Controller based proposed design of black box for multi-parameters

4.3. Flex-Ray protocol for intra vehicular communication application

Virtual Flex-Ray environment is implemented for intra-vehicular communication applications using Flex Ray Protocol using LAB-VIEW. The design of various intra-vehicular applications for Driver and Passenger Doors, AC Control, Oil Temperature Control and External Weather sensing and controlling the activities inside vehicle has been implemented using flex-Ray utilities of LABVIEW platform. The algorithmic sequence of operation is as follows:

- Status of input is pre-setted as true and false for respective vehicle parameter thus tool forms flex-Ray frame for both the conditions of particular vehicle parameter.
- The parameter status is chosen as true or false which goes as input to module designed.
- Module passes the parameter to evaluator and decides output controller action.
- The output is the action taken by flex-Ray controller is displayed along with flexRay output graphic frame.

Figure 7, Figure 8, Figure 9 and Figure 10 shows the designed modules for intra-vehicular communication applications like four doors, AC, Oil Temperature and Weather Conditions.

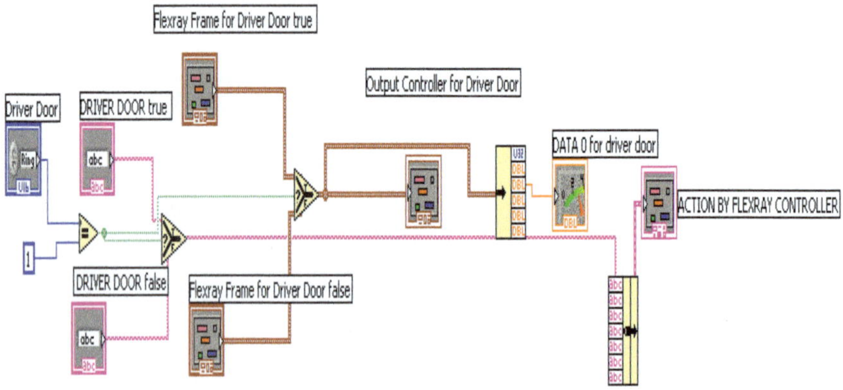

Figure 7. Door Module

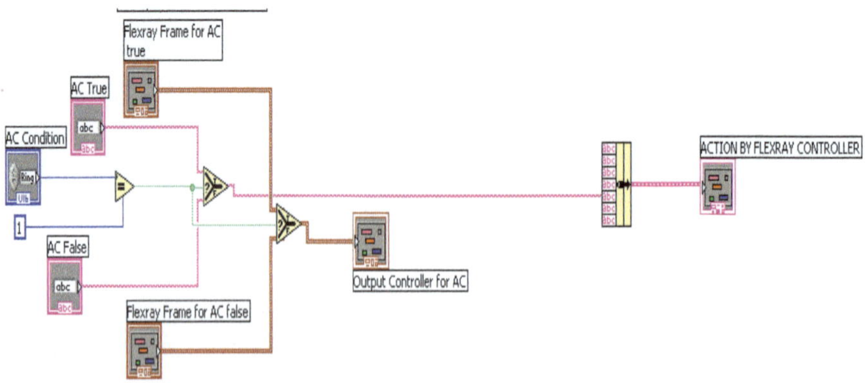

Figure 8. Air Conditioner Module.

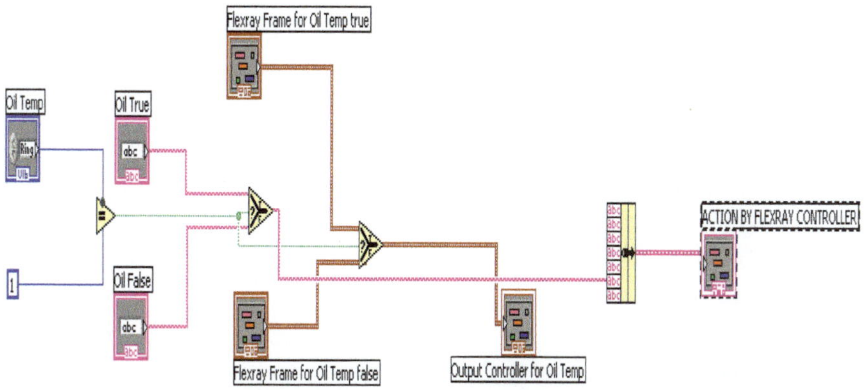

Figure 9. Oil Temperature Module

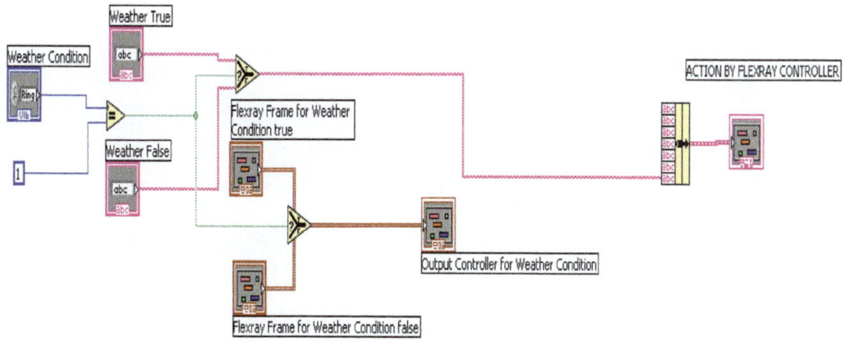

Figure 10. Weather Condition Module

Figure 11 shows design for External weather sensing module. With external weather condition input as rainy to module, the module evaluates the condition and provides the flex-Ray frame for weather condition as true and the output action is taken to turn on the wiper.

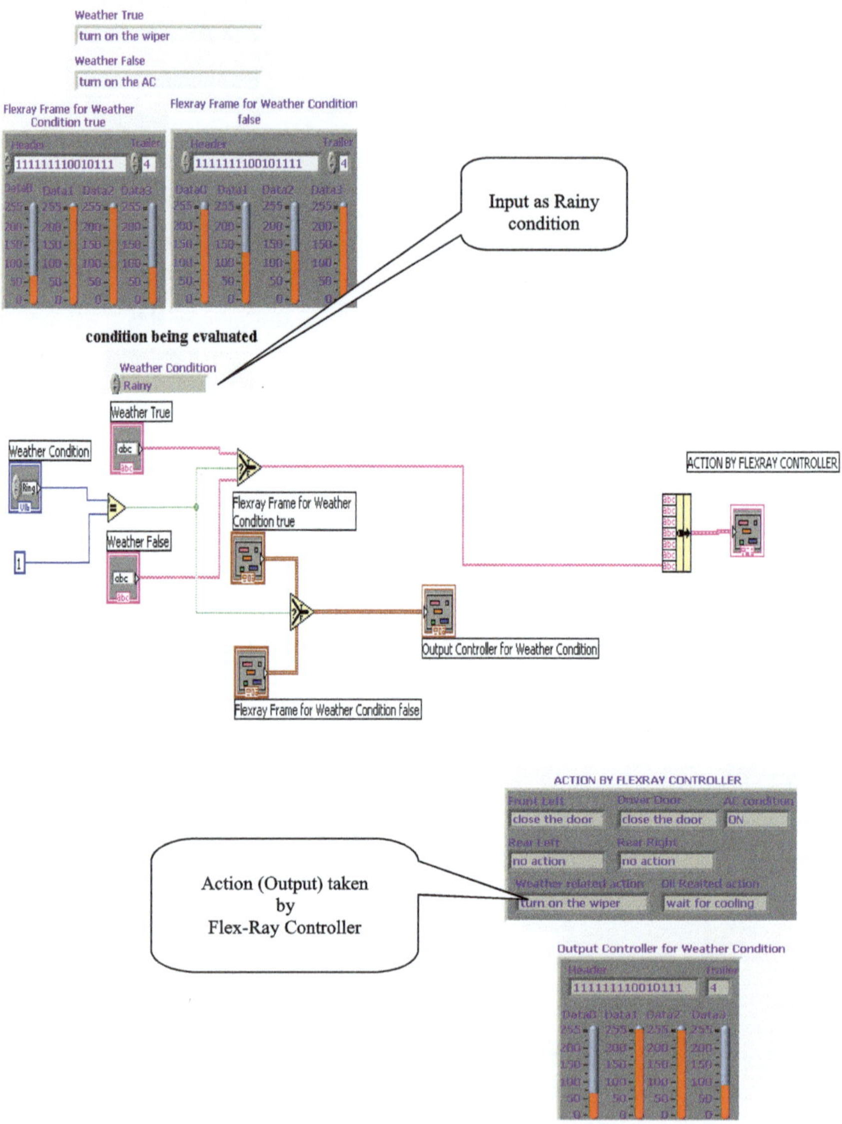

Figure 11. External Weather Sensing Module (input as rainy weather condition)

Figure 12 shows External weather sensing module with external weather condition input as hot, the module evaluates the condition and provides the flex-Ray frame for weather condition as false and the output action is taken to turn on the AC.

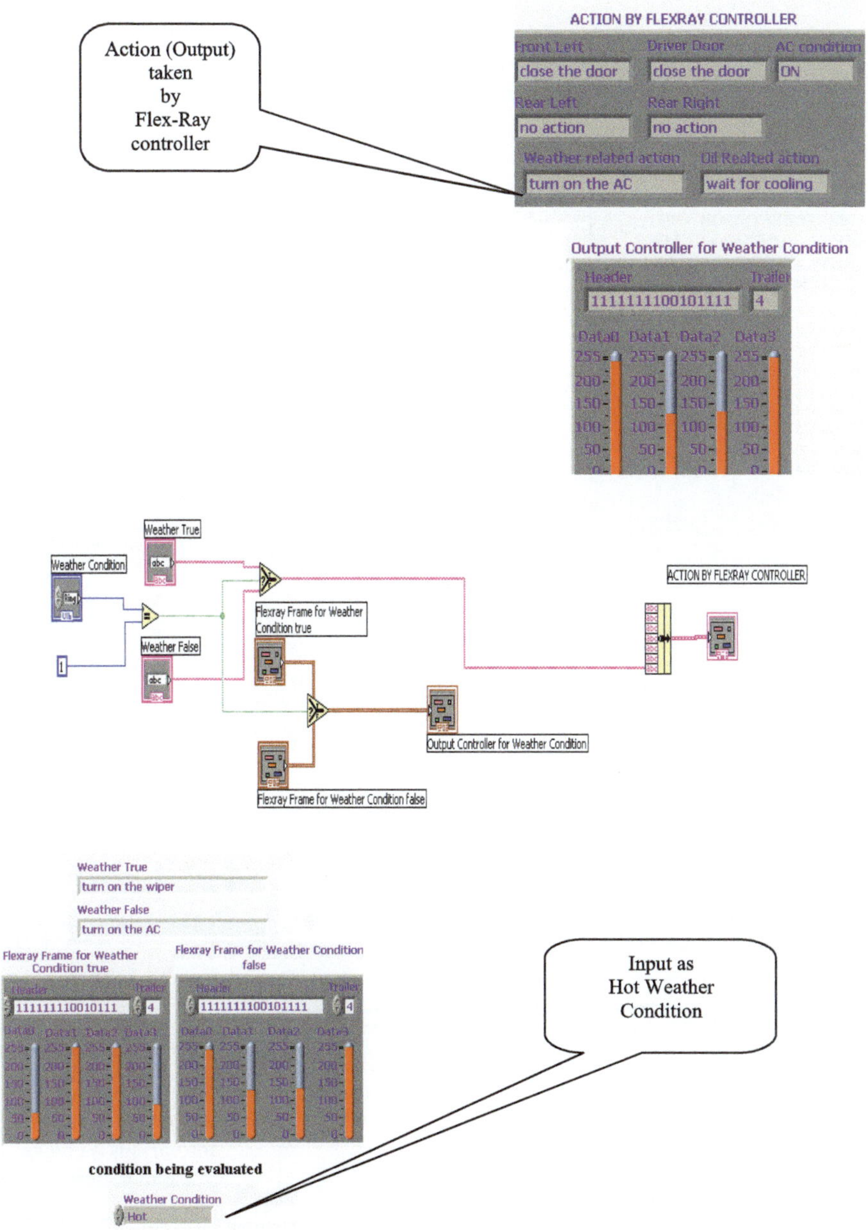

Figure 12. External Weather Sensing Module (input as hot weather condition)

Figure 13 shows results of the designed applications with the possible input conditions and explains the remedial controlling action taken by flex-ray controller on LABVIEW simulation platform.

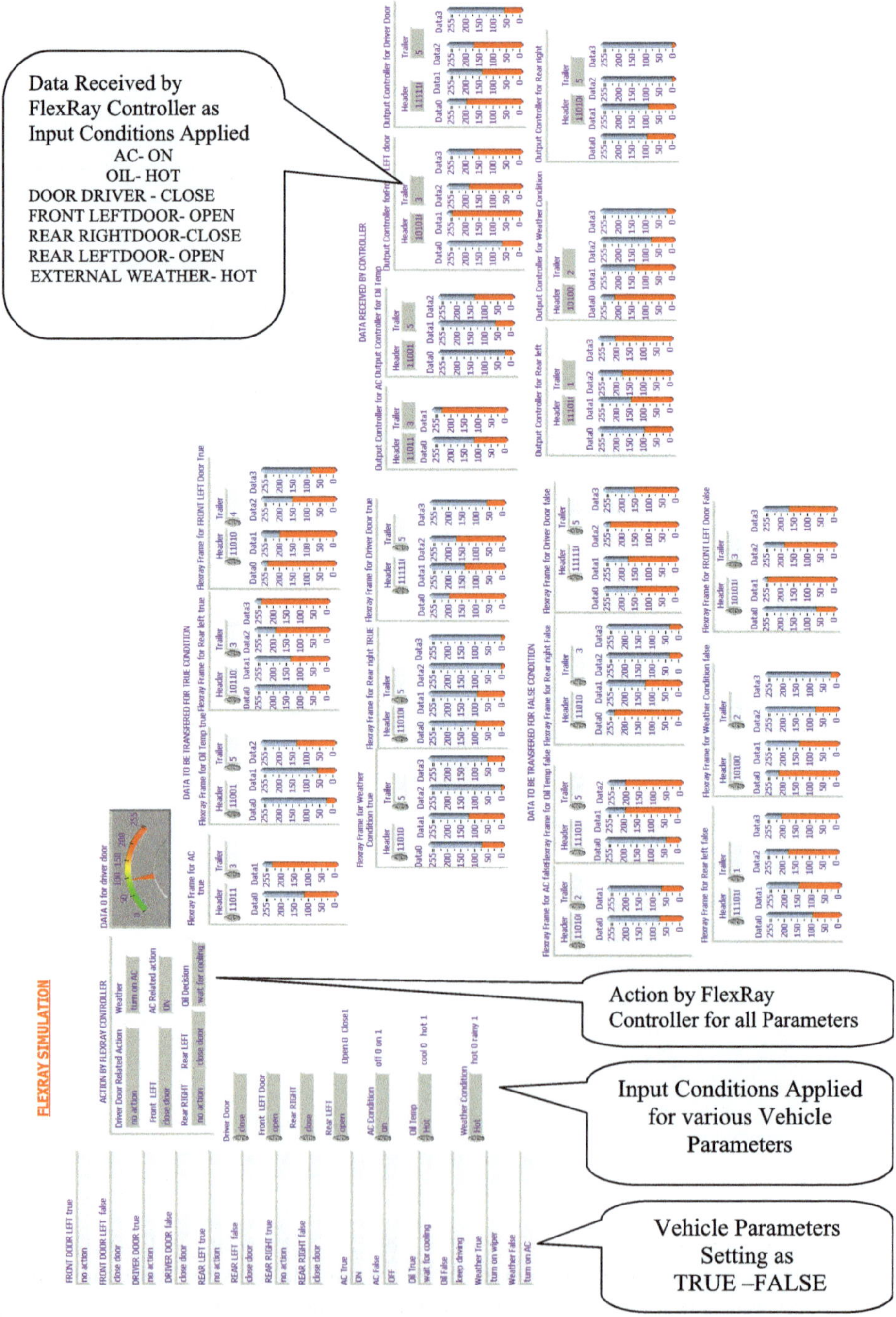

Figure 13. Results for vehicle parameters for Flex-Ray based Design

4.4. SDSCS – AFS Interfacing: Intra-vehicular communication application

The most Versatile Intra-vehicular Communication Application using Automotive Network Controller is Automotive Black-Box. The design approach and implementation of automotive black box is described and discussed below. Automotive Black box is referred to storing of vehicle data at the time of accident and retrieving same for investigation purpose. The main problem while designing of Black Box is deciding the incidence at which vehicle met with accident. The

Figure 14 shows the block diagram of prototype of System for Detecting Sudden Change in Speed (hence accident instance) (SDSCS) of Vehicle for Accident simulation and its interfacing to Adaptive Front Light System (AFS).

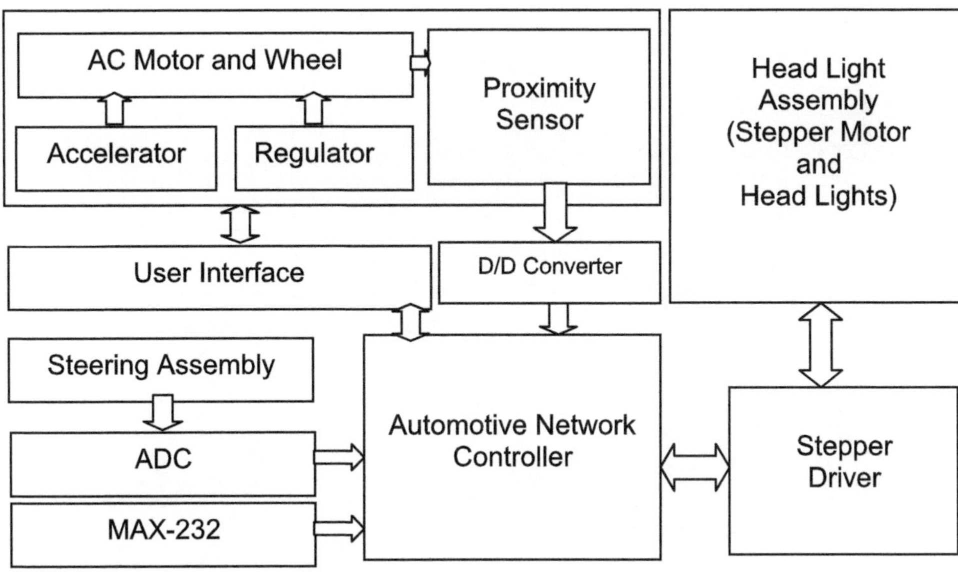

Figure 14. Block Diagram of prototype of SDSCS – AFS Interfacing

Figure 15 shows the details of hardware of the developed prototype. The assembly provides the mechanism for simulation of vehicle speed in form of motor coupled with regulator and breaking –accelerator facility as shown in figure. This motor is in close vicinity of proximity sensor which senses the speed and the same is coupled to controller which calculates the difference at smallest possible interval, thus simulating the accident situation in the vehicle.

AFS- Adaptive Front Lighting System, the name suggests that it is front lighting system of an automobile which will adapt itself according to the need of the curvature of the road.

Once the instance of sudden change in speed of vehicle (instance of accident) is confirmed then the information regarding various activities of vehicle (Vehicle Parameters) can be passed on through automotive controller based system which will take care of recording and storing the vehicle parameters. Intra-Vehicular communication is established between

AFS - SDSCS. The real time steering angle and the real time position of the front lights is available in electrical signal form in AFS. These two real-time output parameters are supplied as real-time inputs to System for Detecting Sudden Change in Speed of Vehicle for accident simulation (SDSCS) for storing purpose.

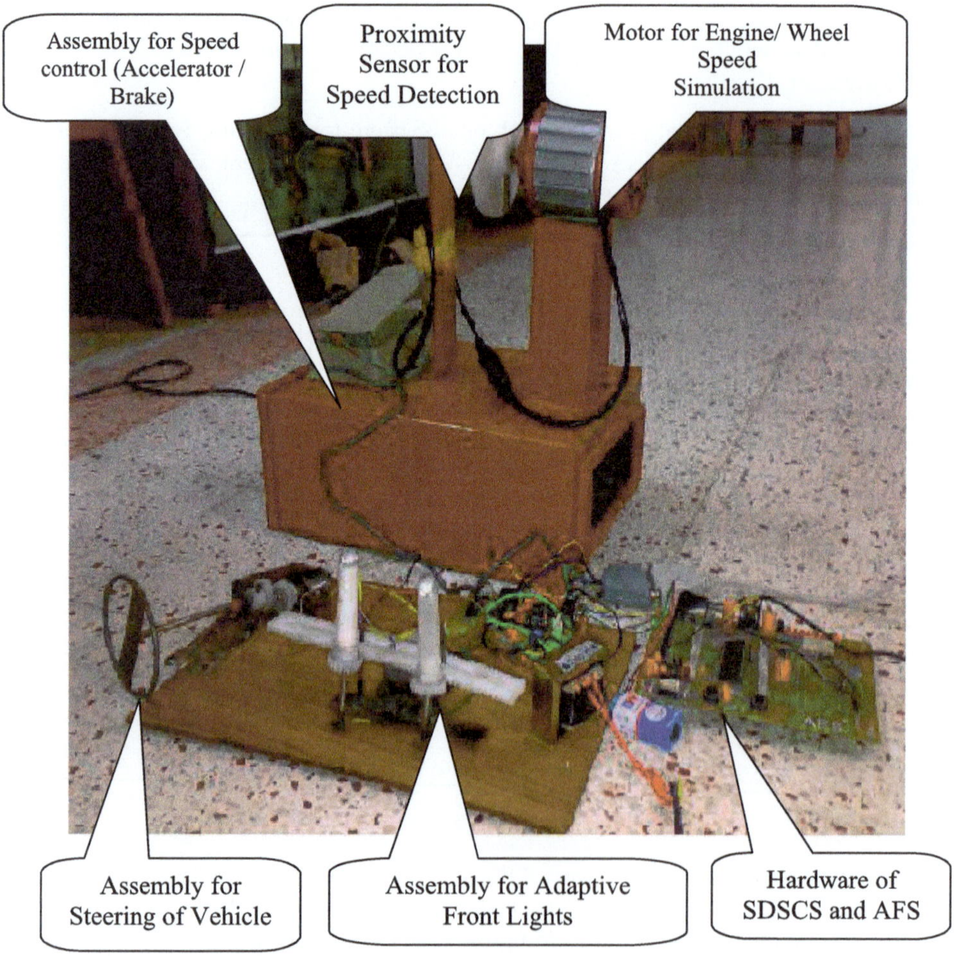

Figure 15. AFS-SDSCS Interfacing

To make the simultaneous movement of the front light in the horizontal plane according to the movement of steering shaft, it needs some range on the movement of the front light. Mainly, the lights should turn to some specific angle and then it should stop moving. As the vehicle takes the steep turn on a road, the direction of the front light should move, if it takes more sharp turn, the front light should rotate deeper in the same direction as that of the steering movement. If in case, the vehicle takes a sudden 'U' turn, the lights should not move directly in the opposite direction, but should be at the extreme of the specified direction.

When steering angle 130^0 the direction of the light system is in the front position as if the vehicle is moving in a straight direction. If Vehicle moves slightly towards the right direction on the curve, the lights system will also move directly from 130^0 to 135^0. If the vehicle takes sharper turn towards the right, let's say from 135^0 to 145^0, the lights will also move in the same direction (extreme right). When the vehicle continues to moves on straight track, the lights will also move in opposite direction from 145^0 to 135^0 and from 135^0 to 130^0.

Range of steering Angle / Variables	X1	X2	X3	X4
125^0 < _ < 135^0	0	0	0	0
135^0 < _ < 145^0	0	0	1	0
145^0 < _	0	0	1	1
135^0 < _ < 145^0	0	0	1	0
125^0 < _ < 135^0	0	0	0	0

Table 1. Right Movement of AFS

The same logic has been applied for the left direction movement of the lights system. If the car slightly turns towards the left direction, the light slightly towards the left direction i.e. from 130^0 to 125^0. if the vehicle takes sharper turn towards the left, accordingly the lights will also move deeper i.e. from 125^0 to 115^0 and if again it moves to the straight direction from the extreme left, it will also make the simultaneous movements from 115^0 to 125^0 and from 125^0 to 130^0

Range of steering Angle / Variables	X1	X2	X3	X4
115^0 < _ < 125^0	1	0	0	0
_ < 115^0	1	1	0	0
115^0 < _ < 125^0	1	0	0	0
125^0< _ < 135	0	0	0	0
115^0< _ < 125^0	1	0	0	0

Table 2. Left Movement of AFS

- **Movement of Front Lights in AFS for different steering positions**
- CASE I: Figure 16 shows position of steering angle is in between 125^0 to 135^0 there will no inclination of lights, the light will be in the neutral direction and the direction of light will be in the front direction.

Range of steering Angle	Variables			
	Over1	Over2	Over3	Over4
$125^0 < \_ < 135^0$	0	0	0	0

Figure 16. AFS (Case I)

- CASE II: Figure 17 shows the light position for right turn. If the vehicle takes right turn and the steering angle is in range of 135 to 145 then the AFS makes lights to move in right.

Range of steering Angle	Variables			
	Over1	Over2	Over3	Over4
$135^0 < \_ < 145^0$	0	0	1	0

Figure 17. AFS (Case II)

- CASE III: If the steering *angle goes beyond* 145 degrees for sharper right turns, Figure 18 shows the logic generated for AFS and the lights will be moved in rightmost.

Range of steering Angle	Variables			
	Over1	Over2	Over3	Over4
$145^{0} <$ _	0	0	1	1

Figure 18. AFS (Case III)

CASE IV: Case III is the rightmost position for lightening system and the lights will again turn back to its original condition, similarly the rotations will take place in the left direction and the same condition will be achieved as in figure 19.

Range of steering Angle	Variables			
	Over1	Over2	Over3	Over4
$135^{0} <$ _ $< 145^{0}$	0	0	1	0

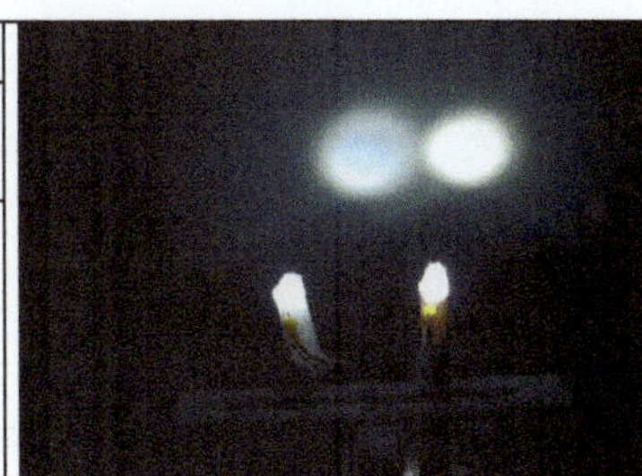

Figure 19. AFS (Case IV)

- **CASE V:** The lights will come back to centre position as in Figure 20.

Range of steering Angle	Variables			
	Over1	Over2	Over3	Over4
$125^{0} < \_ < 135^{0}$	0	0	0	0

Figure 20. AFS (Case V)

- **CASE VI**: If the steering moves in the left direction AFS will move the light towards the left direction, the logic will change as shown in figure 21.

Range of steering Angle	Variables			
	Over1	Over2	Over3	Over4
$115^{0} < \_ < 125^{0}$	1	0	0	0

Figure 21. AFS (Case VI)

- **CASE VII:** Figure 22 shows the left most condition and the vehicle lights will be moved towards extreme left direction.

Range of steering Angle	Variables			
	Over1	Over2	Over3	Over4
_< 115	1	1	0	0

Figure 22. AFS (Case VII)

- **CASE VIII :** Above mentioned was for left most direction now similarly the rotations will take place in the back to the right direction and the same condition will be achieved as it was achieved earlier as in figure 23

Range of steering Angle	Variables			
	Over1	Over2	Over3	Over4
115 < _ < 125	1	0	0	0

Figure 23. AFS (Case VIII)

- **CASE IX:** The logic will change to shift the lights to bring them to straight position as shown in figure 24.

Range of steering Angle	Variables			
	Over1	Over2	Over3	Over4
$125^0 < \_ < 135^0$	0	0	0	0

Figure 24. AFS (Case IX)

Following result windows shows the real time clock for data recording, the set threshold value for deciding the change of sudden change of speed. It also shows the condition of locking. Through menu driven the value of threshold can be altered, re-setted and brakes can be applied for Electro-Mechanical assembly (as explained above). Every time the back history of the data is also recorded. This back history (pre-accident history) can be helpful for post accident analysis.

The parameters recorded and displayed are off line or virtual (i.e. programmable) are status of Wiper, Doors, Air Bags, Seat Belts, Fog Lamps, Side Indicator, Head Lights. In addition to these virtual programmable parameters two real time parameters are also recorded. These parameters are derived from Adaptive Front Light System (AFS) i.e. Range of Steering Angle and The positions of Head lights at the time of data recording. Figure 25 shows data non blocked state of the system.

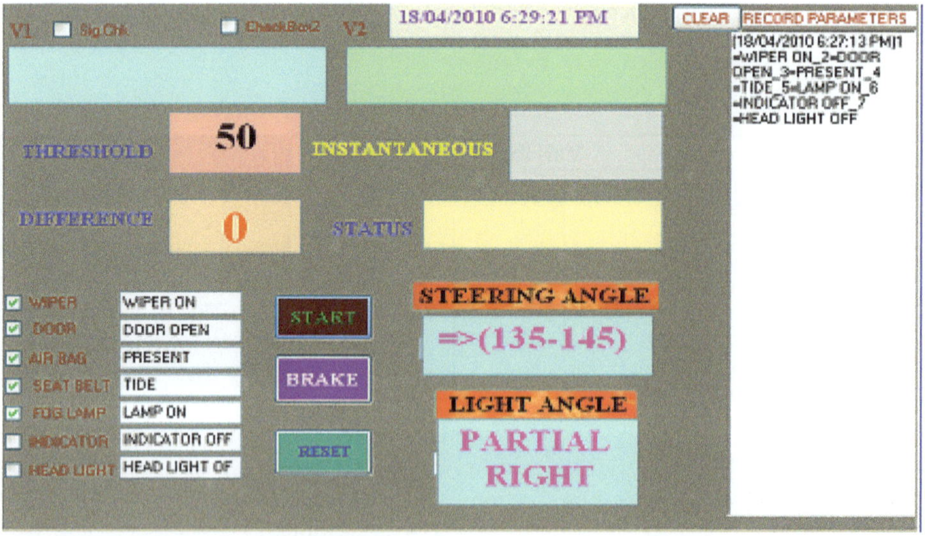

Figure 25. AFS-SDSCS Data not Locked

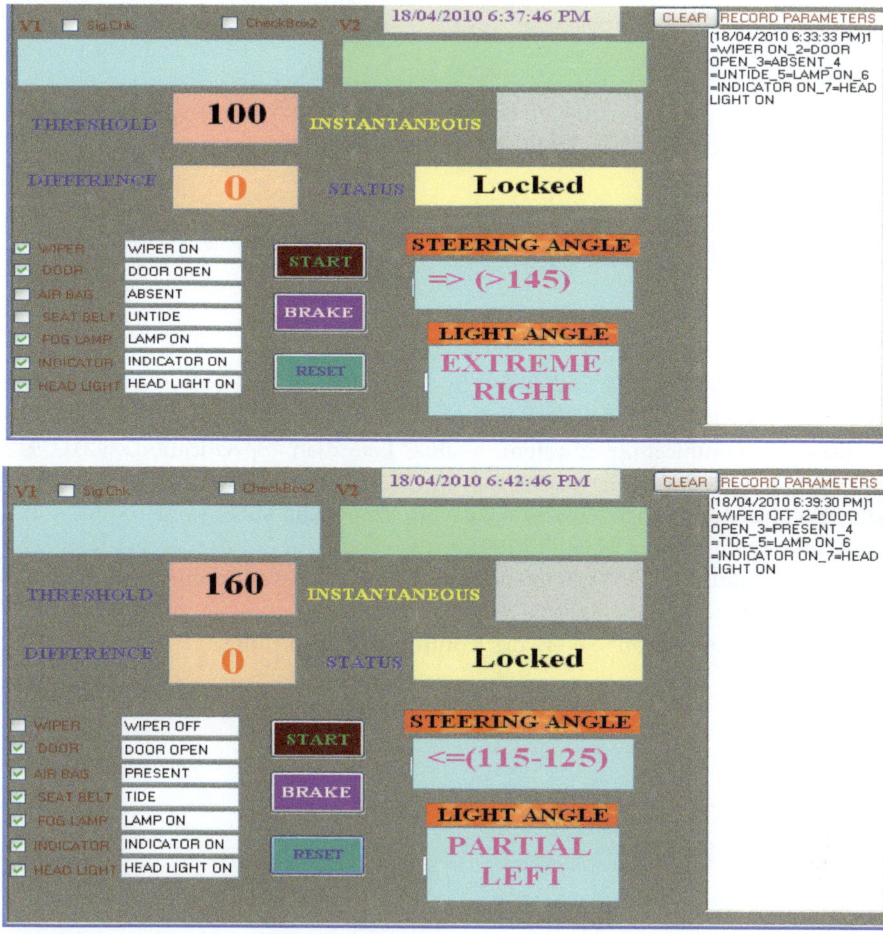

Figure 26. AFS-SDSCS Data Locked and Recorded

Figure 26 shows the blocking state of the system wherein the data is recorded for sudden change of speed that may lead for the unfortunate incident i.e. accident with the vehicle for three sets of virtual and real time parameters.

4.5. Conclusion

The chapter explains various intra-vehicular communication applications implemented through vehicular network controllers such as LIN, CAN, Flex-ray on tools like Modelsim, Xilinx, and Lab View and also some applications are implemented by hardware designing and development. The Applications such as black box, various parameters sensing and controlling respective corresponding actions are implemented and are presented in the chapter. Designed application systems are developed on simulation tools and also prototype is implemented and tested successfully with satisfactory results.

Author details

Preeti Bajaj
Department of Electronics Engineering, India
G.H. Raisoni College of Engineering Nagpur (M.S.), India

Milind Khanapurkar
Department of Electronics and Telecommunication Engineering, India
G.H. Raisoni College of Engineering Nagpur (M.S.), India

5. References

[1] Multiplexed Networks for Embedded Systems CAN, LIN, FlexRay and Safe-by-Wire. Copyright 2007 John Wiley & Sons Ltd.
[2] FlexRay Communications System - Bus Guardian Specification, v2.0, FlexRay Consortium, June 2004.
[3] LIN Specification Package – revision 2.0 – http://www.lin-subbus.org
[4] CAN Specification, BOSCH GmbE 1991.
[5] FlexRay Communications System – Electrical Physical Layer Specification, v2.0, FlexRay Consortium, June 2004.
[6] Publications by Automotive Network Controllers Consortiums

Chapter 10

Optical Techniques for Defect Evaluation in Vehicles

J. P. Carmo and J. E. Ribeiro

Additional information is available at the end of the chapter

1. Introduction

The optical techniques are a powerful tool on situations where either the physical contact or invasive techniques for evaluation are not suitable. Vehicle environments constitute an application field for the optical techniques and are the focus of this chapter. In order to reinforce this kind of techniques, it must be clarified that the idea to manipulate the light backs to the second century before our age, when Archimedes planned to destroy enemy ships using a solar heat ray with an array of actuators to change the shape of a mirror (Bifano T., 2011). Therefore, the field of photonics is the one that offers the possibility to achieve one of the greatest realizations and applications because the light is present in all aspects of the human life and our way of living is impossible without light (Carmo J. P. *et al.*, 2012a). Optical measurement systems are also suitable for harsh monitorization because they are non-contact and full-field techniques. This is the case of Moiré Interferometry, which is used for many optoelectronic applications as displacement measurements (Wronkowski L., 1995), evaluation of microelectronics devices deformation (Xie H. *et al.*, 2004), optical communications (Chen L. *et al.*, 2000), strain measurements with Fiber Bragg Grattings, FBGs, (Silva A. F. *et al.*, 2011) and spectrography (Kong S. H. *et al.*, 2001). In this context, it must be noted that the recent nuclear disaster in Fukushima, Japan, confirms the need of tighter security measures be done within harsh environments (which includes the automobiles) in order to increase both the safety of people and the reliability of vehicles' parts.

2. Principles of full-field optical techniques

2.1. Holographic interferometry

The holography is a method to store and regenerate all the amplitude and phase information contained in the light which is scattered from an illuminated object. This technology offers the possibility of three dimension photography.

Despite the holography requires coherent light, it was discovered by Gabor in 1948 (Gabor D., 1948), more than a decade before the invention of the laser, which in 1960. By means of holography an original wave field can be reconstructed at a later time at a different location. This technique therefore has many potential applications.

In the year of 1965, when the CW lasers became available, Robert Powell and Karl Stetson published the first work on holographic interferometry (Powel R. L. *et al.*, 1965). Using this technique, it is possible record the precise shape and position of a object in two different states, then it is possible compare the two records to measure the movement or deformation, that were displayed as fringes on the image.

In the Figure 1 is shown the optical set-up of holography.

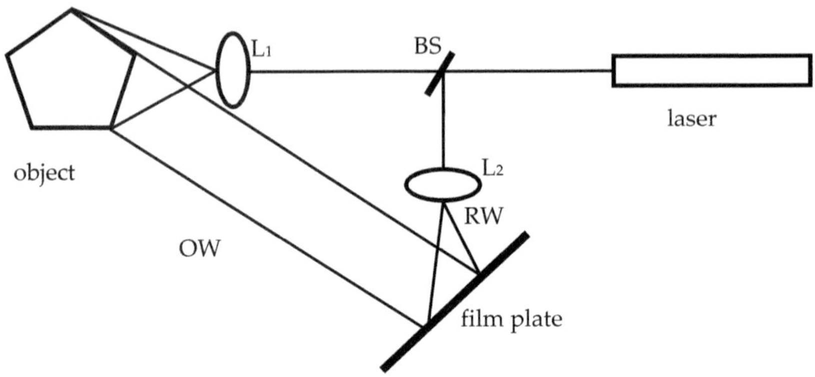

Figure 1. Holography configuration to measure the surface displacement.

The laser beam is divided in two beams using a beam splitter (BS). Each of one is opened by two lenses (L_1 and L_2), on film plate is recorded the interference of two coherent wavefronts: the object (OW) and reference (RW). When the film plate is developed and then illuminated by the reference wavefront, it will be reconstructed a virtual image of the original object. The most used technique for displacement measurement is double exposure or frozen-fringe, which records two holograms on the same film plate, the first one is corresponding to an initial condition of the object and the second one to a displaced or charged condition. The image is reconstructed by a coherent beam illumination; the two reconstructed images will interfere, resulting in a holographic fringe pattern that can be used to measure surface displacements. Each fringe represents a contour of equal displacement of the object, where the value of displacement is one-half the wavelength of the light source used in the recording process (Shang H. *et al.*, 2009).

There are others holographic interferometry techniques in which the most relevant are real-time live-fringe and time-average holographic interferometry. The real-time approach requires that the photographic plate must be physically developed in place without changing its position. The single-exposure hologram is made in the regular way and after it is illuminated by the reference beam while the object is illuminate by the original object beam (Cloud G., 1988). The time-average holographic interferometry approach is used for modal analysis of vibrating bodies. A hologram of the moving body is made with an exposure time lasting over numerous periods of the vibrational motion.

Recent advances in recording materials, lasers and computer allowed more holography techniques being developed, which include: TV holography, digital holography and dynamic holographic interferometry. These techniques open new and challenge fields of research and applications. Nevertheless, this technique has some drawbacks like the displacement components cannot be measured independently in different planes. Another limitation is the strict requirement of vibration (half a wavelength of the illuminating source), if the vibration magnitude is higher occurs decorrelation. As a result, its industrial application as a routine Non Destructive Testing (NDT) technique is restrict, usually these techniques are used in laboratory environment.

2.2. Electronic speckle pattern interferometry

Electronic speckle pattern interferometry is a NDT technique which basic concepts were simultaneously developed Mavsky *et al* (Macovsky A. *et al.*, 1971) in United States and by Butters and Leendertz (Butters J. *et al.*, 1971; Butters N. *et al.*, 1975) in England. Later group pursued the development of the ESPI technique, with Lökberg (Lökberg O. *et al.*, 1976; Lökberg O., 1985) and Beidermann (Beidermann *et al.*, 1975) being important references in this field.

Speckle techniques use the random pattern of dark and bright spots (speckles) that are formed in space when a diffusely reflecting object is illuminated by coherent laser light (laser speckles). A random granular pattern called a speckle pattern is observed when looking, with the eye or with a camera, at a object surface when it is illuminated by a laser - see the Figures 2(a) and (b). The speckles appear only if the surface roughness is greater than the wavelength, λ, of the light. If the object is imaged, each point P on the detector will gain contribution of light coming from a coherence volume, determined by at least the airy spot and the roughness of the surface (Svanbro A., 2005).

The observed speckle pattern unrepeatable for each illuminated area in the sense that the observed pattern is unique for the microstructure of the specific surface area. Another region will give rise to a totally different random speckle pattern. The speckle pattern is records by a CCD camera and could interfere with a reference wave. If the object suffers a deformation caused by an external load, the speckle pattern will change due the variation of wavefront between the object and reference waves. This deformation could be computed using an operation of digital subtracting between the speckle patterns corresponding to before deformation from the one after deformation. This operation will result in an interferogram, which will be displayed on the monitor as a pattern of dark and bright fringes that are called correlation fringes. ESPI allows displacements in different directions to be measured separately with a high-resolution.

ESPI technique can be sensitive to out-of-plane or in-plane displacement, depending of the used optical set-up. For in-plane displacement measurement (Figure 3), the object lies in the

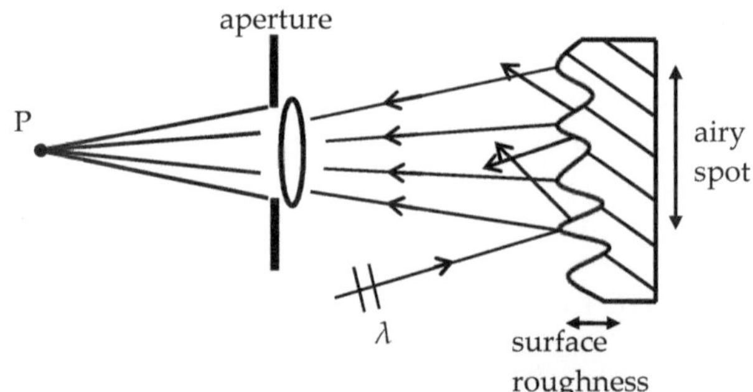

(a)

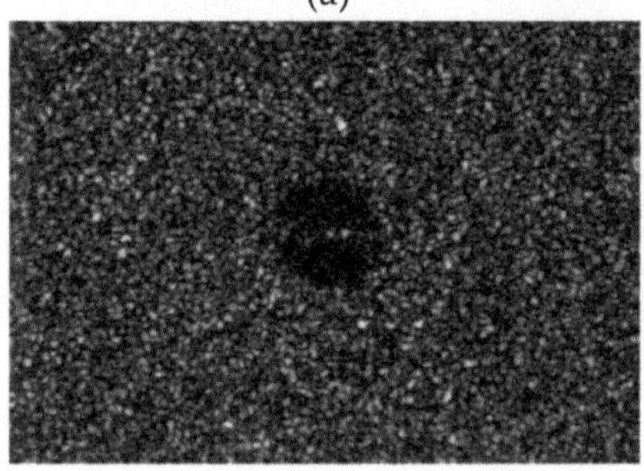

(b)

Figure 2. (a) The schematic representation of speckle formation and (b) a typical speckle pattern.

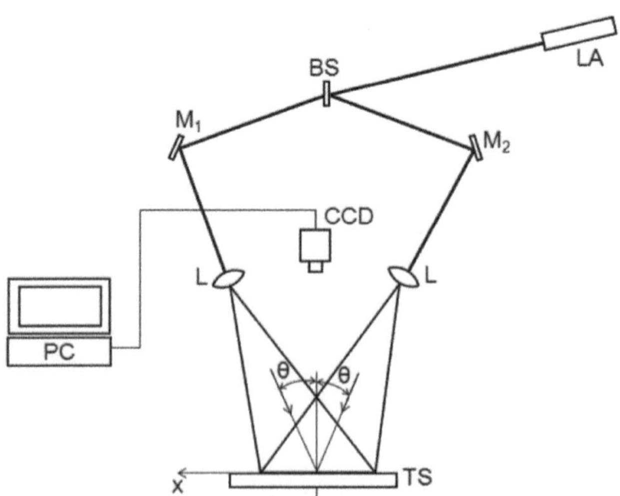

Figure 3. A schematic presentation of the optical set-ups used in-plane ESPI (Jones R. *et al.*, 1974). LA: laser source, BS: beam splitter, M_1 and M_2: plane mirrors, L: lens, TS: test specimen or object.

xy-plane and is illuminated by two plane wavefronts, inclined at equal and opposite angles, θ, to the x-axis surface-normal. No separate reference beam is needed at the image plane. The positive y-axis points out of the page, and the center of the viewing lens aperture lies on the z-axis. The phase change $\Delta\phi$ of the wavefront from the object surface before and after deformation can be expressed as:

$$\Delta\phi = \frac{4\pi}{\lambda}\sin(\theta)U \tag{1}$$

where λ is the wavelength of the laser, θ the incident angle, and U is the in-plane displacement component.

For out-of-plane displacement measurement, the object is illuminated by an object which is viewed by a CCD camera (Figure 4). A reference speckle pattern formed by a reference beam is also observed by the camera. So the resultant image is the interference of these two speckle patterns.

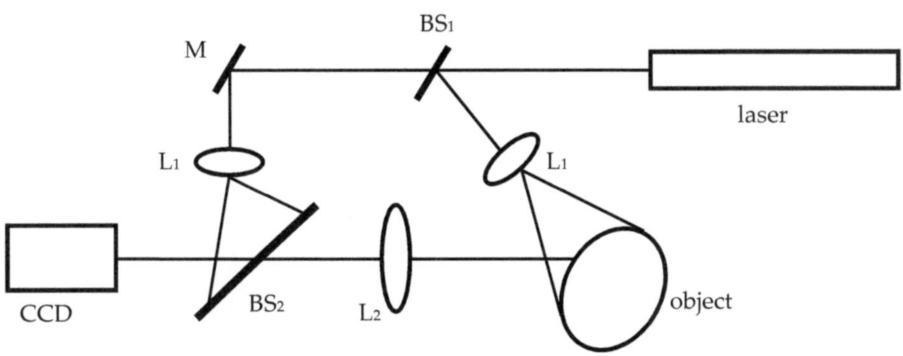

Figure 4. ESPI sensitive to out-of-plane displacement. Where $BS_{1,2}$: beam splitter; M: plane mirror; L_1: concave lens; L_2: convex lens.

The relation between the phase change and the deformation can be expressed as:

$$\Delta\phi = \frac{2\pi}{\lambda}\left[1+\cos(\theta)\right]W \tag{2}$$

where W is the out-of-plane displacement component

2.3. Shearography

Shearography technique is particularly interesting because it enables direct measurements of displacement derivatives which are related to the strains (Hung Y. *et al.*, 1973; Leendertz, J. A. *et al.*, 1973).

The basic principle of shearography, is to carry the rays scattered from one point of the object into interference with those from a adjacent point. To obtained this interference is used a speckle-shearing interferometric camera like the represented in the Figure 5. The optical set-up is similar as that used in common speckle photography, however one half of the camera lens is covered by a thin glass wedge. Thus, the two images focused by each half of the lens are laterally sheared one to other (Gåsvik K. J. *et al.*, 2002). The shearing direction is oriented by the wedge position and it is proportional to its angle. If the shearing is in the x direction, the rays from a point $P(x,y)$ on the object will interfere in the image plane with those from a neighboring point $P(x+\delta x,y)$.

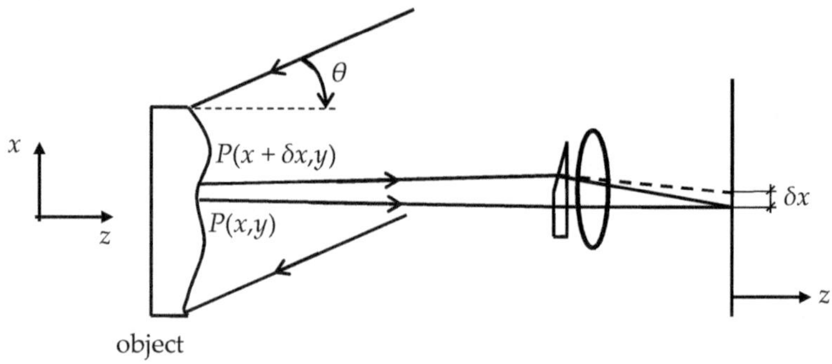

Figure 5. The optical set-up of shearography.

When the object is deformed there is a relative displacement between the two points that produces a relative optical phase change $\Delta\phi$ is given by:

$$\Delta\phi = k \times \left\{ \begin{array}{c} \left[1+\cos(\theta)\right] \times \left[w(x+\delta x, y) - w(x,y)\right] + \\ + \sin(\theta)\left[u(x+\delta x, y) - u(x,y)\right] \end{array} \right\} \tag{3}$$

where θ is the angle of incidence and u and w are the displacement components in the x-and z-directions, respectively. If the shear δx is very small, the relative displacements may be approximated by the displacement derivatives and thus Equation (3) becomes:

$$\Delta\phi = k\{\left[1+\cos(\theta)\right]\frac{\partial w}{\partial x} + \sin(\theta)\frac{\partial u}{\partial x}\}\delta x \tag{4}$$

By rotating the camera 90º around the z-axis, u in Equation (4) is replaced by v, the displacement component in the y-direction.

2.4. Moiré methods

The word Moiré derives from the French term meaning wet silk or fringe patterns produced by the interference of aligned fibers in thin tissues. In Engineering, Moiré refers to a technique for experimental analysis for determining the displacement or deformation from a set of interference fringes resulting from the superposition of two systems - one in the deformed specimen and the other that is undeformed can be used as reference. Moiré techniques have appeared in the field of optics since nineteenth century (Rayleigh L., 1874). The Moiré techniques allow the measurement of deformations which occur when an object is subjected to applied charges. These methods can measure out-of-plane and in-plane displacements. The displacements are measured simultaneously over the whole view field and the measurement data are exposed as contour maps of displacements. One of most important Moiré method is the Moiré interferometry that was performed by Post (Post D. *et al.*, 1994) in 1980 and represents one the most important advances in optical techniques applied to experimental stress analysis in the last decades. The Moiré interferometry principle is based in the interference of two coherent light beams or lasers beams. This interference produces regions in space of constructive and destructive interference where two equal coherent beams intersect. The most important advantages of Moiré interferometry compared with other optical methods are: excellent fringe contrast, high sensitivity and special resolution (typically 0.417 μm per fringe), dynamic range of measurement and high signal-to-noise ratio (Ribeiro J. E., 2006). On the other hand, an important disadvantage of this technique is the need of a high-frequency grating replication onto the sample surface before the measurement begins. The replication of grating could restrict its application in cases where the replicated grating may considerably change the stiffness of the testing sample, another difficulty of this process is the need of a technician with experience implement the replication and it is very laborious.

To measure in-plane displacement with Moiré interferometry are used two symmetrical incident beams of mutually coherent light (Post D. *et al.*, 1994), see the Figure 6.

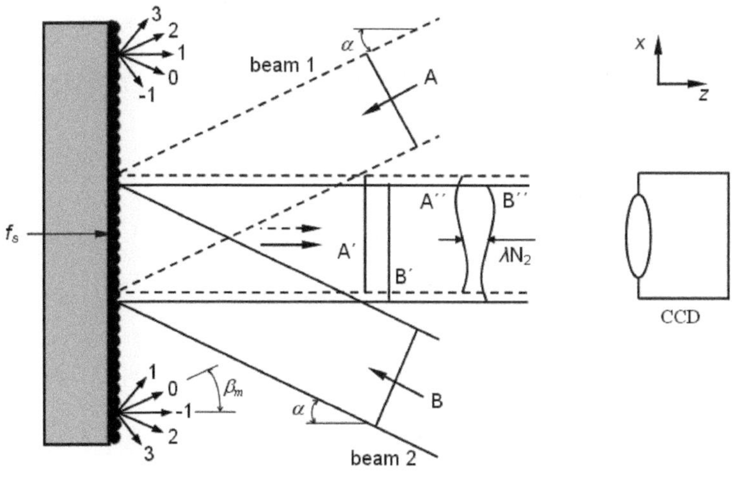

Figure 6. Schematic representation of Moiré interferometry principle (Post D. *et al.*, 1994).

The diffraction equation for the represented grating is:

$$\sin(\beta_m) = \sin(\alpha) + m\lambda f_s \tag{5}$$

where β_m are angles of diffracted beams, α is the angle of incident beam, m^{th} the diffraction order, λ the laser wavelength and f_s is the grating frequency of specimen.

When the equation (5) is satisfied, the ±1 order diffraction beams will emerge normal to the specimen grating, producing a uniform intensity throughout the field (null field) (Shang H. *et al.*, 2009).

The obtained displacement fields allow the calculation of components orthogonal direction (x, y) and they can be expressed as:

$$u = \frac{1}{f} N_x = \frac{1}{2f_s} N_x$$
$$v = \frac{1}{f} N_y = \frac{1}{2f_s} N_y \tag{6}$$

By differentiation of displacements is obtained the corresponding strain components:

$$\varepsilon_x = \frac{\partial u}{\partial x} = \frac{1}{2f_s}\frac{\partial N_x}{\partial x} \approx \frac{1}{2f_s}\frac{\Delta N_x}{\Delta x}$$
$$\varepsilon_y = \frac{\partial u}{\partial y} = \frac{1}{2f_s}\frac{\partial N_y}{\partial y} \approx \frac{1}{2f_s}\frac{\Delta N_y}{\Delta y} \tag{7}$$
$$\varepsilon_{xy} = \frac{1}{2}\left(\frac{\partial u}{\partial y} + \frac{\partial v}{\partial x}\right) \approx \frac{1}{2f_s}\left(\frac{\Delta N_x}{\Delta y} + \frac{\Delta N_y}{\Delta x}\right)$$

where N_x and N_y are fringe orders in x and y direction, respectively.

To measure out-of-plane deformation and obtained the shape of a body is frequently used the well-kwon experimental technique Shadow Moiré (Takasaki H., 1970). The optical set-up is schematically represented in the Figure 7.

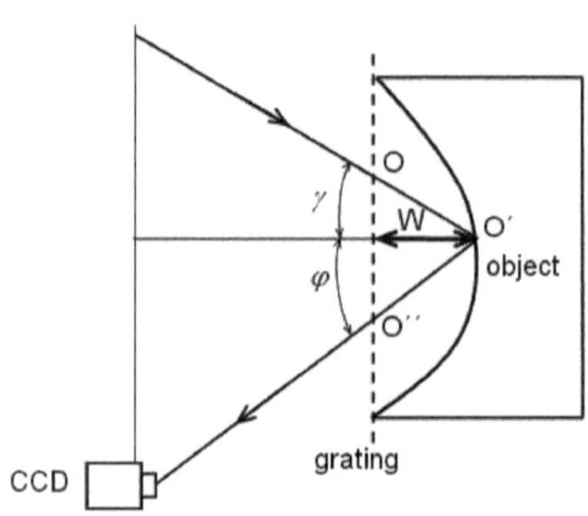

Figure 7. An optical set-up of Shadow Moiré.

The grating lying over the curved surface is illuminated under the angle of incidence γ (measured from the grating normal) and viewed under an angle φ (Gåsvik K. J., 2002). From the Figure 7, it possible to observe that a point O on the grating is projected to a point O′ on the surface which by viewing is projected to the point O′′ on the grating. This is equivalent to a displacement of the grating relative to its shadow equal to:

$$W = \frac{np}{\tan(\gamma) + \tan(\varphi)} \tag{8}$$

where W is the out-of-plane displacement, n is the order of fringe and p is the pitch of the grating.

2.5. Digital Image Correlation (DIC)

Digital image correlation belongs to the class of non-contacting measurement methods that were developed by Sutton (Sutton M. A. *et al.*, 1983; Sutton M. A. *et al.*, 1986; Sutton M. A. *et al.*, 1991; Sutton M. A. *et al.*, 1988) and by Bruck (Bruck A. *et al.*, 1989), in which acquires images of an object, store these images in digital form and perform image analysis to obtain full-field shape, deformation and motion measurements (Sutton M. A. *et al.*, 2009). In this method is used a mathematical correlation to estimate the displacement in the plane surfaces or structures of components subject to mechanical or thermal stresses. This technique is based in the use of random patterns on the surface of the components or structures. The technique compares two images acquired at different states, one before deformation (reference image) and other after deformation (deformed image) (Hu T. *et al.*, 1985).

In this technique the object is illuminated by an incoherent light source and the patterns of intensity results from the surface texture. These patterns of intensity, which would have a random distribution, are subdivided into smaller areas. Each subregion [originally defined in the image recorded] is then compared with images obtained by correlation between two different states of deformation in the object. If $f(x,y)$ is a discrete function that defines the grayscale of pixel on the original image and $f^*(x^*,y^*)$ is the grayscale of pixel on the final image (Marcellierl H. *et al.*, 2001). The relation between the two functions is defined by:

$$f^*(x^*,y^*) = f[x + u(x,y), y + v(x,y)] \tag{9}$$

where u and v represent the displacement field.

The determination of the displacement field is obtained by correlation between the random pattern of the initial image (reference) and its transform (deformed). This operation is performed for all patterns that meet in the center of the virtual grating on the initial image so as to obtain all the displacement field of each grating element.

Considering the displacement field for a random pattern, that is uniform and bilinear along the axes x and y, and given by:

$$u(x,y) = a_u x + b_u y + c_u xy + d_u \tag{10}$$

$$v(x,y) = a_v x + b_v y + c_v xy + d_v \tag{11}$$

The correlation between images of before and after deformation can be expressed as:

$$C = \frac{\sum_{x,y\in S}[f(x,y)f^*(x^*,y^*)]}{\left[\sum_{x,y\in S} f^2(x,y)\sum_{x,y\in S} f^{*2}(x^*,y^*)\right]^{\frac{1}{2}}} \tag{12}$$

where $f(x,y)$ and $f^*(x^*,y^*)$ are intensity distributions of speckle images before and after deformation, respectively. When $C=1$, the two images have the best correlation, i.e., they become identical.

This method has many advantages compared with others, which is possible to detach the follow: the cost of DIC is relatively low when compared to moiré and ESPI, because optical set-up is simple and it is not necessary a coherent light source; the resolution of DIC depends not only on the optical magnification, but also on the image processing (sub-pixel algorithms). Using a zoom lens and high resolution imaging system including CCD, can be reached a resolution of sub-micron.

2.6. Phase measurement techniques

In last decades, have been developed different techniques for quantitative phase measurement from fringe patterns (Robinson D. W. *et al.*, 1993; Creath K. *et al.*, 1994). These techniques are classified in two groups: spatial and temporal. In the spatial phase-measurement techniques, the phase information is extracted from a single interferogram which has a large number of tilt fringes that are used as a carrier frequency. For the temporal phase-measurement techniques, the phase is measured using a single point in an interferogram as the phase difference between the analyzed and reference beams is changed in a controlled way by the means of a piezoelectric device. The N-frame techniques measure a sequence of interferograms with known phase shifts techniques and they are the most popular type of these (Creath K. *et al.*, 1996).

The phase shifts or phase-shifting are the techniques extensively used in automatic fringe analysis of interferogram. It has considerably improved the accuracy of optical techniques, allowing the fast visualization of results. In the phase-shifting technique, the phase change is introduced usually by a calibrated phase steps. The phase of the whole image can be computed by analyzing the intensity patterns taken at different step.

The intensity or irradiance recorded by a detector for a single interferogram can be written as:

$$I(x,y) = I_0(x,y)\{1+\gamma(x,y)\cos[\varphi(x,y)+\alpha_i]\} \tag{13}$$

where $I_0(x,y)$ is background contribution, $\gamma(x,y)$ is the interference fringe amplitude, and $\phi(x,y)$ is the function of fringe phase, α_i the amount of phase shifting, (x,y) the coordinates of the image plane.

The most common phase-shifting technique is the four-point which uses four intensity values with $\pi/2$ relative phase shifts between steps. It is written in the form:

$$\varphi = \operatorname{atan}(\frac{I_4 - I_2}{I_1 - I_3}) \tag{14}$$

where I_1, I_2, I_3 and I_4 are intensities recorded in the detector for four different interferograms with phase shifts of $\alpha=0$, $\pi/2$, π and $3\pi/2$.

Analyzing the equation (15) it is possible to verify that there are discontinuities for the phase values of π and $-\pi$ which results from the atan function. These discontinuities are eliminated by using algorithms developed for this purpose (unwrapping). Thus, the absolute phase maps can be computed by phase unwrapping to reveal the accurate displacement field.

3. Industrial applications of full-field optical techniques

3.1. Vehicle shape measurement

In the automotive industry, it is essential measuring with accuracy the 3-D shapes of objects to ensure product development and manufacturing quality.

Traditionally, the most used technique to measure the vehicle shape is the use of structured method combined with photogrammetry. In this case are used some targets that are fixed on the vehicle body which allow the coordinate transformation from local to global. A structured light is projected on the vehicle surface combined with absolute phase measurement and phase shifting using fringe frequency change to determine the local coordinate pixel by pixel at one view direction (Chen F. *et al.*, 2000). The process is repeated for more than a two hundred views to cover hall vehicle. After these measurements are obtained a cloud of points that are patched together using a least mean squares method.

Recently, have been developed new progresses in 3D shape measurement (Zhang S., 2010), in this process is used a projector to project sinusoidal patterns and is called digital fringe projection technique, if is adopted a phase-shifting algorithm to, this technique is called digital fringe projection and phase-shifting technique.

In the digital fringe projection technique, a computer creates the digital fringe patterns which are composed of vertical straight stripes that are sent to a digital video projector. The fringe images (vertical straight stripes) are projected onto the object, which are distorts by the surface profile, then a CCD camera captures the distorted fringe images for the computer that analyzes the fringe images to obtain 3D shape information based on the deformation using triangulation. In this process is frequently used phase shifting algorithms which improve the optical metrology resolution (Huang P. S. *et al.*, 2006).

3.2. Measurement of residual stresses

The residual stresses can be defined as those which remain in the material or component after the manufacturing process and in the absence of external forces and thermal gradients

(James M. R. *et al.*, 1996). The manufacturing processes are the most common cause of residual stresses and some examples of manufacturing processes that can introduce residual stresses in the produced components are casting, welding and machining, all of them are used in automotive industry. However, the residual stresses can also arise for maintenance or repair operations. Sometimes, these stresses can also be induced in service during installation or by occasional overloads.

The effects of residual stresses can be beneficial or detrimental, depending on its magnitude, sign, and its distribution. In most cases residual stresses are harmful because they overlap with the operating stresses. However, they could be also beneficial, especially by allowing the increased of the fatigue limit in components which are dynamically requested. Another particularly important feature concerning the residual stress is that its presence usually is undetected until the malfunction or failure occurring.

There are currently different techniques for measuring residual stress, such as the contour method (Richter-Trummer V. *et al.*, 2011), the hole-drilling with strain gages or optical processes which use visible radiation of light (Ribeiro J. E., 2006), X-ray and neutrons diffraction, magnetic techniques and ultra sounds (James M. R. *et al.*, 1996). Some of them use mechanical processes to stress release in order to measure the residual stresses while in others it detects its presence by the effect of material properties. However, none of which is universally applicable, despite the most used technique to measure residual stresses is the hole-drilling method.

The hole-drilling method is an experimental technique used for measuring the strains on surface caused by the stress release. These strains can be measured using electrical discrete transducers or calculated from the displacement field on surface. The optical techniques used for measuring the displacement have great advantages compared with other techniques, emphasizing the performing of a global or field measurement and without contact. These techniques allow in-plane and out-of-plane measurements with a resolution that can be variable, ranging from a low resolution (tenths of mm) until very high with a magnitude of used light wavelength (a few tenths of a micrometer). The most used optical techniques combined with hole-drilling method (Ribeiro J. *et al.*, 2009) are the Moiré interferometry (Ribeiro J. *et al.*, 2009) and the ESPI (Cheng P. *et al.*, 2008) to measure residual stresses.

In this chapter a process to measure residual stresses with optical techniques (in-plane ESPI and Moiré Interferometry) is developed and combined with the hole-drilling method. Measurements were carried out on a ring and plug specimen, constructed to produce well known residual stress fields. The calibration coefficients were obtained by numerical simulation with a Finite Element Method (FEM) code. For these measurements were prepared two optical set-ups, one for Moiré Interferometry (Post D. *et al.*, 1994), see in the Figure 8, and the other for double-illumination ESPI (Jones R. *et al.*, 1974), see in the Figure 3. Both were used to measure the in-plane displacements generated by residual stress release. Image processing algorithms, involving filtering, phase calculation and unwrapping, and spatial differentiation were used in data post-processing to transform surface displacement into residual stress fields.

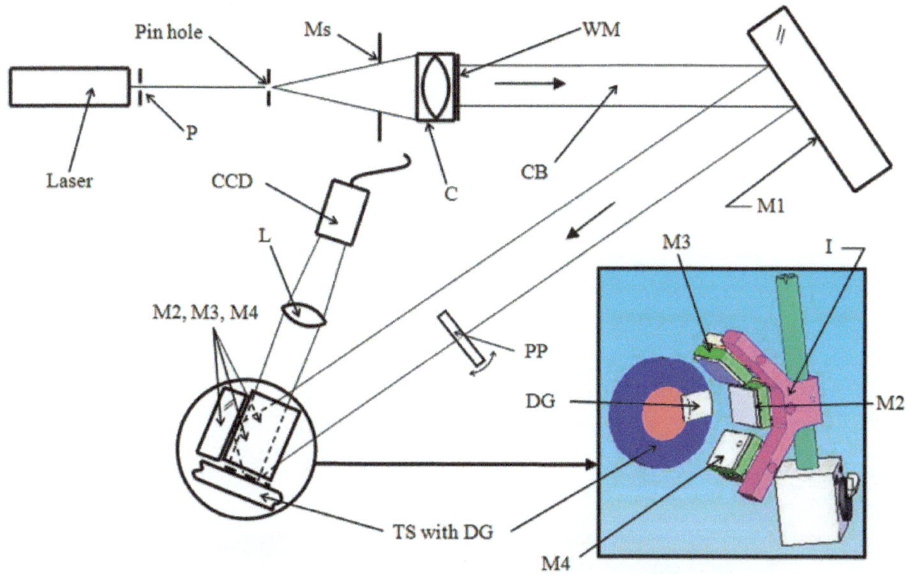

Figure 8. The schematic presentation of the moiré interferometry optical set-up used (Ribeiro J. *et al.*, 2011).

Where LA is the laser source, Ms is a mask, C is the collimator, WM is a window mask, CB is the collimated beam, M1, M2, M3 and M4 are plane mirrors, PP is a parallel plate glass, L is a lens, TS is the test specimen, DG is the diffraction grating, and I is the interferometer.

The stresses were released according to the conventional procedure used with strain gages. Drilling a hole on the surface of a specimen with residual stresses produces a stress relaxation around it. The corresponding deformation was assessed in this work with an optical technique. In the Figure 9 is possible to observe the system used to measure the residual stresses in the ring and plug specimen using the Moiré Interferometry an ESPI.

In the ESPI measurement an initial specklegram is acquired and saved. Then, the drill is placed in front of the test specimen and a small hole is drilled to a given depth. Next, the drill is removed and another specklegram is acquired. The interferogram resulting from the correlation of the two recordings leads to the surface displacements caused by the stress relaxation. In this case the surface information is codified in the speckle patterns.

For the Moiré Interferometry a high frequency grating was previously bonded on the surface of the specimen. The grating used was a 1200 lines/mm, obtained by aluminum vaporization on the top of an epoxy replication of a master grating. The set-up proposed by Post (Post D. *et al.*, 1994) was used in the virtual grating generation by laser interferometry. The first recording was obtained by superposition of the virtual grating over the object replication grating. Then, a hole is drilled at the desired place, and the Moiré fringes due to stress relaxation are obtained and recorded. A tiltable parallel plate glass was used to promote phase modulation with a four image phase calculation algorithm (Creath K. *et al.*, 1996).

In both cases, the residual stress field was computed establishing an appropriate stress–displacement relationship by a FEM code.

(a)

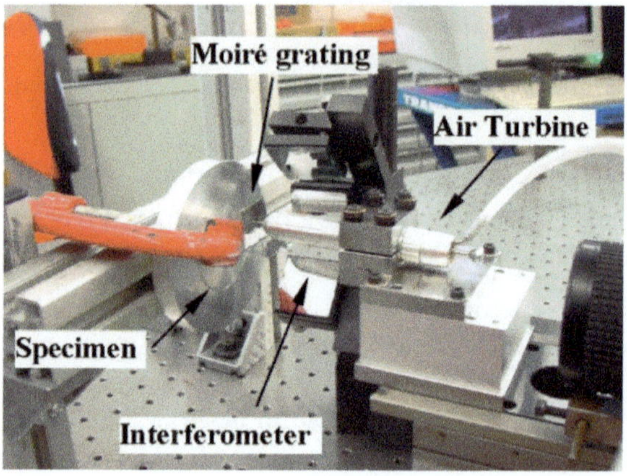

(b)

Figure 9. An apparatus to measure residual stresses using the hole-drilling technique associated to: (a) in-plane ESPI; (b) Moiré Interferometry (Ribeiro J. *et al.*, 2011).

The ring and plug specimen has a closed form solution (Lamé M. G., 1852) for the residual stresses, and relatively simple stress distribution. In the plug the stress is constant, in the ring only depends on the radial position. The specimen was prepared in agree of the reference (Ribeiro J. *et al.*, 2009). After the specimen preparation and its stress state control, several measurements of residual stresses were made using optical techniques and the hole-drilling method. The optical set up used to measure residual stresses with in-plane ESPI and Moiré Interferometry are schematically represented in the Figures 6 and 8, respectively. The Figure 9 shows details of both set-ups with the additional system for drilling of holes.

To calculate the displacement field eight images should be recorded in both techniques, at different phase shift, being half of them before and the remaining after hole drilling. In ESPI the displacement field is calculated by subtracting the speckle phase maps obtained before and after strain relaxation.

The Figure 10 summarizes the results obtained for residual stresses measurements. In this graphic presentation the solid lines represent the stresses calculated by the close form solutions given by Lamé (Lamé M. G., 1852). The experimental data is represented by dots. In all the cases the error was less than 17%, with most of the measurements being around 5% of error.

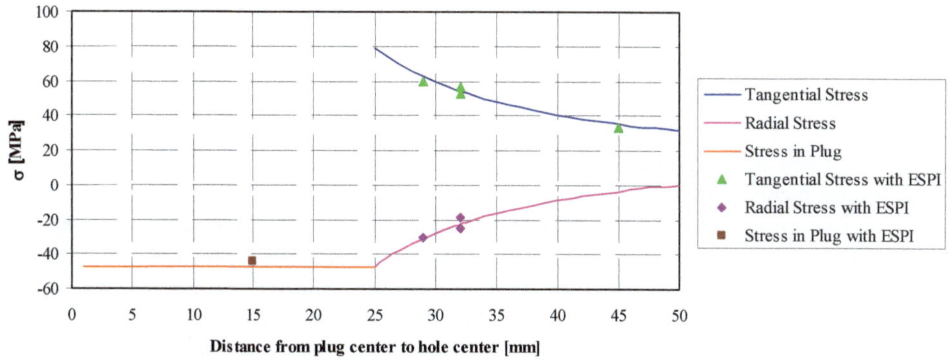

(a)

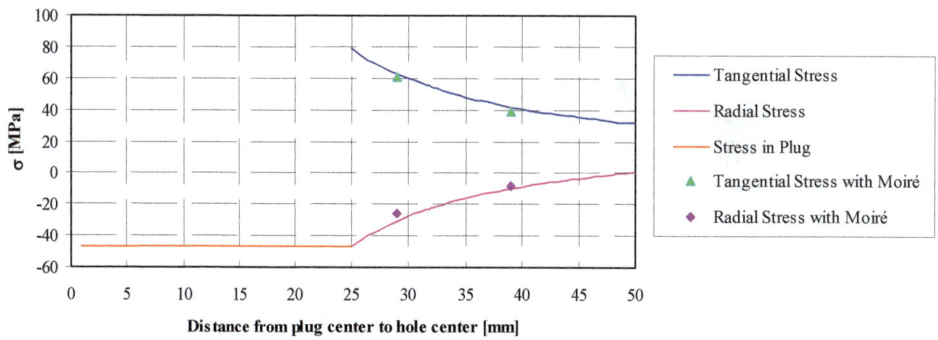

(b)

Figure 10. A closed form solution and experimental measurements with (a) ESPI and (b) Moiré.

In this work an experimental methodology for residual stress assessment is presented using optical techniques with the hole drilling method. These methods were tested with a well known residual stress field in a ring and plug specimen. The experimental results obtained with in-plane ESPI and Moiré Interferometry are in good agreement with the closed form solution. These optical techniques are a very interesting alternative to the traditional hole-drilling method with strain gages, and present some advantages, it is a global measurement, has better resolution and allows measurements closer to the hole edge.

4. Other optical techniques for the automotive industry

4.1. Smart materials based on FBGs

There are available a wide variety of optical fiber sensors, which can be divided in three categories. The first category includes the external type (also known as extrinsic) where the fiber is only used to transfer the measured information from a distant location (Beard P. C. *et al.*, 1996). The intrinsic category includes the optical fibers where at least one optical property (with the consequence of modulating the light) is affected in result of the measured information (Boerkamp M. *et al.*, 2007). Finally, it exists the hybrid category that includes all the situations where the light is transferred for further optoelectronic conversion on a distant receiver (Yau S.-K. *et al.*, 1983). In this context, the Fiber Bragg Gratings (FBGs) are the most suitable sensing elements with an increased usage for structural measurements due to their high performance in terms of the sensitivity and linear response. These sensors differ from conventional optical fiber sensors approaches because their optical signal is not based on power amplitude but instead on spectral changes. This factor is important for embodiment techniques into structures (for example, in the vehicles' structures) since the bends on the optical fiber introduce intensity changes (Kung P., 2009). Moreover, the FBGs are suitable sensing elements for doing physical measurements where a kind of displacement is available. Examples of such applications found in the literature include the measurements of strain (Ling H. Y. *et al.*, 2006), pressure (Peng B. J. *et al.*, 2005), force (Zhao Y. *et al.*, 2005), acceleration (Antunes P. *et al.*, 2011), tilt rotation by an angle (Xie F. *et al.*, 2009), temperature (Gu X. *et al.*, 2006) humidity (Arregui F. J. *et al.*, 2002), magnetic fields (Orr P. *et al.*, 2010), cardiorespiratory function (Silva A. F. *et al.*, 2011), gait function analysis (Rocha R. P. *et al.*, 2011) and integration on wearable garments (Carmo J. P. *et al.*, 2012b).

Looking to the technology point of view, the FBGs consists on periodic changes of the refraction index allong the fiber core. These changes are "shaped" by exposing the fiber to intense ultra violet (UV) light with a suitable interferometer mask (Hill K. O. *et al.*, 1997). The typical lengths of produced gratings are in the range of few millimetres and are characterized by a narrowband resonance spectral reflection. The resonance behaviour depends on grating pitch and on their axial variation because the resonance behaviour strictly follows external actions in the exact proportion as the silica matrix surrounding the gratings (Kersey A. D. *et al.*, 1997). The ultimate effect of the resonances is the reflection by successive and coherent scatterings from the index modulation of a narrow band of the incident optical field injected into the fiber (Hill K. O. *et al.*, 1997). In a FBG, the strongest interaction or mode coupling responsible for the reflected light occurs at a well-known wavelength, λ_B [nm], also known as Bragg wavelength. The Bragg's spectral component depends directly from the grating period of the FBG, Λ [nm], from the modal index, n_{eff}, also known as effective refractive index of the FBG, and is given by:

$$\lambda_B=2n_{eff}\Lambda \qquad (15)$$

The shift in the wavelength, $\Delta\lambda_B$ [nm], with respect to the cross sensitivity with the temperature and the axial strain changes, ΔT [K] and $\Delta\varepsilon$, (respectively) is given by (Wei C. L. *et al.*, 2010; Silva *et al.*, 2010):

$$\frac{\Delta\lambda_B}{\lambda_B} = (1-\rho_\varepsilon)\Delta\varepsilon + (\alpha+\xi)\Delta T \quad (16)$$

where ρ_ε is the photo-elastic coefficient of the fiber, α is the thermal expansion coefficient of the fiber material, and ξ is the thermo-optic coefficient of the fiber material. Typical value for a 1550 nm FBG fabricated in a silica fiber, the temperature and the axial strain sensitivities are respectively $\Delta T \approx 13$ pm.ºC^{-1} and $\Delta\varepsilon \approx 1$ pm.$\mu\varepsilon^{-1}$ (Silva A. F. *et al.*, 2011). It mandatory to compensate the temperature offset when the measurements consist only on strains. The Figure 11 illustrates a FBG structure after being written on an optical fiber. This figure also illustrates that the injection of a broadband pulse on the FBG results in a reflected beam located around the Bragg wavelength, λ_B . It is then possible to determine the exact strain by measuring the reflected spectra and/or the shifting produced in the Bragg wavelength (Zhou W. *et al.*, 2010).

It must also be noted that the absence of mechanical steps on sensor's fabrication results in the possibility to fabricate high sensitivity sensors with high reproducibility of their characteristics (Hill K. O. *et al.*, 1997). However, the most important features that made FBG based systems a wide established technology were their electrically passive operation, electromagnetic interference immunity, compact size, self referencing capability, and more important, inherent multiplexing-ability, which enables a wide number of sensors in a single fiber as well as Bragg a single interrogation system (Wang Q. *et al.*, 2007).

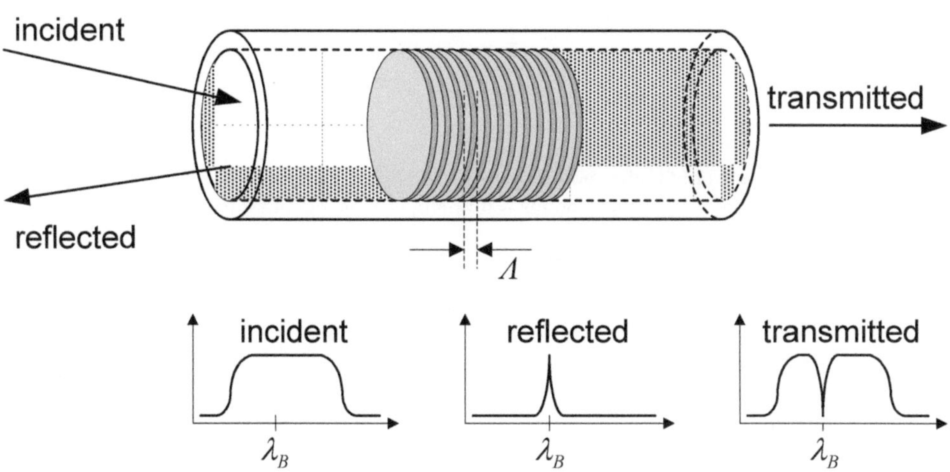

Figure 11. Illustration of working principle of FBGs.

The artwork illustrated in the Figure 12(a) helps to understand how to apply FBGs on vehicles. The use of a polymeric carrier with elastic properties, such as those developed and fabricated by Silva (Silva A. F. *et al.*, 2012) in polychloroethanediyl (polyvinyl chloride, or simply PVC) (Saeki Y. *et al.*, 2002) with one or more embedded FBGs can be used to sense deformations along the vehicle structure. The beauty of all is in the intrinsic simplicity, e.g., the whole system require only the use of a broadband optical source (and centred at the 1550 nm),

(a)

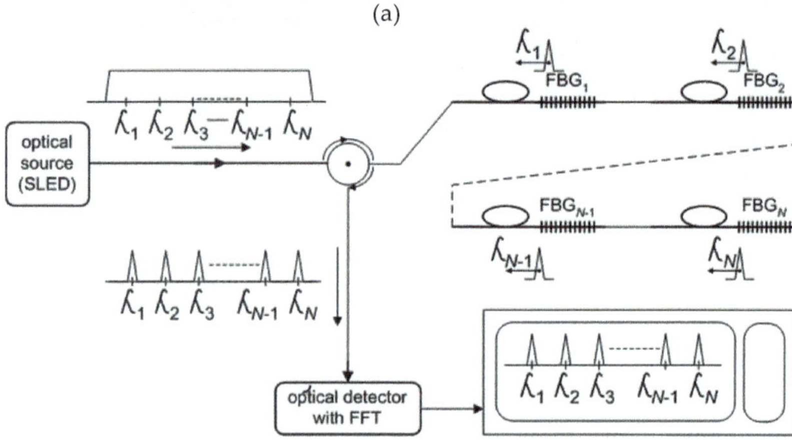

(b)

Figure 12. (a) Application of FBGs for monitoring structural deformations on vehicles, (b) Optoelectronic monitoring system based on FBGs for online/offline monitoring of vehicles.

a single optical circulator, few FBGs centred at different Bragg wavelengths λ_B and a optical receiver with enough spectral resolution R (Carmo J. P. *et al.*, 2012a) spanned along the wavelength range to cover all Bragg wavelengths. The Figure 12(b) illustrates an example for both online (i.e., during the motion of the vehicles) and offline (i.e., with the vehicle parked for maintenance by a mechanic) monitoring. In terms of technology and systems there are available a wide number of companies that offer optical sources, FBGs, optical circulators and optical sources at reasonable costs. Of course, the most challenging task is embedding FBGs into the flexible carriers. However, this problem was already addressed and the complete details can be found on (Silva A. F. *et al.*, 2012). The optical source Superluminiscent LEDs from DenseLight Semiconductors manufacturer can constitute a possibility due to its capability to generate broadband light beams with wavelengths in the range 1530-1570 nm with a maximal power of 8 mW (Denselight, 2012). Additionally, this optical source also presents a full width at half maximum (FWHM) of 60 nm with a maximum ripple of 0.2 dB. The FBGs can be acquired with the FiberSensing company (FiberSensing, 2012), which offers to their clients the possibility to write gratings in hydrogen loaded standard telecommunication fiber (Corning SMF28e+) using the phase mask technique and a pulsed Excimer Laser. The I-MON 80D interrogation monitor from Ibsen Photonics (Ibsen, 2012) is an interesting option to implement the optical receiver because it is especially suitable for real-time spectrum monitoring of signals acquired from FBG sensors. One of the characteristics of this optical detector is operating in the wavelength range of 1529-1561 nm with a resolution of 1 pm for at least 20 dBm of optical power. Moreover, the maximum wavelength drift of this optical detector is around 2 pm per ºC with a dynamic range of 30 dB, an input optical power in the range 10-50 dBm and a power consumption of 250 mW. Finally, the optical circulators can be found in the manufacturer ThorLabs with interesting costs and low losses (ThorLabs, 2012). The 6015-3 optical circulator constitues an interesting option because it operates in a relatively broadband wavelength range with isolation higher than 40 dB (e.g., between 1525 nm and 1610 nm according the manufacturer). Simultaneously, the loss due to the polarization sensitivity of this component is very low (e.g., less than 0.1 dB). Moreover, the delay spread is also very low (e.g., less than 0.05 ps). To finish, this optical circulator provides a high return loss (e.g., higher than 50 dB) for optical powers up to 500 mW .

4.2. Spectroscopy

Spectroscopy is the science investigating the interaction between radiation and matter centered on the wavelengths. The propagation of energy occurs through waves that can be classified either as a function of the frequency υ, or the wavelength λ. The relation between these two properties is given by the following equation: $c=\lambda v$ where c is the speed of light ($\approx 3\times 10^8$ m.s^{-1}). Radiation, within the context presented in this paper, describes how photons propagate through a medium and is given by $E=h.v$ where E is the energy associated with the photons emission and h is the Planck constant and is equal to $6.626/10^{34}$ J.s. With these equations it is possible to correlate the different properties, which are the foundations of the spectrography laws. The all-generic purpose spectroscopic techniques used more often are the X-ray diffraction (XRD, from 10^{-3} to 10 nm), ultraviolet (UV, from 10 to 400 nm), visible (VIS, from 380 to 780 nm) and infrared (IR, from 0.7 to 1000 μm).

Paint is essentially made of three components: binders, pigments (responsible for the color) and solvents. From these components of paint, attention will only be given to the one that define colors by selectively absorbing the visible wavelengths. Pigments are mainly composed by groups of aromatic rings linked by chromophores (may be azo compounds, carbonyl groups, basic, etc.). The mechanism responsible for providing a color to a pigment is the combination of two molecular components: the chromophores and the auxochromes (Zhang Y. *et al.*, 1995). Chromophores are the part of the molecule that gives the necessary conjunction to obtain colors. The auxochromes are a group of atoms linked to the chromophores that complement the action of the latter by performing the necessary changes, within the system's total energy, resulting in the final color. An example in basic dye (where chromophores have thiazine group), is seen in the Figure 13 (Lachheb H. *et al.*, 2002).

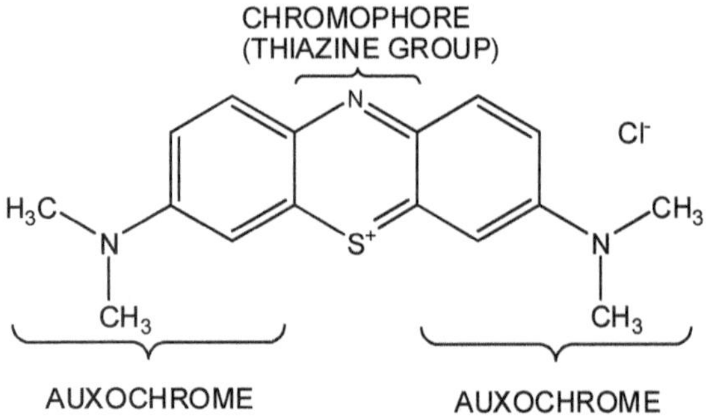

Figure 13. Example of the presence of chromophores and auxochromes in basic blue 9.

From the mentioned spectroscopic techniques, only the UV/Vis and IR, allow the detection of colors from a specific chemical compound by analyzing the intensity level of the absorption band. Therefore, when a chemical compound absorbs within a specific spectrum, it means that it will have the complementary color. As an example, if the most intense absorption band in a UV/Vis spectrum is in the 600 nm, i.e. absorbs the orange color, this means that its color will be the orange's complementary; which is blue. Consequently, when these analyses are done, it is not the color that it is determined directly but, instead, the chemical composition of the compound.

The main performance parameter is the spectral resolution $R=\lambda/\Delta\lambda$, where λ [m] is a particular wavelength setting, and $\Delta\lambda$ is the smallest discernible wavelength difference for the given λ. High-performance spectrometers can reach very high resolutions, greater than $R=10^6$ (Lindblom P. *et al.*, 1990) but are macrosized and expensive. However many times, mainly in the industry due to its intrinsic fast-moving nature, it is not required such a high level of resolution. Microspectrometers on the other hand, because of their small size limitation, have resolutions which are more suitable for rapid color evaluation without the excessively high-quality optical performance for such industrial applications. Spectrometers built using integrated silicon microsystems, have a spectral resolution $R<50$ and for MEMS based systems, $R=100$ (Carmo J. P. *et al.*, 2012a). Color identification is a process being widely used in many different fields. A very interesting and challenging one is to apply microspectrometers in industrial environments. Some industries are more suitable for such color evaluation and measurement procedures such as the: paint, biomedical, textile, ceramic, glass, chemical, etc. Industrial production processes permit the possibility to measure different chemical compounds by spectroscopic methods, which allow studying the colors and according to the needs, different wavelengths spectrum are used. In this context, the color identification and evaluation in the different states of matter is a process with increasing focus by the industry. Within the chemical industry, where is included the textile, polymeric and the ceramics industries, as well as the biomedical, are some of the ones that need more of these kind of characterizations. In the mentioned industries, there is a flagrant need to analyze and detect the presence of different chemical compounds in laboratorial environment, by spectroscopy techniques that allow the study of the different colors. These different spectral analysis techniques demand very expensive, bulky equipments, hardware and software (Wolffenbuttel). A tunable microspectrometer for non-extremely demanding optical resolutions industrial applications is a very interesting solution, especially for end-use customer services offered by the paint industry (Xie F. *et al.*, 2009). Customization of colors can be done quicker, and cheaper, by analyzing on an early stage, the development of any color is having. This procedure can potentially save money and time, which is of extreme value in the industry. Any necessary correction can be performed at any stage of color design. An application in the automotive industry can be seen in Figure 14.

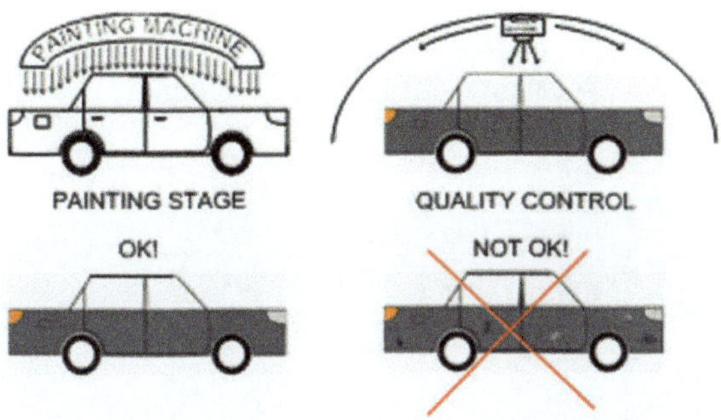

Figure 14. Microspectrometer use in color evaluation in the automotive industry.

The Figure 14 illustrates from the top to bottom and left to right, a practical use for microspectrometers in the automotive industry. First, the cars are painted by an industry standard process and then, the color of the car is evaluated throughout its entire painted areas, by a microspectrometer in scanning-type movements, in real-time. If the automatic color evaluation process does not detect any defect the car may move to the next phase. If that is not the case, then, the correspondent actions can be applied at once without creating any delays or bubbles in the process line, which is always an undesirable and costly event. These principles are usable in many other different fields. The glass, jewelry, paint and textile industries have obvious interests because colors are a direct way to evaluate the progress of the different production steps and final product design and quality. The Figure 15 shows a block diagram for a setup to perform color evaluation in the paint.

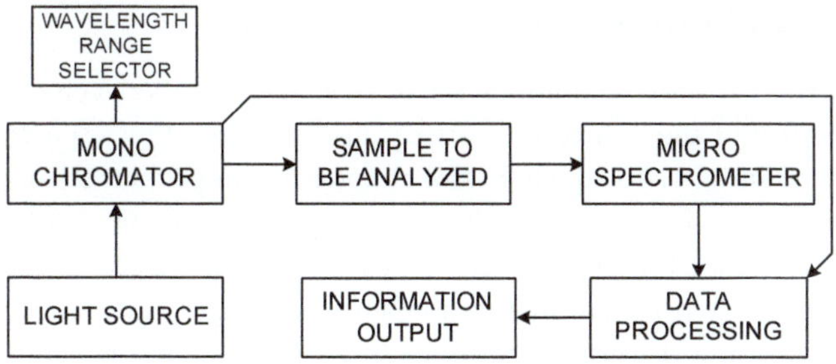

Figure 15. Color evaluation setup.

A light source projecting the full visible spectrum is narrowed down by a monochromator to set the desired wavelength range. Then, the selected spectrum is transmitted through the sample and the output light is recorded and analyzed by a photodetector and adequate software. In the following step, the resulting light is compared with the known selected light reference for improved resolution and sample characterization. By discriminating the difference between these two colors, a more accurate reading is obtained and a specific color pigment can be evaluated, investigated and discriminated. To finish with an example, it is possible to observe a photography showing a microspectrometer made in silicon for analysing the visible wavelength in the Figure 16. A detailed discussion about microspectrometers can be found in (Carmo J. P. et al., 2012a) and in (Wolffenbuttel R. F., 2005).

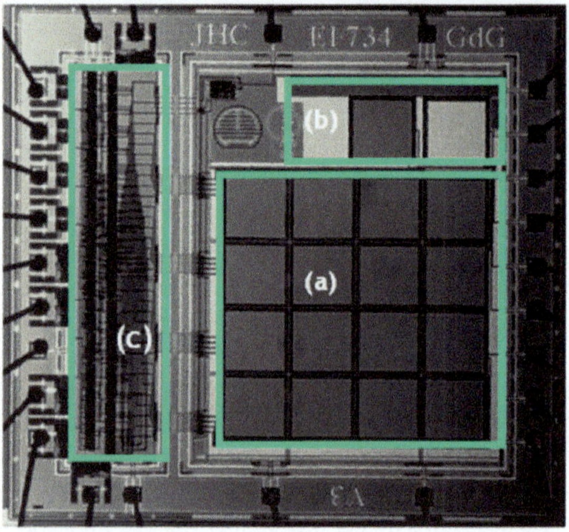

Figure 16. Photograph of a prototype of an array type microspectrometer composed by 16 channels, each one with a CMOS photodiode with a Fabry Perot etalon mounted on top. Three main subsystem blocks can be identified in the photograph: (a) array of 16 Fabry Perot mounted above 16 photodiodes, (b) dark current reduction, and (c) electronics for readout and signal processing (analog to digital conversion and amplification) and a serial bus for communicating with external devices (Carmo J. P. *et al*, 2012a). Reproduced with authorization of Elsevier Science.

5. Conclusions and further directos of research

This chapter presented optic techniques for analysing defects on vehicles. However, and given the importance of the vehicular industry in the world, the research doesn't finishes with one or other technique. To conclude this chapter it is important to reinforce the need of renewable power sources for reducing the greenhouse gas emissions to the atmosphere (Bell L. E., 2008) and to reduce the dependence of fossil fuel combustion sources (Vining C. B., 2008). The current research on production methods for obtaining the next generation algae biofuels is winning support as an even more sustainable alternative to fossil fuel sources (Savage N., 2011). However and despite the claim of experts to promote this upcoming power source to be a sustainable energy, the biofuels don't solve the problem of greenhouse emissions (Wisner R., 2009) and is a threat to forests and agricultural fields (Hudiburg T. W. *et al.*, 2011). Moreover, oils can't be yet produced at large scale from algae and at the same time further/though research in genetics must be done in order to engineer synthetic microorganisms for excreting the desired hydrocarbons (Biello D., 2011). The demand for combining vehicles with renewable power is even more evident when locking the hybrid vehicles available on roads (Boulanger A. G. *et al.*, 2011) or the forthcoming generation of new plug¬ in vehicles that use the grid power to charge their lithium ion battery packs for powering their electric traction motors (Voelcker J., 2011). However and despite the availability of these type of vehicles, these is still much to do at the both the electronics (Cao J. *et al.*, 2011) and batteries level (Kularatna N., 2011). In this sequence of ideas, it is still possible to provide cleaner forms of energy from vehicles based on internal combustion of oil derivatives. The heat released from the vehicle's engine can be scavenged using solid state thermoelectric generators (TEGs) based in the Seebeck effect for generating electricity in a process known as energy harvesting (Carmo J. P. *et al.*, 2010). These TEGs characterizes for a few number of useful features, such as silent operation and without moving parts in its constitution (DiSalvio F. J., 1999). These features are behind the successful use by NASA of radioisotope TEGs (RTGs) for more than three decades of operation in deep space probes (Wolverton M., 2008) and are now under study for application in vehicles (Snyder G. J., 2008).

Author details

J. P. Carmo
University of Minho, Department of Industrial Electronics, Guimarães, Portugal

J. E. Ribeiro
Polytechnic Institute of Bragança, Department of Mechanical Technology, Bragança, Portugal

6. References

Antunes, P., Varum, H., and André, P. (2011). Uniaxial fiber Bragg grating accelerometer system with temperature and cross axis insensitivity", *Measurement*, 44(*1*): 1-5.

Arregui, F. J., Matias, I. R., Cooper, K. L., and Claus, R. O. (2002). Simultaneous measurement of humidity and temperature by combining a reflective intensity-based optical fiber sensor and a fiber bragg grating, *IEEE Sensors Journal*, 2(*5*): 482-487.

Beard, P. C., and Mills, T. M. (1996). Extrinsic optical-fiber ultrasound sensor using a thin polymer film as a low-finesse Fabry-Perot interferometer, *Applied Optics*, 35(*4*): 663-675.

Beidermann, K., and Ek, L. (1975). A recording and display system for hologram interferometry with low resolution imaging devices, *Journal Physics E: Scientific Intruments*, 8(*7*): 571-574.

Bell, L. E. (2008). Cooling, heating, generating power, and recovering waste heat with thermoelectric systems, *Science*, 321: 1457-1462.

Biello, D. (2011). The false promise of biofuels, *Scientific American*, 305: 58-65.

Bifano, T. (2011). MEMS deformable mirrors, *Nature Photonics*, 5: 21-23.

Boerkamp, M., Lamb, D. W., and Lye, P. G. (2007) Using an intrinsic, exposed core, optical fibre sensor to quantify chemical scale formation, *Journal of Physics: Conference Series*, 76(*1*): 1-7. Institute of Physics Paper 012016.

Boulanger, A. G., Chu, A. C., Maxx, S., and Waltz, D. L. (2011). Vehicle electrification:status and issues, *Proceedings of the IEEE*, 99(*6*): 1116-1138.

Bruck, A., McNeil, R., Sutton, M., and Peters, H. (1989). Digital image correlation using Newton-Raphson method of partial differential correction, *Experimental Mechanics*, 29: 261-267.

Butters, J., and Leendertz, J. (1971). Holographic and video techniques applied to engineering measurement, Journal Measurement and Control, 12(4): 349-354.

Butters, N., Jones, R. C., and Wykes, C. (1975). Electronic speckle pattern interferometry in speckle metrology, *Springer-Verlag, New York*: 111-157.

Cao, J., and Emadi, A. (2011). Batteries need electronics, *Industrial Electronics Magazine*, 5(*1*): 27-35.

Carmo, J. P., Goncalves, L. M., and Correia, J. H. (2010). Thermoelectric microconverter for energy harvesting systems, IEEE Transactions on Industrial Electronics, 57(3): 861-867.

Carmo, J. P., Rocha, R. P., Bartek, M., de Graaf, G., Wolffenbuttel, R. F., and Correia, J. H. (2012a). A review of visible-range Fabry-Perot microspectrometers in silicon for the industry, *Optics & Laser Technology*, Accepted: 1–9.

Carmo, J. P., Silva, A. F., Rocha, R. P., and Correia, J. H. (2012b). Application of fiber Bragg gratings to wearable garments, *IEEE Sensors Journal*, 12(*1*): 261-266.

Chen, F., Brown, G. M., and Song, M. (2000). Overview of three-dimensional shape measurement using optical methods, Optical Engineering, 39(1): 10-22.

Chen, L., and Smith, P. (2000). Demonstration of incoherent wavelength-encoding/time-spreading optical CDMA using chirped moire gratings, *IEEE Photonics Technology Letters*, 12(*9*), 1281-1283.

Cheng, P., Kwak, S., and Choi, J. (2008). ESPI combined with hole drilling method to evaluate heat treatment induced residual stresses, Proceedings of the SPIE Fourth International Symposium on Precision Mechanical Measurements.

Cloud, G. (1998). *Optical methods of engineering analysis*, Cambridge University Press.

Creath, K. (1994). *Phase-shifting holographic interferometry*, Holographic Interferometry, Principles and Methods, Springer, Berlin, Ch. 5: 109-150.

Creath, K., and Schmit, J. (1996). N-point spatial phase-measurement techniques for non-destructive testing, *Optics and Lasers in Engineering*, 24: 365-379.

Denselight (2012). Denselight Semiconductors. Online and visited last time on 15th April 2012: http://www.denselight.com.

DiSalvo, F. J. (1999). Thermoelectric cooling and power generation, *Science*, 285: 703-706.

FiberSensing (2012). Online and visited last time on 15th April 2012: http://www.fibersensing.com.

Gabor, D. (1948). A new microscopic principle, *Nature*, 161(*3*): 777-778.

Gåsvik, K. J. (2002). *Optical metrology*, Third Edition, John Wiley & Sons Ltd, England.

Gu, X., Guan, L., He, Y., Zhang, H. B., and Herman, P. R. (2006). High-strength fiber bragg gratings for a temperature-sensing array, *IEEE Sensors Journal*, 6(*3*): 668-671.

Hill, K. O., and Meltz, G. (1997). Fiber Bragg grating technology fundamentals and overview, *IEEE Journal of Lightwave Technology*, 15(*8*): 1263-1276.

Hu, T., Ranson, W., Sutton, M., and Peters, W. (1985). Application of digital image correlation techniques to experimental mechanics, *Experimental Mechanics*, 25: 232-244.

Huang, P. S., and Zhang, S. (2006). Fast three-step phase shifting algorithm, *Applied Optics*, 45(*21*): 5086-5091.

Hudiburg, T. W., Law, B. E., Wirth, C., and Luyssaert, S. (2011). Regional carbon dioxide implications of forest bioenergy production, *Nature Climate Change*, 1: 419-423.

Hung, Y., and Taylor, C. (1973). A tool for measurement of derivatives of surface displacement, *Proceedings of the Society of Photo-Optical Instrumentation Engineering (SPIE)*, 41: 169-75.

Ibsen (2012). Ibsen photonics. Online and visited last time on 15th April 2012: http://www.ibsen.dk.

James, M. R., and Lu, J. (1996). *Handbook of measurement of residual stresses*, The Fairmont Press Inc., United States.

Jones, R., and Leendertz, J. A. (1974). Elastic constant and strain measurement using a three beam speckle pattern interferometer, *Journal of Physics E: Scientific Instruments*, 7(*8*): 653-657.

Jones, R., and Leendertz, J. A. (1974). Elastic constant and strain measurement using a three beam speckle pattern interferometer, *Journal Physics E: Scientific Intruments*, 7(8): 653-657.

Kersey, A. D., Davis, M. A., Patrick, H. J., LeBlanc, M., Koo, K. P., Askins, C. G., Putnum, M. A., and Friebele, E. J. (1997). Fiber grating sensors, *IEEE Journal of Lightwave Technology*, 15(*8*): 1442-1463.

Kong, S. H., Wijngaards, D. D. L., and Wolffenbuttel, R. F. (2001). Infrared micro-spectrometer based on a diffraction grating, *Sensors and Actuators A: Physical Sensors*, 92(*1-3*): 88-95.

Kularatna, N. (2011). Rechargeable batteries and their management, *IEEE Instrumentation & Measurement Magazine*, 14(*2*): 20-33.

Kung, P. (2009). Condition monitoring solutions for the wind turbines, Proceedings of the Canadian Machinery Vibration Association Seminar, Vancouver, Canada.

Lachheb, H., Puzenat, E., Houas, A., Ksibi, M., Elaloui, E., Guillard, C., Herrmann, J.-M. (2002). Photocatalytic degradation of various types of dyes (Alizarin S, Crocein Orange G, Methyl Red, Congo Red, Methylene Blue) in water by UV-irradiated titania, *Applied Catalysis B: Environmental*, 39: 75-90.

Lamé, M. G. (1852). *Leçons sur la théorie mathématique de l'élasticité des corps solids*, bachelier, Paris.

Leendertz, J. A., and Butters, J. N. (1973). An image-shearing speckle-pattern interferometer for measuring bending moments, *Journal of Physics E: Scientific Instruments*, 6(*11*): 1107-1110.

Lindblom, P., Meinander, M., and Olsson, T. (1990). Spectroscopy with the MEGA spectrometer, a very high resolution grating spectrometer, *Review of Scientific Instruments*, 61(*10*): 2546-2548.

Ling, H. Y., Lau, K. T., Cheng, L., and Jin, W. (2006). Viability of using an embedded FBG sensor in a composite structure for dynamic strain, *Measurement*, 39(*4*):328-334.

Lökberg, O. (1985). Electronic speckle pattern interferometry, *Physics in Technology*, 11: 16-22.

Lökberg, O., and Högmoen, K. (1976). Use of modulated reference wave in electronic speckle pattern interferometry, *Journal of Physics E: Scientific Instruments*, 9(*10*): 847-851.

Macovsky, A., Ramsey, S., and Shaeder, L. (1971). Time-laps interferometry and countering using television system, *Applied Optics*, 10: 2722-2727.

Marcellierl, H., Vescovo, P., Varchon, D., Vacher, P., Humbert, P. (2001). Optical analysis of displacement and strain fields on human skin, *Skin Research and Technology*, 7(*4*): 246-253.

Orr, P., and Niewczas, P. (2010). An optical fiber system design enabling simultaneous point measurement of magnetic field strength and temperature using low-birefringence FBGs, *Sensors and Actuators A: Physical Sensors*, 163(*1*): 68-74.

Peng, B. J., Zhao, Y., Yang, J., and Zhao, M. (2005). Pressure sensor based on a free elastic cylinder and birefringence effect on an FBG with temperature-compensation, *Measurement*, 38(2): 176-180.

Post, D., Han, B., and Ifju, P. (1994). High sensitivity moiré: experimental analysis for mechanics and materials, *Springer-Verlag*, New York.

Powell, R. L., and Stetson, K. A. (1965). Interferometric vibration analysis by wavefront reconstruction, *Journal of Optics Society of America*, l55(*12*): 1593-1598.

Rayleigh, L. (1874). On the manufacture and theory of diffraction gratings, *Philosophical Magazine*, 47(*310*): 81-93 & 193-205.

Ribeiro, J. E. (2006). *Caracterização experimental e numérica de campos de tensões residuais provocadas por processos de fabrico*, PhD Thesis, Faculdade de Engenharia da Universidade do Porto. *In Portuguese*.

Ribeiro, J., Monteiro, J., Lopes, and Vaz, M. (2011). Moiré interferometry assessment of residual stress variation in depth on a shot peened surface, *Journal of Strain: An International Journal for Experimental Mechanics*, 47: e542-e550.

Ribeiro, J., Monteiro, J., Vaz, M., and Lopes, H. (2009). Measurement of residual stresses with optical techniques, *Journal of Strain: An International Journal for Experimental Mechanics*, 45: 123-130.

Richter-Trummer, V., Moreira, P., Ribeiro, J., and Castro, P. (2011). The contour method for residual stress determination applied to an AA6082-T6 friction stir butt weld, *Materials Science Forum*, 681: 177-181.

Robinson, D. W., and Reid, G. T. (1993). *Interferogram analysis*, Ch. 5: 141-193, Ch. 4: 94-140, Ch. 6: 194-229. Institute of Physics (IOP), Bristol, 1993.

Rocha, R. P., Silva, A. F., Carmo, J. P., and Correia, J. H. (2011). FBG Sensor for measuring and recording the knee joint movement during gait, Proceedings of the 33rd Annual International Conference of the IEEE Engineering in Medicine and Biology Society (EMBC '11), Boston, Massachusetts, USA.

Saeki, Y., and Emura, T. (2002). Technical progresses for PVC production, *Progress in Polymer Science*, 27(*10*): 2055-2131.

Savage, N. (2011). Algae: the scum solution, *Nature*, 474: S15-S16.

Shang, H., and Gao, J. (2009). Theories and industrial applications of optical interferometric NDT techniques: a review, *Insight*, 51(*5*): 240-251.

Silva, A. F., Carmo, J. P., Mendes, P. M., and Correia, J. H. (2011). Simultaneous cardiac and respiratory frequency measurement based on a single fiber Bragg grating sensor, *Measurement Science and Technology*, 22(*7*): 1-5. Institute of Physics Paper 075801.

Silva, A. F., Goncalves, A. F., Ferreira, L. A., Araujo, F. M., Mendes, P. M., and Correia, J. H. (2011). A Smart skin PVC foil based on FBG sensors for monitoring strain and temperature, *IEEE Transactions on Industrial Electronics*, 58(*7*): 2728-2735.

Silva, A. F., Goncalves, A. F., Ferreira, L. A., Araujo, F. M., Mendes, P. M., and Correia, J. H. (2010). PVC smart sensing foil for advanced strain measurements, *IEEE Sensors Journal*, 10(*6*): 1149-1155.

Silva, A. F., Gonçalves, A. F., Mendes, P. M., and Correia, J. H. (2012). PVC formulation study for the manufacturing of a skin smart structure based in optical fiber elements, *Journal Polymers for Advanced Technologies, Wiley Publisher*, 23(*2*), 220-227.

Snyder, G. J. (2008). Small thermoelectric generators, *The Electrochemical Society Interface*, 17(*3*): 54-56.

Sutton, M. A., Cheng, Q., Peters, H., Chao, J., and McNeill, R. (1986). Application of an optimized digital correlation method to planar deformation analysis, *Image and Vision Computing*, 4: 143-151.

Sutton, M. A., McNeill, R., Jang, J., and Babai, M. (1988). Effects of subpixel image restoration on digital correlation error, *Journal of Optical Engineering*, 27: 870-877.

Sutton, M. A., Orteu, J., and Schreier, H. W. (2009). Image correlation for shape, motion and deformation measurements, *Springer Science+Business Media*, New York.

Sutton, M. A., Turner, L., Bruck, A., and Chae, A. (1991). Full-field representation of discretely sampled surface deformation for displacement and strain analysis, *Experimental Mechanics*, 31: 168-177.

Sutton, M. A., Wolters, J., Peters, H., Ranson, F., and McNeil, R. (1983). Determination of displacements using an improved digital correlation method, *Image and Vision Computating*, 1: 133-139.

Svanbro, A. (2005). *Speckle interferometry and correlation applied to large-displacement fields*, PhD Thesis, Luleå University of Technology.

Takasaki, H. (1970). Moiré topography, *Applied Optics*, 9(*6*): pp 1467-1472.

ThorLabs (2012). Thorlabs GmbH, Germany. Online and visited last time on 15th April 2012: http://www.thorlabs.de.

Vining, C. B. (2008). Desperately seeking silicon, *Nature*, 451: 132-133.

Voelcker, J. (2011). One million plugin cars by 2015?, *IEEE Spectrum INT* (*International Edition*): 9-11.

Wang, Q., Rajan, G., Wang, P., and Farrell, G. (2007). Macrobending fiber loss filter, ratiometricwavelength measurement and application, *Measurement Science Technology*, 18(*10*): 3082-3088.

Wei, C. L., Lai, C. C., Liu, S. Y., Chung, W. H., Ho, T. K., Ho, S. L., McCusker, A. Kam, J., and Lee, K. Y. (2010). A fiber bragg grating sensor system for train axle counting, *IEEE Sensors Journal*, 10(*12*): 1905-1912.

Wisner, R. (2009). Biofuels and greenhouse gas emissions on a collision course, *Oil Mill Gazetter*, 114: 2-20.

Wolffenbuttel, R. F. (2005). MEMS-based optical mini and microspectrometers for the visible and infrared spectral range, *Journal of Micromechanics and Microengineering* 15(*7*): S145-S152.

Wolverton, M. (2008). Stirling in deep space: to cut back on radioisotope fuel, NASA goes back 200 year, *Scientific American*, 298:22.

Wronkowski, L. (1995). Diffraction model of an optoelectronic displacement measuring transducer, *Optics & Laser Technology*, 27(*2*):81–89.

Xie, F., Chen, Z., and Ren, J. (2009). Stabilisation of an optical fiber Michelson interferometer measurement system using a simple feedback circuit", *Measurement*, 42(*9*): 1335-1340.

Xie, F., Chen, Z., and Ren, J. (2009). Stabilisation of an optical fiber Michelson interferometer measurement system using a simple feedback circuit, *Measurement*, 42(*9*): 1335-1340.

Xie, H., Shang, H., Dai, F., Li, B., and Xing, Y. (2004). Phase shifting SEM moiré method, *Optics & Laser Technology*, 36(*4*), 291–297.

Yao, S.-K., and Asawa, C. (1983). Fiber optical intensity sensors, *IEEE Selected Areas in Communications*, 1(*3*): 562-575.

Zhang, S. (2010). Recent progresses on real-time 3D shape measurement using digital fringe projection techniques, *Optics and Lasers in Engineering*, 48: 149-158.

Zhang, Y., Ghosal, S., Casstevens, M, and Burzynski, R. (1995). Bifunctional chromophore for photorefractive applications, *Applied Physics Letters*, 66(*3*): 256-258.

Zhao, Y., Zhao, Y., and Zhao, M. (2005). Novel force sensor based on a couple of fiber Bragg gratings, *Measurement*, 38(*1*): 30-33.

Zhou, W., Dong, X., Ni, K., Chan, C. C., and Shum, P. (2010) Temperature insensitive accelerometer based on a strain chirped FBG, *Sensors and Actuators A: Physical Sensors*, 157(*1*): 15-18.

Section 3

Mechanics

Chapter 11

3D Surface Analysis for Automated Detection of Deformations on Automotive Body Panels

Arjun Yogeswaran and Pierre Payeur

Additional information is available at the end of the chapter

1. Introduction

1.1. Context and motivation

Quality control in the manufacturing industry has traditionally been performed manually by workers. As manufacturing increases in speed and volume through the introduction of automation, the human worker becomes a limiting factor in speed, accuracy, and consistency. In the automotive industry, quality control is critical to ensure that automotive body parts meet predefined standards. Identifying deformations, such as undesired dings and dents on panels, and marking them so that they are repaired while still on the assembly line is essential. In current industrial settings, the procedure for identifying surface defects on automotive body panels often requires a laborious manual surface rubbing operation. This time-consuming process is difficult for a human, especially when dealing with small deformations that require close inspection, and may result in a decreased accuracy when the repetitive task is performed over the course of an entire work shift. Automation of quality control could significantly improve the accuracy and speed of the assembly line, thus increasing the number of panels inspected within an allotted time, maximizing the number of accurately detected defects, and minimizing the number of false detections.

To fully automate this process, a system would have to analyze the surface of the body part to be inspected, determine the position of deformations, and mark those deformations on the body part. This chapter focuses on the analysis of 3D surfaces and automatic detection of deformations. One important contribution of this work comes from the imposed requirement that this system must be able to detect deformations without knowledge of the ideal shape of the part, meaning it cannot use a master work or CAD model for comparison. Some automated deformation detection techniques focus on the difference between the scanned model and an existing ideal model or master work (Newman & Jain 1995; Lilienblum et al. 2000). Certain challenges lie within this approach. The first constraint is that a very precise ideal model must be available, because small faults in the master work can result in erroneously detected defects during execution. The second challenge is related to the registration of the scanned part with the ideal model. Due to vibrations on the assembly line and slight inconsistencies in the acquired model, inaccurate registration may occur resulting in incorrectly detected defects. Also, if panels of different models are processed on the same assembly line, or a new piece is introduced to the system, significant calibration and set up is required to synchronize the master work with the acquired model. Because of these difficulties, this chapter introduces a more generic and robust technique which does not require an ideal model.

1.2. Objectives

This chapter deals primarily with the design of a deformation detection system. Its requirements are to identify deformations of interest over the surface of automotive body parts, with minimal human interaction and independently from the type of acquisition system used. The deformations of interest are dings and dents, where dings are surface deformations which protrude from the surface and dents are depressions into the surface. This chapter focuses on deformation detection when no ideal model of the automotive part is provided, similar to an approach which is alluded to by Döring *et al.* (Döring et al. 2004) and explored by Chen (Chen 2008). Since there is no CAD model of a master work to compare the measured model to, the deformation detection must be done without knowledge of the expected surface and requires certain assumptions to be made based on common characteristics of surface deformations compared to the characteristics of an undeformed surface. However, not all characteristics can be assumed, known, or easily defined. Therefore some basic parameters need to be set by the operator to provide the system with a minimal knowledge of the approximate size or scale of the deformations that the manufacturer wants to detect and eliminate from its products. This is not unrealistic, as the operator generally has a clear idea of the approximate range of sizes for the deformations to be detected.

Given these requirements, a system is proposed which analyzes the digital 3D model of an automotive part collected along the assembly line, determines the locations of only the deformations of interest, and classifies them as dings or dents. Areas of significant surface variation could be deformations. But other features of an automotive body panel such as aesthetic curves and door handles, or inaccurate surface measurements such as acquisition artifacts and noise, also represent surface curves that must not be falsely detected as dings or dents. The deformation detection system is comprised of a surface shape analysis phase to extract areas of interest, a segmentation phase to group areas containing pieces of deformations together into segments, and a classification phase to determine which segments contain deformations and which contain design features.

A deformation detection pipeline is proposed, which combines an enhanced octree-based feature extraction, with segmentation and classification to extract deformations from a 3D mesh of an automotive surface panel. This pipeline supports multi-resolution analysis of 3D models, providing the capability of extracting deformations regardless of the resolution or scale of the model, and relies on intuitively adjustable parameters for the operator to target the feature extraction towards desired characteristics of the deformations.

2. Literature review

The process to automatically determine the location of a defect on an automotive body part requires several steps, some of which are still complex research topics. This section reviews important research that is relevant to the topic of automated deformation detection proposed in this chapter. In order to analyze the surface for automotive body parts, the latter must first be digitally represented as a 3-dimensional object. Various techniques in 3D acquisition are explored in section 2.1. To determine the location of deformations in the digitized 3-dimensional surface, the surface must be analyzed for certain characteristics. Surface shape analysis is discussed in section 2.2.

2.1. 3D acquisition

In order to analyze the surface of a real-world object in 3-dimensions, it must be scanned and converted into digital 3-dimensional data. Laser scanners are very common and highly accurate 3D acquisition tools (Parthasarathy et al. 1982; Sequeira et al. 1995; Marszalec & Myllyla 1997; Gokturk et al. 2004; Blais et al. 2007). The latter are able to produce high resolution, high accuracy scans. However they are usually expensive systems that take a long time to complete a full scan, and often require some mechanical system to move the laser and acquire readings before accumulation into a point cloud. For lower cost, lower scan times, and minimal mechanical complexity, stereoscopic vision systems are a very popular way of digitizing a 3-dimensional scene (Murray & Jennings 1997; Murray & Little 2000; Se et al. 2001). If prominent features are lacking in a scene, such as on the surface of a smooth automotive body part, these construction techniques may fail due to a lack of usable points. One popular technique to overcome the limitations of using traditional stereoscopic imaging is to acquire 3-dimensional models using structured light scanners. This type of sensor projects a set of artificial features onto a model or scene that is being scanned, and then uses a vision system to acquire the model in 3D. Most structured lighting systems use a single camera along with a projector to acquire the 3D points (Rocchini et al. 2001; Zhang et al. 2002), or combine a pattern projector with a standard stereo pair of cameras to avoid calibration with the projector (Payeur & Desjardins 2009).

A recent trend in 3D sensing is the use of Microsoft's Kinect, which is a low-cost portable sensor that provides 3D visualization of a scene. Using structured light principles, an infrared laser projector generates artificial features onto a scene (which are invisible to the human eye), and a CMOS sensor reconstructs a scene through vision techniques. High quality scene reconstruction using the Kinect sensor has been studied (Shahram et al. 2011; Yan & Didier 2011). Relatively high quality reconstruction of real-world scenes can be achieved, yet its accuracy is still too low to detect the slight variations in an automotive panel surface that constitute deformations. Also, due to possible holes and inaccuracies in reconstruction using a single image, several frames of reconstructed scenes must be stitched together and heavily post-processed to provide a full reconstruction of an automotive part. Though preliminary work with the Kinect is promising, more research must be done to adapt its use to detecting the fine contours of an automotive panel for deformation detection. Given the current state of the technology, and the purposes of this work, the laser scanner remains to be the most accurate way to scan 3D models.

2.2. 3D surface shape analysis

The most critical component of a surface deformation detection system is surface analysis to locate the defects in question. In the current context, no ideal model of the automotive part is provided, therefore the algorithm has no *a priori* knowledge of what the surface should look like without deformations. Advanced surface shape analysis techniques must be performed to determine the locations of probable deformations.

Given that the 3-dimensional data can be converted from a range image to a 2-dimensional image where each pixel intensity represents the depth of that point on the object from the viewpoint, features can be extracted and images can be segmented using traditional 2D image processing techniques. Well-known edge detectors, such as the Sobel and Canny operators, can highlight the areas that belong to features (Faugeras 1993). The efficacy of such algorithms varies greatly, since determining the peaks and valleys in histograms with significant noise or varying characteristics is difficult.

The k-means algorithm is a very well-known clustering algorithm that partitions a dataset into a specified number, *k*, of clusters (Plataniotis & Venetsanopoulos 2000). However, selecting the value of *k* is most important, and in the case of an unknown number of deformations, this value cannot be known for sure. Unseeded region growing (Plataniotis & Venetsanopoulos 2000) can overcome some of the problems with k-means algorithms by not requiring any initial knowledge. Similar to the limitations of thresholding in edge detection, gradually changing pixel intensities between actual regions of the image may not be sufficient for accurate segmentation.

These techniques can all be extended to 3 dimensions by using points or voxels instead of pixels, and adjacency can be determined by distance or connectivity in a grid or tree, as is done by Palagyi and Kuba (Palagyi & Kuba 1999). Also, the data being used as the intensity value in an image can be redefined as distance in a range image or 3-dimensional surface deformation metrics such as standard deviation of normals or a curvedness value (Koenderink & Doorn 1992; Dorai & Jain 1997).

Various techniques from the field of 3D data analysis can be used for the purpose of deformation detection. Simple deformations in a mesh can resemble outliers on a smooth surface. Using noise removal techniques to identify areas of noise-like characteristics can be beneficial to determining the location of the defect. Schall *et al.* propose a noise removal method that also provides applications in outlier removal (Schall et al. 2005). The statistical method estimates the density of each area of the point cloud, and uses the neighbouring points to adapt a probable surface to each point. Since points are moved to their most probable location along a surface, the spatial density of the resulting point cloud is relatively consistent throughout the surface of the object. An outlier point exists where the spatial density in a surrounding is too low. Therefore a basic threshold can determine outliers. Though effective, this algorithm is dependent on the sampling density, its parameter selection is not intuitive and may cause unpredictable performance, and will fail when deformations do not resemble outliers.

The moving least-squares surface reconstruction technique proposed by Mederos *et al.* (Mederos et al. 2003) uses a hierarchical segmentation technique that finds redundant points such that the point cloud density can be reduced before surface reconstruction. This segmentation technique results in clusters of points, where the surface variation within each cluster is minimal and the boundaries between those clusters could define a significant deformation on the surface. Though computationally expensive, analyzing the eigenvalues and eigenvectors of the covariance matrix of a cluster of points can estimate local surface properties (Hoppe et al. 1992; Shaffer & Garland 2001). A binary space partitioning tree is used to segment the model into clusters of points that lie on surfaces of low variation, where subdivision is based on the flatness criterion, which represents variation within a group of points, as described by Pauly *et al.* (Pauly et al. 2002). Such an algorithm is effective at determining the characteristics of a model for surface reconstruction or resampling, but requires an extension to be used for efficient feature extraction, since the boundaries must be determined instead of just the clusters. The use of a binary space partitioning tree is very effective to separate the mesh, but tends to become too deep of a tree to traverse efficiently, since each node can only be subdivided into 2 nodes at a time.

Woo *et al.* (Woo et al. 2002) introduce a technique based on octree structures, and use recursive subdivision of the volume of a 3D mesh to identify features. It removes segments of the mesh as the octree is generated, and leaves parts of the mesh that belong to features in the final octree data structure. It requires a model with a reconstructed surface and partitions the model into subsections which represent varying levels, or scales, of features. Surface normal vectors can be calculated for all triangles composing the surface, and ultimately for each point by averaging the normals of the triangles that the point belongs to. Variations in the orientation of the surface within a given region are estimated from the standard deviation of normal vectors within that region. This method facilitates the partitioning process. All of the points that make up the surface of the object are initially added to the root of the tree structure. The standard deviation of their normals is calculated, and compared to a threshold. First the mean normal is computed:

$$\overline{N} = \sum_{i=0}^{n} N_i / n \tag{1}$$

where n is the number of points at the node, $\overline{N}$ is the mean normal, and N_i is the unit normal of point i. Then the standard deviation, σ, of the normals can be estimated as:

$$\sigma = \sqrt{\frac{\sum_{i=0}^{n}(N_i - \bar{N})^2}{n}} = \sqrt{\frac{\sum_{i=0}^{n}(x_i - x)^2 + \sum_{i=0}^{n}(y_i - y)^2 + \sum_{i=0}^{n}(z_i - z)^2}{n}} \quad (2)$$

A threshold must be defined for the subdivision as the maximum standard deviation allowed in a volume before it should be further divided. This threshold is defined by the user. If the standard deviation is larger than the threshold, the volume represented by the root is divided into 8 octants represented by 8 children being added to the root. The points from the root are redistributed based on their spatial location into each of the 8 octants, and thus into each of the 8 children of the root. This process is repeated recursively for each of the children, and for their children, and so on until either there are no children remaining, or a sufficient level of feature details is discovered. The depth of the tree determines how detailed the feature level is. Figure 1 details the recursive process of the feature segmentation at various scales. The final structure provides a tree where the points are distributed amongst the tree nodes. Leaves at greater depths represent finer detailed features of the mesh contained in smaller volumes. Leaves at lesser depths represent larger scale features contained in larger volumes.

Woo *et al.'s* technique is effective, yet because it uses a single threshold value throughout the entire tree, its ability to detect features can be unpredictable. A feature has to sufficiently affect the standard deviation of the surface normals across the selected volume for the method to investigate the mesh at a higher resolution. If this is not the case then the feature is not identified. The criteria for setting the threshold is that it must be high enough such that smooth curvatures and noise are not detected, but low enough such that the deformation features are detected. This remains a subjective criterion that varies with the point cloud. Since standard deviation is used, it is hard to find values which meet the defined criteria.

On the other hand, the technique generates broad shallow trees which are easier to traverse, as opposed to deep narrow trees generated by binary space partitioning methods such as those in (Shaffer & Garland 2001; Mederos et al. 2003). This allows analysis at higher resolutions, with reduced computational load. The octree representation allows features to be represented in the point cloud dataset as well as in a volumetric grid, giving the flexibility of using a variety of techniques for added segmentation.

Pauly *et al.* present a technique that allows feature extraction from a 3D object composed of surfaces, at various detail levels (Pauly et al. 2003). Weights are assigned to each point in the point cloud, representing the amount of local variation in the surface normals. At different scales, different local neighbourhood sizes are used. Introducing the idea of *feature persistence,* a threshold can be selected, such that local maxima weights over that threshold can be considered feature nodes. As a feature persistently exceeds that threshold, across multiple scales, it can be classified as a strong feature, rather than only a small local feature.

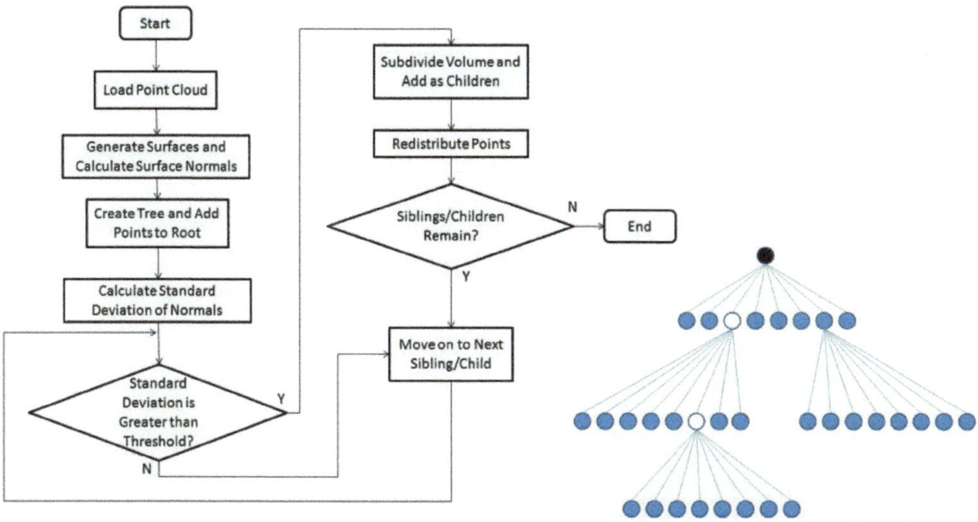

(a) (b)

Figure 1. a) Octree segmentation subdivisioning flowchart, b) an example octree where the top node represents the entire mesh, and nodes at deeper levels represent subdivisions of the mesh at proportionally higher resolutions.

Results show that this method performs well under noise and is effective at identifying prominent features. Also, the idea of feature persistence is interesting, where prominent features appear over multiple scales, and can be very important in using multiresolution information to identify important features.

Vosselman *et al.* (Vosselman et al. 2004) exploit the knowledge of ordered point clouds in the form of scan lines, and combine various techniques to segment point clouds by recognizing geometric shapes and flat smooth surfaces for the analysis of industrial and city scans from LIDAR data. Each scan line is broken into line segments based on orientation and proximity, and a plane-of-best-fit equation is calculated. Adjacent scan lines are compared based on some similarity criterion to be connected as a planar surface or other shapes such as spheres and cylinders. The dependency on ordered point cloud data is a limitation of the technique, since data can come from various sources and may not always be in the form of scan lines. Also, the very distinct shapes that are being extracted are effective in scans of a city or in an industrial setting, but the techniques are less suited to the more curved and variable surfaces of automotive body parts, since such shapes do not fall into the category of basic geometric primitives.

Jagannathan and Miller (Jagannathan & Miller 2007) use a metric known as curvedness (Koenderink & Doorn 1992; Dorai & Jain 1997) for segmentation, to extract regions of the mesh with high curvature. The curvedness is calculated for each point in the mesh. Using iterative graph dilation and filtering of outlier curvedness values, the mesh is broken up into sub-meshes with similar curvedness values. Based on the results shown, the algorithm has great success in segmenting 3D models with very large distinct form changes. However, it might be difficult to predict the success of this algorithm when faced with finding subtle shallow deformations on the surface of a flat or curved mesh, especially when dealing with significant noise and acquisition artifacts.

Döring *et al.* tackle a similar problem to this work, by detecting deformations on car body panels (Döring et al. 2004). The deformation extraction is only briefly explained as finding the differences between the point cloud and an inertial surface approximation of a low polynomial degree. The experiments in this paper work under similar assumptions to this chapter, in that there is no ideal model or *a priori* knowledge to compare to the model being analyzed. Surface deformations must be extracted by analysis of the model surface against what is assumed to be a smooth ideal surface instead of being compared to an existing model of what the surface should look like. This chapter is more concerned with the extraction of surface deformations than the classification, while Döring *et al.'s* work emphasizes the classification of the feature as one of many types of known deformations.

3. Automated surface deformation detection

3.1. General deformation detection framework

The proposed system takes a 3D mesh as an input, and outputs the sections of the mesh which are deformations of interest along with whether they are a ding or a dent. Given that no CAD model of the ideal surface is considered available, the proposed system must locate and classify the deformations of interest using assumptions based on common characteristics of dings and dents. Since some assumptions regarding size and scale of deformations cannot be made without more information, a minimal and intuitive set of parameters must be set by the operator to ensure accurate detection with minimal human intervention. This also ensures that design features of the automotive panel are not accidentally extracted as deformations, since they are generally much larger than the deformations of interest and can easily be separated by size and scale. The outputs of the proposed system are passed onto a robotic deformation marking system briefly discussed in section 3.2.

The proposed system contains 3 major components, as shown in Figure 2.The surface shape analysis component is tasked with dividing the 3D mesh into sections and analyzing each one for the magnitude of the deformation contained in that section. The segmentation component combines sections from the surface shape analysis which seemingly belong to the same deformation. The classification component classifies each segment from the segmentation as either a ding or dent, and removes segments which do not meet the criteria of being a deformation of interest, such as vehicle design features and acquisition noise.

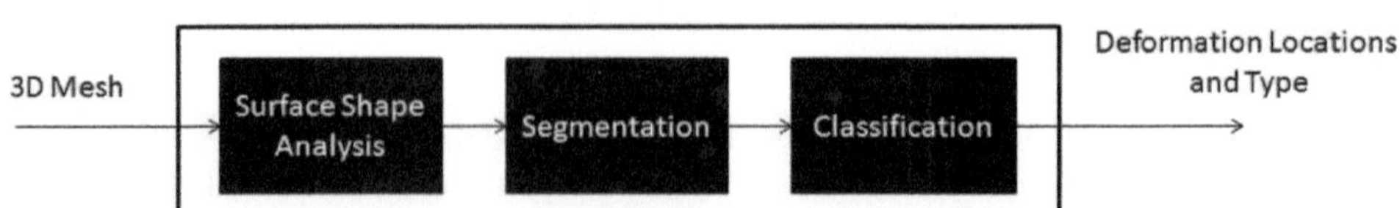

Figure 2. System diagram of proposed deformation detection system.

3.2. Experimental platform and setup

The more extensive research project that this work is part of involves the development of an automated deformation detection and marking system (Borsu et al. 2010). The primary objective is to identify deformations over an automotive panel and physically mark those deformations while the automotive part moves along an assembly line.

The 3D acquisition subsystem provides the deformation detection subsystem with 3D point cloud or mesh information. The deformation detection subsystem, which is the focus of this chapter, analyzes the surface of the automotive body panel and determines the locations and type of all deformations of interest. The robotic marking subsystem tracks the moving automotive panel along the assembly line, and marks the deformations with a robotic arm (Borsu 2010). The relationships between the subsystems are shown in Figure 3.

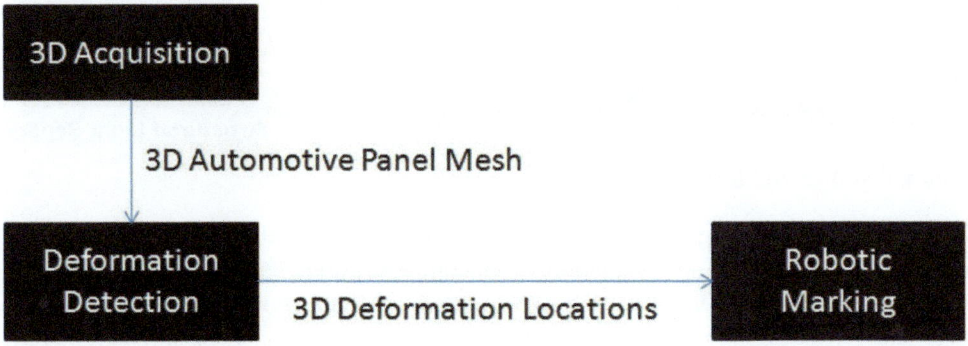

Figure 3. Relationship between subsystems for automated deformation detection and marking.

The automated deformation detection and marking system is created on a smaller scale in a lab setting. This serves as a test bed for the developed techniques, and demonstrates that they can work in a real-world setting. An image of the setup is shown in Figure 4.

To represent the idea of a moving assembly line, a PC-operated sled system is used and simulates a shortened conveyor in a lab setting. One of several real or imitation automotive panels is mounted on the sled system to imitate a real automotive panel. At the beginning of the assembly line, when the automotive panel is static, a structured light sensor is used to generate a dense 3D reconstruction of the surface of the automotive panel (Boyer 2009; Boyer et al. 2009). The deformation detection subsystem processes this 3D data, and acquires the location of the deformations. The panel continues moving along the sled system and is tracked by the robotic marking system (Borsu & Payeur 2009; Borsu 2010). Then, based on the locations automatically provided by the deformation detection subsystem, the robotic manipulator is positioned to smoothly mark deformations on the automotive panel surface.

3.3. Data sets

A 3D acquisition system provides the only input used by the deformation detection system to identify the location of deformations of interest. A detailed discussion of the 3D sensing

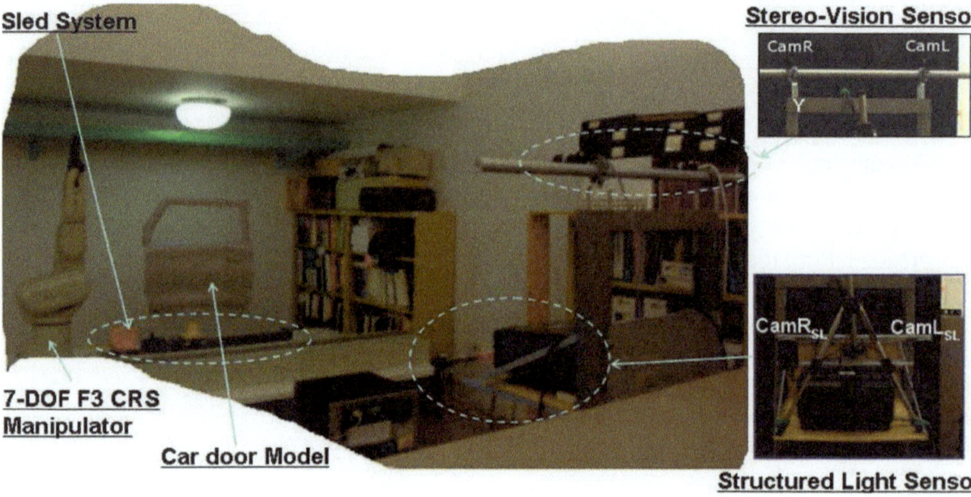

Figure 4. Experimental lab setup.

systems used for acquiring the shape of the automotive body panels is beyond the scope of this chapter. For laboratory evaluation, a custom structured light sensor (Boyer 2009; Boyer et al. 2009; Payeur & Desjardins 2009) is used in combination with slightly enlarged dings and dents deformations artificially affixed on the test panels. Alternatively, higher resolution datasets collected by industrial partners with an active laser range sensor on real automotive panels is also used to demonstrate the capability of adaptation of the proposed approach to different scales and its independence from the 3D acquisition system. Surface reconstruction, performed with the ball-pivoting algorithm proposed by Bernadini *et al.* (Bernardini et al. 1999), is used to generate a mesh triangulation out of the acquired 3D points. The output of this module is provided as input to the deformation detection subsystem, which is the starting point for the original work presented in this chapter.

Real-world test data is important to determine the effectiveness of the approach, since any 3D acquisition system does not provide ideal meshes for this application. The reflective characteristics of the surface, the subtle variations in its shape, and the large distance the panel is positioned from the sensor, cause the acquisition system to introduce an abundance of noise and acquisition artifacts. These real-world meshes serve as test cases for non-optimal acquisition and surface characteristics resulting from the acquisition errors.

The first real-world mesh is a desktop computer casing panel modified by hammering 3 dents into it. Though this is not an automotive part, it simulates real-world deformations on a relatively flat surface. The panel is 20cmx15cm and each dent is circular, with dimensions of approximately 2cm in diameter and 0.25cm in depth. The computer casing panel was scanned at two different resolutions, with the high resolution version containing 14 626 points and the low resolution version containing 3647 points. Since it might be unrealistic to expect an accurate extraction of deformations from this mesh, a filtered version of the low resolution scan is created using a Laplacian smoothing filter to remove the noise while maintaining the deformations. The meshes are shown in Figure 5 with deformations circled. The amount of surface variation, holes along the boundaries, and noise are all visible in the images of the computer casing panel meshes.

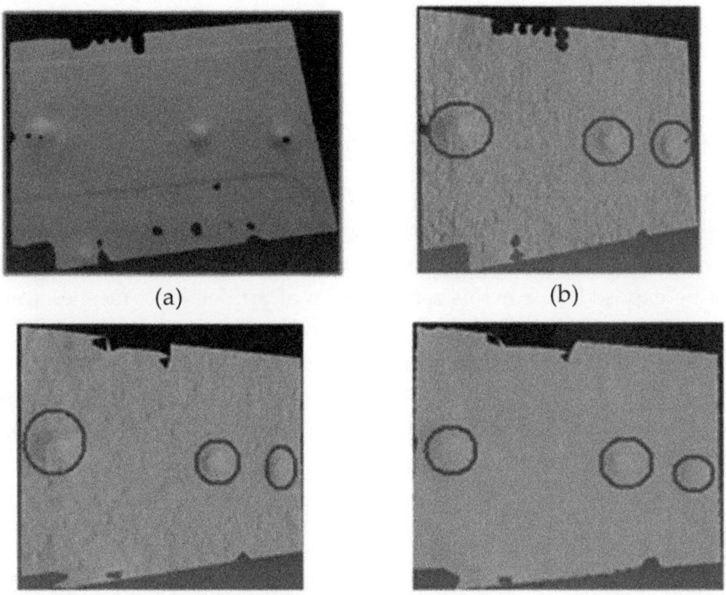

(a) (b) (c) (d)

Figure 5. a) Indented computer casing panel, b) high resolution scan, c) low resolution scan, and d) filtered low resolution scan, with deformations circled.

A mock car door was crafted out of cardboard, consisting of a curved body, a door handle, and a window frame. The door is approximately 70cmx78cm. Three dings, made up of paper were stuck to the door at various positions, where each ding is circular and approximately 1cm in diameter and 1cm in depth. The scanned car door contains 32 202 points. A Laplacian filtered version is also used. The filtered and unfiltered versions are shown in Figure 6 with deformations circled.

The car door is acquired well, with the deformations, door handle, surface variation and window frame all appearing. There is lots of surface variation along the borders due to acquisition errors, and a significant amount of noise, which may interfere with isolating the deformations from parts of the noise as the peaks of the noise are almost as high as the peaks of the deformations. The filtered version reduces noise levels by minimizing the peaks and further separating them from the deformation peaks.

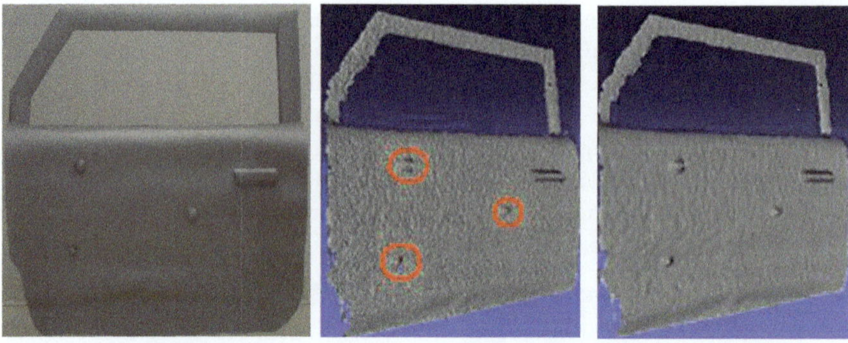

(a) (b) (c)

Figure 6. a) Car door sample, b) unfiltered mesh with deformations circled, and c) filtered mesh.

These real-world meshes are adequate to test the proposed system's behavior when applied upon meshes with noise, acquisition artifacts, and real-world characteristics. However, it does not provide a comprehensive enough set of data to test the various situations that the system might be exposed to. For this reason, a set of artificial test meshes were generated, with characteristics that were not found in the acquired real-world meshes, but that may occur in other real-world meshes. These artificial meshes resemble deformations of interest, of various sizes and scale, under different surface conditions. These meshes also attempt to test the functionality of the system while working under ideal acquisition scenarios where there is no noise and no acquisition artifacts.

A flat mesh with a small dent was created as well as a flat mesh with a large ding, as shown in Figure 7. Similarly, a curved surface mesh with a small dent and a curved mesh with a large ding were created, as shown in Figure 8. The flat meshes are used to determine if the designed algorithms can detect small scale as well as large scale deformations. The curved meshes help determine if the designed algorithms can detect a deformation in spite of a curved or uneven surface around it.

(a) (b)

Figure 7. a) Flat mesh with small dent, b) flat mesh with large ding.

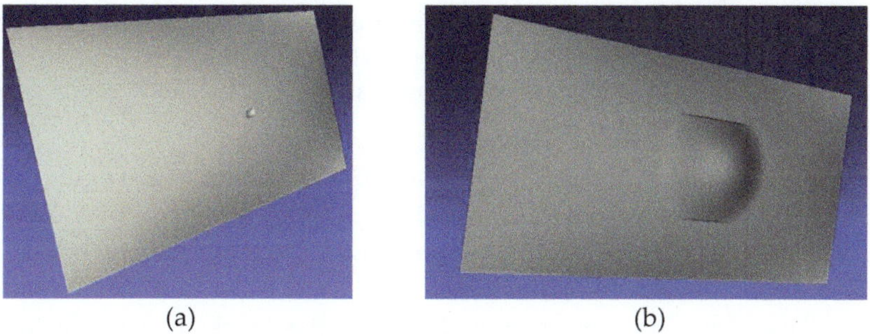

Figure 8. a) Curved surface with small dent, b) curved surface with large ding.

4. Octree-based surface shape analysis

The criticial component of the proposed defect detection system is its surface shape analysis module. The goal of the latter is to break the mesh up into pieces and determine which of those pieces likely belong to the defects in question, as shown in Figure 9.

The outcome of the methods presented in this section is the labeling of all parts of the mesh as either belonging to a feature or not. This module does not determine what collection of mesh pieces define a deformation, however it will determine which mesh pieces likely contain part of a deformation. The output of the shape analysis module is therefore passed to the segmentation phase, as depicted in Figure 2, to determine which collection of mesh pieces defines a deformation.

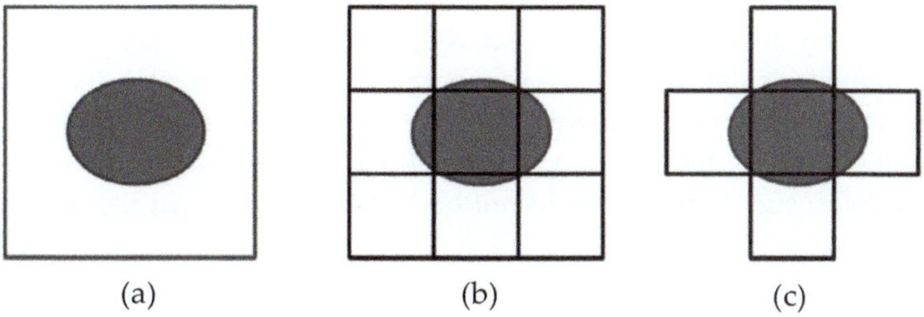

Figure 9. a) Mesh with oval feature, b) mesh broken into pieces, and c) feature extraction results.

This chapter proposes an original surface shape analysis technique that is based on octrees for the automated deformation detection framework. The octree-based technique divides the mesh into cubic volumes and analyzes the mesh contained in those volumes to determine if they belong to a feature. Taking inspiration from the octree-based segmentation method proposed by Woo *et al.* (Woo et al. 2002), as explained in section 2.2, a number of improvements are proposed.

The original technique represents the entire volume surrounding a point cloud in the root node of a tree. Then, by evaluating the standard deviation of the point normals, σ, against a threshold, it determines whether there should be a subdivision of that volume into eight octants. When subdivided, each of the resulting octants is a volume, and is represented by a child node. The points in the original volume are redistributed into the new octants, based on their position, and stored in the child node which represents the new volume it belongs to. This process is repeated for each node, until a tree of sufficient depth is generated or no more volumes require subdivision.

The original technique is designed to find significant changes in a 3D point cloud, such as sharp edges. However, deformation detection for dings and dents over automotive body panels requires detection of slight variations over a smooth surface, which may not be consistent over multiple resolutions. For this reason, the original technique must be revisited. Two major aspects are introduced in this work to enhance the algorithm's flexibility and performance for the purposes of the application considered in the present work: *i*) using a triangle-based analysis rather than a point-based analysis of surface shape, and *ii*) defining non-uniform weighting of surface normals. These enhancements will be discussed in sections 4.1 and 4.2, respectively. A third improvement is also proposed, that uses the octree to aggregate multi-resolution information into performing the feature extraction after the tree generation is complete. This will be discussed in section 4.3.

4.1. Triangle-based analysis

The original method (Woo et al. 2002) operates directly on the 3D point cloud, with knowledge of the reconstructed surface, to calculate the appropriate values for subdivision of the octree. The calculation of the point normal uses all the triangles surrounding the point, and the subdivision of the octree relies on the standard deviation of the point normals contained inside each node. The triangles surrounding a point provide several pieces of information about the surface, yet reducing this information to a point normal using an averaging calculation acts as a smoothing filter, inherently inducing a loss of valuable surface information as shown in Figure 10. In the context of the present work where deformations to be detected are of a small size compared to the remainder of the panel surface curves, such a filter should not be included directly in the feature extraction

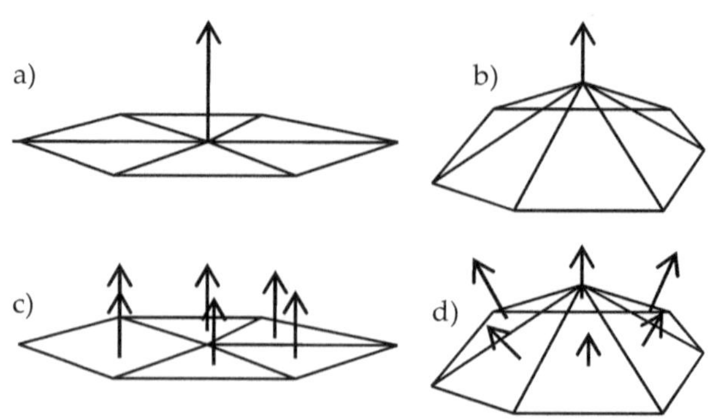

Figure 10. a) Point normal describing a flat surface, b) same point normal describing a non-flat surface, c) triangle surface normals describing a flat surface, and d) different triangle surface normals describing a non-flat surface.

technique. For these reasons, a first improvement that is proposed consists of using more information to describe the surface of a model by relying on the triangles for surface calculations rather than on points.

4.2. Non-uniform weighting of surface normals

The second strategy to enhance the performance of the algorithm proposed in (Woo et al. 2002) consists in using information about the size of the triangles represented by the surface normals to improve the standard deviation, σ, calculation from Eq. 2.

When using 3D scanning that provide a non-uniform sampling density, the original technique will assign equal weight to every point. As a result, there might be more points to describe a certain region within a volume, and the variation in that region is more strongly accounted for in the σ calculation than in other regions, even if there is no disparity in the surface area represented by these regions, as shown in Figure 11.

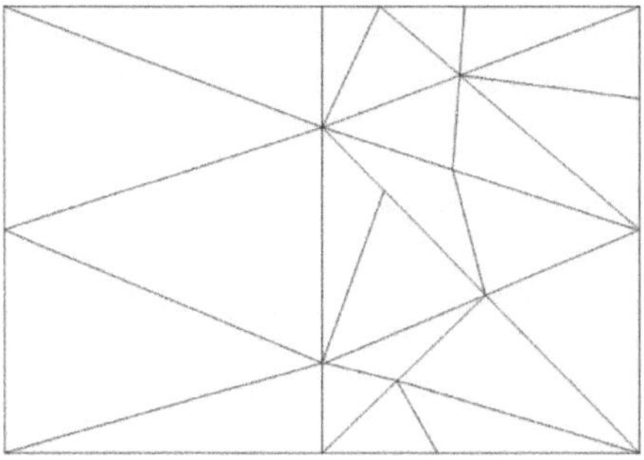

Figure 11. Non-uniformly distributed scan points. The right half of mesh contributes more to the standard deviation value than the left half due to higher density of points/triangles.

A solution to this problem is achieved by using the area of each triangle as a weight to calculate the mean normal and σ values. This approach helps to minimize the effect of small noisy areas, to overcome the effect of non-uniformly distributed points, and to provide a more accurate representation of the surface variation over the region being analyzed.

First, the area of each triangle is calculated:

$$a_i = \frac{1}{2} |v_{i1} \times v_{i2}| \tag{3}$$

where v_{i1} and v_{i2} are any two of the edge vectors that define a triangle, T_i. Then the weighted average normal is calculated over all the triangles that are contained within a given node of the octree, with the area of each triangle serving as a weight:

$$A = \sum_{i=0}^{n} a_i \tag{4}$$

$$\overline{N} = \sum_{i=0}^{n} \frac{a_i}{A} N_i \tag{5}$$

where n is the number of triangles contained in the volume represented by the node of interest in the octree, $\bar{N}$ is the weighted average normal, and N_i is the unit normal of each triangle. Finally, the weighted standard deviation, σ, of the normals can be estimated as:

$$\sigma = \sqrt{\frac{\sum_{i=0}^{n} a_i (N_i - \bar{N})^2}{A}} = \sqrt{\frac{\sum_{i=0}^{n} a_i (x_i - x)^2 + \sum_{i=0}^{n} a_i (y_i - y)^2 + \sum_{i=0}^{n} a_i (z_i - z)^2}{A}} \quad (6)$$

As in the original method, if σ is greater than the defined threshold for that resolution, the volume is subdivided for further investigation at a higher resolution, until the entire tree has been generated. This improvement makes the standard deviation values more accurately represent the amount of deformation within a volume. Figure 12 compares non-uniform weighting to uniform weighting. Detected regions containing deformations are marked in red over surface maps corresponding to various levels of resolution in the octree. Deeper levels in the octree correspond to finer details in the 3D surface mesh.

It can be seen that uniform weighting (upper line) is not as effective at extracting the complete set of deformations. By octree resolution level 10, none of the deformations are extracted with uniform weighting. The non-uniform weighting scheme, with similar thresholding, more consistently extracts the deformations across all resolution scales.

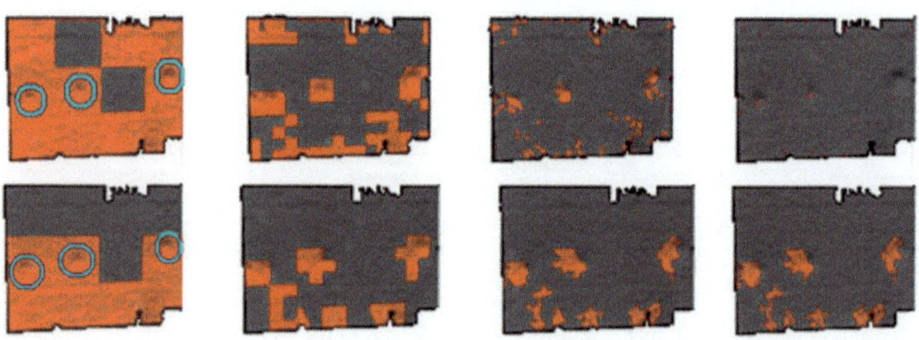

Figure 12. Deformation detection performance at octree resolution levels 4, 6, 8, and 10 under: uniform weighting (top), and non-uniform weighting (bottom). Actual deformation locations are circled on left hand-side images, and detected deformations are marked in red.

4.3. Aggregate standard deviation

Even with the aforementioned improvements of triangle-based analysis and non-uniform surface normal weighting, deformations extraction using an octree-based distribution of 3D scan points requires the determination of the appropriate thresholds. Based on the thresholds alone, it is difficult to predict what features will remain and what features will be removed by the time the deeper and higher resolution levels of the octree are reached. A slight threshold change can produce drastically different results. In general, a more robust technique is required to deal with meshes with varying characteristics, involving an intuitive relationship between the threshold and the results.

The proposed aggregate standard deviation variation of the octree-based technique allows the generation of the tree in its entirety, without limiting subdivision to only nodes that meet a certain threshold. Then, all nodes at selected resolutions of the fully generated tree can be extracted and those with low surface variation can be removed. A new metric is introduced to measure multi-resolution surface variation.

Before the new metric is detailed, it is important to show how the algorithm can be applied using the standard deviation, σ, as the main metric. Analyzing the histogram of the σ values can help in selecting a proper threshold to isolate feature nodes. Histograms are computed to separate the range between the lowest σ and highest σ values at a given octree resolution level into 256 equally divided bins. In the histograms, the x-axis represents the bins, where bin 1 is the lowest range of σ values and bin 256 is the highest, and the y-axis represents the number of nodes with σ values in each bin. In Figure 13, standard deviation values are mapped as a grayscale representation. Black pixels represent low deviation, and white pixels represent large deviation. Therefore, the nodes containing deformations appear in the higher bins. Over most surfaces, the majority of the nodes are mapped to non-feature nodes, such that the bulk of the normal distribution gets classified as non-feature nodes, that is nodes with low σ values. Using the computer casing panel scan as an example, Figure 13 represents the σ values on the mesh as intensity values, along with its histogram, a selected threshold, and the thresholding results. Nodes with σ values above the threshold are extracted as features, which leads to the extraction of the three deformations along with some extra noisy areas.

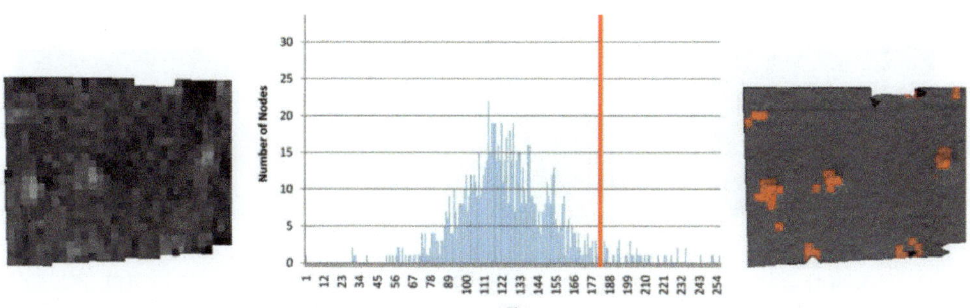

(a) (b) (c)

Figure 13. a) Image of surface shape standard deviation mapped as intensities for the low resolution computer casing panel sample encoded in octree at resolution level 6, b) corresponding histogram of σ values with selected threshold, and c) extracted features including three deformations.

Relying on σ values alone does make use of the multi-resolution capabilities of the octree structure and places a lot of emphasis on the proper selection of the threshold at a given resolution. As described in section 2.2, Pauly *et al.* (Pauly et al. 2002) proposed the idea of multi-resolution feature persistence, where a strong feature can be retained only if it is persistently detected across multiple adjacent scales. In order to combine some of the key concepts of multi-resolution feature persistence with the octree-based feature extraction technique, it is proposed in this work that the characteristics of nodes are accumulated across multiple resolution levels of the octree. The accumulated standard deviation of the surface normals for an octree node is estimated as follows:

$$\sigma_{children} = \sum_{i=0}^{n} \frac{\sigma_i}{n} \tag{7}$$

$$s = (\sigma)(\sigma_{parent})(\sigma_{children}) \tag{8}$$

where s is the aggregate standard deviation value, σ is the standard deviation value of surface normals in the current node, σ_{parent} is the standard deviation value of surface normals in the parent node, σ_i is the standard deviation value of surface normals in the ith child, and n is the number of children that are not empty, such that only nodes that contain 3D points are considered not to bias the metric. $\sigma_{children}$ is calculated as the average standard deviation value of the node's non-empty children. Note that σ values are calculated using non-uniform weighting as detailed in section 4.2. At any given scale, each node will contain a value representing the accumulated standard deviation, s, of a certain volume of the mesh located under itself in the octree structure.

Using the computer casing panel scan as an example, Figure 14 shows how thresholding the s values at a single resolution can be successful in isolating areas of interest. The histogram shown is in the same format as the histogram of Figure 13, but maps s values into pixel intensities instead of σ values.

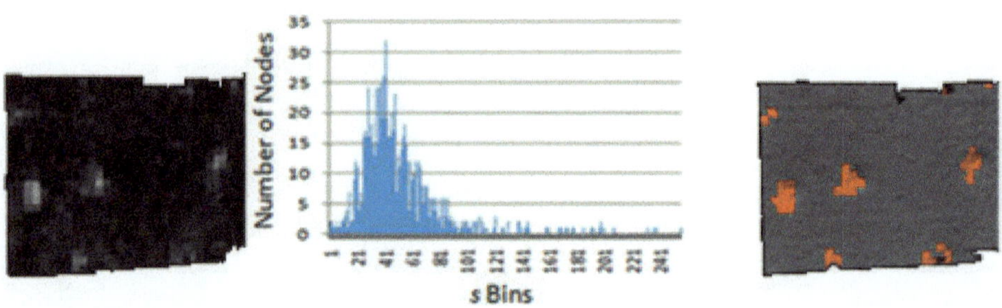

(a) (b) (c)

Figure 14. a) Intensities corresponding to aggregate standard deviation, s, in low resolution computer casing panel sample at resolution level 6 of the octree, b) s histogram with threshold, and c) extracted features including three deformations.

The aggregate standard deviation, s, value provides a greater separation between feature nodes and non-feature nodes than the local standard deviation, σ, only. A more accurate separation also adds tolerance to non-optimal thresholds. To compare the tolerance to non-optimal thresholds when using the σ value against using the s value for feature extraction, the algorithm is applied on the filtered low resolution computer casing panel mesh with both metrics at octree resolution level 5. A suitable threshold was determined such that the results are comparable between using the σ and s values respectively. Then, using the histogram, thresholds which are 50 bins in either direction of the selected threshold are used to determine how that affects the extraction results. When using the σ metric, a threshold of 0.223 is used, which corresponds to bin 163. Then, a threshold of 0.157, corresponding to bin 113, and a threshold of 0.289, aligned with bin 213 are also applied. When using the *s* metric, a threshold of 0.295 is used, which corresponds to bin 176. Then, a threshold of 0.214, corresponding to bin 126, and a threshold of 0.377, aligned with bin 226 are also applied. Figure 15 shows the results using σ for feature extraction, and Figure 16 shows the results using *s* for feature extraction.

(a) (b) (c)

Figure 15. a) Deformed regions detected over computer casing panel at resolution level 5 of the octree with a) optimal σ threshold, b) optimal σ threshold minus 50 bins, and c) optimal σ threshold plus 50 bins.

(a) (b) (c)

Figure 16. a) Deformed regions detected over computer casing panel at resolution level 5 of the octree with a) optimal *s* threshold, b) optimal *s* threshold minus 50 bins, and c) optimal *s* threshold plus 50 bins.

This case demonstrates that a change in threshold setting affects the deformation detection method more extensively when using the local standard deviation, σ, value as a metric than when using the proposed aggregate standard deviation, *s*, value. When using the σ value as a metric, the surface analysis captures more noise and transient features when the threshold is lowered, and removes all of the deformations when the threshold is increased. When using *s* as a metric, the outcome of the surface analysis does not change significantly with the different thresholds, as the deformations are all still present and no significant additional surface variation is detected. As a consequence, when thresholds need to be selected from experimentation, the expected results with aggregate standard deviation are much less sensitive to changes in threshold setting than with the enhancement described in section 4.2 alone. The increase in robustness when compared to non-optimal thresholding, and the significant separation between non-deformation areas and deformation areas in the *s* metric, justify the use of *s* over σ as a metric.

Also, since only three levels of the octree need to be analyzed for thresholding, in accordance with Eq. 8, only the level of the octree at which the nodes are being extracted and thresholded, along with the levels immediately above and below, must be generated. By selectively generating only the necessary levels of the tree from the 3D points cloud, the efficiency of the algorithm is improved significantly. The increase in intuitiveness of the thresholding parameter setting and the efficiency over the original octree-based method of Woo *et al. (Woo et al. 2002)*, which thresholds nodes during tree generation as opposed to after tree generation, also support the development and application of the algorithm introduced in this work.

To demonstrate the effectiveness of the proposed algorithm, in addition to the results presented above on the computer casing panel, the method is applied to the more challenging unfiltered mesh of the car door with the aggregate standard deviation threshold set at 0.155, for the octree resolution level 6. The results are shown in Figure 17. The same algorithm is also applied to the artificial curved meshes, with the threshold set at 0.002, at resolution level 5 of the octree, and the results are shown in Figure 18.

(a) (b)

Figure 17. a) Intensity map corresponding to s values on unfiltered car door octree at resolution level 6, and b) features extracted including the three deformations of interest.

(a) (b)

Figure 18. Feature extracted, at resolution level 5, on artificial curved mesh with a) small dent, and b) large ding.

In terms of deformation extraction effectiveness, provided that the correct threshold is selected, the algorithm performs similarly on both the filtered and unfiltered car door meshes. It also extracts many of the edges around the door and window frame. These edges are very rough areas in the meshes due to the limitations of the acquisition system, and generate large s values in their surroundings, resulting in them being extracted. Despite these small issues with the noisy data, the algorithm isolates the deformations well while increasing robustness and decreasing memory usage when compared to using only the previous octree-based method enhancements.

5. Segmentation and classification

When using the octree-based surface shape analysis technique described in section 4, each node records information determining whether it belongs to a feature or not. Among the nodes that correspond to a deformation, those that contain pieces of a same deformation must eventually be grouped together to segment the deformation from the rest of the panel surface scan, as shown in Figure 19.

Finally, before the entire system outputs the segments that contain deformations, each of them must be classified as a ding or a dent as that is one of the primary objectives of the proposed solution. The octree-based feature extraction requires the classification component to handle this task by receiving the segments containing the deformations of interest as an input and labeling them as dings or dents for the output. A two-step segmentation and classification strategy is proposed to achieve this goal.

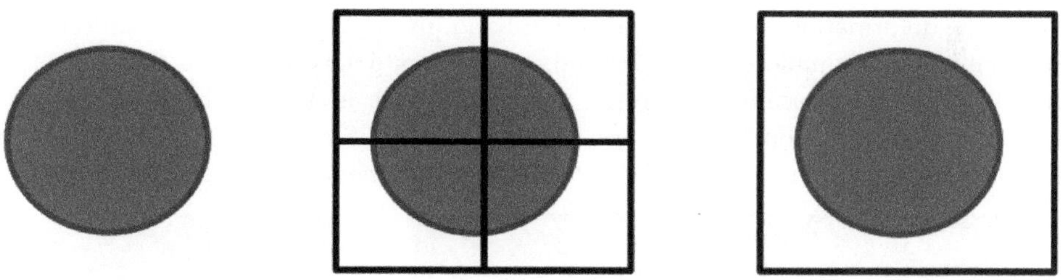

(a) (b) (c)

Figure 19. a) Original deformation, b) octree-based surface shape analysis results, and c) octree-based segmentation.

5.1. Single-resolution segmentation based on octrees

If the scale of the desired deformations is known, an appropriate resolution of the octree can be selected to extract those deformations. Since different depths of the octree correspond to different spatial resolutions, selecting all nodes at a certain depth (defined as octree levels in the previous section) will provide a voxel representation of the object at that scale. However, the appropriate resolution level to segment the deformation must generally be lower than the resolution level considered for the surface shape analysis described previously. Indeed small discontinuities in the deformation should not be detected and segmented as individual deformations based on the connectivity between nodes in the higher resolution version of that deformation. On the other hand, the segmentation resolution must be sufficiently high to avoid deformations being grouped with non-deformations, and to reduce the size of small segments defining features such as noise, in order to avoid confusion with the actual deformations during the classification phase.

After the feature extraction removes all non-feature voxels at the desired resolution, grouping is performed. These remaining voxels are denoted as *feature voxels*. Sets of feature voxels are grouped together to define a deformation based on adjacency, since each voxel contains a piece of a deformation. Since the voxels are cells of a 3-dimensional grid, adjacent voxels can be determined based on their coordinates in the 3-dimensional grid and looking for voxels at adjacent coordinates to the current one. By extending the idea of blob extraction, which is a well-known 2-dimensional image processing algorithm, to three dimensions, adjacent feature voxels can be grouped together. The final result is a set of voxel groups, where each group represents a segment containing a deformation.

The proposed octree-based method is applied on the flat mesh with a small deformation of Figure 7a to extract the small deformation until octree resolution level 8. The segmentation results are shown in Figure 20. Figure 21 shows the segmentation applied at octree resolution level 6 on feature extraction results on the indented computer casing panel high resolution surface mesh.

These results demonstrate that the segmentation can group the required voxels to properly define the deformations. On the artificial mesh, applying segmentation at resolution level 6 segments the deformation clearly. On the other hand, the segmentation at resolution level 4 shows that the deformation is still located, but covers a larger surface than the actual deformation. This is because the resolution considered is lower, therefore the voxel containing the deformation is larger and entirely marked. Similarly successful results are achieved on the computer casing panel, with all deformations being successful grouped.

(a) (b) (c)

Figure 20. a) Octree-based feature extraction, at octree resolution level 8, on the flat mesh with small deformation, b) bounding box of segmented deformation at octree resolution level 6, and c) bounding box of segmented deformation at octree resolution level 4.

Figure 21. Bounding boxes defining the areas segmented as actual deformations on the filtered high resolution computer casing panel surface mesh at resolution octree level 6.

5.2. Classification

Classification represents the final phase of the proposed deformation detection process. It helps determine whether the identified segments are dings or dents. It also provides abilities to ensure that the extracted segments are indeed deformations of interest. Ideally, the previous steps of surface shape analysis and segmentation have already removed most non-deformation areas. But in case some erroneous deformation areas remain, the classification phase provides the necessary filtering stage to remove those areas and reduce false positives. Complex classification methods, such as using neural networks as proposed by Döring *et al.* (Döring et al. 2004), or spin image signatures as attempted by Assfalg *et al.* (Assfalg et al. 2007), could be implemented at this stage. However this work focuses on simpler and more computationally efficient solutions that take advantage of the fact that accurate results have already been obtained by the surface shape analysis and segmentation components.

5.2.1. Classification of the type of deformation

To measure the shape characteristics of the segments, a basic understanding of the orientation of the segment must be determined. A least-squares plane-of-best-fit fitted to the 3D points contained in a segment, specifically the boundary points, is used to determine the orientation of the shape represented by a given segment. Since the boundary points are on the outside edges of the segment, they would more likely belong to the regular surface of the automotive panel than to the deformation. This leads to a plane best fitted to the surface of the automotive panel around the deformation, and determines the general orientation of the surface shape that is contained in the segment. A descriptor, called the point-count descriptor, uses the number of points that share a similar positional relationship to the plane-of-best-fit in estimating the direction of variation of the surface contained in the segment. If a majority of the points contained in the segment are above the plane-of-best-fit, that is, in the direction of the normal vector, the deformation is classified as a ding. If a majority of the points are below the plane-of-best-fit, that is in the opposite direction to the normal vector, the deformation is classified as a dent.

In any classification, a certainty measure is also important. The percentage of the points that are above the plane-of-best-fit in the case of a ding, or below the plane-of-best-fit in the case of a dent, provides the certainty measure on the classification. This way, if there are a similar number of points that are above and below the plane-of-best-fit, the certainty measure is close to 50%, indicating uncertainty.

To test the classification technique's ability to determine whether a deformation is a ding or a dent, it is applied on deformation segments of every mesh using the point-count descriptor, and the results are compared. Non-ideal extraction and segmentation results are presented in Figure 22, while the resulting classifications are presented in Table 1.

Over the artificial flat and curved meshes, it can be seen that the classification is correct. These results show that the classification behaves well on artificial models, corresponding to an acquisition system with minimal noise and acquisition artifacts. For the real world meshes (car door and computer panel), the descriptor accurately classifies each of the dents on the computer casing except for one, which is recognized as a ding rather than a dent. This can be attributed to the non-ideal feature extraction, as classification is dependent on the quality of the latter step. However, the certainty measure reflects the inaccuracy of the classification by being close to 50%, lower than that of the correctly classified deformations. Overall, it can be seen that the classification provides proper results even when feature extraction and segmentation results are non-ideal.

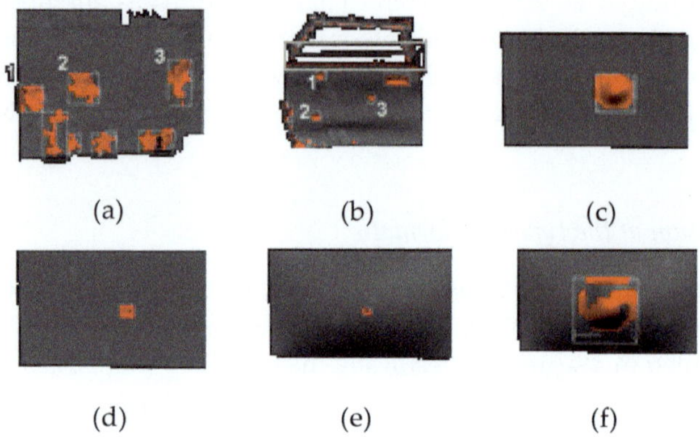

Figure 22. Octree-based feature extraction and single-resolution segmentation applied on a) computer casing panel mesh with dent segments labeled, b) car door mesh with ding segments labeled, c) flat mesh with large ding segment labeled, d) flat mesh with small dent segment labeled, e) curved mesh with small dent segment labeled, and f) curved mesh with large ding segment labeled.

Model	**Actual**	**Point-Count Descriptor Estimates**	
	Type	*Type*	*Certainty*
Car Door			
Def 1	Ding	Ding	0.636
Def 2	Ding	Ding	0.644
Def 3	Ding	Ding	0.609
Computer Casing Panel			
Def 1	Dent	Ding	0.502
Def 2	Dent	Dent	0.511
Def 3	Dent	Dent	0.546
Flat Mesh			
Small	Dent	Dent	0.583
Large	Ding	Ding	0.896
Curved Mesh			
Small	Dent	Dent	0.667
Large	Ding	Ding	0.558

Table 1. Comparison of actual deformation characteristics and the results of classification following octree-based feature extraction and segmentation.

5.2.2. Additional classification

Though it is not the primary goal of the classification stage, the latter also allows fine-tuning of certain parameters by the operator as a final effort to ensure that only deformations of interest are outputted as marked segments. In the process of analyzing descriptors to determine whether a segment is a ding or a dent, segments that do not meet the characteristics of either can be removed. Certain well-known characteristics of deformations can be taken into account to remove non-deformation areas that remain.

The combined surface area of the mesh contained in a segment makes a suitable descriptor of deformations of interest. Thresholding the surface area is another effective strategy for the removal of noise and acquisition artifacts, as those erroneous extracted segments typically cover only very small surface area. Similarly, thresholding surface area also proves effective in removing large surface features that do not correspond to deformations expected over an automotive body panel at the assembly stage, such as actual door handles or aesthetic curves. The latter cover much larger surface areas than dings and dents.

After the orientation of the shape contained in the segment is determined, as detailed in section 5.2.1, and the plane-of-best-fit provides the shape its own local coordinate system, characteristics such as the deformation size in the x, y, and z directions can be measured relative to the shape's local coordinate system to provide an estimate of the shape's height, width, and depth. Noise typically has a small depth, while features like door handles have a larger depth and width relative to the deformations of interest. Applying thresholds on these parameters further increases the reliability of isolating only segments that contain actual deformations.

Though a certain dependency on the setting of threshold values remains, the combination of these descriptors that provide a large amount of information about the various shapes contained in the extracted segments proves to be an excellent technique to improve the reliability of the feature extraction process. In order to demonstrate the relevance of using the final classification phase to further refine the selection of actual deformation segments, here focusing on dings and dents over smoothly curved surface meshes, some cases of poor feature extraction scenarios are artificially created using the experimental models described in section 3.3, but with non-optimal parameters for feature detection and segmentation. The resulting non-deformation areas that get included into detected segments are used to test the classification system's ability to distinguish between actual deformations and non-deformations.

Figure 23 depicts a non-optimal case where many false positives are detected and segmented as potential deformation areas over the car door surface mesh, and shows how the classification is able to remove them. By setting the parameters of the additional classification to remove depths less than 8.5mm or greater than 17mm, most of the broad curvatures on the panel, the door handle, and the small noisy areas are removed. Also, setting the parameters to remove segments with surface area less than 2800 mm^2 or segments with surface area greater than 5200 mm^2 results in further non-deformation areas being removed. In spite of the non-optimal tuning of the quality inspection system, only one additional area, influenced by the boundary effect around the lower left-hand side of the panel, is preserved as a potential deformation segment in Figure 23b. This represents a major improvement from the 40+ erroneous segments initially identified in Figure 23a.

(a) (b)

Figure 23. a) Deformation and non-deformation areas initially segmented on the car door mesh, b) additional classification removes most non-deformation areas.

6. Conclusion

In this chapter, an original feature detection, segmentation and classification framework is proposed to process 3D point clouds and corresponding surface meshes in order to meet the advanced requirements of an automated deformation detection system for use in the context of automotive panels quality control over an assembly line. The requirements in place are that such a system must be able to detect deformations of interest, using 3D analysis, without knowledge of the ideal surface and without any comparative CAD models. The deformations must also be classified as dings or dents. The proposed approach assumes that the operator possesses a minimal knowledge about the approximate size and scale of these deformations of interest in the context of the specific application. The proposed technique then makes optimal use of this additional information to refine the deformation isolation process which leads to an accurate separation of ding and dent deformations from desirable aesthetic design features that typically appear over automotive panels.

A variety of techniques were reviewed for the deformation detection pipeline. An octree-based technique is revisited and refined for surface shape analysis. A single-resolution segmentation method is also presented to refine the location of deformations. Finally, a classification approach is proposed and a complete experimental evaluation is performed on every stage of the surface inspection procedure. The complete pipeline is effective in identifying the location of deformations of interest, and classifying them as dents or dings when presented with a 3D mesh of a surface.

Experiments were conducted on both artificial and real-world test data, offering a set of meshes encompassing various characteristics. These experiments demonstrated that the proposed approach can be used in both ideal circumstances, such as finding a large deformation over a flat, noiseless mesh, as well as in more complex circumstances, such as finding small deformations over a noisy, smoothly curved surface, with acquisition artifacts and holes. The experimental results demonstrate that the proposed framework is scalable, effective and robust to meshes with noise and acquisition artifacts, along with non-ideal surfaces containing shape variations other than the deformations of interest. The proposed technique is therefore suitable for integration in an automated deformation detection and marking system for quality control on automotive panels assembly lines.

Author details

Arjun Yogeswaran and Pierre Payeur
University of Ottawa,
Electrical Engineering and Computer Science (EECS),
Canada

7. References

Assfalg, J., Bertini, M., Bimbo, A. D. & Pala, P. (2007). "Content-based Retrieval of 3-D Objects Using Spin Image Signatures." IEEE Transactions on Multimedia 9(3).

Bernardini, F., Mittleman, J., Rushmeier, H., Silva, C. & Taubin, G. (1999). "The Ball-Pivoting Algorithm for Surface Reconstruction." IEEE Trans. Visualization and Computer Graphics 5(4): 349-359.

Blais, F., Taylor, J., Cournoyer, L., Picard, M., Borgeat, L., Godin, G., Beraldin, J. A., Rioux, M. & Lahanier, C. (2007). Ultra High-Resolution 3D Laser Color Imaging of Paintings: the Mona Lisa by Leonardo da Vinci. Proceedings of the 7th International Conference on Lasers in the Conservation of Artworks, Madrid, Spain.

Borsu, V. (2010). Pose and Motion Estimation of Parts Exhibiting Few Visual Features for Robotic Marking of Deformations, University of Ottawa.

Borsu, V. & Payeur, P. (2009). Pose and Motion Estimation of a Moving Rigid Body with Few Features. Proceedings of the IEEE International Workshop on Robotic and Sensor Environments, Lecco, Italy.

Borsu, V., Yogeswaran, A. & Payeur, P. (2010). Automated Surface Deformations Detection and Marking on Automotive Body Panels. Proceedings of the 6th IEEE International Conference on Automation Science and Engineering, Toronto, Ontario, Canada.

Boyer, A. (2009). Adaptive structured light imaging for 3D reconstruction and autonomous robotic exploration, University of Ottawa.

Boyer, A., Curtis, P. & Payeur, P. (2009). 3D Modeling from Multiple Views with Integrated Registration and Data Fusion. Proceedings of the 6th Canadian Conference on Computer and Robot Vision, Kelowna, BC, Canada.

Chen, H. (2008). Automatic Dent Detection on Car Bodies, Hong Kong University of Science and Technology.

Dorai, C. & Jain, A. (1997). "A Representation Scheme for 3D Free-Form Objects." IEEE Trans. Pattern Analysis and Machine Intelligence 19(10): 1115-1130.

Döring, C., Eichhorn, A., Girimonte, D. & Kruse, R. (2004). "Improving Surface Detection for Quality Assessment of Car Body Panels." Mathware & Soft Computing 11: 163-177.

Faugeras, O. (1993). Three-Dimensional Computer Vision, MIT Press.

Gokturk, S. B., Yalcin, H. & Bamji, C. (2004). A Time-Of-Flight Depth Sensor - System Description, Issues and Solutions. Proceedings of the 2004 Conference on Computer Vision and Pattern Recognition Workshop, Washington D.C., USA.

Hoppe, H., DeRose, T., Duchamp, T., McDonald, J. & Stuetzle, W. (1992). Surface Reconstruction from Unorganized Points. Proceedings of SIGGRAPH '92, Chicago, Illinois.

Jagannathan, A. & Miller, E. (2007). "Three-dimensional Surface Mesh Segmentation using Curvedness-based Region Growing Approach." IEEE Trans. Pattern Analysis and Machine Intelligence 29(12): 2195-2204.

Koenderink, J. & Doorn, A. v. (1992). Surface Shape and Curvature Scales. Image and Vision Computing: 557-565.

Lilienblum, T., Albrecht, P., Calow, R. & Michaelis, B. (2000). Dent Detection in Car Bodies. Proceedings of the 15th International Conference on Pattern Recognition (ICPR), Barcelona, Spain.

Marszalec, J. & Myllyla, R. (1997). "Shape Measurements Using Time-Of-Flight-Based Imaging Lidar." Proceedings of the SPIE Conference on Three-Dimensional Imaging and Laser-based Systems for Metrology and Inspection III 3204: 14-15.

Mederos, B., Velho, L. & Figueiredo, L. H. D. (2003). "Moving Least Squares Multiresolution Surface Approximation." Proceedings of SIBGRAPI 2003 - XVI Brazilian Symposium on Computer Graphics and Image Processing: 19.

Murray, D. & Jennings, C. (1997). Stereo Vision Based Mapping and Navigation for Mobile Robots. Proceedings of the IEEE International Conference on Robotics and Automation (ICRA '97), Albuquerque, New Mexico, USA.

Murray, D. & Little, J. J. (2000). "Using Real-Time Stereo Vision for Mobile Robot Navigation." Autonomous Robots 8(2): 161-171.

Newman, T. S. & Jain, A. K. (1995). "A System for 3D CAD-Based Inspection using Range Images." Pattern Recognition 28(10): 1555-1574.

Palagyi, K. & Kuba, A. (1999). "Directional 3D Thinning Using 8 Subiterations." Discrete Geometry for Computer Imagery 1568: 325-336.

Parthasarathy, S., Birk, J. & Dessimoz, J. (1982). "Laser Rangefinder for Robot Control and Inspection." Proceedings of SPIE Robot Vision 1982 336: 2-11.

Pauly, M., Gross, M. & Kobbelt, L. P. (2002). "Efficient Simplification of Point-Sampled Surfaces." Proc. of IEEE Visualization: 163-170.

Pauly, M., Keiser, R. & Gross, M. (2003). "Multi-scale Feature Extraction on Point-Sampled Surfaces." Computer Graphics Forum 22(3): 281-289.

Payeur, P. & Desjardins, D. (2009). "Structured Light Stereoscopic Imaging with Dynamic Pseudo-random Patterns." Proceedings of the International Conference on Image Analysis and Recognition (ICIAR 2009) 5627: 687-696.

Plataniotis, K. N. & Venetsanopoulos, A. N. (2000). Color Image Processing and Applications, Springer Publishing.

Rocchini, C., Cignoni, P., Montani, C., Pingi, P. & Scopigno, R. (2001). "A Low Cost 3D Scanner Based on Structured Light." Computer Graphics Forum (Eurographics 2001 Conf. Issue) 20(3): 299-308.

Schall, O., Belayev, A. & Seidel, H. (2005). Robust Filtering of Noisy Scattered Point Data. Proceedings of Point-Based Graphics, 2005, Eurographics/IEEE VGTC Symposium, Stony Brook, New York, USA.

Se, S., Lowe, D. G. & Little, J. (2001). "Vision-Based Mobile Robot Locatization and Mapping Using Scale-Invariant Features." Proceedings of the International Conference on Robotics and Automation 2001: 2051-2058.

Sequeira, V., Goncalves, J. & Ribeiro, M. (1995). "3D Environment Modelling Using Laser Range Sensing." Robotics and Automation 16(1): 81-91.

Shaffer, E. & Garland, M. (2001). "Efficient Adaptive Simplification of Massive Meshes." IEEE Visualization '01.

Shahram, I., Richard, A. N., David, K., Otmar, H., David, M., Steve, H., Pushmeet, K., Jamie, S., Andrew, J. D. & Andrew, F. KinectFusion: real-time dynamic 3D surface reconstruction and interaction. ACM SIGGRAPH 2011 Talks. Vancouver, British Columbia, Canada, ACM.

Vosselman, G., Gorte, B., Sithole, G. & Rabbani, T. (2004). "Recognising Structure in Laser Scanner Point Clouds." International Archives of Photogrammetry, Remote Sensing and Spatial Information Sciences 46: 33-38.

Woo, H., Kang, E., Wang, S. & Lee, K. H. (2002). "A new segmentation method for point cloud data." International Journal of Machine Tools and Manufacture 42(2): 167-178.

Yan, C. & Didier, S. 3D shape scanning with a Kinect. ACM SIGGRAPH 2011 Posters. Vancouver, British Columbia, Canada, ACM.

Zhang, L., Curless, B. & Seitz, S. M. (2002). Rapid Shape Acquisition Using Color Structured Light and Multi-pass Dynamic Programming. Proceedings of Int. Symposium on 3D Data Processing Visualization and Transmission, Padova, Italy.

Chapter 12

Structural Health Monitoring in Composite Automotive Elements

Hernani Lopes and João Ribeiro

Additional information is available at the end of the chapter

1. Introduction

The composite materials have demonstrated an improvement in some properties, like the weight, the durability, the corrosion resistance, and the sound and warmth insulation, relatively to the classical metallic materials. In addition, the low cost and flexibility of the structures manufacturing process with composite materials has motivated there growing in automotive engineering. With the advent of composite materials, lighter and with specific resistance higher than the metallic, the elements with lower responsibility in vehicles were gradually replaced by these new materials. Nowadays, the energetic crises, with the increase of oil prices, have forced the automotive industry to go further and creating a new generation of more efficient vehicles. One of the key elements in this strategy is to build new light weight vehicles, and the best option to achieve this goal is increasing the use of composite materials. This means that basic structural elements have to be constructed in composite materials. In these applications, the structural elements are highly demanded and work near of its mechanical strength limit, with high safety requirements. Also, these structures usually present a high strength/weight ratio. Accordingly, it requires a low tolerance to damage and therefore requires a tighter control of the integrity of the components by periodic inspections with non-destructive techniques. In those circumstances, a low tolerance to damage is required and, therefore, a tight control of the components integrity by periodic inspections with non-destructive techniques. Despite its higher strength / weight ratio, the composite elements are more sensitive to internal damages and present types of defects and/or damages are different than the metallic. The main damages in composite laminates are the interlaminar debonding, micro-cracks, micro-buckling and inclusions. These internal damages usually result from the manufacturing process and/or external stresses during service. The interlaminar unbound or delamination is the kind of invisible damage and, therefore, more severe and more common in structural components. Such damage appears essentially in laminated structures, like plate or shells with low curvature, and usually results in a substantial loss of structural performance by reducing its load capacity.

In engineering there is broad interest in structural health monitoring, looking for the early damage detection. The risk of human lives loss resulting from structural unpredictable failure, as in the airplanes crashes, bridges or buildings, have motivated the investigation of the scientific community of the various branches of engineering. Similarly, the superior performance requirement of the materials has stimulated the development and improvement of experimental techniques with application in monitoring of the structural integrity. The earlier damage identification is intended to prevent structural failure and the programming the replacement of damaged element. The main technical inspection methods can be divided into global and localized methods (Alamos, 1996). The available inspection global techniques are based in the sound or ultrasound propagation, magnetic field variations, radiation inspection, electric current, the thermal emissivity and visual inspection techniques. All these techniques assume that the behavior of the material in the vicinity of the damage is known. In addition, a large majority of these techniques is intended exclusively for research of damage in metallic structures. Moreover, the detection of delamination from the change of the mechanical characteristics and the static or dynamic structural response has been extensively referred in the technical literature as preferable. The use of composites elements in the automotive market (Altenbach, 2004) is increasing and replacing the traditional ones, this trend shows the importance in the characterization of mechanical properties (Gibson, 2012) and health monitoring (Boller, 2009) during their time lives.

2. Damage inspection techniques

The development of global methods for damage detection in composite structures has been primarily motivated by applications to the aviation and aerospace industries (Lopes, 2011). In technical literature are presented different methodologies for damage characterization in composite materials. These are usually based in experimental measurements of located and /or global structure parameters. Generally, the damage identification methods can be classified into four levels of increased detection (Rytter, 1993): Level 1: Structural integrity; Level 2: Damage localization; Level 3: Damage quantification and Level 4: Prognosis of remaining service life. The first three levels of damage detection are related to methodologies directly supported in experimental measurements. Otherwise, a more complete characterization of damage requires the use of analytical and numerical time to estimate the remaining life, fourth level of damage characterization. The actually experimental techniques don't allow the proper quantification damage in composite structures. Indeed, the fourth level of characterization of the damage require the information from three previous levels, this explains why there aren't any numerical model capable to predict the remaining life of such components.

2.1. Structural integrity

The first work referenced in the literature addressing the damage detection in composites structures was made by Adams et al. (Adams, 1975). The proposed methodology is based on the principle of dynamic stiffness decreased and the damping increased due to the presence of structural damage. The change in stiffness, both local and global, leads to the decrease of natural vibration frequencies. As such, the non-uniform distribution of internal forces in each mode shape produced different variations into the natural frequencies. These changes are directly related to the location of the damage in structure. On the other hand, there is an increase in the structural damping caused by the growth of the vibratory energy dissipation in the region of damage (Peroni, 1991). The comparison of corresponding frequencies or damping ratio, before and after the structure is put to use, allow their integrity to be evaluated. The main advantage of using this technique is the simplicity of measuring dynamic structural properties, natural frequency and modal damping, which are global parameters and don't dependent on the measuring points, could be obtained by sparse measurements.

2.1.1. Methods based on natural frequencies and modal damping

The natural frequencies variation principle in one-dimensional structure was tested with introduction of a single damage, by removing the equivalent to 1% of its cross-section (Adams, 1978). The structural damage was successfully detected by the decrease of their natural frequencies. However, this methodology was insufficient to locate and quantify the damage severity, and also, shown the need for a more complete structural characterization. Similarly, experimental results on a bridge of a motorway had proven the effectiveness in detection of damage by a decrease in the natural vibration frequencies (Biswas, 1990). The same methodology was applied to offshore the structures monitoring (Loland, 1976; Vandiver, 1975). Subsequently, the decrease of natural frequencies and increased damping were investigated for delamination detection in composite structures (Lai, 1995). The experimental results show a higher sensitivity in variation of natural frequencies than in the modal damping, due to the low resolution and the instability of measurements.

2.1.2. Methods based on frequency response functions

The use of frequency response functions (FRF) was considered by many other authors as a solution for the detection of the structural integrity (Tsai, 1988; Mannan, 1990; Samman, 1991; Biswas, 1994; Samman, 1994a; Samman, 1994b). The experimental measurement of FRFs in a laboratory bridge model, allowed the identification of a 3 mm cut in one of the tested bars (Mannan, 1990). The analysis of the risk failure in trusses structures was investigated by using poles changes in the FRFs (Manning, 1994).

2.2. Damage localization

The methods dedicated to the damage localization are based on physical principle of the reduction local structural stiffness. Indirectly, they can be identified from the localization of disturbances or discontinuities in the experimental structural response, like the: displacement, rotation, bending moment or strain fields. Another approach is based on the analysis of the structural stiffness or flexibility changes, which are identified from experimental modal parameters.

2.2.1. *Methods based on modal response*

The extension to two-dimensional structures, plate geometry, of the methodology proposed by Adams et al. (Adams, 1975) was presented by Cawley (Cawley, 1979a; Cawley, 1979b; Cawley, 1980). As an alternative to experimental modal analysis, is proposed to apply the sensitivity method to the modes shapes to deduce the location of damage in plates (Cawley, 1979b). The effect produced by the damage into the modes shapes depends on the size and stress state of the damaged region. For the analyzed structures is observed that the stress varies along the plate thickness, being zero in middle plane. The sensitivity method was applied to a damage numerical model, where was consider stiffness and mass variation negligible in the damage elements. This analysis was used in localization of a sequence of damage introduced into a laminated composite plate (Cawley, 1979b). The procedure required a large computational effort to achieve a good agreement between the numerical model and the experimental measurements.

In the laminates structures, their dynamic characteristics are strongly dependent on the stack and orientation of the laminate layers (Saravanos, 1996). The experimental analysis performed in graphite/epoxy (T299/823) of a clamped-free beams which damage dimensions less than 10% of the beam's length, demonstrated that the global modal parameters, frequency and damping, were not sufficient to localize the damage and other local parameters should be also used (Saravanos, 1996).

The effect produced by damage in the fundamental modal vibration of a clamped-free beam was investigated by Yuen (Yuen, 1985). His numerical study shows the influence of the damage in the modal response. Its effect in the amplitude of the mode shapes was also analyzed by Chen (Chen, 1988). In this case, the distributions of the kinetic and potential energy were used as indicators to localize the damage.

The methodology for damage localization from the disturbances or discontinuities analysis of the modal rotation field was proposed by Abdo et al. (Abdo, 2002). The numerical study performed with a finite element model of a plate with different boundary conditions, show that the modal rotation field is more sensitive than the modal displacement field on the damage localization. This study also proved that was possible to localize damages up to 5% reduction of local stiffness using the perturbation analysis of the modal rotation field.

The sensitivity analysis of the FRF for damage localization in beams was presented and later improved by Lin (Lin, 1990; Lin, 1994). Recent results using experimental data demonstrated the good performance of this method (Maia, 2006).

2.2.2. *Methods based on modal curvature*

The modal curvature field analysis method was proposed by Pandey (Pandey, 1991). His method is based on the perturbation analysis of the model curvature field between undamaged and damage state. The curvatures are computed by applying the second-order central differences method to the modal displacements field. Contrary to the natural mode shapes, the perturbations in the modal curvature field are coincident with the damage region and its magnitude is proportional to its severity. A finite element model was used to calculate the natural mode shapes of a beam for clamped-free and simply supported boundary conditions. The damage was created by locally reducing the beam's flexural stiffness. The analysis shows that the change on curvature locates correctly the damage but not quantify its severity.

It was also shown that the parameters COMAC (Co-ordenate Modal Assurance Criteria) (Lienven, 1988) and MAC (Modal Assurance Criteria) (Wolff, 1989), calculated from modes shapes displacements, are effective only for identification of significant damages, and the smaller damage are masked by the superposition process. Likewise, the frequencies and deviations in the modal curvature field were used to estimate the location of deep cracks (Dimarogonas, 1996). According to the theory of elasticity for thin beams and plates bending, the deformation at the surface is proportional to its curvature (Timoshenko, 1959). The measurement of the modal deformation field was suggested to identify the damages position. The deformation field is a better indicator of damage location than the modal displacement field (Yao, 1992; Chen, 1994; Yan, 1996). Same conclusion was presented by Chang (Chang, 1993), who compared the sensitivity of the several modal parameters.

The formulation of the model of damage index is based on the modal curvature field information. The damage index correlates the curvature field before and after the introduction of damage at each of the structure segments (Stubbs, 1995). Its average deviation from the normal distribution of the damage index is used as an indicator to identify the most likely of damage region.

The extension of the curvature method at all frequencies was proposed by Sampaio et al. (Sampaio, 1999). The changes in curvature of the frequency response functions (FRF), before and after the introduction the damage, are used to identify its location. The numerical simulations show that the method is most effective for the frequency band up to the first natural frequency or anti-resonance. Its comparison with the amplitude difference of the curvatures and the damage index method proved the superior performance of the method. The procedure was tested using experimental data measure from a concrete bridge, where was created four levels of damage in four different positions. Despite the greater effectiveness of the method, only the most severe case of damage was identified. Similar results were obtained with other methods.

Based on modal curvatures field analysis, Ratcliffe (Ratcliffe, 1997) developed a new procedure that doesn't require the previous knowledge of the structure behavior. The calculation of the modal curvature field is performed by applying the Laplacian operator to natural modes shapes. The localization of the damage is identified from the discontinuity or perturbation in the computed curvature field. Numerical simulations derived from finite element analysis allowed to identify the damage for the case of 10% thickness reduction in the section of a beam. In order to improve the sensibility of this technique, the authors presented a modified version of this method, which called the damage detection gapped smoothing method (Yoon, 2005). The difference between the smooth and the original modal curvatures profile is used, being smooth profile obtained by adjusting a third degree polynomial to the experimental data.

The application of this new technique to numerical modes shapes made possible the localization of damage up to 0.5% reduction of the beam thickness. It was also concluded that the effectiveness of the damage location is greatest for the fundamental mode shape and improves with increasing spatial resolution data. The experimental demonstration of the procedure was performed on a steel beam with a local damage, created by a cut along the cross section of the beam direction and with half thickness depth. For the first two natural modes was successfully identified the location of the damage. However, the low accuracy in the results led the authors to suggest the electrical strain gages to be used as an alternative technique for measuring the curvature. Direct measurement of the curvature of the natural modes of vibration of beams was proven to have superior performance in the localization of damages (Change, 1994).

The damage detection gapped smoothing method was applied to experimental mode shapes of a composite beam in order to locate delamination (Ratcliffe, 1998). The high sensitivity of this method was also demonstrated by localizing the damage in a steel beam, equivalent to 0.8% reduction in thickness (Ratcliffe, 2000). The application of hybrid techniques for the extraction of the smoothing curvature field, allowed the sensitivity improvement of this method (Yoon, 2001). The comparison between the experimental mode shapes and corresponded analytical ones, allows the localization of the damage (Yoon, 2001). Later, this method was applied to locate defects or delaminations in laminated composite plates (Yoon, 2005). In this case, the structural irregularity index, used to localized damage, derives from damage detection gapped smoothing method. The procedure can be applied to a response at any fixed frequency, but it is preferable to use the information from a frequency band. In this latter case, the structural irregularity index is analyzed statistically to serve as reference in the identification of damage position. The numerical simulations using the finite element method proved the effectiveness of this procedure. However, only the border of the damage position can be identified. This technique was tested on damage laminated plates. The methodology was unable to find the delaminations created artificially on the plates during manufacturing process by the introduction sheet of Teflon layer. The results show a superior performance of frequency band response method when compared with single modal response method, as result of canceling the measurement errors through the accumulative process.

2.2.3. Methods based on the dynamic measurement of stiffness or flexibility

As variant of the curvature method and having the change of local stiffness due to the damage, it was suggested the use new methodologies based on measurement of the structure dynamic stiffness (Yoon, 2005; Change, 1994). The differences in the structural stiffness matrix between undamage and damage cases is use to detect and locate cracks in structures [48]. Further, the method of the stiffness matrix error, defined by the difference of the stiffness matrix between the analytical/numerical model and experimental data, it was proposed to detect damage in the case of a large variation of stiffness (Park, 1988). For small variations, the same author proposed the use of a weight function by including others modal parameters. However, a large number of mode shapes are needed to increase the effectiveness of this method (Gysin, 1986). Indeed, in experimental modal analysis only the lowest frequency modes are measured. Furthermore, the analytical stiffness matrix should be representative of the experimental model. The combination of these two limitations will affect the precision of the method and conditions on its practical application. To overcome these limitations, it was proposed the differences flexibility method (Pandey, 1994; Pandey, 1995). This normalized model of the flexibility matrix was successfully tested for damages localization in underwater platforms (Rubin, 1983). The advantage of using the flexible matrix is in the accuracy of the estimating this matrix coefficients using a small number of mode shapes (Pandey, 1994). The flexibility matrix is defined as the inverse of the stiffness matrix. Thus, reduction in rigidity produce increased flexibility in the structure. Indeed, an approximation of the flexible matrix can be obtained from the experimental modal analysis.

The structural damage can be detected and located from disturbances in the matrix of flexibility. Numerical and experimental results obtained in beams established the effectiveness of this methodology (Pandey, 1995). In this work, the damage was identified based on local maximum analysis, computed from the difference between the flexibility matrices of the original and damage structure. Sequences of five damage cases, created in two different locations of the beam, were located and gradual evolution of its severity identified. However, for the case of multiple damages it was only possible to identify the location of the most dominant.

2.2.4. *Methods based on wavelets transform*

The methods based on natural modes shapes and its spatial derivatives prove its effectiveness using numerical data. However, the success of these techniques is affected by the noise present in experimental data (Gentile, 2003). A new research domain of structural damage is the application Wavelets transform to extract the signal spatial derivatives components. This has the advantage of identifying small changes or discontinuities in the signal, without the propagation of noise, such as in the common differentiation techniques.

The identification of singularities in the distribution of the signal components can be used to detect the location of damages. The study of the most appropriate technique for finding cracks in beams based on Wavelets signal processing was presented by Rucka and Wilde (Rucka, 2006). The proposed technique allows the position identification of the damage without the previous knowledge of the structure behavior or the use of mathematical models. Several damages with different degrees of severity were investigated based on the optical measurement of the beams bending profile. The damage is located by identifying the local maxima of the signal components for each profile (Rucka, 2006). The Gaussian and Coifet Wavelet functions show to be the most effective in localizing the slot, up to 27% section reduction of the beam thickness.

The comparison of differentiation process among Wavelets transform and several other differential operators, to calculate the curvature mode shape and sequent location of damage in beams, was presented by Messina (Messina, 2004). The differential operators integrate a low-pass filter to reduce the unwanted high frequency noise. This study reveals the perturbation on the low frequencies signal produced by the use of strong filter, which becomes clearer for higher order derivatives. The Fourier transform filter, the weighted least squares, Lanczos's differentiator filters and Gaussian Wavelets differentiator filter techniques were investigated to calculate the curvatures from the modal displacement field contaminated by Gaussian noise. For the computed curvatures field is observed similar results using all the techniques. However, the profile curvature obtained using the Gaussian Wavelet transform presents coarse result for the undamaged region. In this case, the curvature fields will present greater number of disturbances, making difficult the damage localization.

2.3. Quantification of the damage severity

The ultimate level of the damage characterization is the quantification of stiffness decrease and estimation of the damage real dimensions. The procedure requires a high accuracy in evaluating the structural response. The quantification stiffness in the damage region can be estimated from the local variation of the curvature or using mathematical models. With respect to area affected by the damage, this can be assessed by analyzing the contours of local disturbances, normally, requires the use of dedicate digital image processing techniques.

2.3.1. Methods based on the sensitivities of modal parameters

The sensitivities method of the modal parameters was used to localize and quantify the severity of damage in a discrete system with multiple degrees of freedom (Zhu, 2005). The sensitivity of the natural frequencies and the modal displacements, modal rotations and modal curvature fields were compared in order to evaluate the effectiveness in the damage localization. The numerical simulations of a mass-spring model with 10 degrees of freedom showed that the modal curvature field is the more sensitive to the damage, while the modal rotation is a better indicator of its position. A procedure defined in two steps was proposed to locate and quantify the severity of damages. First, the damage is located from the perturbations analysis of the modal curvature field. Next, the damage severity is estimate using a limited number of measured frequencies. The methodology was investigated by using the experimental analysis of different damage scenarios in a periodic model of a building with three degrees of freedom. The results are considered satisfactory for medium damage severity (13.12% - 26.74%). This author concluded that the quantification of the damage severity increases with the number of natural frequencies used in the calculations. Deviations in the results are pointed to experimental measurement errors.

2.3.2. Methods based on the measurement of modal curvatures fields

The effectiveness of the methods based on the modal curvature field is determined by the quality measurements imposed. Typically, the modal curvature is obtained through the application numerical differentiation techniques to experimental modal displacement field. As a result, the high frequency experimental noise is amplified and propagates through this process and has an strong impact on the final quality of the results. Alternatively, the direct measurement of the curvature has the advantage of avoiding the numerical differentiation of the data with consequent improvement in the efficacy of the methods.

The method for damage localization and quantify its severity on a sandwich beam, by measuring the curvature of the mode shapes using piezoelectric transducers, was presented by Lestari et al. (Lestari, 2005). The procedure is based on curvature difference between the original and damaged structure, measured directly using 31 piezoelectric sensors (polyvinylidenefluoride film) glued and equally spaced on the structure surface. The natural frequencies and modal curvature field of a clamped-free sandwich beam with a local damage caused artificially: the first – by removing the nucleus (to simulate the debonding between the core and the skins) and, second – by crushing the lower interface core / skin (to simulate the crushing of the core). The difference between curvature (damage factor) and the sum of the differences between curvatures, allow identifying the approximate of the damage location. The results also show that the crushing produces greater reduction in structural stiffness than separation between core and skin. The estimative of the damage region stiffness variation was obtained from the difference of first six modal curvature fields. It was observed a local stiffness reduction for the delamination damage between 30% and 60% and for the crushing damage of 40% to 90%. The disparity in some of the values is justified by errors associated to the measurement of curvature field.

2.3.3. *Methods based on full-field measurement of displacement or rotation field by interferometric techniques*

The optical interferometry techniques have been widely investigated in the last four decades, robust tools and have proven to be very effective in non-destructive inspection of structures (Lee, 1991; Sirohi, 1993; Hung, 1997a; Hung, 1999; Hung, 1998; Sirohi, 1999; Gomes, 2000; Santos, 2004)). Its advantages are undeniable compared to classical techniques for the inspection of composite materials (Lee, 1991). The ESPI (Electronic Speckle Pattern Interferometry) and Shear (Shearography) techniques are two examples currently used for nondestructive inspection of composite structures. These are full-field techniques for measuring the information on a surface and allow easily locating disturbance in structural response. The ESPI technique measures the absolute value of the displacements of the surface, including the rigid body. The principle of the Shear technique was first demonstrated by Leendertz and Butters (Leendertz, 1973) through the construction of the Michelson optical interferometer. The Shear technique is only sensitive to the object displacements gradient, which can assumed, for the case of small displacements, as a good approximation to the surface rotation field (Kreis, 2005). Due to their properties and insensitivity to rigid body movement, this technique is often used for the localization of delaminations in composite structures. Comparative analysis of the measurements quality between the ESPI and Shear techniques for internal damage identification was studied by several authors (Hung, 1997; Gomes, 2000). In ESPI technique, the fringes produced by rigid body motion makes difficult its interpretation and may mask the presence of damage. Consequently, the Shear technique is more suitable for damage localization. In this case, the internal damage is reveal by appearing two juxtaposed lobes of concentric fringes representative of local disturbance.

Contrary to ESPI technique, the Shear measures directly the surface rotation field and eliminates the need of the numerical differentiation. Another major advantage is the simplification of the optical setup, which allow the use of low coherence length Lasers. Also, the optical setup can be built in a very compact form, giving a greater stability to the measurements and isolation to external disturbances (Gomes, 2000). In addition to the damage localization in composite structures, several other applications can be found in the literature including: the static and dynamic measurement of the rotation field and the measurement of surface residual stress. (Santos, 2004; Kreis, 2005; Hung, 1978; Lùkberg, 1997; Hung, 1997b; Devesa, 2002; Pedrini, 1997).

The damage identification using shearography technique is based on comparison of two states of the object deformation. The type of excitation will depend on the type of defect and the material used. The success of such techniques will be influence by several factors: the material properties, type of defect and method of excitation. The thermal, the vacuum or transient excitations are most effective techniques to the reveal the inter-layer defects (Hung, 1997a; Gomes, 2000; Santos, 2004; Ambu, 2006). However, the best choice is typically defined by a heuristic process in which the previous experience influences the selection of the excitation method. The analysis of the fringes obtained by ESPI techniques was used to investigate damages in laminated thin plates (Ambu, 2006). Damage was induced by the impact of metallic spheres with different masses. The thermal excitation with infrared lamp was used and the plate out-of-plane displacement field was measured. The damage size and position were estimated from the raw fringes analysis, using digital image processing techniques. The results revealed that the ESPI technique is less sensitive to damage in relation to the holographic techniques. Both of these optical techniques can identify well the damage location. However, the comparison with ultrasound (C-scan) measurements reveals that the interferometric techniques are inadequate to located damages with depths greater than 0.7 mm. The introduction of the phase calculation methods on the classic interferometric techniques permit to improve the spatial resolution of the measurements. The phase map and the corresponding raw fringes, measured with the ESPI technique, were used to locate and quantify delaminations in a fiberglass reinforced polyester plates (Richardson 1998). The damages created by impact were analyzed by C-Scan and by sectioning the matrix. Both methods were used as reference for the analysis of other techniques. The results show a good correlation between optical techniques and reference techniques.

The propagation elastic waves in plates and pipes using holographic interferometry techniques were proposed by several authors (Aprahamian, 1971; Fallstrom, 1989a; Fallstrom, 1989b; Olofsson, 1994; Olofsson, 1996; Fallstrom, 1996; Fallstrom, 1998). A double pulse Laser technique is used to measure the displacements in isotropic plates (Aprahamian, 1971; Olofsson, 1994), anisotropic plates (Fallstrom, 1989a; Fallstrom, 1989b; Olofsson, 1994; Fallstrom, 1996; Fallstrom, 1998) and anisotropic tubes (Olofsson, 1994; Olofsson, 1996). The holographic interferometry and the flexural waves propagation were also used to investigate the debonding in the interface areas of ceramic-metal plate (Conrad, 2001). The quality of the bond between the two materials influences the wave propagation of transients bending waves. These are produced by a piezoelectric exciter mounted near of plate's surface. Distinct damage models were introduced into three plates, simulating the interface discontinuities and cracks in the ceramic plate. A YAG pulse Laser with double cavity was used to generate two pulses with adjustable intervals between 1 and 80 μs. The two interferometric pulses were recorded on a holographic plate by changing the angle of incidence. Afterword, the interferograms were reconstructed and the phase map is extracted by using the phase-shift technique. The damages located by identifying perturbations in the phase map of the wave propagation. However, the complexity of fringes distribution observed in the phase map doesn't permit the interpretation of the damaged area.

The sandwich panels removed from the wing of an airplane model were used to investigate the damage, created by low-speed impact (Ruzek, 2006). The skin made of carbon fiber and core honeycomb, suffered several low energy impacts (10J - 40J). The Shear and C-scan techniques were used for non-destructive inspection of the panels and with the purpose of visually comparing the results. Based on the measurements acquired with Shear technique were possible to identify very well the damage location and size. By contrast, the C-Scan technique, due to their operating principle has proved to be less suitable for such structures. In addition to be more time consuming, presents difficulties in dealing with the multiple discontinuities in the material, rupture / indentations in the skin and distortion of the honeycomb core produced by the impact.

The application of ESPI and Shear techniques to the analysis of debondings in the interface region of thin coatings was investigated by Gomes et al. (Gomes, 2000). The thermal excitation was used to reveal the damage position and its size. The phase maps present similar results for both techniques. However, a superior fringe contrast was observed in the measurements with Shear technique. A technical review of Shear technique and its several applications to composite materials was presented by Hung et al. (Hung, 1999). The measurement of residual stresses, strain field and damage localization are some of the mentioned applications using this technique.

A modified version of the Mach-Zehnder optical interferometer was developed by Pedrini et al. (Pedrini, 1996) for measuring the modal rotation field in plates. The two images created by the interferometer are sheared and rotated before being combined, in order to later compute the phase map. A double pulse Ruby Laser was used to record modal rotation field of a circular plate. The same technique was then used in localizing damages in a sandwich plate with Nomex core and fiberglass skins (Santos, 2004). The two damages were artificially created by removing 2.5 cm diameter of skin and 1 cm diameter of core. An impact hammer with electromagnetic drive was used to produce transient excitation of the structure. The time plate response was recorded for posterior damage localization. The smallest damage was located based on fringes concentration analysis. However, larger damages produce a higher number of fringes, becoming difficulty to distinguish them form fringes produce by the natural vibration of the structure. These and other difficulties have led other authors to develop alternative procedures for nondestructive inspection of damage. The high speed measurements using phase shift technique was considered in order to reduce phase errors and improve the quality of the data (Davila, 2003). The procedure is based on the technical implementation of the temporal phase and the application of phase unwrapping algorithm for recording in time the structure dynamic response.

The combination of ultrasound excitation technique with pulsed interferometric technique for investigating internal damages in the plates was developed by Cernadas et al. (Cernadas, 2001). The ultrasonic elastic waves (Lamb waves) can penetrate deep into the material to reveal the surface damages. These are detected through the use of high resolution optical techniques. The measurement by real-time holography has already been used to identify minor cuts and holes in plates (Schroeder, 1996). The ESPI technique and pulsed Laser method can be used to measure freeze in time the bending wave propagation (Mast, 2001). However, the high speed and small amplitude of the ultrasonic elastic waves requires a good insulation of exterior disturbances. The use of the ESPI technique with double pulse laser has been suggested to solve the stability problems in the measurement (Cernadas, 2001). A piezoelectric transducer coupled to the plate surface is used to generate surface acoustic waves, Rayleigh waves that propagate along the structure surface. These waves can also be generated remotely by strong Laser pulses. However, to generate this Rayleigh waves with amplitude in the measurement range of the ESPI technique, requires the application of high energy on the surface and its protection to prevent being damage. In this study, two types of damage were introduced into an aluminum plate. A blind hole and cross section cut were generated artificially in the plate. The sequence of two images, separated by 1.5 μs, was used to measure the plate displacement field caused by a propagation chain of the Rayleigh waves, introduced through the fourth Laser pulses. Through the disturbances fringes analysis was possible to identify the damages. The introduction of the phase calculation into the displacement measurement allows improving the quality of the results (Cristina, 2003). This new methodology was used to measure the real and imaginary components of the displacement field generated by the propagation of Rayleigh waves. Based on these two components was possible identified damage the position and size of damage, even in the situation of poor quality fringes.

Author details

Hernani Lopes
Polytechnic Institute of Bragança, Department of Applied Mechanics, Bragança, Portugal

João Ribeiro
Polytechnic Institute of Bragança, Department of Mechanical Technology, Bragança, Portugal

3. References

Abdo, M. A. Hori, M. (2002). A numerical study of structural damage detection using changes in the rotation of mode shape, *Journal of Sound and Vibration*, 251(2): 227-239.

Adams, R. D., Cawley, P., Pye, C. J., Short, D. (1978). A vibration technique for non-destructively assessing the integrity of structures, *Mechanical Engineering Science,* 20(2): 93-100.

Adams, R. D., Walton, D., Flitcroft, J. E., Short, D. (1975). Vibration testing as a non-destrutive test too for composite materials, Composite Reliability, ASTM STP(580): 159-175.

Alamos, L. (1996). Damage Identification and Health Monitoring of Structural and Mechanical Systems from Changes in Their Vibration Characteristics: A Literature Review, Energy Citations Database.

Altenbach, H., Altenbach. J., Kissing, W. (2004). Mechanics of composite structural elements, Edited by Springer-Verlag, Berlin and New York.

Ambu, R., Aymerich, F., Ginesu, F., and Priolo, F. (2006). Assessment of NDT interferometric techniques for impact damage detection in composite laminates, *Composites Science and Technology,* 66 (2): 199-205.

Aprahamian, R., Evensen, D. A., Mixson, J. S., Jacoby, J. L. (1971). Holographic Study of Propagating Transverse Waves in Plates, *Experimental Mechanics,* 11(5): 357-362.

Biswas, M., Pandey, A. K., Samman, M. M. (1990). Diagnostic experimental spectral/modal analysis of a highway bridge, *The International Journal of Analytical and Experimental Modal Analysis,* 40(1): 22-31.

Biswas, M., Samman, M., Pandey, A., Bluni, S. (1994). Modified chain code computer vision techniques for interrogation of vibration signatures for structural fault detection, *Journal of Sound and Vibration,* 175: 89-104.

Boller, C., Chang, F., Fujino, Y. (2009). Encyclopedia of Structural Health Monitoring, Wiley, 2960.

Cawley, P., Adams, R. D., (1979a). A vibration technique for non-destrutive testing of fibre composite structures, *Jounal of Composite Materials,* 12(2): 161-175.

Cawley, P., Adams, R. D. (1979b). The Location of Defects in Structures From Measurements of Natural Frequencies, *Journal Strain Analysis,* 14(2): 49-57.

Cawley, P., Adams, R.D. (1980). Defect location in structures made from advance composite materials, in Proceedings of Third International Conference on Composite Materials: 973-983.

Cernadas, D., et al. (2002). Non-destructive testing with surface acoustic waves using double-pulse TV holography, *Measurement Science and Technology,* 13(4): 438–444.

Chang, K. C., S. Z., Lee, G. C. (1993). Modal analysis technique for bridge damage detection, in American Society of Civil Engineers Proceedings: 1083-1088.

Change, J., Tomlinson, G. R., Worden, K. (1994). A simply approach to the numerical and experimental modelling of the dynamics of a cracked beam, in Proceedings of the 12th International Modal Analysis Conference: 778-795.

Chen, J. C. (1988). On-orbit damage assessment for large space structures, *American Institute of Aeronautics and Astronautics Journal,* 26(9): 1119-1126.

Chen, Y., Swamidas, A. S. (1994). Dynamic characteristics and modal parameters of a plate with a small growing surface cracks, in Proceedings of the 12th International Modal Analysis Conference: 1155-1161.

Conrad, M., Sayir, M. (2001). Composite Ceramic-metal Plates Tested with Flexural Waves and Holography, *Experimental Mechanics,* 41(4): 412-420.

Cristina, T., et al. (2003). Measurement of the complex amplitude of transient surface acoustic waves using double-pulsed TV holography and a two-stage spatial Fourier transform method, *Measurement Science and Technology,* 14(12): 2127–2134.

Davila, A., Ruiz, P., Kaufmann, G., Huntley, J. (2003). Measurement of sub-surface delaminations in carbon fibre composites using high-speed phase-shifted speckle interferometry and temporal phase unwrapping, *Optics and Lasers in Engineering,* 40: 447–458.

Devesa, L., et al. (2002). Detection of impact defects in laminated composites by holographic interferometry, *Advanced Materials Forum I,* 230(2): 279-282.

Dimarogonas, A. D. (1996). Vibration of cracked structures-a state of the art review, *Engineering Fracture Mechanics,* 5: 831–857.

Fallstrom, K. E., Lindblom, O. (1998). Transient bending wave propagation in anisotropic plates, *Journal of Applied Mechanics-Transactions of the Asme,* 65(4): 930-938.

Fallstrom, K. E., Gustavsson, H., Molin, N. E., Wahlin, A. (1989a). Transient Bending Waves in Plates Studied by Hologram Interferometry, *Experimental Mechanics,* 29(4): 378-387.

Fallstrom, K.E., Lindgren, L.E., Molin, N.E., Wahlin, A. (1989b). Transient Bending Waves in Anisotropic Plates Studied by Hologram Interferometry, *Experimental Mechanics,* 29(4): 409-413.

Fallstrom, K.E., Olofsson, K., Saldner, H. O., Schedin, S. (1996). Dynamic material parameters in an anisotropic plate estimated by phase-stepped holographic interferometry, *Optics and Lasers in Engineering,* 24(5-6): 429-454.

Gentile, A., Messina, A. (2003). On the continuous wavelet transforms applied to discrete vibrational data for detecting open cracks in damaged beams, *International Journal of Solid and Structures,* 40(2): 295-315.

Gibson, F. R. (2012). Principles of Composite Material Mechanics, Edited by Taylor & Francis Group, USA.

Gomes, J. F., Monteiro, J. M., Vaz, M. A. (2000). NDI of interfaces in coating systems using digital interferometry, *Mechanics of Materials,* 32(12): 837-843.

Gysin, H. P. (1986). Critical application of the error matrix method for localisation of finite element modeling inaccuracies, in Proceedings of the 3th International Modal Analysis Conference: 1339-1351.

Hung, M. Y., Long, K. W., Wang, J. Q. (1997b). Measurement of residual stress by phase shift shearography, *Optics and Lasers in Engineering,* 27(1): 61-73.

Hung, Y. M., Dahuan, S. (1998). Technique for rapid inspection of hermetic seals of microelectronic packages using shearography, *Optical Engineering,* 37(5): 1406-1409.

Hung, Y. Y., Hovanesian, J. D. (1978). Surface Slopes Measurement by a Multisource Shearing Interferometry, *Journal of the Optical Society of America,* 68(10): 1391-1391.

Hung, Y. Y. (1997a). Digital shearography versus TV-holography for non-destructive evaluation, *Optics and Lasers in Engineering,* 26(4-5): 421-436.

Hung, Y. Y. (1999). Applications of digital shearography for testing of composite structures, *Composites Part B-Engineering,* 30(7): 765-773.

Kreis, T., (2005). Handbook of holographic interferometry: optical and digital methods, Weinheim: Wiley-VCH.

Lai, J. Y., Young, K. F. (1995). Dynamic of graphite/epoxy composite under delamination fracture and environmental effects, *Jounal of Composite Structures,* 30(1): 25-32.

Lee, S.M., (1991). International Encyclopaedia of Composites, 4, VCH Publishers.

Leendertz J., Butters, J. N. (1973). An image-shearing speckle-pattern interferometer for measuring bending moments, *Journal of Physics E: Scientific Instruments,* 6 (11): 1107-1110.

Lestari, W., Qiao, P. (2005). Damage detection of fiber-reinforced polymer honeycomb sandwich beams, *Composite Structures,* 67(3): 365–373.

Lienven, N. A., Ewins, D. (1988). Spatial correlation of modal shapes, the Co-ordinate Modal Assurance Criterion (COMAC), in Proceedings of the 6th International Modal Analysis Conference: 690-695.

Lin, C.S. (1990). Location of modeling errors using modal test data, *AIAA Journal,* 28(9): 1650–1654.

Lin, R. M., Ewins, D. J. (1994). Analytical model improvement using frequency response functions, *Mechanical Systems and Signal Processing,* 8(4): 437-458.

Loland, O., Dodds C. J., (1976). Experiences in developing and operating integrity monitoring system in North sea, in Proceedings of the 7th Annual Offshore Technology Conference: 313-319.

Lopes, H., Santos J., Soares, C., Guedes, R., Vaz, M. (2011). A numerical-experimental method for damage location based on rotation fields spatial differentiation, *Computers & Structures,* 89(19-20): 1754-1770.

Lùkberg, O., Slettemoen, G. (1997). Basic electronic speckle pattern interferometry. Applied Optics and Optical Engineering, New York: Academic Press.

Maia, N. M., Santos, J. V., Sampaio, R. P., Soares, C. M. (2006). Damage Identification Using Curvatures and Sensitivities of Frequency-Response-Functions, in Proceedings of The Third European Workshop On Structural Health Monitoring: 547-554.

Mannan, M. A., Richarson, M. (1990). Detection and location of structural cracks using FRF measurements, in Proceedings of the 7th International Modal Analysis Conference: 652-657.

Manning, R. A. (1994). Structural damage detection using active members and neural networks, *American Institute of Aeronautics and Astronautics Journal,* 32(Technical Notes): 1331-3.

Mast, T. D., Gordon, G. A. (2001). Quantitative flaw reconstruction from ultrasonic surface wavefields measured by electronic speckle pattern interferometry, *IEEE Trans. Ultrason. Ferroelectr. Freq. Control,* 48(2): 432-44.

Messina, A. (2004). Detecting damage in beams through digital differentiator filters and continuous wavelet transforms, *Journal of Sound and Vibration,* 272(1-2): 385–412.

Olofsson, K. (1994). Pulsed Holographic Interferometry for the Study of Bending Wave Propagation in Paper and in Tubes, Doctoral Thesis, Luleã University of Technology - Sweden.

Olofsson, K., Fallstrom, K.E., Palagyi, P. (1996). Laser generated and recorded transient bending waves in composite tubes, *Experimental Mechanics,* 36(3): 224-231.

Pandey, A. K., Biswas, M. (1994). Damage detection in structures using changes in flexibility, *Journal of Sound and Vibration,* 169(1): 3-17.

Pandey, A .K., Biswas, M. (1995). Experimental-Verification of Flexibility Difference Method for Locating Damage in Structures, *Journal of Sound and Vibration,* 184(2): 311-328.

Pandey, A. K., Biswas, M., Samman, M. M. (1991). Damage Detection from Changes in Curvature Mode Shapes, *Journal of Sound and Vibration,* 145(2): 321-332.

Park, Y. S., Park, H. S., Lee, S. S. (1988). Weighted-error-matrix application to detect stiffness damage by dynamic-characteristic measurement, *Journal of Analytical and Experimental Modal Analysis,* 3(3): 101-107.

Pedrini, G., Zou, Y., Tiziani, H. J. (1996). Quantitative evaluation of digital shearing interferogram using the spatial carrier method, *Pure and Applied Optics: Journal of the European Optical Society Part A,* 5(3): 313–321.

Peroni, I., Paolozzi, A. (1991). Effect of debonding damage on modal damping of sandwich panel, in Proccedings of 9th International Modal Analysis Conference: 1617-1622.

Ratcliffe, C. P. (1997). Damage detection using a modified laplacian operator on mode shape data, *Journal of Sound and Vibration,* 204(3): 505-517.

Ratcliffe, C. P. (2000). A frequency and curvature based experimental method for locating damage in structures, *Journal of Vibration and Acoustics-Transactions of the Asme,* 122(3): 324-329.

Ratcliffe, C. P., Bagaria, W. J. (1998). Vibration technique for locating delamination in a composite beam, *American Institute of Aeronautics and Astronautics Journal,* 36(6): 1074–1077.

Richardson, M. O., et al. (1998). ESPI non-destrutive testing of GRP composite materials containing impact damage, *Composites Part A,* 29(A): 721-729.

Rubin, S., Coppolino, R. N. (1983). Flexibility monitoring of offshore jacket platforms, in Proceedings of the 15th Annual Offshore Technology Conference: 201-208.

Rucka, M., Wilde, K. (2006). Crack identification using wavelets on experimental static deflection profiles, *Engineering Structures,* 28(2): 279-288.

Ruzek, R., Lohonka, R., Jironc, J. (2006). Ultrasonic C-Scan and shearography NDI techniques evaluation of impact defects identification, *NDT&E International,* 39(2): 132-142.

Rytter, A. (1993). Vibration Based Inspection of Civil Engineering Structures, Doctoral Thesis, Aalborg University -Denmark.

Samman, M. M., Biswas, M. (1994a). Vibration testing for nondestructive evaluation of bridges. I: theory, *Journal of Structural Engineering,* 120(1): 269-289.

Samman, M. M., Biswas, M. (1994b). Vibration testing for nondestructive evaluation of bridges. II: results, *Journal of Structural Engineering,* 120(1): 290-306.

Samman, M. M., Biswas, M., Pandey, A. K. (1991). Employing pattern recognition for detecting cracks in a bridge model, *The International Journal of Analytical and Experimental Modal Analysis,* 6(1): 35-44.

Sampaio, R. P., Maia, N. M., Silva, J.M. (1999). Damage detection using the frequency-response-function curvature method, *Journal of Sound and Vibration,* 226(5): 1029-1042.

Santos, F., Vaz, M., Monteiro, J. (2004). A new set-up for pulsed digital shearography applied to defect detection in composite structures, *Optics and Lasers in Engineering,* 42(2): 131-140.

Saravanos, D. A., Hopkins, D. A. (1996). Effects of delaminations on the damped dynamic characteristic of composite laminates: analysis and experiments, Journal *of Sound and Vibration,* 195(5): 977-993.

Schroeder, F. D. Crostack, H. A. (1996). Real-time holography of ultrasonic surface waves, in Proceedings of SPIE: 290-295.

Sirohi, R. S. (1993). Speckle Metrology, Marcel Dekker.

Sirohi, R., Tay, C., Shang, H., Boo, W. (1999). Nondestructive assessment of thinning of plates using digital shearography, *Optical Engineering,* 38(9): 1582-1585.

Stubbs, N., Kim, J. (1995). Field verification of a non destructive damage localization and severity estimation algorithm, in Proceedings 13th International Modal Analysis Conference, Nashville, U.S.A.

Timoshenko, S., Woinowsky-Krieger, S., (1959). Theory of plates and shells, 2d ed. Engineering societies monographs, New York,: McGraw-Hill.

Tsai, W. H., Yang, J. C. (1988). Nondestructive evaluation of composite structures using system identification technique, *Journal of Engineering Materials and Technology,* 110(2): 134–139.

Vandiver, J. K. (1975). Detection of structural failure on fixed platforms by measurements of dynamic response, in Proceedings of the 6th Annual Offshore Technology Conference: 243-252.

Wolff, T., Richarson, M. (1989). Fault detection in structures from changes in their modal paramenters, in Proceedings of the 7th International Modal Analysis Conference: 87-94.

Yan, L. H., Leung, T. P., Xue, K. Z. (1996). Theoretical and experimental study of modal strain energy, *Journal of Sound and Vibration,* 191: 251-260.

Yao, G. C., Chang, K. C., Lee, G. C. (1992). Damage diagnosis of steel frames using vibrational signature analysis, *Journal of Engineering Mechanics,* 118(9): 1949-61.

Yoon, M. K., et al. (2001). Local damage detection using a global fitting method on mode shape data, in A Conference on Structural Dynamics: 231-237.

Yoon, M. K., et al. (2005). Local damage detection using the two-dimensional gapped smoothing method, *Journal of Sound and Vibration,* 279(1-2): 119-139.

Yuen, M. M. (1985). A numerical study of eigen parameters of a damaged cantilever, *Journal of Sound and Vibration,* 103: 301-310.

Zhu, H. P., Xu, Y. L. (2005). Damage detection of mono-coupled periodic structures based on sensitivity analysis of modal parameters, *Journal of Sound and Vibration,* 285(1-2): 365–390.

Chapter 13

Semi-Active Suspension Control Considering Lateral Vehicle Dynamics Due to Road Input

Takama Suzuki and Masaki Takahashi

Additional information is available at the end of the chapter

1. Introduction

An automotive performance has improved from the demand of ride comfort and driving stability. Many research have proposed various control system design methods for active and semi-active suspension systems. To improve vehicle response by steering, a roll angle control [1], a distribution control of the suspension stiffness of front and rear [2], and pitch angle control [3][4] have been proposed. Furthermore, to improve vehicle response due to disturbance on the road, a tire vertical load control [4-6] caused by road disturbance was proposed. In recent year, it is reported that roll and lateral vehicle motions are affected by the road input. These motions are affected by not only road displacement but also the change of tire side force [8-11]. As shown in Figs. 1 and 2, the tire side force is caused by a toe change and a scuff due to the roll motion and tire side force. However, there are few research about a suspension control method which takes into consideration with a suspension characteristic that is change of tire side force caused by suspension stroke and tire side force.

In this research, we designed a suspension control system which reflects a suspension characteristic, the change of tire side force caused by suspension stroke and tire side force. A new semi-active suspension control method is proposed to reduce the vehicle vibration and vehicle lateral motion due to the road input.

To design such a system, the time delay of the road input from the front wheel to the rear wheel needed to be modeled. In the vehicle control, there are several control systems which take into consideration with this delay [12-16]. However, the purpose of these research is to find a way to reduce the vehicle vertical vibration. Oraby evaluated motion of the vehicle lateral direction [7]. However, the dynamics of the lateral direction do not be considered in the control system design.

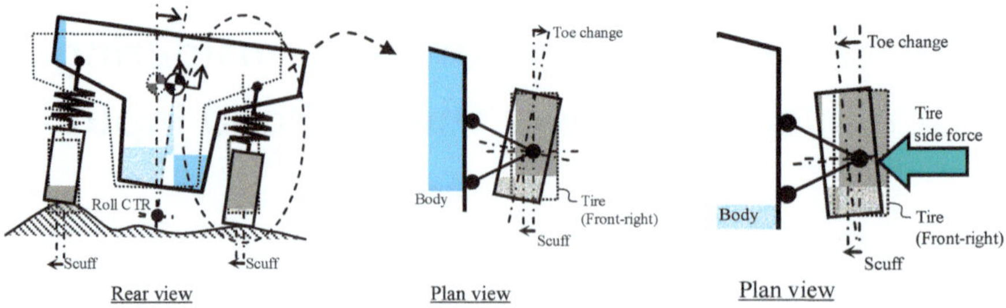

(a) Roll motion. (b) Tire side force

Figure 1. Toe change and scuff caused by roll motion and tire side force.

Symbol	Value		Symbol	Value	
M_b	1900	kg	K_t	260×10^3	N/m
M_t	50	kg	l_f	1.34	m
I_r	600	kgm^2	l_r	1.46	m
I_p	3000	kgm^2	T_f	1.50	m
I_y	3200	kgm^2	T_r	1.50	m
K_f	33×10^3	N/m	h_r	0.45	m
K_r	31×10^3	N/m	h_p	0.53	m
C_f	Discuss in section 3.3		h_{rf}	0.62	m
C_r	Discuss in section 3.3		h_{rr}	0.11	m

Table 1. Specification of vehicle model

In order to design the controller, a vehicle model including the tire side force change caused by road disturbance is constructed. Moreover, the road input from the front wheel to the rear wheel time delay is modelled with Pade approximation. These suspension characteristics and the time delay are modelled with liner model. The disturbance accommodating H_∞ control which considers the vehicle model including the suspension characteristic is proposed. In the control, we set the lateral acceleration of the vehicle to one of controlled outputs. New semi-active suspension control method is proposed to reduce the vehicle vibration and vehicle lateral motion due to the road input. To verify the feasibility of the proposed method, several numerical simulations are carried out.

2. Modeling

2.1. Modeling of the vehicle

Figure 2 shows a full vehicle model which is equipped with semi-active suspension between each wheel and the vehicle body. The weight of the vehicle body is supported by the spring. We assume that a vehicle model is a generic sedan car as shown in Table 1. The equations of motion which are, lateral, bounce, roll, pitch, yaw, and each unsprung motion are as follows:

$$
\begin{aligned}
&(M_b + 4M_t)V\left(\frac{d\beta}{dt} + \gamma\right) = F_{y1} + F_{y2} + F_{y3} + F_{y4} \\
&M_b \ddot{Z}_{cg} = -K_f z_{s1} - C_f \dot{z}_{s1} + F_1 - K_f z_{s2} - C_f \dot{z}_{s2} + F_2 \\
&\qquad\qquad - K_r z_{s3} - C_r \dot{z}_{s3} + F_3 - K_r z_{s4} - C_4 \dot{z}_{s4} + F_4 \\
&I_r \ddot{\phi} = {}^{T_f}\!/_2 \left[-K_f (z_{s1} - z_{s2}) - C_f (\dot{z}_{s1} - \dot{z}_{s2}) + (F_1 - F_2) \right] \\
&\quad + {}^{T_r}\!/_2 \left[-K_r (z_{s3} - z_{s4}) - C_r (\dot{z}_{s3} - \dot{z}_{s4}) + (F_3 - F_4) \right] \\
&\quad + G_m (F_{y1} + F_{y2})(h_{cg} - h_{rf}) + G_m (F_{y3} + F_{y4})(h_{cg} - h_{rr}) + M_b g H_r \phi \\
&I_p \ddot{\theta} = -L_f(-K_f z_{s1} - C_f \dot{z}_{s1} + F_1 - K_f z_{s2} - C_f \dot{z}_{s2} + F_2) \\
&\quad + L_r(-K_r z_{s3} - C_r \dot{z}_{s3} + F_3 - K_r z_{s4} - C_4 \dot{z}_{s4} + F_4) + M_b g H_p \theta \\
&I_y \frac{d\gamma}{dt} = (F_{y1} + F_{y2}) l_f - (F_{y3} + F_{y4}) l_r \\
&M_{ti} \ddot{z}_{ui} = K_t z_{ti} - F_i - K_f z_{si} - C_i(t) \dot{z}_{si} \quad (i = 1,2) \\
&M_{ti} \ddot{z}_{ui} = K_t z_{ti} - F_i - K_r z_{si} - C_i(t) \dot{z}_{si} \quad (i = 3,4)
\end{aligned}
\tag{1}
$$

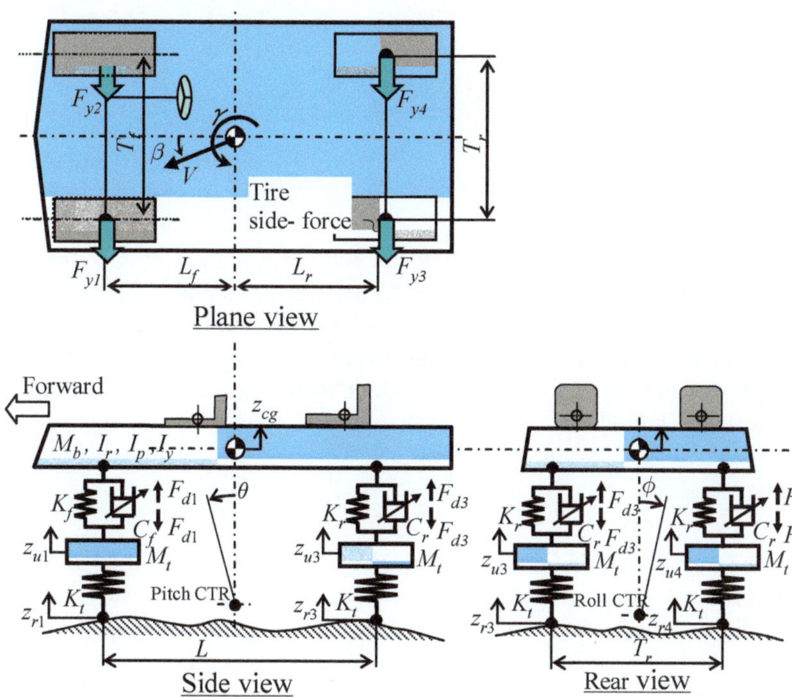

Figure 2. Vehicle model

where H_r is the distance from a roll center to the CoG of the vehicle body, and H_p is the distance from a pitch center to the CoG of the vehicle body. These parameters are constant.

The spring coefficients of each wheel are different from each other, and were set to $K_{1,2} = K_f$, $K_{3,4} = K_r$, F_{yi} means a tire side force, z_{si} means a suspension stroke of each wheel, z_{ti} means deformation of the each tire.

$$\begin{aligned} z_{s1} &= z_{cg} + T_f/2\phi - L_f\theta - z_{u1} \\ z_{s2} &= z_{cg} - T_f/2\phi - L_f\theta - z_{u2} \\ z_{s3} &= z_{cg} + T_r/2\phi + L_r\theta - z_{u3} \\ z_{s4} &= z_{cg} - T_r/2\phi + L_r\theta - z_{u4} \\ z_{ti} &= z_{ri} - z_{ui} (i = 1...4) \end{aligned} \tag{2}$$

As shown in Fig.2, tire side force is caused by the toe change and the scuff. The toe change and the scuff are caused by the suspension stroke and the tire side force. The tire side force of each tire, F_{yi}, is as follows:

$$\begin{aligned} F_{y1} &= K'_{Cf}\left(-\beta - \frac{L_f}{V}\gamma - K_{sf}z_{s1} + \frac{2h_{rf}}{T_f V}\dot{z}_{s1}\right) \\ F_{y2} &= K'_{Cf}\left(-\beta - \frac{L_f}{V}\gamma + K_{sf}z_{s2} - \frac{2h_{rf}}{T_f V}\dot{z}_{s2}\right) \\ F_{y3} &= K'_{Cr}\left(-\beta + \frac{L_r}{V}\gamma - K_{sr}z_{s3} + \frac{2h_{rr}}{T_r V}\dot{z}_{s3}\right) \\ F_{y4} &= K'_{Cr}\left(-\beta + \frac{L_r}{V}\gamma + K_{sr}z_{s4} - \frac{2h_{rr}}{T_r V}\dot{z}_{s4}\right) \end{aligned} \tag{3}$$

where, a value in the parenthesis in Eq.(3) means the tire slip angle of each tire. K_{sf} and K_{sr} are the coefficient of toe angle change caused by suspension stroke. The third term in the parenthesis is the toe angle change caused by the suspension stroke of each wheel, as shown in Table 2 (A). The fourth term in the parenthesis is the tire slip angle caused by the scuff per unit time as shown in Table 2 (B). The equivalent cornering stiffness which includes the compliance steer is as follows:

$$K'_{Ci} = \frac{K_{Ci}}{1 - K_{Ci}K_{SFi}} \quad (i = 1,...,4) \tag{4}$$

where, K_{Ci} is the cornering stiffness of each tire. K_{SFi} is the coefficient of the toe angle change caused by the tire side force, as shown is Table 2 (C). Finally, the characteristic shown in Table 2 (D) is derived from a first order delay of the equivalent cornering stiffness (Eq. (4)).

Input \ Output	Toe-angle change	Scuff
Suspension stroke	Modeling (A)	Modeling (B)
Tire side-force	Modeling (C)	Modeling (D)

Table 2. Modeling of suspension characteristics

Symbol	Value		Symbol	Value	
K_{sf}	0.13	rad/m	K_{SFf}	8.0×10^{-6}	rad/N
K_{sr}	0.13	rad/m	K_{SFr}	1.0×10^{-6}	rad/N
K_{Cf}	77000	N/rad	K_{PFf}	6.7×10^{-6}	m/N
K_{Cr}	71000	N/rad	K_{PFr}	6.7×10^{-6}	m/N

Table 3. Specification of suspension characteristics

$$G_{di}(s) = \frac{1}{T_i s + 1}$$
$$T_i = \frac{K_{Ci} K_{PFi}}{V\left(1 - K_{Ci} K_{SFi}\right)} \quad (i = 1,\ldots,4) \tag{5}$$

where, K_{PFi} is the coefficient of the scuff caused by the tire side force. The characteristics of the tire side change is modelled from Equations (3) to (5). As shown in Table 3, each parameter is determined that stability factor was set to the usual value, 1.43×10^{-3} s^2/m^2 based on the reference [18].

2.2. Vehicle response caused by antiphase road disturbance

The state equation of the vehicle model which reflects change the tire side force is defined as Eq. (6). This model is an LTI model when the vehicle velocity is fixed.

$$\begin{aligned}
&\dot{x} = A_P x(t) + B_{P1} w(t) + B_u u(t) \\
&x(t) = [z_{u1}\ z_{u2}\ z_{u3}\ z_{u4}\ z_{cg}\ \phi\ \theta\ \dot{z}_{u1}\ \dot{z}_{u2}\ \dot{z}_{u3}\ \dot{z}_{u4}\ \dot{z}_{cg}\ \dot{\phi}\ \dot{\theta}\ \beta\ \gamma\ F_{y1}\ F_{y2}\ F_{y3}\ F_{y4}]^T \in \mathbb{R}^{20} \\
&w(t) = [z_{r1}\ z_{r2}\ z_{r3}\ z_{r4}]^T \\
&u(t) = [F_1\ F_2\ F_3\ F_4]^T
\end{aligned} \tag{6}$$

We confirmed that the vehicle responsed to an antiphase road disturbance. In this research, the phases of road input of left-right are zero and π. Zero means a coordinate phase, and π means the antiphase. In the antiphase road, the road displacement at the vehicle velocity, V, is as follows.

$$z_{r2} = -z_{r1},\ z_{r3} = e^{-\frac{L}{V}s} z_{r1},\ z_{r4} = -e^{-\frac{L}{V}s} z_{r1} \tag{7}$$

The transfer function of the vehicle responce caused by the antiphase road disuturbance at 33.3 m/s (120 km/h) is shwon in Fig. 3. Here, Figure 3 (a) is the date of previous study. Figure 3 (b) is the data of our study. In these figures, specification of the vehicle model of each study is different. However, from the result showing the same tendency for vehicle responses, we found that the our model is appropriate.

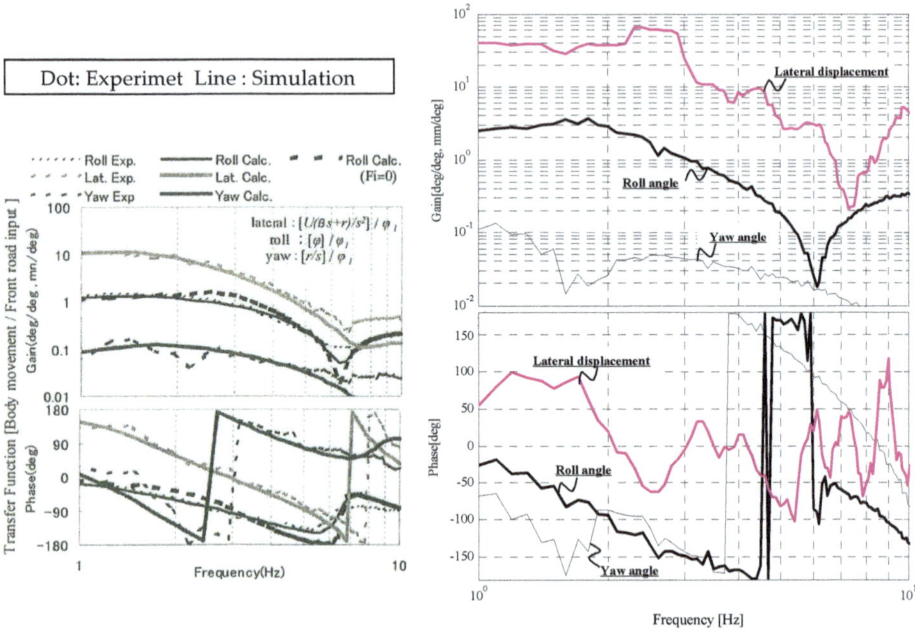

(a) Previous study (Koumura, 2008) (b) Our study (Simulation)

Figure 3. Vehicle responce coused by antiphase road disuturbance

2.3. Design of damper coefficient

In general, the performance of the controller is affected by the damping coefficients of the suspension. In this section, we design the damping coefficient for the evaluation values which are the vertical acceleration at the coordinate phase road and the lateral acceleration at the antiphase road. In the next section, we design the control system which uses the damping coefficients designed in this section. The road condition is shown in Figs. 4 and 5. The road displacement is assumed such that the Power Spectral Density (PSD) characteristic of the road surface is C class defined by ISO [19]. The lateral accelerations are shown in Fig. 8 and the lateral acceleration of vehicle body are derived from a geometric relation as follows.

$$\ddot{y}_{cg} = \ddot{y}_{RC} - H_r \ddot{\phi} \tag{8}$$

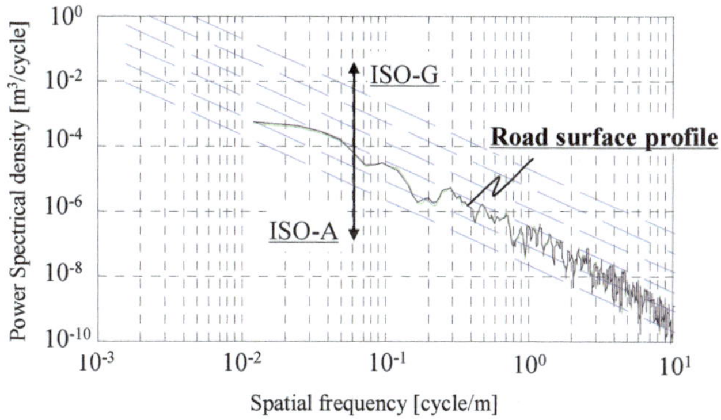

Figure 4. Power spectral density of road surface profile[19]

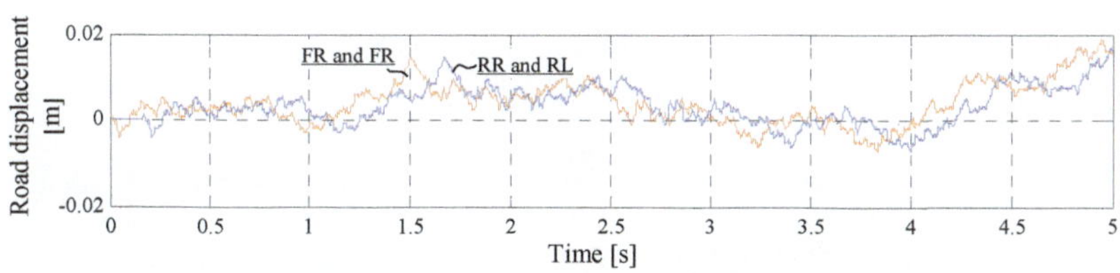

(a) Coordinate phase

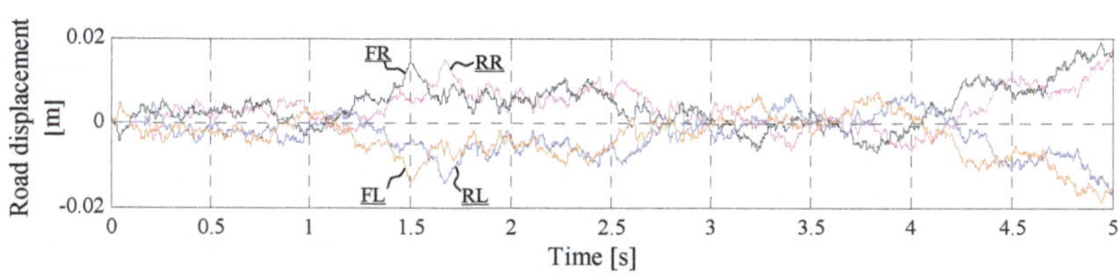

(b) Antipahse

Figure 5. Road displacement

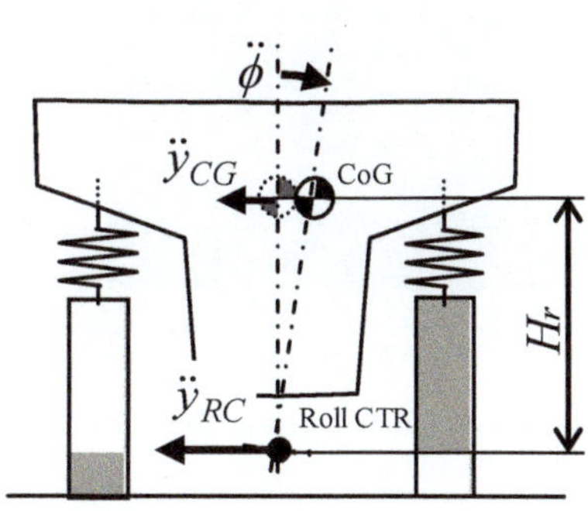

Figure 6. Lateral acceleration (Rear view)

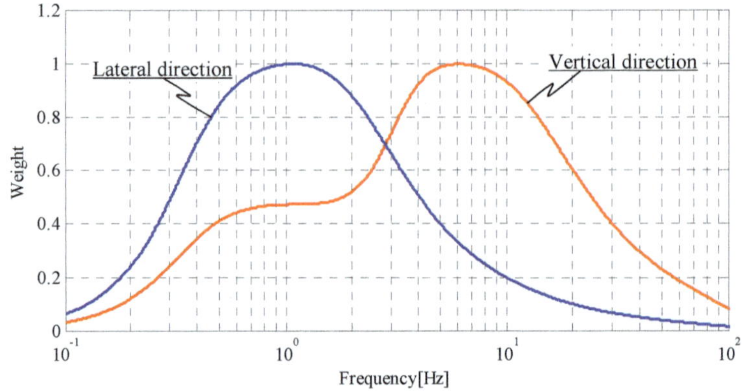

Figure 7. Sensitivity curves of lateral and vertical acceleration [20]

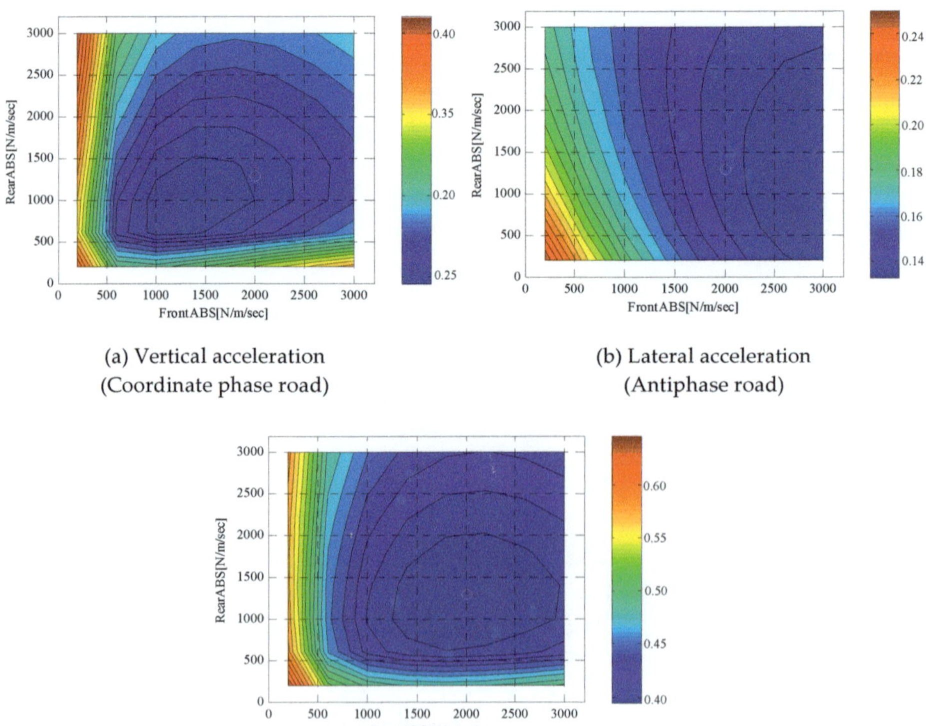

(a) Vertical acceleration (Coordinate phase road)

(b) Lateral acceleration (Antiphase road)

(c) Sum of Vertical and lateral acceleration ((a) + (b))

Figure 8. Contour diagram of RMS value of accelerations (unit : m/s², O : minimum point of (c))

The sensitivity curves of the lateral and the vertical acceleration are defined by ISO [20]. We designed the damping coefficient for the suspension to minimize the Root Mean Square (RMS) which considers the sensitivity curves. A frequency response of the filter [21] which modeled sensitivity curve with the transfer function is shown in Fig. 7. We calculated the RMS values using the time history of the vertical and lateral accelerations which pass this filter. The simulation results when both front and rear damping coefficients are changed are shown in Fig. 8, which is a contour diagram of the RMS values. There are minimum point in Fig. 8 (a), (b) and (c). The evaluation value is the sum of vertical and lateral acceleration of RMS shown in Fig. 8. (c). From the simulation results, the damping coefficients of the suspension are 2000 and 1300 N/m/s.

3. Controller design

We used the controller designed based on the linear H_∞ control theory to a semi-active suspension. The control system of the semi-active suspension is shown in Fig. 9. An actuator force command F_{li} is calculated with the linear H_∞ controller. The damping force command F_{di} is translated with a liner-bilinear converter.

3.1. Approximation of time delay of road disturbances

When the front and rear tire passes the same path, road disturbance affects the rear tire has the delay to the front tire. As shown in Fig. 10, the road input from the front wheel to the rear wheel is the delayed. The delay is modeled with a third-ordered Pade approximation. As shown in Fig. 10 (b), we consider the effect of vehicle velocity in the control system design by using a linear model. To check whether the Pade approximation is correct, we carried out the numerical simulations. The response of the vehicle model which includes the Pade approximation when the vehicle runs on the antiphase road disturbance is shown in Fig. 11. The vehicle velocity is 16.7 m/s (60km/h). In the lower frequency of the resonance frequency of the tire, we found that the result of the simulation model which uses the time delay and Pade approximation are almost the same.

3.2. Disturbance-accommodating control

We found that feedforward control of disturbance information in the finite frequency range and feedback control improve performance [22]. The power spectral density of the actual velocity of disturbances had flat characteristics in a low frequency, and decreased according to frequency at a region of high frequency. We assumed that it regarded as the colored noise formed by shaping filter which has a transfer function with low-pass characteristics. This filter of the each wheel is based on the road condition which defined by ISO [19]. The filter is as follows:

$$
\begin{aligned}
&Q_w \begin{cases} \dot{x}_{wi}(t) = A_{wi}x_{wi}(t) + B_{wi}w_{gi}(t) \\ w_{wi}(t) = c_{wi}x_{wi}(t) \quad (i = 1,\ldots,4) \end{cases} \\
&\frac{w_{wi}(s)}{w_{gi}(s)} = \frac{\varpi_d^{\,2}}{s^2 + 2\xi_d\varpi_d s + \varpi_d^{\,2}} \quad (i = 1,\ldots,4)
\end{aligned}
\tag{9}
$$

where, w_{gi} is road input of the each wheel, w_i is road input of the vehicle model of the general model as shown in Fig. 7. It was referred to as $\varpi_d = 50 \times 2\pi$ and $\xi_d = 0.706$.

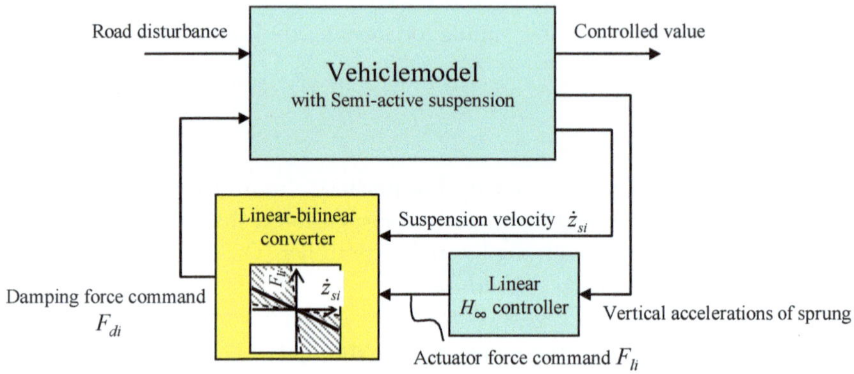

Figure 9. Control system of semi-active suspension

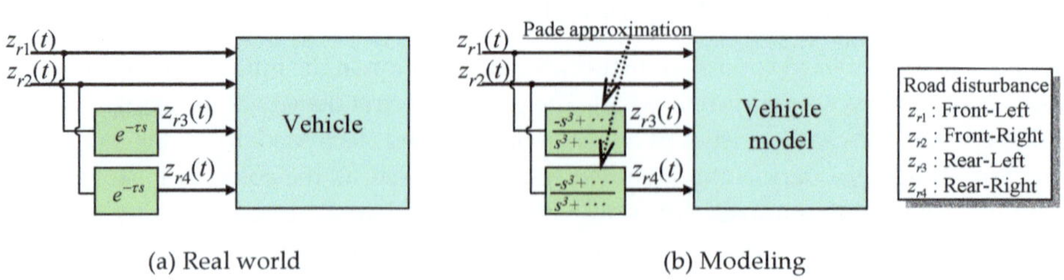

Figure 10. Approximation of time delay of road disturbances from front to rear wheels.

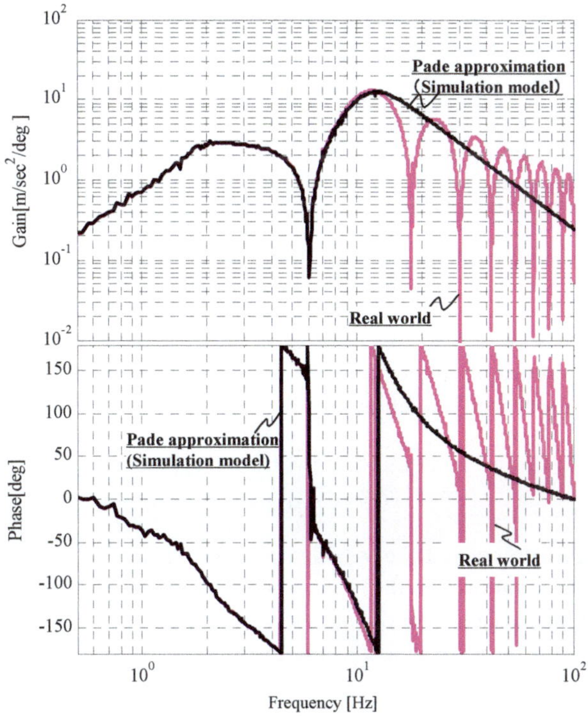

Figure 11. Vehicle response of vehicle model which includes the Pade approximation caused by antiphase road disturbance

3.3. Disturbance-accommodating H_∞ control

The feedforward control of disturbances resulted in worse accuracy outside the assumed frequency [22]. Furthermore, because each resonance frequency of the vehicles and tire differs, the control system design considering each resonance frequency is needed. Therefore, the control system was designed by using the H_∞ method in the control theory.

We integrated each state variable of the road disturbance model and frequency weights for controlled values. The frequency weights are as follows:

$$
\begin{aligned}
Q_{zc} &\begin{cases} \dot{x}_{zc}(t) = A_{zc}x_{zc}(t) + B_{zc}x(t) \\ z_{zc}(t) = c_{zc}x_{zc}(t) \ + D_{zc}x(t) \end{cases} \\
Q_{zi} &\begin{cases} \dot{x}_{zi}(t) = A_{zi}x_{zi}(t) + B_{zi}u(t) \\ z_{zi}(t) = c_{zi}x_{zi}(t) \ + D_{zi}u(t) \end{cases}
\end{aligned}
\tag{10}
$$

where, z_{zc} is controlled value of the vehicle motion, z_{zi} is controlled value of the control input. Figure 7 shows a block diagram of the generalized plant to design the controller, and the state-space form of the generalized plant is as follows:

$$\begin{cases} \dot{x}_g(t) = A_g x_g(t) + B_{g1} w_g(t) + B_{g2} u(t) \\ z_g(t) = C_{g1} x_g(t) + D_{g11} w_g(t) \ \ + D_{g12} u(t) \\ y_g(t) = C_{g2} x_g(t) \ + D_{g22} u(t) \end{cases} \tag{11}$$

H_∞ norm of the transfer function $G_{z_g w_g}$ from disturbance $w_g(t)$ to controlled value $z(t)$ is expressed by the following equation.

$$\underset{u}{miu} \left\| G_{Z_g W_g} \right\|_\infty = \sup_{w} \frac{\left\| z_g \right\|_2}{\left\| w_g \right\|_2} =: \gamma^* \tag{12}$$

where, γ^* is a minimum of H_∞ norm of the generalized plant realized with H_∞ controller. The controller is the following equation [23].

$$\begin{cases} \dot{x}_k(t) = A_k x_k(t) + B_k y_g(t) \\ u(t) = C_k x_k(t) \end{cases} \tag{13}$$

In this reserch, we assumed that the vehicle speed is constant. The measured outputs, $y(t)$, are four vertical accelerations of the wheel position of the vehicle body. The controlled values, $z(t)$, are vertical acceleration, lateral acceleration, vertical velocity of the vhicle body, tire deformation, and actuating force. Frequency weight W_i, shown in Fig. 13, was determined by trial and error.

A bandpass filter, W_1, that had a peak frequency equal to the resonance frequency of the passenger's head was used based on sensitivity curves [20], such as that being standardized by ISO and shown in Fig. 4. The low pass filter, W_2, was used based on sensitivity curves of lateral direction. In order to prevent the increase of response in each resonance, a low pass filter W_3 and a bandpass filter W_4 are used. Moreover, to prevent steady control input, a high pass filter, W_5, was used.

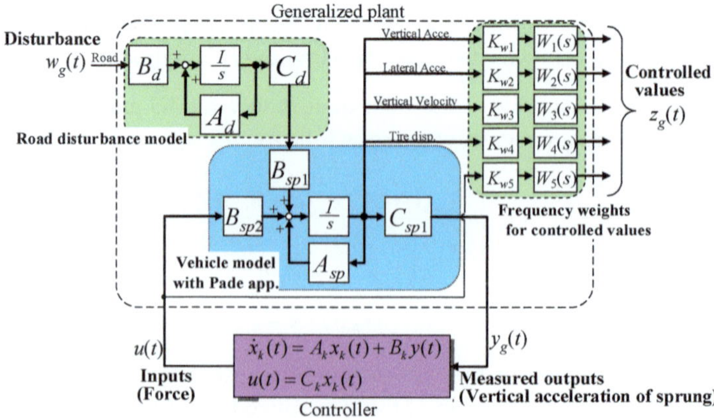

Figure 12. Generalized plant

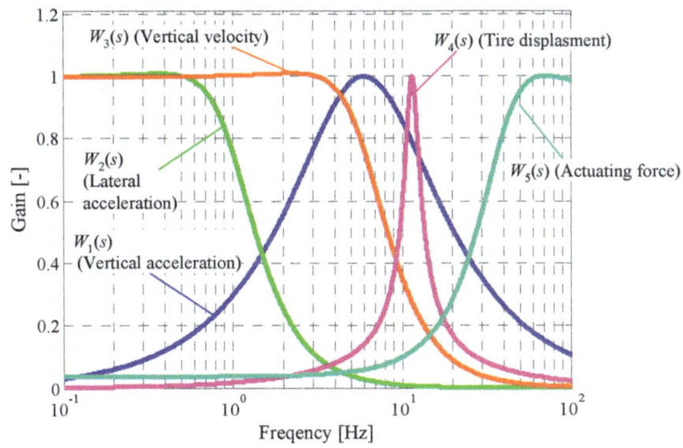

Figure 13. Frequency weight

3.4. Comparison of the response of the frequency weight K_{w2}

To check the effect of the suspension model which includes the tire side force caused by the road disturbance, we compared two control systems, one which considers the tire side force and the control system, and one which does not consider one. The design of two general control methods changed the controlled value of the lateral acceleration of the vehicle body. Frequency weights, W_1, W_2, W_3, W_4, $K_{w1} = 500$, $K_{w3} = 500$, $K_{w4} = 2000$, and $K_{w5} = 20$, are the same value in the proposed method and general methods. We combined the control system and vehicle model which includes the tire side force shown in Chapter 2. We used MATLAB (The Math Work Inc.) to calculate the Runge-Kutta method for the differential equations. The computational time step is 1 ms. Vehicle velocity is 16.7 m/s (60 km/h). The road surface conditions are shown in Fig. 5(b). We calculated the RMS values from the time history of lateral acceleration which passes the filter shown in Fig.7.

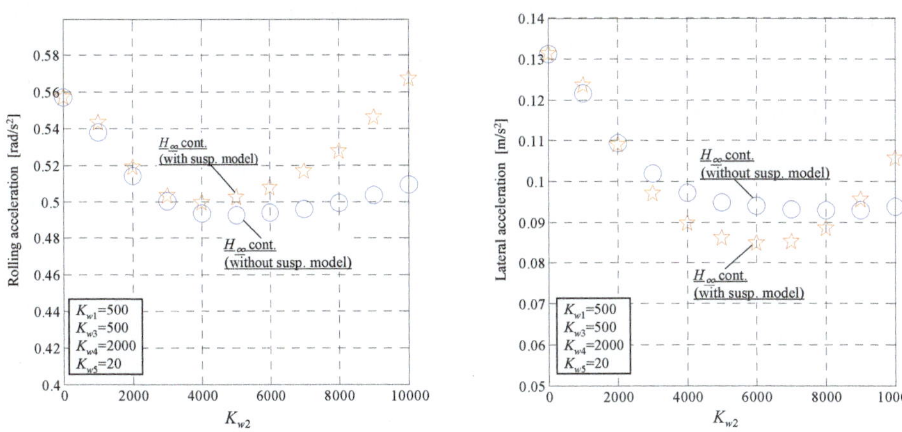

(a) Rolling acceleration (b)Lateral acceleration

Figure 14. RMS value changed by evaluation function of accelerations (Antiphase road)

The RMS value changed by the frequency weight, K_{w2} is shown in Fig. 14. For a frequency weight, K_{w2}, is 1 to 4000, when K_{w2} increases, the RMS value of rolling and lateral accelerations decreased. Furthermore, when K_{w2} increases, RMS of lateral acceleration decreased. From the result of Fig. 14.(b), it was confirmed that the proposed method can reduce the lateral acceleration due to the road disturbance in comparison with the general method. We compare the control performance of two controllers which are designed using the frequency weight shown in Fig. 14.(b). In two controllers, K_{w2} is set the smallest RMS value in 14.(b).

4. Simulation

4.1. Simulation condition

To clarify the effect of the suspension model which includes the tire side force, we compare the two control systems. The two control systems are designed in Cheater 3. As for the control system which inflects the tire side force (H_∞ cont. (with susp. model)), the frequency weight of lateral acceleration, K_{w2}, is 4000. The control system which does not consider the tire side force (H_∞ cont. (without susp. model)), the frequency weight of lateral acceleration, K_{w2}, is 6000.

In the simulation, we combine the control system and the vehicle model which includes the tire side force shown in Chapter 2. The vehicle velocity is 16.7 m/s (60km/h). Moreover, we assume that the range of the damping coefficient of the semi-active suspension is 100-10000 N/m/s in Fig. 15. The semi-active damper has the first order delay. The cut off frequency is 10 Hz. We used MATLAB (The Math Work Inc.) to calculate the Runge-Kutta method for the differential equations. There were two road conditions. One is the coordinate phase road in Fig. 5 (a), and the other is the antiphase road in Fig. 5 (b). To demonstrate the control effect, we show the responses which does not control he semi-active damper (Non control).

4.2. Simulation results

In check whether the proposed method reduced vertical and lateral motion, we did the numerical simulations. The time history of the vertical acceleration caused by the coordinate phase and the lateral acceleration caused by the antiphase road is shown in Fig. 14. The Lissajous figure for the suspension velocity and the damping force of the front suspension are shown in Fig. 15. The PSD of the vertical acceleration and the lateral acceleration is shown in Fig. 16.

In the coordinate phase road, there are few difference in the time history and the Lissajous figure. As for the PSD, two control systems can reduce the vibration better than the non control one near the resonance frequencies of the vertical direction of the vehicle body and vertical direction of the human head. On the other hand, in the antiphase road, there are differences in the time history and the Lissajous figure. The PSD of lateral acceleration, two control systems can reduce the vibration near the frequency range with a high level of lateral acceleration sensitivity. Additionally, the control system which includes suspension model of the tire side force change caused by road disturbance can reduce the vibration better than the control system which does not include the tire side force change in the same frequency range. The simulation results confirmed that using the proposed control system reduces the vertical and the lateral motion of the vehicle caused by the disturbance in the road.

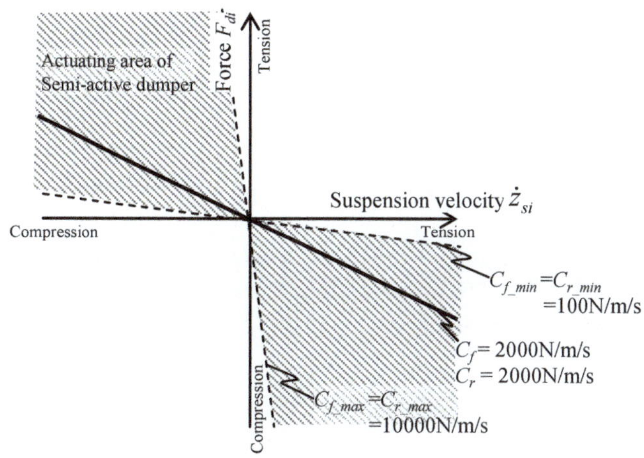

Figure 15. Semi-active dumper model

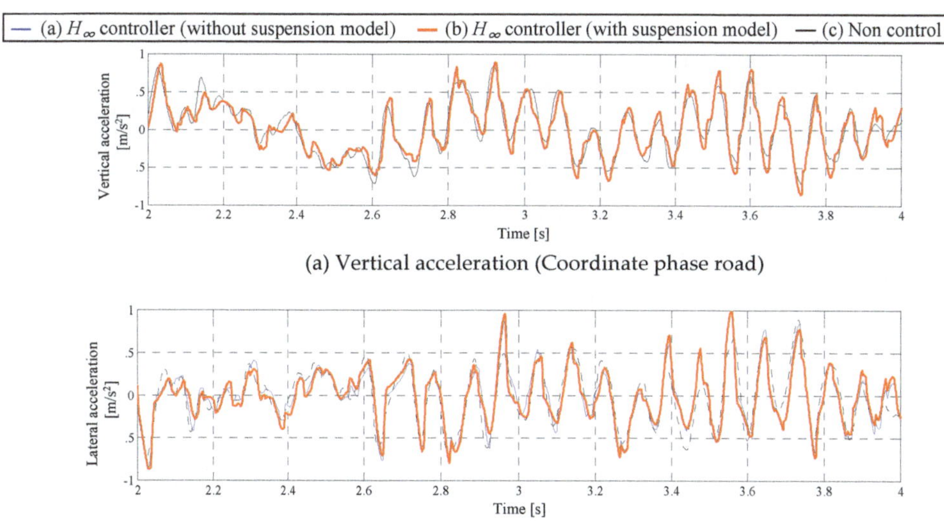

(b) Latera acceleration (Antiphase road)

Figure 16. Time history

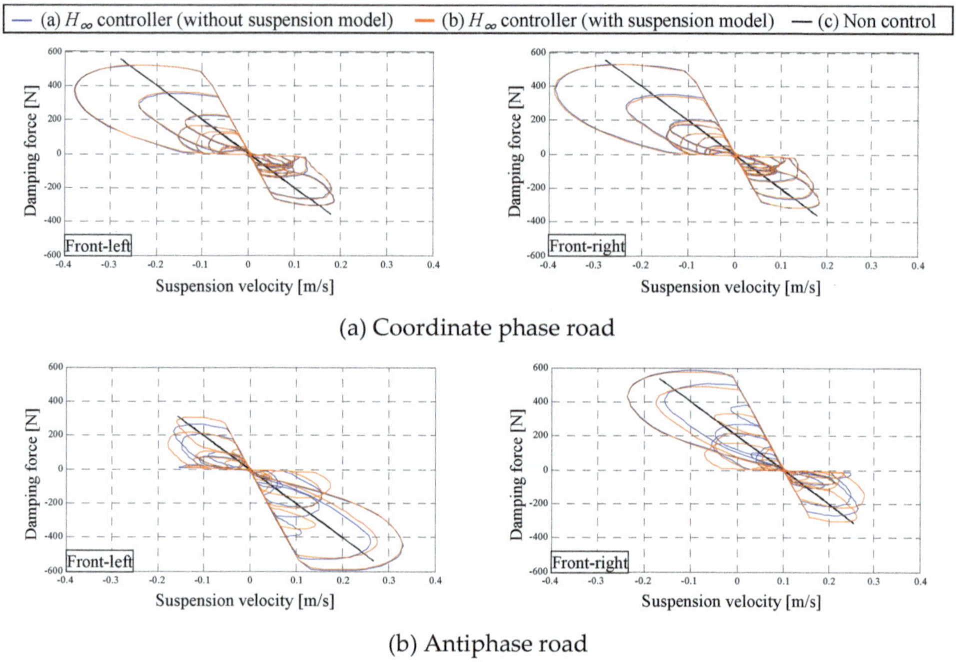

Figure 17. Lissajous figure (Suspension velocity and damping force)

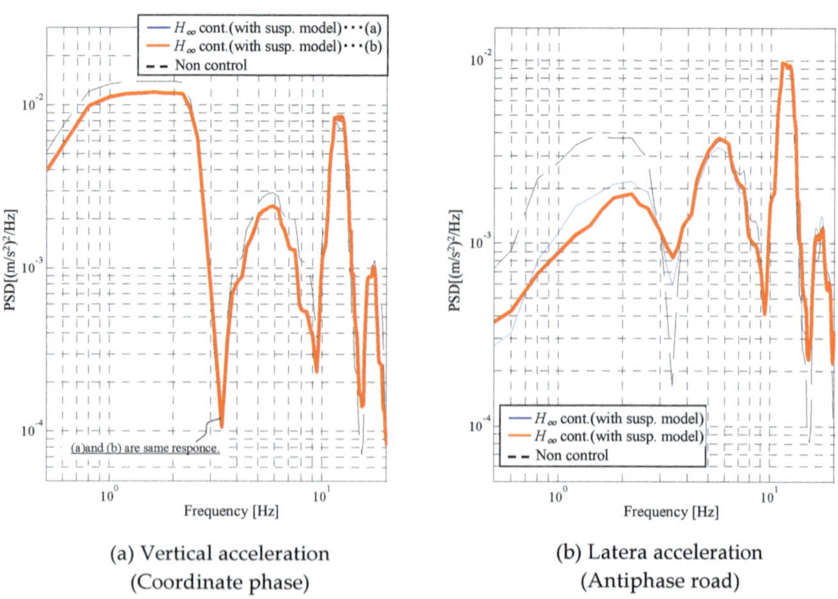

Figure 18. PSD values

5. Conclusion

This research proposed disturbance accommodating H_∞ control which considers the vehicle model in order to reduce the vehicle vibration and vehicle lateral motion due to the road input. We formulate the vehicle model including the suspension characteristic that is the toe change and the scuff caused by suspension stroke and tire side force. In order to verify the feasibility of the proposed method, the numerical simulations were carried out. From the result, it was confirmed that the proposed control system is effective for reducing not only the vertical vibration but also the lateral vehicle motion due to the road disturbance in comparison with the control method without considering the suspension characteristic.

Author details

Takama Suzuki and Masaki Takahashi
Keio University, Japan

6. References

[1] Tanaka T, Harada M, Takizawa S, Tatemoto M, Active Controlling System of Suspension, J Japan Society of Automotive Engineering, 1998, 42(1): 92-101, (in Japanese).

[2] Kawarasaki Y, Fukunaga Y, Hasegaw H, Okuyama Y, Kurozu K, Iijima T, Development of a Hydraulic Active Suspension by Nissan, Proc Japan Society of Automotive Engineering Annual Congress, 1989, 892: 193-196, (in Japanese).

[3] Yonekawa T, Ohnuma T, Mori Y, Gotoh T, Buma S, Effect of Active Controlled Suspension System on Vehicle Dynamics, Trans Japan Society of Automotive Engineering, 1991, 22(3): 63-68, (in Japanese).

[4] Hanamura Y, Mori R, Araki Y, Harada H, Influence of Attitude Control by Active Controlled Suspension on Stability of Vehicle, Trans Japan Society of Mechanical Engineers, Series C, 1998, 64(617): 163-168, (in Japanese).

[5] Hanamura Y, Nakajo K, Araki Y, Oya M, Harada H, Control of Vehicle Maneuverability and Stability of Means of Attitude Control with Vertical Load Control, Trans Japan Society of Mechanical Engineers, Series C, 1998, 64(624): 242-247, (in Japanese).

[6] Hanamura Y, Fujita K, Araki Y, Oya M, Harada, H, Control of Vehicle Maneuverability of 4 Wheeled Vehicle by Active Suspension Control with Additional Vertical Load Control, Trans Japan Society of Mechanical Engineers, Series C, 1999, 65(629): 236-243 (in Japanese).

[7] Hamahira M, Hanamura Y, Araki Y, Oya M, Active Vehicle Suspension Control with Vertical Load Control by H_∞ Controller, Proc Dynamics and Design Conference 1999, M108: 41-46 (in Japanese).

[8] Ikuo K, Eiich Y, Shunichi D, An Analysis of Pitch and Bounce Mouton, Requiring High Performance of Ride Comfort, Vehicle System Dynamics, 2004, 41: S83-92.

[9] Koumura S, Analysis of Roll Lateral Vehicle Behavior by Road Input, Proc Japan Society of Automotive Engineering, Seminar of Vehicle dynamics, 2008, 08(46): 31-48, (in Japanese).
[10] Koumura S, Ohkita T, Analysis of Roll Lateral Vehicle Behavior by Road Input, Trans Japan Society of Automotive Engineering, 2008, 39(4): 23-29, (in Japanese).
[11] Koumura S, Ohkita T, Ride Comfort Evaluation through Analysis of Roll and Lateral Vehicle Behaviors Due to Rod Input, J Society of Automotive Engineering, Passenger cars – Mechanical system,2008, 2008-01-0581: 472-479.
[12] Antonio M, Nagai M, Optimal Preview Control of Rear Suspension Using Nonlinear Networks, Proc Int Symp Advanced Vehicle Control, 1992, 020: 117-122.
[13] Kimura T, Akatsu Y, Tabata H, Fumuyama K, Application of Preview Control to an Active Suspension System for Vehicle Vibration Control, Trans Japan Society of Automotive Engineering, 1994, 25(3): 93-97, (in Japanese).
[14] Araki Y, Harada H, Oya M, Preview Control of Active Suspension Using Disturbance Information of Front Wheel, Trans Japan Society of Mechanical Engineers, Series C, 1994, 60(578): 172-177 (in Japanese).
[15] Aida K, Naganawa A, Suda S, Shimomura H, Mizuguchi Y, Fundamental Study on H_∞ Preview Control in Active Suspensions for Railway Vehicles, Trans Japan Society of Mechanical Engineers, Series C, 1998, 64(621): 58-64, (in Japanese).
[16] Yu F, Zhang JW, Crolla DD, A study of a Kalman filter vehicle suspension system using correlation of front and rear wheel road input, Trans I MechE, Series D, 2000, 214: 493-502.
[17] Oraby WAH, Aly MA, Ei-Demerdash SM, Selim AM, Influence of Active Suspension Preview Control on the Vehicle Lateral Dynamics, Trans Society of Automotive Engineering[CD-ROM], 2007, No.2007-01-2347.
[18] The Japan society of Automotive engineering, Automotive technology series, Technology for improvement of vehicle dynamics, Asakura-Shoten, 1998, 43-44, (in Japanese).
[19] ISO-8608:1995, Mechanical vibration -Road surface profiles- Reporting of measured data, International Organization for Standardization, 1995.
[20] ISO-2631-1:1997, Mechanical vibration and shock - Evaluation of human exposure to whole-body vibration-, International Organization for Standardization, 1997.
[21] Rimell AN and Mansfield NJ, Design of Digital Filters for Frequency Weightings Required for Risk Assessments of Workers Exposed to Vibration, Proceedings of the Industrial Health 2007, 2007, 45: 512-519.
[22] Okamoto B, Yoshida K, Bilinear Disturbance-Accommodating Optimal Control of Semi-Active Suspension for Automobiles, Trans Japan Society of Mechanical Engineers, Series C, 2000, 66(650): 3297-3304 (in Japanese).
[23] Glover K, Doyle JC, State-space Formula for All Stabilizing Controllers that Satisfy an H_∞-norm Bound and Relations to Risk Sensitivity, J Systems & Control letters, 1988, 11: 167-172.

Chapter 14

Development of a Dimensionless Model for Predicting the Onset of Cavitation in Torque Converters

Darrell Robinette, Carl Anderson and Jason Blough

Additional information is available at the end of the chapter

1. Introduction

The presence of cavitation in any turbomachine can be detrimental to hardware durability, operating performance or noise and vibration characteristics. In an automatic transmission, the torque converter has the potential to cavitate under specific operating conditions resulting in a degradation in torque transfer, irregular engine speed regulation, blade damage or noise complaints depending on the degree of cavitation. Recent trends in planetary automatic transmissions are towards an increased number of fixed gear states and an expanding area of torque converter lockup clutch operation to reduce fuel consumption. The result is a reduction in the packaging volume available for the torque converter torus as indicated in Fig. 1 and the requirement to accommodate the increasing specific torque output of down sized, boosted engines. This combination of torus design and engine matching increases the opportunity for cavitation with higher toroidal flow velocities, higher pressure differentials across blade surfaces and greater power input to the fluid during normally inefficient operating points.

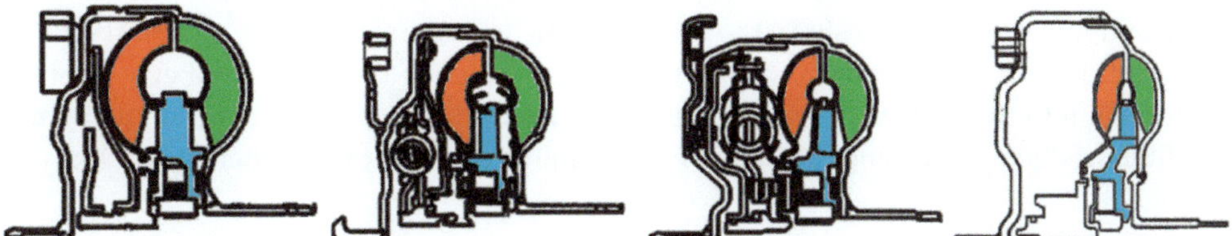

Figure 1. Recent design trend in automotive torque converter torus dimensions; pump (green), turbine (red) and stator (blue)

The onset of cavitation can be readily identified by a distinct change in radiated sound from the torque converter during particular driving maneuvers. Advanced stages of cavitation are distinguished by a loss of torque transfer through the torque converter and unexpected engine speed regulation due to the presence of large scale multi-phase flow structures. The focus of this chapter will be on incipient cavitation and detecting the onset of cavitation rather than on advanced stages of cavitation. The chapter will detail an experimental technique for acoustically detecting onset of cavitation in torque converters of varying designs and performance characteristics over a range of operating points. The effects of torque converter design parameters and operating conditions on cavitation are quantified through dimensional analysis of the test results. The objective is to develop a tool for designing converters with minimal cavitation by fitting power product and response surface models to the dimensionless data. Initially, torque converter designs following exact geometric similitude are analyzed. The constraint of similitude are relaxed in subsequent model iterations to allow application of the dimensional analysis and model fitting process to a broader range of converter designs. Empirical power product and response surface models were produced capable of predicting the onset of cavitation with less than 10% and 7% error, repectively, for a large population of torque converters.

2. The torque converter and cavitation

2.1. Torque converter design and performance

The torque converter is a particular class of turbomachine that utilizes a pressurized fluid circulating through multiple bladed elements that add or extract energy from the fluid. The torque converter is the powertrain component that couples the engine crankshaft to the gear system of the automatic transmission, allowing the transfer of power. It serves a number of other purposes, including decoupling the engine and transmission, allowing engine idle without the aid of a clutch, multiplying torque at low engine speeds to overcome poor self-starting capability of the internal combustion engine and regulation of engine speed during vehicle acceleration.

The torque converter pump, connected to the engine crankshaft, rotates at engine speed and torque, increases the angular momentum of the toriodal fluid as it flows from inner to outer torus radius. The toroidal fluid then flows into the turbine where its angular momentum is decreased and torque is transferred to the automatic transmission via the turbine shaft. The toriodal fluid exits the turbine and enters the stator which redirects the fluid at a favorable angle back into the pump. The redirection of the fluid increases the angular momentum of the fluid reducing the amount of increase required by the pump, thereby multiplying torque.

Dimensionless parameters are frequently used in the turbomachinery field to relate fundamental design characteristics to performance such as speed, torque, flow or head. The torque converter is no exception and uses a semi-dimensionless parameter known as K-factor,

$$K = \frac{N_p}{\sqrt{T_p}} \tag{1}$$

where N_p and T_p represent pump speed and pump torque, respectively, to characterize a specific design of the torus and bladed elements. A dimensionless form of K-factor can be formulated by including fluid density and torus diamter to form unit input speed,

$$U = K\sqrt{\rho D^5}\ . \tag{2}$$

Unit input speed is the inverse square of the torque coefficienct, which is more commonly found in turbomachinery. The value of K-factor or unit input speed are single value functions of the ratio of turbine (output) speed to pump (input) speed, referred to as speed ratio,

$$SR = \frac{N_t}{N_p}\ . \tag{3}$$

Torque ratio, the ratio of turbine to pump torque (see Eq. 4), quantifies torque multiplication of a design and is a single value of function of speed ratio.

$$TR = \frac{T_t}{T_p} \tag{4}$$

Efficiency of the torque converter is also used to quantify performance and is the product of speed ratio and torque ratio. Figure 2 summarizes the typical hydrodynamic performance attribtues of K-factor, TR and efficiency as a function of SR. At zero turbine speed (SR=0), referred to as stall, K-factor and TR are constant regardless of pump speed. As SR increases TR decreases monotonically until the coupling point is reached and is unity thereafter. K-factor in general will remain approximately constant until a threshold SR where it begins to increase gradually up to coupling and then asymptotically approaches infinite capacity as SR nears unity.

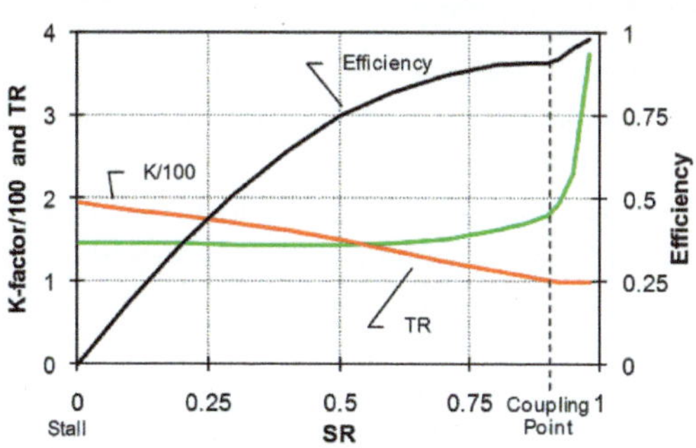

Figure 2. Typical dimensionless and semi-dimensionless performance characteristics of a hydrodynamic torque converter

2.2. Cavitation in torque converters

The change in angular momentum across the pump, turbine or stator produces a torque acting on individual blades that is equal to the product of the local static pressure and radius from converter centerline integrated over the entire surface area of the blade. A pressure differential exists between low and high pressure surfaces of a blade and in order for an increase in torque transfer to occur, the pressure differential across the blade surface in each element must increase. At a critical torque level, the localized pressure on the low pressure side of a blade will drop below the vapor pressure of the toroidal fluid, causing the nucleation of cavitation bubbles. Stall, SR=0, is most susceptible to the occurrence of cavitation due to the combination of high toroidal flow velocities, high incidence angle at element inlets and the thermal loading of the fluid. Cavitation may also occur while climbing a steep grade, repeated high throttle launches or during initial vehicle launch towing a heavy load. The discussion in this chapter will be focued on cavitation while operating at the stalled condition.

Incipient cavitation produces a negligible effect on overall torque converter performance due to a small volume of two phase flow in a localized region of the toroidal flow circuit. Continued operation at moderate to heavy cavitation will begin to cause a departure from the idealized K-factor relationship as an increased volume of toroidal flow is displaced by vapor. Previous testing by [1] found that heavy and sustained cavitation will cause an approximate 3% decrease in pump torque at stall. The onset of cavitation could more precisely be detected by a noted increase in the fluctuating component of pressure using in-situ pressure taps and a microwave telemetry technique, see [1]. The testing conducted at stall showed that a sudden increase in the ensemble averaged fluctuating pressure measurements signified the onset of cavitation for a particular charging pressure. Pressure measurements at the leading edge of the stator blade by [8] showed the fluctuating component of pressure to damp out as cavitation occurred as the cavitation structure formed at the leading edge of the stator blade. These results were confirmed by CFD work performed through [3], showing that the leading edge of the stator is indeed the initial nucleation site for cavitation at stall. As pump speed increased further, the size of the cavitation region grew at the leading edge of the stator blade.

The collapse of cavitation bubbles produces a broadband noise and may become objectionable to vehicle occupants. The radiated noise can also be used to identify the onset of cavitation as found by [4]. Using an acoustically treated test fixture, the onset of cavitation was readily found by an abrupt increase in filtered sound pressure level (SPL) above a critical pump speed. Utilizing a non-contact measurement technique facilitates testing of multiple torque converter designs, however, the size of the test fixture limited the testing to a single diameter torus and element designs spanning a narrow 23 K-factor point range. A dimensionless model for predicting pump speed at onset of cavitation was developed by [5] for this data, however, was not practical to predict cavitation for a diverse torque converter design population. The goal of this investigation was to utilize a similar nearfield acoustical measurement technique and test a general design population of torque converters at stall to determine the operating threshold at incipient cavitation. Through dimensional analysis, this data would be used to develop a comprehensive design tool capable of predicting the onset of cavitation with reasonable accuracy.

3. Experimental setup

3.1. Dynamometer test cell

A torque converter dynamometer test cell was constructed with an acoustically treated test fixture to acquire nearfield acoustical measurements to identify the onset of cavitation. The dynamometer setup consisted of four major subsystems as shown in Fig. 3; a hydraulic supply system, drive dynamometer, acoustical test fixture, and an absorbing dynamometer with an optional stall plate to prevent rotation. The hydraulic system has setpoint control of the charge pressure and back pressure imposed upon the torque converter as well as regulation of the temperature of the automatic transmission fluid at the inlet to the test fixture. A 215 kW direct current drive dynamometer was used to simulate typical engine speeds and torques. The 225 kW absorbing dynamometer was connected to the stall plate to prevent rotation and induce the stall operating condition. The acoustical test fixture was sized to accommodate at least 150 mm of acoustical foam to improve measurement coherence and allow torque converter designs of widely varying torus dimensions. Two data acquisition systems (not shown) were used to acquire data. One for operating data (speed, torque, pressure, temperature, flow, etc.) at a 1 kHz sample rate and one for nearfield microphone and pump speed data at a sample rate of 51.2 kHz.

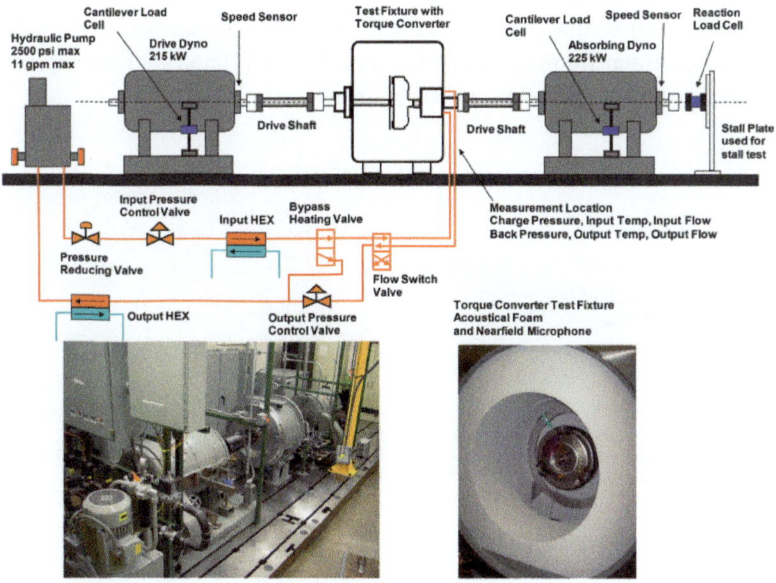

Figure 3. Torque converter dynamometer test cell and acoustically treated test fixture with nearfield microphone measurement

3.2. Stall cavitation test

A standardized test procedure for the stall operating condition was developed to transition each torque converter tested from a noncavitating to cavitating condition. Starting at 500 rpm, pump speed was sweep at 40 rpm/s until a speed sufficiently higher than onset of cavitation (~ 1200 to 2300 rpm). Fourteen stall speed sweep tests at various combinations of charge pressures ranging from 483 to 896 kPa, delta pressures between 69 to 345 kPa and input temperatures of 60, 70 and 80 °C were completed for each torque converter tested. The range of operating points tested is contained in Table 1.

	Low	High
Charge Pressure	483 kPa	896 kPa
Back Pressure	69 kPa	345 kPa
Input Temperature	60 °C	80 °C

Table 1. Range of operating point pressures and temperatures tested

3.3. Acoustical detection of incipient of cavitation

A sample nearfield acoustical measurement during a stall speed sweep test is shown in Fig. 4, in both the time domain (a) and frequency domain (b). The subtle change in SPL near 1400 rpm and frequency characteristic above 6 kHz correlates to the collapse of incipient cavitation bubbles. All torque converters tested exhibited similar behaviour when onset of cavitation occurs. Similar findings have been reported by [2] and [7] for incipient cavitation and acoustical detection techniques in rotating machinery.

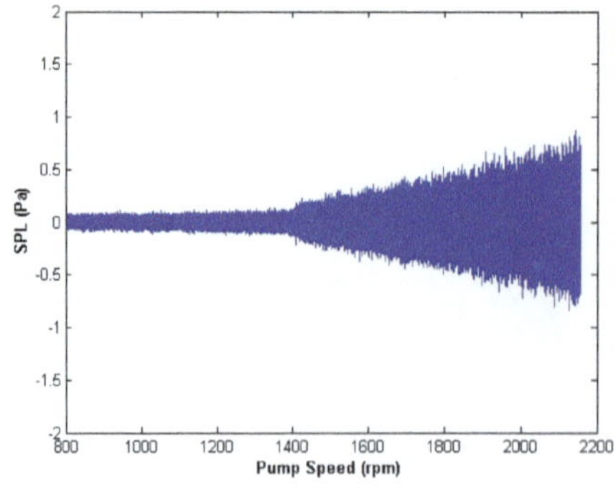

a) Time domain

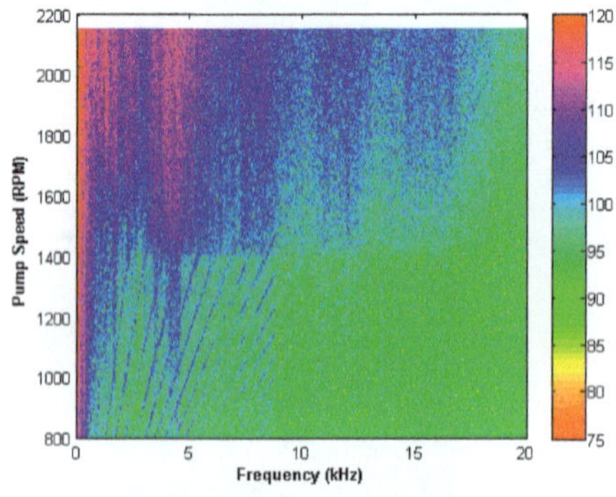

b) Frequency domain

Figure 4. Time (a) and frequency (b) domain nearfield acoustical measurement for standardized stall speed sweep cavitation test

To eliminate subjectivity in identifying the exact pump speed at onset of cavitation, a standardized signal processing technique was developed to reduce the nearfield acoustical measurements shown in Fig. 4 into a signal that more readily indicates the onset of cavitation. A 6 kHz Kaiser window high pass filter was first applied to the nearfield microphone data to remove noise unrelated to cavitation. The filtered SPL was then time averaged at 40 equally spaced points throughout the stall speed sweep test, resulting in Fig. 5a. Post processing the data in this manner showed an abrupt increase in filtered SPL to occur at a particular pump speed that is taken as the onset of cavitation for this investigation. A metric referred to as slope^2, see Fig. 5b, was computed and used in an automated algorithm to identify the exact pump speed at which cavitation occurred using a threshold criteria of a 12.5% increase. This value was found to give an accuracy of +/- 40 rpm with 95% confidence across the broad range of converter designs and operating points tested. The details regarding the experimental setup, post processing and identifying the exact pump speed at onset of cavitation can be found in [10].

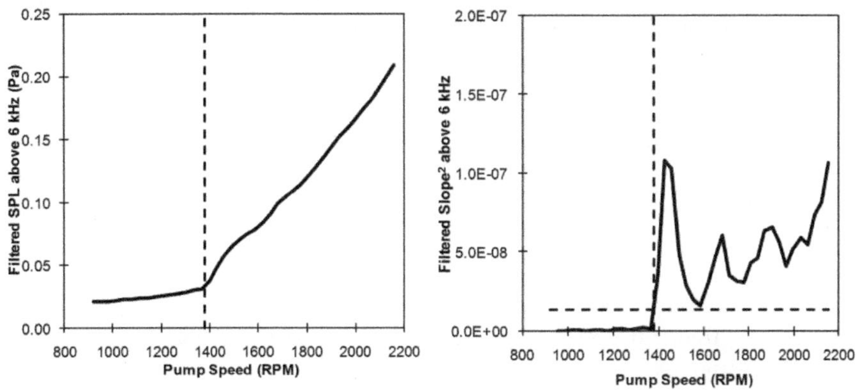

a) Filtered SPL   b) Filtered slope^2

Figure 5. Filtered SPL (a) and slope^2 (b) post processed from nearfield acoustical measurements to identify pump speed at onset of cavitation

3.4. Experimental torque converters

Fifty one torque converters were tested for this investigation of varying design characteristics that would satisfy the performance requirements of a wide range of powertrain applications. Although each torque converter differed geometrically in some respect, each design followed a common philosophy of torus and bladed element design to achieve certain performance attributes. The torque converter designs were arranged into four major populations, from exact geometric similitude to a general design population including all 51 torque converters with no similitude. Six diameters were tested in this investigation and will be denoted D_1 through D_6. D_1 corresponds to the smallest diameter at slightly smaller than 240 mm and D_6 the largest diameter at slightly larger than 300 mm. The details of the torque converters within each design population will be discussed later in the chapter as dimensionless quantities when the empirical models for predicting the onset of cavitation are presented. The exact designs characteristics will be omitted or presented as dimensionless quantities for proprietary reasons.

4. Dimensional analysis

4.1. Non-dimensionalizing onset of cavitation

Analysis of geometrically similar torque converter designs was performed to determine if the onset of cavitation could be non-dimensionalized using diameter and other relevant parameters. Such a dimensionless quantity would allow experimentally determined incipient cavitation thresholds for a given diameter, element design and operating point to be scaled to multiple diameters when exact dimensional scaling and operating point are preserved. The two torque converters considered are of exact geometric scaling with diameters D_2 and D_6, with the same stall unit input speed of 140 and stall torque ratio of 2.0. The designs only differ in pump blade count, which was necessary to maintain identical unit input speed.

Due to the relationship between speed and torque, either quantity could be used to describe the critical operating threshold at the onset of cavitation. Torque, however, is a more relevant quantity than speed as torque capacity is a more useful quantity when performing torque converter and powertrain matching and is fundamentally related to the pressure drop across the blades of the pump, turbine and stator. Stator torque at the onset of cavitation, $T_{s,i}$ was selected over pump or turbine torque for comparing critical cavitation stall operating thresholds as previous research by [3] and [8] showed cavitation to originate at the leading edge of the stator blade. Stator torque can generically be written as,

$$T_s \propto n_{sb}\left(A_{blade}\Delta p\right)r \tag{5}$$

where n_{sb} is the number of stator blades, A_{blade}, the surface area of an individual blade, Δp, the pressure differential between high and low pressure sides and r, the radius from centreline. In general, smaller diameter torque converters will begin cavitating at higher pump speeds, but lower elements torques relative to a larger geometrically scaled torque converter design. This can readily be seen in the plots of filtered SPL versus pump speed (a) and stator torque (b) in Fig. 6.

The critical stator torque at incipient cavitation will primarily depend on design parameters and fluid properties, most notably a characteristic size dimension and the fluid pressure imposed upon the toroidal flow. Figure 7 is a plot of $T_{s,i}$ versus the average of the controlled charge and back pressures for all fourteen operating conditions. It is clear from Fig.7 that diameter and average pressure significantly influence the onset of cavitation. The stator torque threshold at cavitation was found to be 95 to 118 Nm greater than the smaller diameter with 95% confidence, as reported by [10].

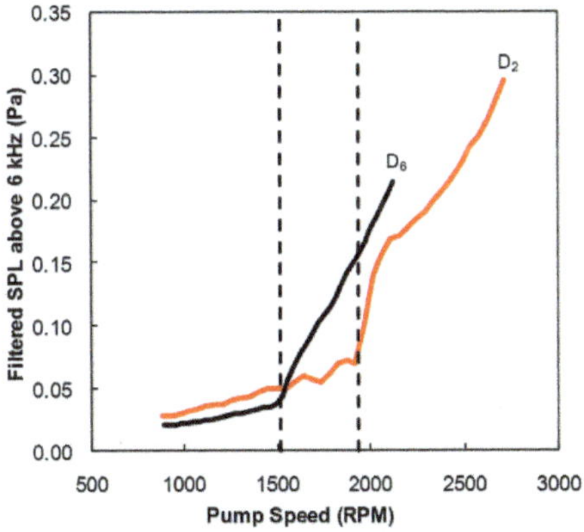

a) Pump speed

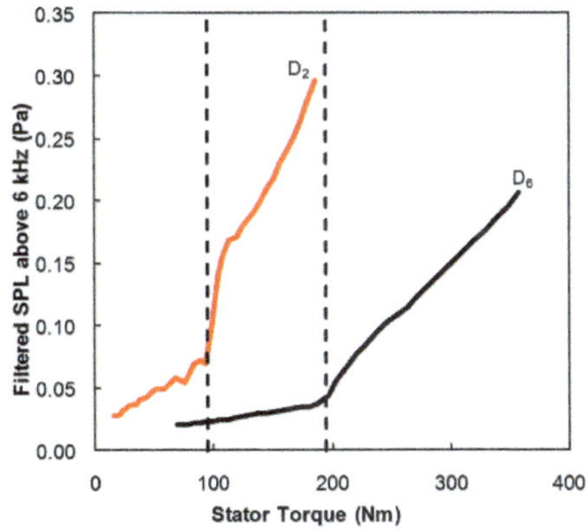

b) Stator torque

Figure 6. Filtered SPL versus pump speed (a) and stator torque (b) at the onset of cavitation

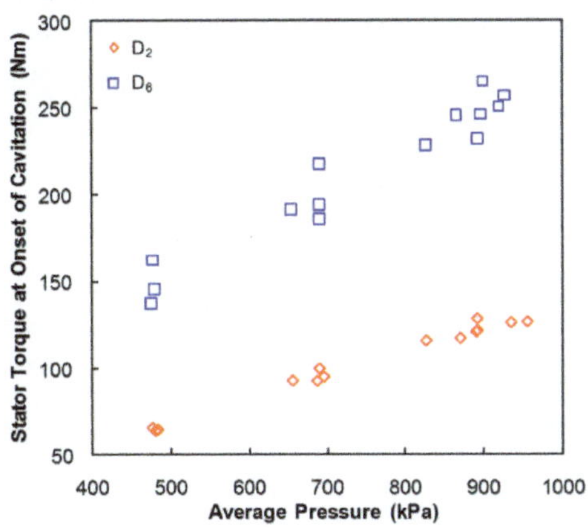

Figure 7. Stator torque at onset of cavitation versus average absolute pressure imposed upon the toroidal flow for geometrically scaled torque converters

These observations lead to the development of a dimensionless quantity by [11] incorporating stator torque at the onset of cavitation, $T_{s,i}$, diameter, D and average pressure, p_{ave} as given by Eq. 6.

$$\frac{T_{s,i}}{D^3 p_{ave}} \tag{6}$$

This quantity will be referred to as dimensionless stator torque and represents a similarity condition when geometric similitude is observed. Thus, stator torque cavitation threshold can be scaled to another diameter when the same stall operating point is maintained. The similarity condition for dimensionless stator torque is expressed by,

$$\left(\frac{T_s}{D^3 p_{ave}}\right)_1 = \left(\frac{T_s}{D^3 p_{ave}}\right)_2 \tag{7}$$

where the subscript 1 represents the known cavitation threshold and diameter and subscript 2 represents the cavitation threshold and diameter that are to be scaled. For Eq. 7 to hold true the average pressure, the combination of charge and back pressures, must be equivalent to properly scale the stator torque threshold. When the ratio of average pressures becomes unity it can subsequenctly be dropped from Eq. 7. The similarity condition for dimensionless stator torque is illustrated clearly in Fig. 8 when filtered SPL is plotted versus dimensionless stator torque. For the geometrically scaled designs at the the same operating point, onset of cavitation reduces to approximately the same value of dimensionless stator torque, as indicated by the vertical dashed lines. A difference of 1.7% between dimensionless stator torque values at incipient cavitation was found.

Alternatively pump speed could be used to formulate a similarity condition using dimensionless pump at onset of cavitation, Eq. 8, as proposed by [1]. This dimensionless quantity incorporates the critical pump speed threshold at cavitation, $\omega_{p,i}$, with diameter, D, the charging pressure imposed upon on the toroidal flow, p_c, and automatic transmission fluid density, ρ. The onset of cavitation occurs at nearly the same value of dimensionless pump speed as seen in Fig. 9.

$$\Omega_{\mathrm{i,p}} = \frac{\omega_{p,i} D}{\sqrt{p_c/\rho}} \tag{8}$$

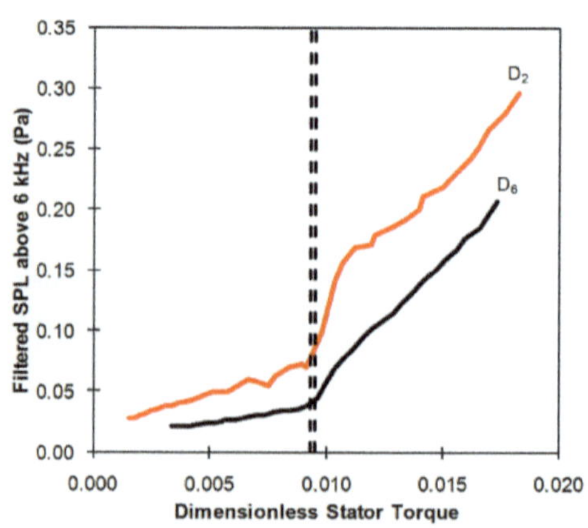

Figure 8. Filtered SPL versus dimensionless stator torque for geometrically scaled torque converters, dashed lines indicate onset of cavitation

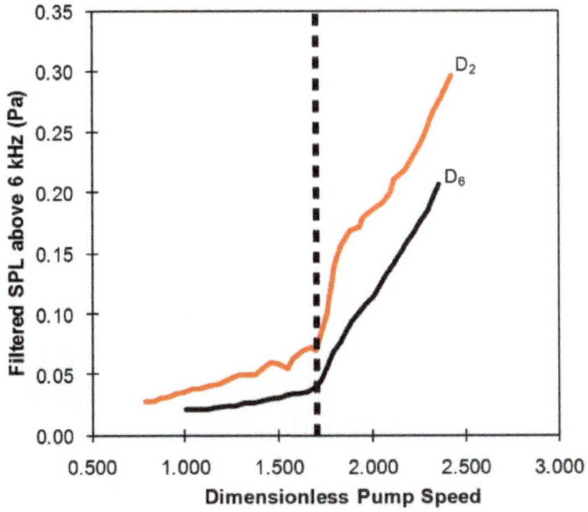

Figure 9. Filtered SPL versus dimensionless pump speed for geometrically scaled torque converters, dashed lines indicate onset of cavitation

4.2. Pi group formulation

The dimensional analysis and formation of pi groups for onset of cavitation in torque converters is initiated by formulating a list of regressor variables that will be fit to the response variable in developing empirical models. The response variable, stator torque at the onset of cavitation, can be described by the regressor variables that determine the boundary value problem of the toroidal flow for incipient cavitation at stall. The list of variables will include those that describe the torque converter design and those related to the operating conditions of the fluid. The design variables that principally determine the stator torque cavitation threshold at stall include the diameter, D, axial length, L_t, and element blade designs comprising the pump, turbine and stator. Element blade design manifests itself as the integrated quantities of K-factor and TR. Both of these quantities correlate to cavitation behaviour at stall, see [10], and their simplicity is favored over a lengthy list of individual blade design parameters. The number of stator blades as well as their shape can strongly influence performance and therefore the onset of cavitation. The airfoil shape of the stator blade is described by the ratio of blade maximum thickness, t_{max}, and chord length, l_c, and has been shown to have a statistically significant effect on the stator torque cavitation threshold, see [10]. The pressures imposed upon the torque converter are included in the list of variables for the dimensional analysis. As found by [10], increasing charging pressure can alter the stator torque cavitation threshold at stall by as much as 150 Nm for a particular design. Secondary variables that affect the cavitation threshold are those describing the thermal conditions of the toroidal flow, namely cooling flow rate, Q, and the automatic transmission fluid properties of density, ρ, viscosity, μ, specific heat, C_p, and thermal conductivity, k. Rather than using charge and back pressures, average pressure, p_{ave}, and pressure drop, Δp, will be used as they implicitly comprehend the cooling flow rate through the toroidal flow. Equation 9 gives the list of primary and secondary varies that determine $T_{s,i}$ in a torque converter,

$$T_{s,i} = f\left(D, L_t, K, TR, n_{sb}, t_{\max}, l_c, p_{ave}, \Delta p, \rho, \mu, C_p, k\right). \tag{9}$$

The dimensionless form of Eq. 9 was resolved into dimensionless stator torque as a function of dimensionless design parameters and dimensionless operating point parameters using the PI Theroem with repeating variables of D, p_{ave}, ρ and C_p. The exponents for the repeating variables contained in each pi group are found using,

$$\Psi = -\mathbf{Q}\mathbf{P}^{-1} \tag{10}$$

where $\mathbf{\Psi}$, $\mathbf{Q}$ and $\mathbf{P}$ are 10x4, 10x4 and 4x4 matrices. The rows of $\mathbf{P}$ are the repeating variables D, p_{ave}, ρ and C_p, while the rows of $\mathbf{Q}$ are the remaining variables, $T_{s,i}$, L_t, K, TR, n_{sb}, t_{max}, l_c, Δp, μ, and k from Eq. 9. The columns of both $\mathbf{P}$ and $\mathbf{Q}$ represent the dimensions of M, L, t and θ. A $\mathbf{\Pi}$ group matrix, Eq. 11, is formed by combining the $\mathbf{\Psi}$ matrix and a 10x10 identity matrix, $\mathbf{I}$, in which the diagonal represents the remaining variables, $T_{s,i}$, Lt, K, TR, n_{sb}, t_{max}, l_c, Δp, μ, and k raised to the power of 1. Each row of Eq. 11 represents a Π group, while the columns are the variables from Eq. 9 in the following order, D, p_{ave}, ρ, C_p, $T_{s,i}$, L_t, K, TR, n_{sb}, t_{max}, l_c, Δp, μ, and k. The cells of the matrix in Eq. 11 are the exponents assigned to each variable. Equation 12 lists the Π groups found using this method.

$$\mathbf{\Pi} = [\,\mathbf{\Psi}, \mathbf{I}\,] = \begin{array}{c} \\ \Pi_1 \\ \Pi_2 \\ \Pi_3 \\ \Pi_4 \\ \Pi_5 \\ \Pi_6 \\ \Pi_7 \\ \Pi_8 \\ \Pi_9 \\ \Pi_{10} \end{array} \begin{array}{c} \begin{array}{cccc|cccccccccc} D & p_{ave} & \rho & C_p & T_{s,i} & L_t & K & TR & n_{sb} & t_{max} & l_c & \Delta p & \mu & k \end{array} \\ \left[\begin{array}{cccc|cccccccccc} -3 & -1 & 0 & 0 & 1 & 0 & 0 & 0 & 0 & 0 & 0 & 0 & 0 & 0 \\ -1 & 0 & 0 & 0 & 0 & 1 & 0 & 0 & 0 & 0 & 0 & 0 & 0 & 0 \\ 2.5 & 0 & 0.5 & 0 & 0 & 0 & 1 & 0 & 0 & 0 & 0 & 0 & 0 & 0 \\ 0 & 0 & 0 & 0 & 0 & 0 & 0 & 1 & 0 & 0 & 0 & 0 & 0 & 0 \\ 0 & 0 & 0 & 0 & 0 & 0 & 0 & 0 & 1 & 0 & 0 & 0 & 0 & 0 \\ -1 & 0 & 0 & 0 & 0 & 0 & 0 & 0 & 0 & 1 & 0 & 0 & 0 & 0 \\ -1 & 0 & 0 & 0 & 0 & 0 & 0 & 0 & 0 & 0 & 1 & 0 & 0 & 0 \\ 0 & -1 & 0 & 0 & 0 & 0 & 0 & 0 & 0 & 0 & 0 & 1 & 0 & 0 \\ -1 & -0.5 & -0.5 & 0 & 0 & 0 & 0 & 0 & 0 & 0 & 0 & 0 & 1 & 0 \\ -1 & -0.5 & -0.5 & -1 & 0 & 0 & 0 & 0 & 0 & 0 & 0 & 0 & 0 & 1 \end{array}\right] \end{array}. \tag{11}$$

$$\begin{aligned}
\Pi_1 &= \frac{T_{s,i}}{D^3 p_{ave}} \\
\Pi_2 &= \frac{L_t}{D} \\
\Pi_3 &= K\sqrt{\rho D^5} \\
\Pi_4 &= TR \\
\Pi_5 &= \frac{t_{\max}}{D} \\
\Pi_6 &= \frac{l_c}{D} \\
\Pi_7 &= n_{sb} \\
\Pi_8 &= \frac{\Delta p}{p_{ave}} \\
\Pi_9 &= \frac{\mu}{D\sqrt{\rho p_{ave}}} \\
\Pi_{10} &= \frac{k}{D\sqrt{\rho p_{ave}} C_p}
\end{aligned} \tag{12}$$

A new dimensionless quantity, stator blade thickness ratio, is formed by combining Π_5 and Π_6. The Prandlt number is formed with the combination of Π_9 and Π_{10} to quantify thermal effects. The list of variables that effect stator torque at the onset of cavitation found in Eq. 9 are now made dimensionless as summarized in Eq. 13 as a function of dimensionless stator torque,

$$\frac{T_{s,i}}{D^3 p_{ave}} = f\left(\frac{L_t}{D}, U, TR, n_{sb}, \frac{t_{\max}}{l_c}, \frac{\Delta p}{p_{ave}}, \mathrm{Pr}\right) \tag{13}$$

where the five dimensionless design parameters are torus aspect ratio, L_t/D, unit input speed, U, stall torque ratio, TR, number of stator blades, n_{sb}, and stator blade thickness ratio, t_{max}/l_c. The two dimensionless operating point parameters are dimensionless operating pressure, $p_{ave}/\Delta p$, and the Prandlt number. These seven dimensionless parameters will be used to form empirical models on dimensionless stator torque.

Not all the dimensionless design parameters of Eq. 13 are necessary when considering the various populations of the torque converter designs considered in this investigation. When exact geometric similitude is observed, all five dimensionless design parameters are constant and only the dimensionless operating point parameters are used to fit a model. Those dimensionless design parameters that remain constant are removed from Eq. 13 before a model is fit to the data. As the constraint of geomertric similitude is relaxed, dimensionless design parameters are added back into Eq. 13 before proceeding fit a model to the data.

5. Dimensionless models

5.1. Model functions

The power product method (PPM) and response surface methodology (RSM) were used to develop relationships between dimensionless stator torque and the dimensionless design and operating point parameters of Eq. 13. The power product method has been traditionally used as the standard model function for most dimensionless prediction models found in fluid mechanics and heat transfer. The response surface method, however, is a statistical approach to developing a linear regression. An empirical power product and response surface model will be developed for each of the four torque converter populations of varying geometric similitude.

The power product model function is given by Eq. 14, with $\hat{y}$ equal to predicted dimensionless stator torque and Π representing the dimensionless parameters on the right hand side of Eq. 13. The model error, ε, is the difference between measured and predicted dimensionless stator torque. The exponents, b_i, and coefficient, b_0, were found using a Gaussian-Newton numerical search technique based upon minimizing ε; see [11] for more details.

$$\hat{y} = b_0 \left(\prod_{i=1}^{k} x_i^{\,b_i} \right) + \varepsilon \tag{14}$$

A second order function was assumed for the RSM used in this investigation, which contains linear, quadratic and two factor interactions as used by [12]:

$$\hat{y} = b_0 + \sum_{i=1}^{k} b_i \Pi_i + \sum_{i=1}^{k} b_{ii} \Pi_i^2 + \sum_{i=1}^{k-1} \sum_{j=i+1}^{k} b_{ij} \Pi_i \Pi_j + \varepsilon \,. \tag{15}$$

The response parameter, as with the PPM, is dimensionless stator torque and the regressors, Π, are the dimensionless design and operating point parameters of Eq. 13. The coefficients for the RSM were found using least squares method by minimizing the error between experimentally measured, y, and predicted, $\hat{y}$,dimensionless stator torque. Equation 15 can be written in matrix form as

$$\hat{\mathbf{Y}} = \mathbf{Xb} + \boldsymbol{\varepsilon} \tag{16}$$

where $\hat{\mathbf{Y}}$ is a vector of predicted dimensionless stator torque, $\mathbf{X}$ is a matrices of the dimensionless regressors, $\mathbf{b}$ is a matric of the regression coefficients and $\boldsymbol{\varepsilon}$ is a vector of error. The regression coefficents are found using Eq. 17, where $\mathbf{Y}$ replaces $\hat{\mathbf{Y}}$ and $\boldsymbol{\varepsilon}$ is omitted.

$$\mathbf{b} = \left(\mathbf{X}^T \mathbf{X} \right)^{-1} \mathbf{X}^T \mathbf{Y} \tag{17}$$

A stepwise regression technique was used to produce a response surface model with the minimum number of dimensionless regressors that were statisically significant as determined by a 95% joint Bonferroni confidence interval and *t*-test. The initial model only included the b_0 coefficient. During the stepwise regression, regressors were added or removed from the model dictated by the statistical criterion described. The reduction in model error realized by the addition of any one regressor was dependent upon the regressors already included in the model. Use of the stepwise regression procedure can greatly reduce the number of regressors, thereby complexity of the model while increasing the amount of variation in the response explained. It should be noted that the dimensionless parameters of *U* and Pr were divided by 1000 for the RSM so that the estimated regression coefficients were of roughly the same magnitude. For a more detailed description of the linear and stepwise regression techniques utilized, the reader is referred to [6] and [9].

The accuarcy and goodness of fit for both the PPM and RSM models were determined by computing the root mean square error (RMSE) and the linear association between the dimensionless response and regressors. RMSE is an estimator of a models standard deviation and for this investigation was computed as a percentage, denoted as %RMSE. Equation 18 defines %RMSE, where n is the number of data points and p, the number of regressors in the model. A value of %RMSE of 10% or less can generally be regarded as providing an empirical dimensionless model with acceptable accuracy.

$$\%RMSE = \sqrt{\frac{\sum_{i=1}^{n}\left(\frac{y_i - \hat{y}_i}{y_i} * 100\right)^2}{n-p}} \tag{18}$$

A measure of the proportionate amount of variation in the response explained by a particular set of regressors in the model is the adjusted coefficient of multiple determination, R^2_a:

$$R_a^2 = 1 - \frac{\sum_{i=1}^{n}\left(y_i - \hat{y}_i\right)^2 \Big/ (n-p)}{\sum_{i=1}^{n}\left(y_i - \overline{y}\right)^2 \Big/ (n-1)}. \tag{19}$$

This metric will calculate to a value between 0 and 1, with values of 0.85 or higher signifying a model that accurately represents the data. R^2_a is generally preferred over R^2, the coefficient of multiple determination, as it is a better evaluator of model variation and the number of regressors present. Whereas R^2 will always increase in value when additional regressors are added to a model, R^2_a may increase or decrease depending on whether or not the additional regressors actually reduce the variation in the response.

5.2. Exact geometric similitude

Two torque converters of exact geometric scaling with diameters D_2 and D_6 were considered for developing a dimensionless prediction model for onset of cavitation at stall. The functional form of Eq. 13 contains only the dimensionless operating point parameters since the dimensionless design parameters are equivalent for both diameters. Equations 20 and 21 are the PPM and RSM models for torque converters with exact similitude and are graphically represented in Fig. 10a and 10b, respectively. The values of the dimensionless design parameters for the two torque converters considered for exact geometric similitude are provided in Fig 10.

$$\frac{T_{s,i}}{D^3 p_{ave}} = b_0 \left[\left(\frac{\Delta p}{p_{ave}} \right)^{b_1} (\mathrm{Pr})^{b_2} \right] \tag{20}$$

$$\frac{T_{s,i}}{D^3 p_{ave}} = b_0 + b_1 \left(\frac{\Delta p}{p_{ave}} \right)^2 + b_2 (\mathrm{Pr})^2 \tag{21}$$

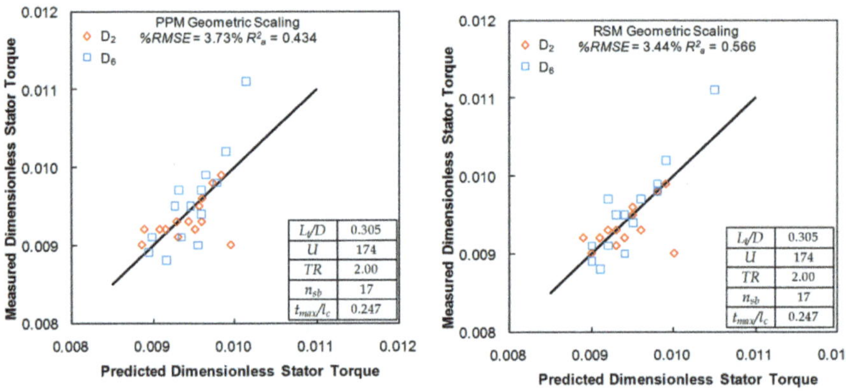

a) PPM b) RSM

Figure 10. Dimensionless stator torque PPM (a) and RSM (b) for exact geometric scaling of a torque converter design for a range of operating points

The model diagonistics provided at the top of each model plot shows %RMSE below 4% and R^2_a above 0.43. The RSM model performance is slightly better than that of the PPM with a 9% decrease in %RMSE and a 23% increase in R^2_a. Although both the PPM and RSM models have an %RMSE below 4%, R^2_a does not meet the 0.85 criteria for a good model fit with sufficient explanation of response variation for the regressors in the model. Use of dimensionless stator torque results in a constant value when geometric similitude is observed and an identical operating point is considered. For the range of dimensionless operating points tested, see Table 2, a small variation in dimensionless stator torque resulted which does not particularly lend itself to empirical modeling, requiring a greater variation in the dimensionless quantities to adequately define a curve. However, even with relatively low values of R^2_a, the prediction accuracy for scaling stator torque cavitation thresholds for a given torus and set of element designs is high.

	Low	High
Average Pressure	481 kPa	963 kPa
Pressure Drop	69 kPa	345 kPa
Prandlt Number	116	300

Table 2. Range of dimensionless operating point parameters tested

5.3. Torus scaling

For the torus scaling population, three torque converters with varying L_t/D ratios were used to investigate the ability to apply dimensional analysis and develop a predictive model with reasonable accuracy when geometric similitude is not observed. The element designs of the three converters were maintained, with the exception of stator blade count for the D_5 converter, such that unit input speeds remained constant. The PPM and RSM models are given by Eq. 22 and 23 and include the dimensionless design parameters of torus aspect ratio and number of stator blades to account for the design variations:

$$\frac{T_{s,i}}{D^3 p_{ave}} = b_0 \left[\left(\frac{L_t}{D} \right)^{b_1} \left(n_{sb} \right)^{b_2} \left(\frac{\Delta p}{p_{ave}} \right)^{b_3} \left(\mathrm{Pr} \right)^{b_4} \right] \tag{22}$$

$$\frac{T_{s,i}}{D^3 p_{ave}} = b_0 + b_1 \left(\frac{\Delta p}{p_{ave}} \right) + b_2 \left(\frac{L_t}{D} \right)^2 + b_3 \left(\frac{L_t}{D} \right) \left(\frac{\Delta p}{p_{ave}} \right) + b_4 \left(\frac{L_t}{D} \right) (\mathrm{Pr}) + b_5 \left(n_{sb} \right) \left(\frac{\Delta p}{p_{ave}} \right). \tag{23}$$

Figure 11a and 11b contain the PPM and RSM models, respectively, for the torus scaling popualtion and include the model performance metrics and dimensionless design parameters. An increase in the range of dimensionless stator torque can be noted over the exact geometric similitude population shown in Figure 10a and 10b. The PPM model experienced a negligible increase in %RMSE and a doubling of R^2_a when compared with the exact geometric scaling PPM model. A decrease in %RSME from 3.44% to 2.87% and an increase in R^2_a from 0.566 to 0.926 was realized compared to the geometric scaling RSM model. Both model functions increase the amount of variation in dimensionless stator torque accounted for by the regressors in the model. The RSM model has a slight advantage over the PPM model due to the statistical nature of determining which dimensionless parameters to include. This can be seen grpahically in Fig. 11 in that the data fits the RSM model closer for all three diameters versus the PPM model. The utility of these particular models for predicting $T_{s,i}$ are limited to the element blade design (fixed U and TR) tested in Fig. 11, but can be sacled to other diameters as long as the L_t/D ratio remains between 0.29 and 0.317.

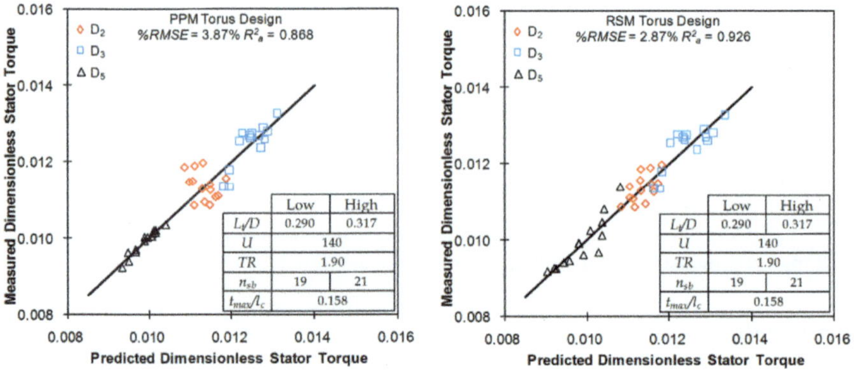

a) PPM b) RSM

Figure 11. Dimensionless stator torque PPM (a) and RSM (b) for torus geometry scaling of a torque converter design for a range of operating points

5.4. Pump and stator

The dimensionless PPM and RSM models for a population in which pump and stator designs varied for a fixed L_t/D are given by:

$$\frac{T_{s,i}}{D^3 p_{ave}} = b_0 \left[(U)^{b_1} (TR)^{b_2} (n_{sb})^{b_3} \left(\frac{t_{\max}}{l_c} \right)^{b_4} \left(\frac{\Delta p}{p_{ave}} \right)^{b_5} (\Pr)^{b_6} \right] \tag{24}$$

and

$$\frac{T_{s,i}}{D^3 p_{ave}} = b_0 + b_1 \left(\frac{t_{\max}}{l_c} \right) + b_2 (U)^2 + b_3 (TR)^2 + b_4 (n_{sb})^2 + b_5 \left(\frac{\Delta p}{p_{ave}} \right)^2 + b_6 (\Pr)^2 + b_7 (U)(TR) + \ldots$$
$$\ldots + b_8 (U) \left(\frac{t_{\max}}{l_c} \right) + b_9 (n_{sb}) \left(\frac{t_{\max}}{l_c} \right). \tag{25}$$

Sixteen torque converter designs with an L_t/D of 0.3 were formed from various combinations of 5 pump and 11 stator designs. The dimensionless design parameters of U, TR, n_{sb} and t_{max}/l_c are required to characterize the effect pump and stator design have on toriodal flow and incipient cavitation, while L_t/D was eliminated as it remained fixed. The %RMSE for both the PPM and RSM models increased approximately 2.25% over the torus scaling population models. This is to be expected as more data points are included, departing further from geometric similitude. Although the torque converter designs are more varied, R^2_a increased for both model functions indicating that the additional dimensionless design parameters helped to explain the variation in dimensionless stator torque. With %RMSE's of roughly 6% and 5% and R^2_a greater than 0.92, both models are accurate for design purposes. Predicted dimensionless stator torque from either model can be scaled to another diameter if the L_t/D ratio of 0.3 is maintained and the elements design falls within the scope of the dimensionless design parameters used to develop the model.

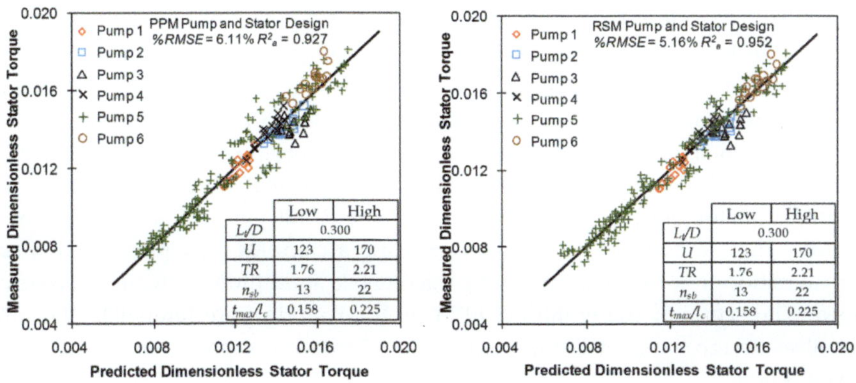

a) PPM b) RSM

Figure 12. Dimensionless stator torque PPM (a) and RSM (b) for various pump and stator geometry for a given torque converter torus design for a range of operating points

5.5. General design

For the general design population of torque converters, variations in torus and element blade geometries were considered to develop a PPM and RSM model for predicting $T_{s,i}$ at stall. All of the dimensionless design parameters in Eq. 13 were required to develop the PPM and RSM models, as all varied for the matrix of designs tested. The PPM model, Eq. 26, contains 7 dimensionless parameters and 8 regression coefficients, while the RSM model, Eq. 27, contains 18 dimensionless parameters as determined by the stepwise regression procedure and 19 regression coefficients.

$$\frac{T_{s,i}}{D^3 p_{ave}} = b_0\left[\left(\frac{L_t}{D}\right)^{b_1}(U)^{b_2}(TR)^{b_3}(n_{sb})^{b_4}\left(\frac{t_{max}}{l_c}\right)^{b_5}\left(\frac{\Delta p}{p_{ave}}\right)^{b_6}(\Pr)^{b_7}\right] \tag{26}$$

$$\frac{T_{s,i}}{D^3 p_{ave}} = b_0 + b_1\left(\frac{L_t}{D}\right) + b_2(U) + b_3(TR) + b_4(n_{sb}) + b_5\left(\frac{t_{max}}{l_c}\right) + b_6\left(\frac{L_t}{D}\right)^2 + b_7(n_{sb})^2 + \ldots$$

$$\ldots + b_8\left(\frac{t_{max}}{l_c}\right)^2 + b_9\left(\frac{\Delta p}{p_{ave}}\right)^2 + b_{10}(\Pr)^2 + b_{11}\left(\frac{L_t}{D}\right)(n_{sb}) + b_{12}\left(\frac{L_t}{D}\right)\left(\frac{t_{max}}{l_c}\right) + b_{13}(U)(TR) + \ldots \tag{27}$$

$$\ldots + b_{14}(U)(n_{sb}) + b_{15}(U)\left(\frac{t_{max}}{l_c}\right) + b_{16}(TR)(n_{sb}) + b_{17}(TR)\left(\frac{t_{max}}{l_c}\right) + b_{18}(n_{sb})\left(\frac{t_{max}}{l_c}\right)$$

Figure 13a and 13b are the PPM and RSM models for a general population of torque converter desgins plotted as measured versus prediceted dimensionless stator torque. The model diagonistics of %RMSE and R^2_a are included in Fig 13 along with the range of dimensionless design parameters of the converter designs. The range of the dimensionless operating parameters remained the same as those reported in Table 2.

Either form of the general design population empirical model, PPM or RSM, demonstrates the capability to be used in the design process to predict onset of cavitation for a torque converter at stall. %RMSE is below and R^2_a is above the threshold criteria of what was deemed acceptable for an accurate and useable model. The RSM model, with it's statistically determined functional form and increased number of dimensionless regressors, resulted in a closer curve fit model than the PPM as seen in Fig. 13, particularly at the extreme values of dimensionless stator torque. The RSM model's %RMSE of 6.52% and R^2_a of 0.936 demonstrates that an accurate model for predicting a complex flow phenomenon such as onset of cavitation in a turbomachine when geometric similitude is greatly relaxed can be developed. The scope of the empirical PPM or RSM models are limited to the range of dimensionless design parameters reported in Fig. 13 to achieve prediction $T_{s,i}$ values within the %RMSE's noted.

Figure 14 is a histogram of the residual errors, ε, for the general torque converter design PPM and RSM models. For both empirical models the errors do not significantly deviate from normalcy and follows that of a normal distribution. Although both models have error well distributed around zero, the RSM shows a narrower and taller distribution, indicating that the RSM model prediction capability exceeds that of the PPM model. This is reinforced by the %RMSE and R^2_a model diagnostics.

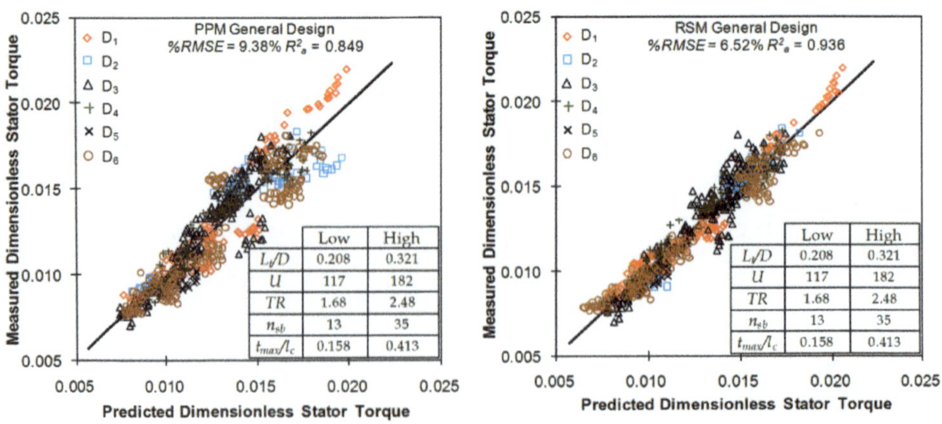

a) PPM b) RSM

Figure 13. Dimensionless stator torque PPM (a) and RSM (b) for a general design population of torque converters for a range of operating points

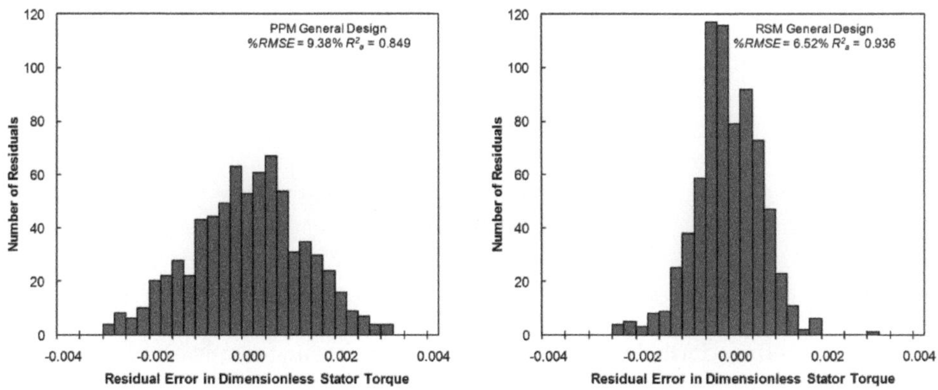

a) PPM

b) RSM

Figure 14. Histogram for general design PPM (a) and RSM (b) model residual error

The span of the experimental data for the 6 diameters, encompassing 51 torque converter designs are presented in Fig. 15 as pump power at the onset of cavitation. As Fig. 15 shows, the onset of cavitation occurs within a consistent range of pump power, between 10 and 65 kW, across the range of diameters tested. In general, a torque converter design with a low K-factor and high TR operating at a high average pressure will transition to cavitation at a higher value of pump power. It is worth noting that incipient cavitation does not occur at moderate or heavy pump power, but rather at moderate to low power levels. This indicates that any given torque converter design has a high potential to cavitate lightly or moderately during typical driving conditions at or near stall. As discussed previously, incipient or light cavitation does not hinder performance or component durability. The objective then of the torque converter engineer is to design the torque converter to operate in a wider speed-torque range cavitation free to prevent conditions of heavy and sustained cavitation during atypical driving conditions.

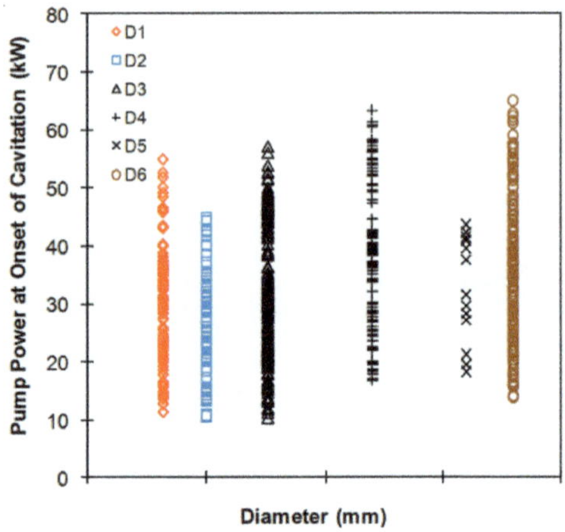

Figure 15. Power at onset of cavitation for general design population of torque converters

Pump torque is the principal parameter used when performing torque converter and engine matching as it is directly equivalent to engine torque. During the design phase of a torque converter and subsequent engine matching, Eq. 28 would be used to take the predicted value of stator torque at the onset of cavitation from Eq. 26 or 27 to calculate pump torque at onset of cavitation. This enables the torque converter engineer to better formulate a design that balances performance requirements and minimize cavitation potential when matching to the torque characteristics of a specific engine.

$$T_{p,i} = \frac{T_{s,i}}{TR - 1} \tag{28}$$

5.6. Model summary

Table 3 compares each PPM and RSM model developed for the four torque converter design populations. As geometric similitude decreases the number of dimensionless regressors required to realize a highly accurate model increases for either the PPM or RSM model. This is an expected trend as the numerous interactions between design parameters become increasingly important in determining stator torque at the onset of cavitation while stalled. The increase in R^2_a confirms this conclusion as the variation in the data is nearly completly explained by the addition of the dimensionless design parameters. The gradual increase in %RMSE from exact geometric similitude to a general design population is not substantial enough to qualify either the PPM or RSM models to be in accurate or deficient for design purposes. The regression coefficients for all of the PPM or RSM models have been purposefully omitted from this chapter as they were derived from proprietary torque converter designs whose performance attributes could otherwise be extracted.

			RSM			PPM		
Population	Similitude	Data Points	Regressors	RMSE (%)	R^2_a	Regressors	RMSE (%)	R^2_a
Exact Similitude	Exact	28	3	3.34	0.566	3	3.73	0.434
Torus Scaling	Elements	42	6	2.87	0.926	5	3.87	0.868
Pump and Stator Design	Torus	224	10	5.16	0.953	7	6.037	0.929
General Design	None	714	19	6.52	0.936	8	9.38	0.849

Table 3. Summary of PPM and RSM models predictive capability by torque converter design population

6. Conclusions

The chapter presents experimentally obtained values for stator torque at the onset of cavitation for torque converters of greatly varying geometric similitude nondimensionalized and used as dimensionless reponse and regressors for developing empirical models. Power product method (PPM) and response surface method (RSM) models were curve fit using regression techniques to dimensionless stator torque as a function of dimensionless design and operating point parameters. PPM and RSM models created from data sets of decreasing geometric similitude showed that R^2_a values above 0.85 are achieved even with greatly relaxed geometric similitude. The %RMSE values resulting for each PPM or RSM model of decreasing geometric similitude were not substantial enough to indicate inadequacies in the dimensional analysis or data modeling methodology. A RSM model was presented which is capable of predicting $T_{s,i}$ for general torque converter design with a %RMSE of 6.52%. This error is deemed sufficiently low and its scope of prediction large enough to rate the RSM model a valuable tool for optimizing torus and element geometries with respect to the onset of cavitation at stall in three element torque converters. The next phase of this research will be to expand testing to include non stall conditions to determine the desinent (disappearance) point of cavitation and non-dimensionalize the test data into an equivalent design tool.

Appendix

Nomenclature

C_p	Specific heat (kJ/kg·K)
D	Diameter (m)
K	K-factor (rpm/Nm$^{0.5}$)
L_t	Axial length (m)
N_p	Pump speed
Pr	Prandlt number
Q	Volumetric flow rate (m^3/s)
R^2_a	Adjusted coefficient of multiple determination
$T_{p,i}$	Pump torque at onset of cavitation (Nm)
$T_{s,i}$	Stator torque at onset of cavitation (Nm)
SR	Speed ratio
TR	Torque ratio
U	Unit input speed
%RMSE	Percent root mean square error
b	Regression coefficient
k	Thermal conductivity (W/m·K)
l_c	Chord length (m)
n_{sb}	Number of stator blades
p_{ave}	Average pressure (kPa)
p_b	Back pressure (kPa)
p_c	Charge pressure (kPa)
Δp	Delta pressure (kPa)
t_{max}	Maximum blade thickness (m)
$\hat{y}$	Predicted dimensionless response
y	Measured dimensionless response
ε	Model residual error
ρ	Density (kg/m^3)
μ	Viscosity (N·s/m^2)
$\omega_{p,i}$	Pump speed at onset of cavitation (rad/s)
θ	Temperature (°C)

Author details

Darrell Robinette
General Motors LLC, USA

Carl Anderson and Jason Blough
Michigan Technological University,
Department of Mechanical Engineering – Engineering Mechanics, USA

7. References

[1] Anderson, C., Zeng, L., Sweger, P. O., Narain A., & Blough, J., "Experimental Investigation of Cavitation Signatures in an Automotive Torque Converter Using a Microwave Telemetry Technique," in Proceedings of the 9th International Symposium on Transport Phenomena and Dynamics of Rotating Machinery (ISROMAC '02), Honolulu, Hawaii, USA, February 2002

[2] Courbiere, P., "An Acoustical Method for Characterizing the Onset of Cavitation in Nozzles and Pumps," *in Proceedings of the 2nd International Symposium on Cavitation Inception,* New Orleans, La, USA, December 1984

[3] Dong, Y., Korivi, V., Attibele, P. & Yuan, Y., "Torque Converter CFD Engineering Part II: Performance Improvement Through Core Leakage Flow and Cavitation Control," *in Proceedings of the SAE World Congress and Exhibition,* Detroit, Mich, USA, March 2002, 2002-01-0884

[4] Kowalski, D., Anderson, C. & Blough, J., "Cavitation Detection in Automotive Torque Converters Using Nearfield Acoustical Measurements," *in Proceedings of the SAE International Noise and Vibration Conference and Exhibition (SAE '05),* Grand Traverse, Mich, USA, May 2005, 2005-01-2516

[5] Kowalski, D., Anderson, C. & Blough, J., "Cavitation Prediction in Automotive Torque Converters," *in Proceedings of the SAE International Noise and Vibration Conference and Exhibition (SAE '05),* Grand Traverse, Mich, USA, May 2005, 2005-01-2557

[6] Kutner, M. H., Nachtsheim, C. J. & Neter, J., *Applied Linear Regression Models,* McGraw-Hill, New York, NY, USA, 4th edition, 2004

[7] McNulty, P. J. & Pearsall, I., "Cavitation inception in pumps," *Journal of Fluids Engineering,* vol. 104, no. 1, pp. 99–104, 1982

[8] Mekkes, J., Anderson, C. & Narain, A., "Static Pressure Measurements on the Nose of a Torque Converter Stator during Cavitation," *in Proceedings of the 10th International Symposium on Transport Phenomena and Dynamics of Rotating Machinery (ISROMAC '04),* Honolulu, Hawaii, USA, February 2004

[9] Montgomery, D. G., *Design and Analysis of Experiments,* 6th Edition, John Wiley and Sons Inc., 2005

[10] Robinette, D., Anderson, C., Blough, J., and Johnson, M., Schweitzer, J. & Maddock D., "Characterizing the Effect of Automotive Torque Converter Design Parameters on the Onset of Cavitation at Stall," *in Proceedings of the SAE International Noise and Vibration Conference and Exhibition (SAE '07),* St. Charles, Ill, USA, May 2007, 2007-01-2231

[11] Robinette, D., Schweitzer, J., Maddock D., Anderson, C., Blough, J., & Johnson, M., "Predicting the Onset of Cavitation in Automotive Torque Converters Part I: Designs with Geometric Similitude," *International Journal of Rotating Machinery,* Vol. 2008

[12] Robinette, D., Schweitzer, J., Maddock D., Anderson, C., Blough, J., & Johnson, M., "Predicting the Onset of Cavitation in Automotive Torque Converters Part II: A Generalized Model", *International Journal of Rotating Machinery*, Vol. 2008

Section 4

Manufacturing

Chapter 15

Performance Measurement in Supply Chains: A Study in the Automotive Industry

Mário Sacomano Neto and Sílvio R. I. Pires

Additional information is available at the end of the chapter

1. Introduction

Transformation and growth on the automobile sector has opened a vast research field involving supply chain performance measurement. The change on the pattern in the relationship of the automakers with suppliers and dealers is accompanied by a need for modifications on the supply chain performance measurement systems.

In the last years, several and new productive arrangements were implanted in the Brazilian automobile industry, among them the modular consortium and the industrial condominiums. Those new arrangements are characterized by a high degree of outsourcing, long-term contracts, co-production of components, transfer of information and support to the suppliers. These new arrangements have also modified the relationship of the automakers with the suppliers (Helper, 1991; Lamming, 1993; Lambert, Cooper & Pagh, 1998; Croxton *et al.*, 2001; Pires e Sacomano Neto, 2008).

There are two basic consequences of those changes to auto parts suppliers: (1) significant increase on the demands of the automakers for quality, just-in-time delivery, global sourcing, follow sourcing, research and development of products, supply of complete functions (systems, sub-systems or modules), financial and technological training; (2) concentration of auto part suppliers in the hands of large international companies groups (Carvalho et al, 2000).

The intense change in the relationship with auto part suppliers opens a vast field of studies regarding the stage and characteristics of supply chain performance measurement. The need for low levels of stocks and the high complexity of the logistic flow management made the automaker outsource important parts of those processes to logistic service integrators. Consequently, the base of the suppliers' consolidation and the outsourcing of internal processes made automakers establish new mechanisms of supply chain performance measurement.

In order to understand this set of modifications, this article aims verify the current stage of the performance measurement system within a relevant supply chain of the Brazilian automotive industry. The study characterizes the main changes of automaker relationship with the auto part suppliers and dealer influenced the supply chain performance measurement.

In order to accomplish the objective of this research, interview were realized with managers of the automaker (industrial condominium), with one systemist supplier (installed inside the automaker's condominium) and other three auto parts suppliers and one dealer. Besides studying four relationships of the automotive chain, that group of companies allowed to understand the performance measurement in highly cooperative relationships (as in the systemist supplier), "more traditional" relationships in the chain (other suppliers) and commercial relationships (dealers).

2. Performance measurement systems

The measurement mechanisms and performance evaluation were incorporate to the business atmosphere since the beginning of the mass production. The need for measuring the performance has intensified since the establishment and growth of large companies. The first type of established measure system was created by Taylor in the work rationalization process, presenting the first scales of "evaluation of merit" (Brandão & Guimarães, 2001). That period was the beginning of the establishment, in several ways and systems, of performance measurement in companies.

After that, the accounting administration, which was composed by measures that had as objective to measure the financial return, turned up. That type of measure turned up in response to the change of the market nature and to the perspective of larger profits through centralized production (Waggoner et al., 1999).

However, the measurement system revolution began when managers felt the need for instruments or systems to assure the maintenance of organization strategies (Eccles, 2000). From 1980 to1990, the performance measures appeared related to quality and customer's satisfaction, respectively (Eccles, 2000). The necessity to control the strategy, the quality and the customers' satisfaction performance turned up because of external factors, such as customers and competitors, interfered significantly on the strategic position of the organizations. The picture 1 illustrate the supply chain indicators. The picture presupposes the existence of an individual group of indicators (represented by ind. 1, ind. 2 and ind. 3), used in each one of the business units. Besides, there are some common indicators to the whole productive chain and these common indicators will determine the supply chain performance.

There is a significant set of concepts on the performance measurement systems. For the FPNQ (1996), a measurement system and evaluation on business performance should be guided according to the aspirations of all the interested parts (stakeholders), including customers, employees, shareholders, suppliers, partners, society and community.

According to Neely et al. (1995) the performance measurement process involves three central concepts: (1) performance measurement: it can be defined as the process of quantifying the efficiency and the effectiveness of the actions; (2) performance measured: metric used to quantify the efficiency and/or the effectiveness of the action and (3) system of performance measurement: it is the unity of metric used to quantify the efficiency and the effectiveness of the actions simultaneously.

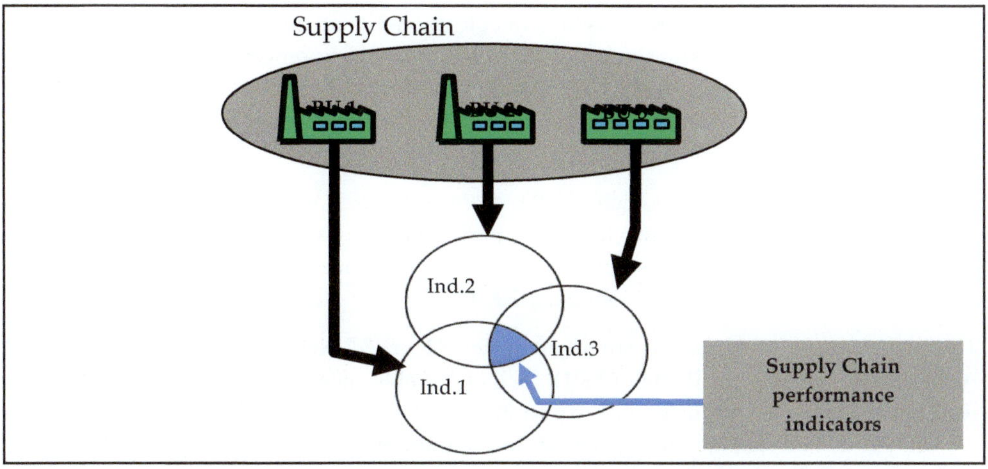

Figure 1. Illustrate the supply chain indicators (Source: Pires e Aravechia, 2001)

The performance measurement is one of the central elements in the management systems, and the performance consists on the result of decisions taken. Such decisions are made through proportional information by the system (Kaydos, 1991). Therefore, the level of performance has direct influence on quality and amount of information.

The objective of a performance measurement system is to establish the level of evolution, stagnation or decadence in certain processes. Performance measurement system also supply refined information to the decision taken process on preventive and/or corrective actions (Nauri, 1998). The performance measurement is essential to maintain high levels of quality and productivity, not only to provide a good control and administration, but also to develop and motivate the organization (Kaydos, 1991).

With high levels of outsourcing in the automakers activities, systems and metrics to supply chain performance should be developed. A great amount of costs in a vehicle is about materials, parts and systems from suppliers. Considering that, the need to control is fundamental for the systems optimization. Due to competition within the supply chain, it is important to think over a way to built knowledge, as well as to, incorporate a new performance measurement system considering the supply chain.

3. Supply chain performance measurement

The Supply Chain Management - SCM is an original term in logistics literature, specifically in purchases and stocks management (Trienekens, 1999). However, the Council of Logistics Management itself defines logistics as "a part of supply chain management that drifts, implements and controls the flows efficiently, storage of products, services and information correlated, starting from the origin to the consumption point, aiming to assist the customers' needs" (Lambert, Cooper & Pagh, 1998 p.3). The authors point out that logistics has a functional role to deal with the information and matter flows in supply chain.

The Supply Chain Management is a strategic model that goes beyond logistics when integrating business operations and the administration of key processes in the supply chain from the development of products to the marketing. Therefore, it can be said that the Supply Chain Management is multifunctional, involving production management areas, marketing, purchases and logistics (Saucer, 2004).

According to Lambert and Pohlen (2001), many measures used in the supply chain are internal and it does not measure the chain performance as a whole. Several authors point out that there are a few studies about performance measure in the supply chain (Beamon, 1999; Gunasekaram et al., 2001; Holmberg, 2000).

There are few performance measurement systems focused exclusively on the supply chain. According to Lambert & Pohlen (2001), lack of suitable measures to the supply chain can result in flaws to get the customer's expectations, sub-optimizations in the chain, barriers to get competitive advantages and generate conflicts among participant companies. According to Beamon (1999), the choice for performance measures in the chain is complex because it depends on many factors, such as size, culture, needs, location, etc.

There are some relevant characteristics that should be considered to analyze the chain performance measurement system: inclusion (measure of all pertinent aspects); universality (to allow comparison among different levels); measurement (the necessary data can be measured) and consistence (solid measures with strategic goals). Those characteristics guarantee consistence and reliability to the measurement system (Beamon, 1999).

The performance measurement integration in the whole supply chain allows the global performance to be evaluated and to identify the internal improvements that should be accomplished in order to have a great impact in competitiveness (Croxton et al., 2001).

There is a tendency to use specific performance measures in the chain, because they are more attractive to the implementation and monitoring point of view (Beamon, 1999). Functionality and effectiveness in this type of measure is considered very limited because they show the chain performance in just one aspect, and serious problems can be hidden. The type of measure most used in this context is related to costs because it is a basic measure used in any measure system, even when it takes into consideration a unit of business in an isolated way. However, the problem is that only this type of measure does not translate the real performance in the chain. It is necessary to include other types of measures to guarantee the effectiveness of the system for the whole chain.

The supply chain performance measures can also be classified in qualitative and quantitative measures. The quantitative measures are easier to measure than the qualitative ones, because they are more subjective, and, therefore, more complex to measure. The measures that can be classified as qualitative are: customer's satisfaction, flexibility, integration of material and information flow, effective risk management and the supplier's performance. The measures that are framed as quantitative are the ones related to cost and the customer's satisfaction (Beamon, 1998).

According to Steweart (1995), there are four areas that should be developed to guarantee the improvement of the chain performance measurement system. These areas are: (1) performance of delivery, (2) flexibility and customer's feedback, (3) logistics costs and (4) performance of assets.

The performance measures provide a great amount of information inside to enlarge the integration of the company and alongside the supply chain. The performance measures have, as objective, not only to monitor the performance system but also to accomplish diagnoses, to propose improvement of performances, to stimulate performances along the chain and customers' service.

One aspect to be considered as the performance measurement system limitation is the lack of connection between the elaboration of the performance measures and the final satisfaction of the customers. That fact is extremely relevant because the performance measures, in a business unit or even in a supply chain, should be unfolded starting from their strategic objectives. Considering that the strategic objectives should be directly linked to the factors that determine the customer's satisfaction, the focal company (or that one which represents the chain strongest relationship) should direct actions which reflect an efficient management and service of demand.

Because of this, the research has incorporated the analysis from second tier suppliers to car dealers, delimiting a richer and not common to research area.

4. The research methodology

This case study has involved one automaker (assembler) using an industrial condominium (focal company), two first tier suppliers (companies A and B), two second tier suppliers (companies C and D) and a vehicle dealer. Figure 2 delimits the supply chain studied, including upstream and downstream companies.

The research was exploratory and descriptive. The collection of data was accomplished through interviews (followed by a semi-structured questionnaire developed previously) with the responsible for the production areas and logistics in the automaker and in the auto part suppliers, always also visiting the productive, operations and logistics areas. At the dealer, the commercial director was interviewed.

An interesting aspect of the studied chain, is that, the supplier A is a "systemist" supplier installed inside the automaker plant. One aspect that was verified and explored in this study is the difference between the relationship (connections) of the automaker with a systemist and the automaker with traditional suppliers. That fact has generated different types of connections and measures performance in the same supply chain. That aspect will be explored later, when analyzing the research results.

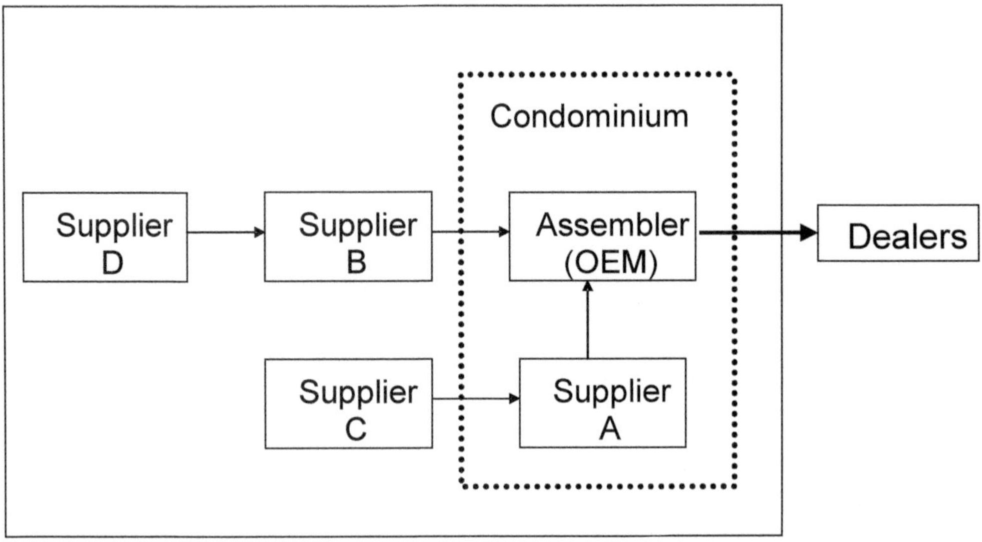

Figure 2. Delimitation of the supply chain studied

5. Study cases

That item presents in a descriptive way data about the automaker, the suppliers and the vehicle dealer.

5.1. Automaker and the industrial condominium (focal company)

Its original manufacturer was inaugurated in 1957 in São Paulo metropolitan area and it represents one of the automotive industry icons and Brazilian industrialization. In the beginning of 2008 it would hired 20.545 employees (automaker: 15.000; outsourced: 5.545) and more than a thousand engineers in the engineering center. Its production capacity was of 1.600 vehicles a day. The figure 3 illustrate an industrial condominium plant, were the suppliers are beside the car manufacture installation.

In 2002 the plant was refurbished to become an industrial condominium. As it is explained by Pires (2004), in the industrial condominium the key suppliers are installed physically beside the automaker in order to supply the productive system. According to the Automotive Business Data (2005), the company invested R$ 2 billion in a structure with high level of automation and modernity. Laser welds, robot islands in the frame, automation in the painting line, use of palm tops to control the production in real time, modular system assembly and truck transporters adapted to the employee's height were some improvements and innovations accomplished in the new plant.

The systemist suppliers are physically located inside the automaker condominium and they have a close and collaborative relationship with the automaker on the product development and the production plan. The levels of trust in this type of relationship are intense, involving high investment from both.

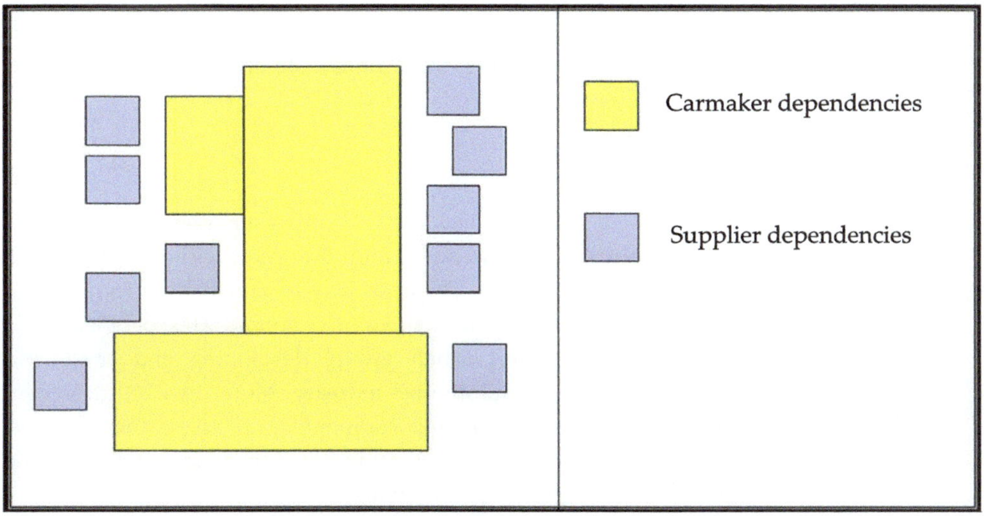

Figure 3. Illustrate an industrial condominium plant

The company logistic has been of high complexity. There are different kinds of suppliers: those responsible for supplying the line in the point of use, being, therefore, responsible for affairs involving transport and storage, and those not responsible for the transport because the automaker takes charge of this activity through a logistics consolidator.

Another responsible factor which makes the complexity of the system worse is the number of components, around 4.000 by platform. This number can vary because the automobile design suffers constant changes. Considering all the platforms present in the plant, the total of components to be managed by the automaker is around 25.000. If there are no systemists, that number would be approximately 40.000 components as mentioned by one of the directors of the automakers.

To work with low levels of stock and high flexibility on the production system, the automaker uses logistics practices to guarantee the production synchronization and the logistics efficiency. Examples of practices that contribute to a better result are the milk run, the cross docking and the just in sequence activities.

The milk run is a provisioning system with pre-defined itinerary and schedules to collect materials in the suppliers, whose main objective is to reduce the provisioning logistics costs via scale savings and rationalization of the routes.

The cross-docking can be defined as a practice that seeks to avoid unnecessary storages (that represent great sources of waste) in distribution centers or in places that work as such. That practices is accomplished by the automaker during the delivery of the components.

The just in sequence is a contemporary logistics practice involving the provisioning process. Its objective is to place the item according to automaker production sequence. That allows stocks and the provisioning time to be optimized. That practice can be considered an evolution within the provisioning process inside the just in time logistics, which points out that the customer should be given the right thing, the right amount and at the right moment (PIRES, 2004).

5.2. Suppliers and customers

- **Supplier A** - is a systemist installed inside the automaker plant which hires 40 employees. The supplier belongs to a German group that deals with the automobile branch, supplying auto parts and systems to trucks, chassis and motors (powertrain systems). In Brazil the group has 22 subsidiaries and it hires around 9.000 people.
- **Supplier B**: supplier B belongs to a German group, developing and producing mechanical and electronic locks to the automotive industry. More than 200 designers work in Germany, USA and Korea developing components to access the cars and immobilization systems known as "passive entrance". In Brazil the plant is settled in the state of São Paulo and supplies directly to the automaker.
- **Supplier C**: supplier C also belongs to a German group that deals with the automobile segment, manufacture of tubes, mechanical engineering and distribution. At the automotive branch the company develops and produces complete modules, components for trucks, chassis and motors. This auto part supplier came to Brazil in 1999 in order to established a flow sourcing with two automakers.
- **Supplier D**: supplier D is a national company that provides printings and cut tools to supplier B. It also supplies to other companies; in its great majority, to the automobile industry suppliers. The company was founded in 1983 in the state of São Paulo and today it hires 19 people.
- **Dealer**: the dealer studied belongs to a business group that operates in São Paulo and Minas Gerais states. The group has, as a whole, 6 dealers.

6. Results

In this item it is shown how the automaker deals with the evaluation of suppliers and dealer. It is also shown the company internal performance system and its deployment in the supply chain.

6.1. Evaluation of suppliers

The evaluation of suppliers is considered to be a relevant issue to the automaker, according to several reports taken along the research. This evaluation is accomplished in a multifunctional form, having different weighs of measurement in the following areas: logistics, engineering, quality, finance and commercial.

Logistics is responsible for evaluating the operational aspects of the suppliers, such as: easiness on the information flow, kinds of packing, ways of transport and storage and service level related to the execution of the scheduling and on-time deliveries. Logistics is also responsible for the placement of the product at the use point as well as for the service of the master program in production. Logistics, operating in a very efficient way, guarantees both a synchronized production and an appropriated level of stock.

The engineering area verifies the development potential in products and processes, quality of the process and it makes possible ways to control the analysis and to select the suppliers. The most technical processes related to the evaluation of suppliers are found in this area.

The quality area takes care of the system making possible the quality process (internal process, means of control, checking of tools and maintenance). This area has got the veto power to evaluate the suppliers because this requirement is considered to be a request qualifier. A qualifier competitive priority of request means that the requested characteristic should be intrinsic to the product or component. The two more relevant aspects in this area are: the product conformity and the specification service of the automaker.

The Finance area also participates in the evaluation of the suppliers analyzing whether the latter has the necessary profitability and "financial health" to be kept in the business. This is an important issue because if the suppliers do not have financial balance, aspects, such as, the quality of the product or the delivery performance can be compromised.

The last area to participate in this process is the commercial one, which studies and monitors the intensity and relationship forms between the suppliers and the company. This also analyzes the contribution of the supplier productivity to reduce the end product prices.

If an area vetoes a supplier, yet it is verified that the supplier has a supply potential, plans for actions are taken to help it.

The five areas that compose the performance evaluation centralize the information in weekly meetings called “the supplier definition forums". These meetings happen twice a week, with the representations of each area and it isq held by the director of the area of Purchasing.

In order to define new suppliers for a new project, two meetings are held inside "the supplier definition forums". The purchasing sector indicates the capable suppliers and in the first meeting, they raise up the quotations of prices on the Internet, which pre-define the suppliers taking part in the process. In the second meeting, the suppliers are appraised for each area and one and only one decision is made to choose the best supplier.

Subjects, such as, quality and delivery performance are important characteristics to evaluate the suppliers. The cost is very often the competitive priority that assumes the position of requested winner.

Suppliers also have internal performance measurement systems that are deployed in external measures. Supplier A has several performance indicators, such as: physics of sales, the customers' complaints, field flaws, stop of the customer's line, medium time of flaws, waste control and refuse, raw material blocked by the suppliers, the product auditing and maintenance control. The criteria directly related to the automaker supply chain are the customer's complaints and stop of the customer's line. Supplier A considers that it has an internal performance measurement system more rigorous than used by the automaker to measure the supplier performance. It considers that the automaker measures, in fact, are the quality and an eventual and unwilling stop at the end of the assembly line

Supplier B also uses several internal performance indicators, such as: delivery performance, number of faults, relationship with the customer, revenue, cost of labor, volume of sales, turn of inventories, percentage of purchases on invoicing, cost of the non conformity, quality, punctuality and the deliver performance, productivity, matter loss excels and industrial efficiency. Those criteria generate information regarding the logistics performance, quality and finances. However, the company considers that the automaker controls, mainly, the quality and the delivery performance (punctuality, flexibility, transport, package, service and accuracy).

Similarly, supplier C is evaluated in quality matters and the delivery performance by the automaker. Every six months, supplier C receives an evaluation from the automaker informing its delivery performance and monthly an evaluation of its quality. The performance measures are more intense in areas, such as, quality and logistics because the engineering was evaluated in begin of the project.

Quality and delivery are the two main appraised criteria (for supplier B) in supplier D. Types of exchanged information limit issues involving the project and programming (types of parts, quantities and dates). A key problem for supplier D is the production reprogramming, which generates problems with planning, generating, very often, delivery delays and harming the upstream companies.

The demands from supplier B to supplier D involve: specification of the invoice request number, quantity and types of parts.

In the chain studied, the relationship with supplier D is considered to be of less intensity among companies because it supplies a commodity with little aggregated value that reflects significantly in the relationship among the companies. Figure 4 tries to characterize the relationship between the aggregated value and the need for integration on the processes and performance measures verified in the chain.

It is noticed that the larger the aggregated value of the supplied parts is, the larger the need for integration of processes and performance evaluation will be, as it happens in the relationship with supplier A, where there has been not only close partnership but also high performance criteria.

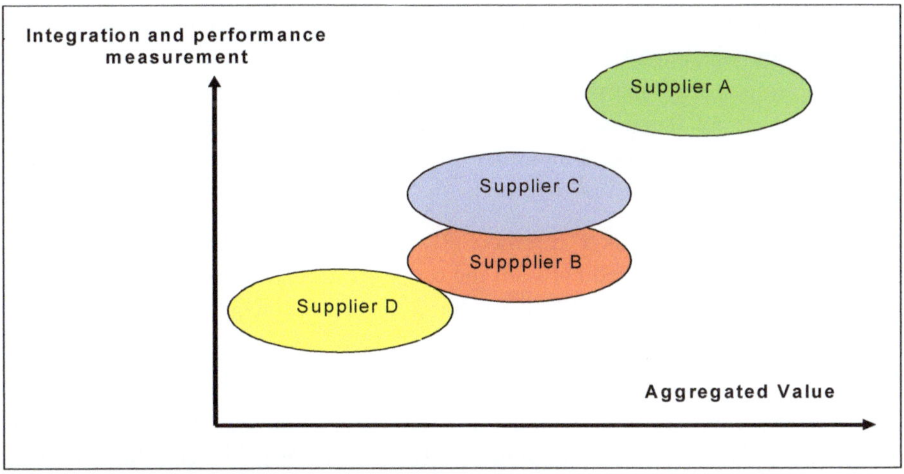

Figure 4. Relationship of the joined value and the integration

6.2. Evaluation of dealers

The automaker's commercial area makes the connection between the automaker and the dealers objecting to assist the final customer's demand regarding the conformity performance and of the delivery performance. The commercial area that has been analyzed is responsible for the region that includes São Paulo city, and it is represented by 40 dealers. Inside the state of São Paulo the automaker studied includes more than 100 dealers. In the whole, the company is represented by 6 regional that involve all its commercial areas (sales) in the country. The interviews done with the automaker commercial area tried to answer questions regarding the relationship among the commercial, manufacturing, logistics areas and dealers.

According to the Brazilian law, the automaker cannot sell to the final consumer. Considering that the final customer is the most important element of the entire chain, dealers are the distribution channels to assist that demand. Therefore, the automaker has great interest in monitoring and evaluating the dealer performance because it checks how the customers are being assisted and how the trademark (brand) is worked.

The relationship between the commercial and the manufacturing areas defines the models to be produced. In the past, the relationship between the commercial and manufacturing areas was represented by the duality of goals. While the manufacturing area wanted to simplify production, the commercial area tried to sell "what the customer wanted". That caused critical cross-sector problems. Nowadays, that relationship is more integrated, effective and with resolution of problems taking together.

The automaker imposes to the dealers a minimum batch of purchase of basic vehicles and another batch of vehicles "ordered" by the dealer. The objective of the commercial area is both to intermediate the management on requests and to intermediate the connection between a viable production and the demand.

There is an integrating exchange and sharing information among the automaker, commercial area and dealers. The dealer can access the information on the automaker production programs in the following five weeks, and the automaker can get information about the cars that are available at the dealer through this same information system. That integration of information permits to optimize stocks and to elevate the level of service at the dealers.

Integration between the automaker commercial area and the dealers is necessary and evident. Monthly meetings are held between the commercial area and big dealers to define the market policies and strategies. Each dealer has a different strategy because it is located in a specific area with different target customer.

In order to improve the service to the customer (demand) several initiatives were implemented by the automaker in the last few years. One of the main measures was the restructuring of the dealer base. As consequence, the relationship between the automaker (performed by the commercial area) and dealers are increasing. The automaker's commercial area does not only measure the dealer performance but also interferes in management policies . The automaker has nominate consultants to dealers both to analyze and to propose changes, since administrative activities, until the customer's loyalty.

There are two ways in which the automaker gives technical and administrative support to dealers: back-office and front-office. The back-office is characterized by consultants that analyze and propose plans of actions that go from the subjects related to the service level to the dealer management. Consultants can be classified in three types: market, distribution and business. São Paulo regional front-office is composed by four managers who run and monitor 40 dealers under the regional responsibility. And they can be considered an element *in loco* to support the back-office at the dealers.

There are several indicators to measure the dealer performance. The three main indicators analyzed by the commercial area are: sales performance, service performance (customers) and monthly business profitability.

The sales performance indicates the potential growth and development of the dealer. That indicator is usually linked with the service indicators to the customer and profitability.

The commercial area spends strong efforts to keep the service levels in order to please the dealer customers. Currently, it is not interesting to the automaker that only the dealer commercializes high quantities of cars. This also interests the automaker, as the customer is being assisted and so are the business levels of profitability. Dealers with short life cycles can harm the vehicle brand. The automaker commercial area force to be bought by larger groups.

Several studies are showing that the automaker should control and give more attention to the after-sale. The automaker measures the customer's satisfaction through a so called QIS (Quality Information System). This system is responsible for measuring aspects related to customer's satisfaction in terms of sales and services at dealers.

The customer's satisfaction is measured by QIS through a questionnaire sent directly to the customer's home. Through QIS the dealer and the automaker identify whether the customer is satisfy. If he/she is not, diagnoses are accomplished and performance plans are elaborated to improve this indicator. The indicators considered in the QIS are: the customer's general satisfaction, satisfaction with sales and services, characteristics of the show-room, the salesperson's evaluation, price evaluation given to second hand cars, attendance after purchase, repair quality , treatment given to the customer, repair price, period of delivery and offers. Through that after-sale monitor the automaker makes a diagnosis to verify the critical points to be improved at the dealers.

6.3. Performance measurement of the supply chain studied

A single systematic way was not verified to measure the supply chain studied. The focal company has internal performance measure systems, which are deployed to the supply chain, both downstream (suppliers) and upstream (dealers) directions.

The suppliers also have internal performance measures. Logistics supplier performance measures are strongly influenced by the automaker demands. In the studied chain, the chain measurement system is in development. Most of the characteristics on the automaker measurement system maintain the characteristics of the traditional systems, that is, of the systems practiced internally in the companies.

The automaker does not have an exclusive measurement system, but it addresses the whole measurement process to the chain, exercising the governance (coordination) of the performance measurement criteria. Despite that characteristic, the automaker internal system permits to consolidate several pieces of information in order to measure the performance along the chain. That group of measures comes from the automaker internal indicators and also the suppliers' definition forums.

The automaker measures and evaluates the internal performance through a solid group of KPIs (Key Performance Indicators) and it includes all the areas and internal collaborators. Criteria and goals are set starting from corporate strategies, for specific areas, such as departments, sectors, etc. The performance criteria are signaled by colors green, yellow and red with very defined limits of control. The area that has consolidated those numbers is the finances (controller), which is responsible for shooting the results of the indicators, both financial and performance of other areas.

The automaker logistics area also makes use of a key group of 10 blocks of indicators which are: percentage of lost production, bill of material, scrap, cripple (number of missing units / incomplete vehicles), levels of inventories, deliveries of replacement parts, overtime, logistics costs (purchase of inputs), overhead, volume and fidelity of the inventory. Similarly, other functional areas in the company have that same backup through the same procedures of the system and each one of their employees has his/her KPI sheet and criteria that add value to the area.

Some logistics performance measures are copied from the automaker to the suppliers. One of the typical performance measures is the cripple, which indicates that when the car is being assembled some parts are lacking and, consequently, measures flaws along the supply chain. Therefore, even indirectly, KPIs end up reflecting in the suppliers. The reflections of KPIs in the suppliers systemists are discussed daily at the automaker premises through meetings where preventive and corrective actions are solved.

The automaker performance measure system is similar to Balanced Scorecard model, mainly, regarding the definition of the measures, starting from the strategic unfolding. The corporation draws the strategy and, from this, the objectives that should be reached are delimited. The strategic objectives of the corporation are transferred to the functional areas. According to the results of the research, Hronec (1994) points out that the performance measures should be developed on the corporate level and be related to the strategy. That happens with the automaker studied, although the alignment of the strategy and the chain performance measures involves great complexity.

According to Pires (2004) there exist some difficulties to implement a strategy along the chain: (1) in each relationship what really prevails is the company strategy, which is derived from its performance in the market and internal competence, and it is not necessarily linked to a strategy that belongs to the chain; (2) the strongest company's strategy in the chain can influence all the other strategies linked to the company; (3) the participation of several links in several chains hinders the implementation of a single strategy along the chain. Those aspects hinder both the alignment of the strategies and also the performance measurement development among companies in the chain.

Dealers working close to the final customer can identify either new needs or strategic issues along the chain. For this reason it is possible to explain the constant attendance of the operational design and service to the customers in the dealers. The customer's satisfaction is a multidimensional measure that evaluates several aspects, for example, quality (under several perspectives), cost, flexibility and even, intangible factors, such as trust.

Measures related to the final customer's satisfaction were thoroughly seen at dealers. Dealers can inform the levels of demand and stocks and transmit it to the automaker that can inform the other relationships in the chain, trying to reduce impacts on the variation of the demand.

Through the interviews done at the automaker, the concern was to verify just measuring companies which were considered strategic to the automaker. The second tier suppliers participate to the performance evaluation just when some serious problems with the product are detected either by the automaker or the first tier supplier. What motivates the evaluation of the second tier supplier is usually the lack of quality regarding conformity and specifications. The Table 1 illustrates the downstream and upstream performance measures in the focal company.

UPSTREAM PERFORMANCE MEASURES	DOWNSTREAM PERFORMANCE MEASURES
Logistics areas, engineering, quality, finances and commercial compose the supplier selection forum. The automaker has got other criteria to choose the systemists supplier: such as: quality, co-design and worldwide performance. The upstream performance measures involve operational and technical criteria.	sales performance; customer service; business profitability; satisfaction with the services, show-room characteristics, salesperson's evaluation, price evaluation on second-hand cars, vehicle delivery, attendance after purchase, repair quality, treatment given to the customer, repair price, period of delivery and car offering. The downstream performance measures seek to measure the dealer performance through the customers' satisfaction.

Table 1. Downstream and upstream performance measures in the focal company.

The group of criteria downstream and upstream is intense, mainly for the companies on the immediate chain. The results of the research show qualitative and quantitative measures, as pointed out by Beamon (1998). Although there is not a unique system for the chain, the measures have shown good inclusion, universality and measuring capacity.

According to Slack (1993), in order to gain in effectiveness, in every supply chain it is mostly necessary to measure only the companies that are part of the immediate chain. Therefore, academically as well as in practice, it was verified that the measure tendency in the chain is to happen in the immediate chain. For Gasparetto (2003) if all the companies are involved in the same efforts, even if it is just with their partners from the immediate chain, the whole chain will be involved in dyadic relationships and the improvement in competitiveness in many companies will be permitted, although the relationship composed by the systemist presents both larger process integration and larger specificity on the performance measures.

Regarding the performance measure selection, Beamon (1999) describes that some companies use just one indicator of performance to evaluate the chains. But, the same author criticizes that position because she believes that a single indicator does not translate, under the operational and strategic perspectives, the total chain performance. The studied supply chain has shown a reasonable level of evolution in this sense because besides having indicators regarding costs, it also has indicators that express the logistics performance, the suppliers' evaluation , the customers' satisfaction and other factors related to the chain.

An important characteristic for the performance measure system is its flexibility. This fact can also be proven in this study because besides measuring the performance of the immediate chain, the automaker determines new performance plans. A fact seen in the case studied, which proves the dynamic character of the system, is the automaker capacity to set certain behaviors. The focal company offers a consultancy service to the dealers seeking to improve the dealer performance in terms of profitability and service to the customer. In doing that, the company improves the supply chain performance using downstream information, such as the demand and stock performance.

The capacity to stimulate performances, according to Neely et al. (1995), consists on one of the greatest benefits in a measurement system. In this case, the measurement system plays an important role to reach certain strategic goals proposed by the automaker. These roles may come from an individual contribution for the system and it should not be faced as a control. However, the characteristics of control are still much stronger in these performance measure systems than in most traditional relationships. This can be explained by the strong influence that the automaker imposes on the chain governance.

The automaker coordinates and controls the provisioning and distribution channels. That role influences the automaker capacity to exercise governance in the chain. For Suzigan, Garcia and Furtado (2002) governance refers to the hierarchy degree and command (or, alternatively, collaboration and cooperation) among the actors. Storper and Harrison (1990, p.408) understand governance as the "actor that has the power to affect the system development". Those relationships can be ruled either by market mechanisms or by cooperation processes supported by mechanism of control and performance.

The automaker has two ways to exercise governance and coordination in the chain: one is accomplished by traditional market relationships (where price and quality are the basic criteria) and another is got through high cooperation (ruled relationships in trust and high change of resources), mainly with the systemists and large dealers. The automaker uses both forms to guarantee the relationships with suppliers. For the commodities suppliers the selection is accomplished by "suppliers' auctions" registered through the internet, where the basic criteria are the price and the quality (examining criteria). Yet, with the systemists suppliers there is an intense cooperation in the activities to improve the levels of productivity in both plants. In that relationship, it has to be taken into consideration the inter-business high interdependence and costs.

Even with more cooperative relationships in the chain, the automaker governance can be classified as core-ring with the leader firm according to Storper and Harrison (1990). In that system the leader companies can be completely independent from the suppliers. The negotiation power is asymmetrical and there is considerable hierarchy. In this system, the leader company is dominant, that is, the participants depend on the leader's strategy. That part of the research is important, because the governance structure involving the relationships of authority and power among the firms and it determines the financial resource allocation, materials and labor along the chain (Gereffi, 1999).

It can be said that the automaker governance, for some suppliers, is based strongly on imposition, because it exercises a great economical power on the other suppliers in the chain. Yet, in other relationships with systemists, the governance is strongly based on cooperation, but with intense performance measurement system. The relationship with the systemists supplier is soft and with cooperation routines. Therefore, the governance legitimacy is accomplished through a combination of coercion and cooperation systems.

7. Final considerations

The necessity of controlling the strategy performance and the customers' satisfaction interferes significantly in the strategic positioning of the organizations and the way it measures the performance.

As mention previously, it was not found a single performance measurement system in the supply chain studied. The automaker internal performance measures deployed to both the suppliers and to dealers. The automaker internal systems, added to the suppliers and dealers evaluation and the suppliers' definition forum are enough to generate several performance indicators to increase the immediate chain. The automaker is responsible for the whole coordination of the chain measure process.

Starting from the performance indicators used by the companies and from the focal company perspective (automaker) in the studied supply chain, it can be seen that the downstream measures are addressed strongly to quality, production and delivery performance. On the other hand, the upstream measures present a strategic and market characteristics and also receiving special attention from the automaker although the downstream processes tend to be more complex and are subjected of unpredictable events.

It is also noticed that the performance indicators related the chain space besides the traditional indicators regarding costs, also presenting indicators that include aspects related to the logistics performance, the suppliers' evaluation, the customers' satisfaction, among others. Additionally, which one can call a "measurement system" of the studied chain (same that he/she comes a form no structured and formalized), it is still for the most part contained in the immediate supply chain. That corroborates the authors' work, such as Lambert and Pohlen (2001), who point out that the companies usually end up just measuring the performance in the immediate chain and avoiding measuring the performance in the whole chain.

Considering what was pointed out by Beamon (1999): inclusion (measure of all the pertinent aspects); universality (to allow the comparison among the different levels); measuring (the necessary data can be measured) and consistence (solid measures with strategic goals), it is possible to state that it has adherence to the studied system. However, the measures are not integrated in a unique system.

As it is expected the supply chain performance measurement tends to be more intense in most cooperative relationships. The systems supplied by the systemists are of high aggregated value, with high specificities and also involving characteristics on the vehicle safety. According to that, the automaker exercises high control on the performance measurement evaluation. However, one of the automaker logistics executives has mentioned how easy it is to deal with systemists suppliers, because of geographic proximity and professionalism. The supplier, installed physically inside the automaker plant, presents a high performance in all the requirements measured by the automaker. That fact reinforces the idea that the supply chain configuration is a decisive factor in the way it dimensions and controls the chain performance measures.

Another important subject to be pointed out is about supply chain governance. This research, takes into consideration the chain asymmetric power. A car industry is typically run by the producer, i.e., industries such as: metal mechanics, electronics, chemistry, development of products and processes. On the other hand, it is clear their progresses towards a cooperation with key suppliers, as it can be seen with systemists suppliers and the distribution channels.

Sharing information was always a fundamental element to build a cooperative relationship. In this research, sharing information is more intensity within the automaker and immediate chain. The automaker can have access to several types of information, so much in its inbound as in the outbound. The opposite, most of the time, is not true. In general, it can be considered that sharing information along the supply chain is fragmented. That means that the automaker works with content and frequency of information not necessarily similar to the relationships of its immediate chain, although the basic information on its master production schedule is shared with all its immediate suppliers.

Finally, it is important to emphasize the fact that the immediate supply chain studied has presented a supply chain configuration including a dealer and systemists suppliers which provided a very contemporary and interesting research for other studies involving the supply chain performance measurement.

Author details

Mário Sacomano Neto and Sílvio R. I. Pires
Methodist University of Piracicaba (UNIMEP), Rodovia do Açúcar, Piracicaba (SP), Brazil

8. References

Beamon. B. N. (1998), Supply chain design and analysis: Models and methods. International Journal of Production Economics, Vol. 55, No. 3, pp. 281-294.

Beamon, B. M. (1999). Measuring supply chain performance. International Journal of Operations & Production Management, Vol.19, No. 3, pp. 275-292.

Brandão, H. P., Guimarães, T. A. (2001). Gestão de Competências Gestão do Desempenho: tecnologias distintas ou instrumentos de um mesmo constructo? Revista de Administração de Empresas, Vol.41, No.1, pp. 8-15.

Carvalho, R. Q. et al. (2000). Globalização e reestruturação da cadeia produtiva na indústria automotiva: qual é o papel do Mercosul? Convênio IPEA-DPCT/IG/UNICAMP. Relatório Final. FUNCAMP, 252p.

Croxton, K. L., Sebastián, J. G. D., Lambert, D. M., Rogers, D. S. (2001). The Supply Chain Management Processes. International Journal of Logistic Management. Vol. 12, No. 2, pp. 13-36.

Eccles, R. G. (2000). Manifesto da mensuração do desempenho . In: Medindo o Desempenho empresarial. Rio de Janeiro: Campus Publishing.

FPNQ (1990). Fundação para o prêmio nacional da qualidade - Critérios de excelência: o estado da arte da Gestão da Qualidade Total. São Paulo, 76 p.

Gasparetto, V. (2003). Proposta de uma sistemática para avaliação de desempenho em cadeias de suprimentos. Florianópolis. 248 p., Thesis Doctorate, UFSC.

Gereffi, G. (1999). International trade and industrial upgrading in the apparel commodity chain. Journal of International Economics, Vol . 48, No, 1 pp.37-70.

Gunasekaran, A., Patel, C., Tirtiroglu, E. (2001). Performance measures and metrics in a supply chain environment. International. Journal of Operations & Production Management, Vol. 21, No. 1/2, pp. 71-87.

Helper, S. (1991). How much has really changed between U.S. automakers and their suppliers? Sloan Management Review, Vol. 32, No. 4, pp.15-28.

Holmberg, S. (2000). A systems perspective on supply chain measurements. International Journal of Physical Distribution & Logistics Management. Vol. 30, No.10, pp. 847-868.

Hronec, S. M. (1994). Sinais Vitais - Usando Medidas de Desempenho da Qualidade, Tempo, e Custos para traçar a rota do futuro da sua empresa. São Paulo: Makron Books.

Kaydos, W. (1991). Measuring, managing and maximizing performance. Portland: Productivity Press.

Lambert, D.M., Cooper, M.C., Pagh, J. D. (1998). Supply Chain management: Implementation Issues and Research Opportunities, The International Journal of Logistics Management, Vol. 9, No. 2, pp.1-19.

Lambert, D., Pohlen, T. (2001). Supply Chain Metrics. The International Journal of Logistics Management, Vol.12, No.1, pp. 1-19.

Nauri, M. H. C. (1998). As medidas de desempenho como base para a melhoria continua dos processos: o caso da Fundação de Amparo à Pesquisa e Extensão Universitária (FAPEU). Thesis Master Degree, UFSC.

Neely, A. (2000). Performance measurement system design: developing and testing a process-based approach. International Journal of Operations & Production Management. Vol. 20, No. 10. pp. 1119-1145.

Pires, S. R. I., Sacomano Neto, M. (2008). New Configurations in Supply Chains: The Case of a Condominium in Brazil's Automotive Industry. Supply Chain Management: an International Journal, Vol. 13, pp. 37-48.

Pires, S. R. I. (2004). Gestão da Cadeia de Suprimentos: conceitos, estratégias, práticas e casos. São Paulo, Atlas Publishing, 310p.

Pires, S. R. I.; Aravechia, C. H. M (2001). Measuring Supply Chain Performance. POMS, 2001, Orlando.

Slack, N. (1993). Vantagem Competitiva em Manufatura. São Paulo, Atlas Publishing.

Stewart, G. (1995). Supply chain performance benchmarking study reveals keys to supply chain excellence. Logistics Information Management, Vol. 8, No. 2, pp. 38-44.

Storper, M.; Harrison, B. (1991). Flexibility, hierarchy and regional developments: the changing structure of industrial production systems and their forms of governance in the 1990s. Research Policy, North-Holland, Vol. 20, No. 5.

Trienekens, J. (1999). Management of Process in chains: a research framework. Thesis. Wageningen University.

Suzigan, W.; Garcia, R.; Furtado, J. (2002). Governança de Sistemas de MPMES em *clusters* Industriais. Text presented at the International Symposium: "Políticas para Sistemas Produtivos Locais de MPMES", Rio de Janeiro.

Waggoner, D. B. Nelly, A. D., Kennerley, M. P. (1999) The forces that shape organizational performance measurement systems: An interdisciplinary review. International Journal of Production Economics. Vol. 60, No. 61, pp. 53-60.

Permissions

The contributors of this book come from diverse backgrounds, making this book a truly international effort. This book will bring forth new frontiers with its revolutionizing research information and detailed analysis of the nascent developments around the world.

We would like to thank João Paulo Carmo, for lending their expertise to make the book truly unique. They have played a crucial role in the development of this book. Without their invaluable contribution this book wouldn't have been possible. They have made vital efforts to compile up to date information on the varied aspects of this subject to make this book a valuable addition to the collection of many professionals and students.

This book was conceptualized with the vision of imparting up-to-date information and advanced data in this field. To ensure the same, a matchless editorial board was set up. Every individual on the board went through rigorous rounds of assessment to prove their worth. After which they invested a large part of their time researching and compiling the most relevant data for our readers. Conferences and sessions were held from time to time between the editorial board and the contributing authors to present the data in the most comprehensible form. The editorial team has worked tirelessly to provide valuable and valid information to help people across the globe.

Every chapter published in this book has been scrutinized by our experts. Their significance has been extensively debated. The topics covered herein carry significant findings which will fuel the growth of the discipline. They may even be implemented as practical applications or may be referred to as a beginning point for another development. Chapters in this book were first published by InTech; hereby published with permission under the Creative Commons Attribution License or equivalent.

The editorial board has been involved in producing this book since its inception. They have spent rigorous hours researching and exploring the diverse topics which have resulted in the successful publishing of this book. They have passed on their knowledge of decades through this book. To expedite this challenging task, the publisher supported the team at every step. A small team of assistant editors was also appointed to further simplify the editing procedure and attain best results for the readers.

Our editorial team has been hand-picked from every corner of the world. Their multi-ethnicity adds dynamic inputs to the discussions which result in innovative outcomes. These outcomes are then further discussed with the researchers and contributors who give their valuable feedback and opinion regarding the same. The feedback is then collaborated with the researches and they are edited in a comprehensive manner to aid the understanding of the subject.

Apart from the editorial board, the designing team has also invested a significant amount of their time in understanding the subject and creating the most relevant covers. They scrutinized every image to scout for the most suitable representation of the subject and create an appropriate cover for the book.

The publishing team has been involved in this book since its early stages. They were actively engaged in every process, be it collecting the data, connecting with the contributors or procuring relevant information. The team has been an ardent support to the editorial, designing and production team. Their endless efforts to recruit the best for this project, has resulted in the accomplishment of this book. They are a veteran in the field of academics and their pool of knowledge is as vast as their experience in printing. Their expertise and guidance has proved useful at every step. Their uncompromising quality standards have made this book an exceptional effort. Their encouragement from time to time has been an inspiration for everyone.

The publisher and the editorial board hope that this book will prove to be a valuable piece of knowledge for researchers, students, practitioners and scholars across the globe.

List of Contributors

Horst Hintze-Bruening
BASF Coatings GmbH, Germany

Fabrice Leroux
Clermont University, CNRS, France

Evripidis Lois and Panagiotis Arkoudeas
National Technical University of Athens, School of Chemical Engineering, Greece

Mohsen Mohseni, Bahram Ramezanzadeh, Hossein Yari and Mohsen Moazzami Gudarzi
Amirkabir University of Technology, Iran

Amaya Igartua and Xana Fdez-Pérez
Ik4 Tekniker, Spain

Iñaki Illarramendi
CIE- Tarabusi, Spain

Rolf Luther and Jürgen Rausch
FUCHS, Germany

Mathias Woydt
Bam, Germany

Vítor Monteiro, Henrique Gonçalves, João C. Ferreira and João L. Afonso
Centro Algoritmi, University of Minho,Guimarães, Portugal

Niels Koch
Altran GmbH & Co.KG, Germany

Ruben Ivankovic, Jérôme Cros, Mehdi Taghizadeh Kakhki, Carlos A. Martins and Philippe Viarouge
Laval University, Canada

Amedeo Troiano, Eros Pasero and Luca Mesin
Politecnico di Torino - Dipartimento di Elettronica, Italy

Preeti Bajaj
Department of Electronics Engineering, India
G.H. Raisoni College of Engineering Nagpur (M.S.), India

Milind Khanapurkar
Department of Electronics and Telecommunication Engineering, India
G.H. Raisoni College of Engineering Nagpur (M.S.), India

J. P. Carmo
University of Minho, Department of Industrial Electronics, Guimarães, Portugal

J. E. Ribeiro
Polytechnic Institute of Bragança, Department of Mechanical Technology, Bragança, Portugal

Arjun Yogeswaran and Pierre Payeur
University of Ottawa, Electrical Engineering and Computer Science (EECS), Canada

Takama Suzuki and Masaki Takahashi
Keio University, Japan

Darrell Robinette
General Motors LLC, USA

Carl Anderson and Jason Blough
Michigan Technological University, Department of Mechanical Engineering - Engineering Mechanics, USA

Mário Sacomano Neto and Sílvio R. I. Pires
Methodist University of Piracicaba (UNIMEP), Rodovia do Açúcar, Piracicaba (SP), Brazil